TODAY'S TECHNICIAN

Shop Manual for

# Automotive Suspension and Steering Systems

TODAY'S TECHNICIAN

# Shop Manual for
# Automotive Suspension and Steering Systems

Don Knowles

Knowles Automotive Training
Moose Jaw, Saskatchewan
CANADA

Jack Erjavec

*Series Advisor*

Columbus State Community College
Columbus, Ohio

Delmar Publishers
ITP An International Thomson Publishing Company

Albany • Bonn • Boston • Cincinnati • Detroit • London • Madrid • Melbourne
Mexico City • New York • Pacific Grove • Paris • San Francisco • Singapore • Tokyo
Toronto • Washington

## NOTICE TO THE READER

Publisher does not warrant or guarantee any of the products described herein or perform any independent analysis in connection with any of the product information contained herein. Publisher does not assume, and expressly disclaims, any obligation to obtain and include information other than that provided to it by the manufacturer.

The reader is expressly warned to consider and adopt all safety precautions that might be indicated by the activities described herein and to avoid all potential hazards. By following the instructions contained herein, the reader willingly assumes all risks in connection with such instructions.

The publisher makes no representations or warranties of any kind, including but not limited to, the warranties of fitness for particular purpose or merchantability, nor are any such representations implied with respect to the material set forth herein, and the publisher takes no responsibility with respect to such material. The publisher shall not be liable for any special, consequential or exemplary damages resulting, in whole or in part, from the readers' use of, or reliance upon, this material.

**COVER PHOTO:** Courtesy of Chevrolet Motor Division, General Motors Corporation
**PHOTO SEQUENCES:** Photography by Rod Dixon Associates and Michael A. Gallitelli, Metroland Photo
Portions of materials contained herein have been reprinted with permission of General Motors Corporation, Service Technology Group.

**DELMAR STAFF**
Senior Administrative Editor: Vernon Anthony
Developmental Editor: Catherine Eads
Project Editor: Eleanor Isenhart
Production Coordinator: Karen Smith
Art/Design Coordinator: Heather Brown



Printed in the United States of America

For information, contact:

Delmar Publishers
3 Columbia Circle, Box 15015
Albany, New York 12212-5015

International Thomson Editors
Campos Eliseos 385, Piso 7
Col Polanco
11560 Mexico DF Mexico

International Thomson Publishing Europe
Berkshire House 168-173
High Holborn
London, WC1V7AA
England

International Thomson Publishing GmbH
Königswinterer Strasse 418
53227 Bonn
Germany

Thomas Nelson Australia
102 Dodds Street
South Melbourne, 3205
Victoria, Australia

International Thomson Publishing Asia
221 Henderson Road
#05-10 Henderson Building
Singapore 0315

Nelson Canada
1120 Birchmont Road
Scarborough, Ontario
Canada M1K 5G4

International Thomson Publishing Japan
Hirakawacho Kyowa Building, 3F
2-2-1 Hirakawacho
Chiyoda-ku, Tokyo 102
Japan



5 6 7 8 9 10 XXX 00 99 98 97

**Library of Congress Cataloging-in-Publication Data**

Knowles, Don.
Automotive suspension and steering systems / Don Knowles.
p. cm. --(Today's technician)
Includes index.
Contents: v. 1. Classroom manual -- v. 2. Shop manual.
ISBN 0-8273-6185-8 (set)
1. Automobiles-- Springs and suspension--Maintenance and Repair--Problems, exercises, etc. 2. Automobiles--Steering-gear--Maintenance and repair--Problems, exercises, etc. I. Title. II. Series.
TL257.K59 1995 94-12957
629.24'3--dc20 CIP

# CONTENTS

CHAPTER 4

## Wheel and Tire Service and Balancing 67



CHAPTER 5

## Shock Absorber and Strut Diagnosis and Service 91



CHAPTER 6

## Front Suspension System Service 111

# Photo Sequences

# PREFACE

Unlike yesterday's mechanic, the technician of today and for the future must know the underlying theory of all automotive systems and be able to service and maintain those systems. Today's technician must also know how these individual systems interact with each other. Standards and expectations have been set for today's technician, and these must be met in order to keep the world's automobiles running efficiently and safely.

The *Today's Technician* series, by Delmar Publishers, features textbooks that cover all mechanical and electrical systems of automobiles and light trucks. Principal titles correspond with the eight major areas of ASE (National Institute for Automotive Service Excellence) certification. Additional titles include remedial skills and theories common to all of the certification areas and advanced or specialized subject areas that reflect the latest technological trends.

Each title is divided into two manuals: a Classroom Manual and a Shop Manual. Dividing the material into two manuals provides the reader with the information needed to begin a successful career as an automotive technician without interrupting the learning process by mixing cognitive and performance-based learning objectives.

Each Classroom Manual contains the principles of operation for each system and subsystem. It also discusses the design variations used by different manufacturers. The Classroom Manual is organized to build upon basic facts and theories. The primary objective of this manual is to allow the reader to gain an understanding of how each system and subsystem operates. This understanding is necessary to diagnose the complex automobile systems.

The understanding acquired by using the Classroom Manual is required for competence in the skill areas covered in the Shop Manual. All of the high priority skills, as identified by ASE, are explained in the Shop Manual. The Shop Manual also includes step-by-step instructions for diagnostic and repair procedures. Photo Sequences are used to illustrate many of the common service procedures. Other common procedures are listed and are accompanied with fine-line drawings and photographs that allow the reader to visualize and conceptualize the finest details of the procedure. The Shop Manual also contains the reasons for performing the procedures, as well as when that particular service is appropriate.

The two manuals are designed to be used together and are arranged in corresponding chapters. Not only are the chapters in the manuals linked together, the contents of the chapters are also linked. Both manuals contain clear and thoughtfully selected illustrations. Many of the illustrations are original drawings or photos prepared for inclusion in this series. This means that the art is a vital part of each manual.

The page layout is designed to include information that would otherwise break up the flow of information presented to the reader. The main body of the text includes all of the "need-to-know" information and illustrations. In the side margins are many of the special features of the series. Items such as definitions of new terms, common trade jargon, tools lists, and cross-references are placed in the margin, out of the normal flow of information so as not to interrupt the thought process of the reader. Each manual in this series is organized in a like manner and contains the same features.

Jack Erjavec, Series Advisor

# Classroom Manual

To stress the importance of safe work habits, the Classroom Manual dedicates one full chapter to safety. Included in this chapter are common safety practices, safety equipment, and safe handling of hazardous materials and wastes. This includes information on MSDS sheets and OSHA regulations. Other features of this manual include:

## Cognitive Objectives

These objectives define the contents of the chapter and define what the student should have learned upon completion of the chapter.
*Each topic is divided into small units to promote easier understanding and learning.*

## Marginal Notes

Page numbers for cross-referencing appear in the margin. Some of the common terms used for components, and other bits of information, also appear in the margin. This provides an understanding of the language of the trade and helps when conversing with an experienced technician.

## Cautions and Warnings

Throughout the text, cautions are given to alert the reader to potentially hazardous materials or unsafe conditions. Warnings are also given to advise the student of things that can go wrong if instructions are not followed or if a nonacceptable part or tool is used.

## References to the Shop Manual

Reference to the appropriate page in the Shop Manual is given whenever necessary. Although the chapters of the two manuals are synchronized, material covered in other chapters of the Shop Manual may be fundamental to the topic discussed in the Classroom Manual.

## A Bit of History

This feature gives the student a sense of the evolution of the automobile. This feature not only contains nice-to-know information, but also should spark some interest in the subject matter.

## Summaries

Each chapter concludes with summary statements that contain the important topics of the chapter. These are designed to help the reader review the contents.

## Terms to Know

A list of new terms appears next to the Summary. Definitions for these terms can be found in the Glossary at the end of the manual.

## Review Questions

Short answer essay, fill-in-the-blank, and multiple-choice type questions follow each chapter. These questions are designed to accurately assess the student's competence in the stated objectives at the beginning of the chapter.

# Shop Manual

To stress the importance of safe work habits, the Shop Manual also dedicates one full chapter to safety. Other important features of this manual include:

## Performance Objectives

These objectives define the contents of the chapter and define what the student should have learned upon completion of the chapter. These objectives also correspond with the list of required tasks for ASE certification. *Each ASE task is addressed.*

Although this textbook is not designed to simply prepare someone for the certification exams, it is organized around the ASE task list. These tasks are defined generically when the procedure is commonly followed and specifically when the procedure is unique for specific vehicle models. Imported and domestic model automobiles and light trucks are included in the procedures.

## Photo Sequences

Many procedures are illustrated in detailed photo sequences. These detailed photographs show the students what to expect when they perform particular procedures. They also can provide a student a familiarity with a system or type of equipment, which the school may not have.

## Marginal Notes

New terms are pulled out and defined. Common trade jargon also appears in the margin and gives some of the common terms used for components. This allows the reader to speak and understand the language of the trade, especially when conversing with an experienced technician.

## Cautions and Warnings

Throughout the text, cautions are given to alert the reader to potentially hazardous materials or unsafe conditions. Warnings are also given to advise the student of things that can go wrong if instructions are not followed or if a nonacceptable part or tool is used.

## References to the Classroom Manual

Reference to the appropriate page in the Classroom Manual is given whenever necessary. Although the chapters of the two manuals are synchronized, material covered in other chapters of the Classroom Manual may be fundamental to the topic discussed in the Shop Manual.

## Customer Care

This feature highlights those little things a technician can do or say to enhance customer relations.

## Tools Lists

Each chapter begins with a list of the Basic Tools needed to perform the tasks included in the chapter. Whenever a Special Tool is required to complete a task, it is listed in the margin next to the procedure.

### Service Tips

Whenever a special procedure is appropriate, it is described in the text. These tips are generally those things commonly done by experienced technicians.

### Case Studies

Case Studies concentrate on the ability to properly diagnose the systems. Each chapter ends with a case study in which a vehicle has a problem, and the logic used by a technician to solve the problem is explained.

### Terms to Know

Terms in this list can be found in the Glossary at the end of the manual.

### Diagnostic Chart

Chapters include detailed diagnostic charts linked with the appropriate ASE task. These charts list common problems and most probable causes. They also list a page reference in the Classroom Manual for better understanding of the system's operation and a page reference in the Shop Manual for details on the procedure necessary for correcting the problem.

### ASE Style Review Questions

Each chapter contains ASE style review questions that reflect the performance objectives listed at the beginning of the chapter. These questions can be used to review the chapter as well as to prepare for the ASE certification exam.

## Instructor's Guide

The Instructor's Guide is provided free of charge as part of the *Today's Technician Series* of automotive technology textbooks. It contains Lecture Outlines, Answers to Review Questions, Pretest and Test Bank including ASE style questions.

## Classroom Manager

The complete ancillary package is designed to aid the instructor with classroom preparation and provide tools to measure student performance. For an affordable price, this comprehensive package contains:

Instructor's Guide
200 Transparency Masters
Answers to Review Questions
Lecture Outlines and Lecture Notes
Printed and Computerized Test Bank
Laboratory Worksheets and Practicals

## Reviewers

Special thanks to the following instructors for reviewing this material:

Fred Raadsher
British Columbia Institute of Technology

Alan Penuela
Ventura College

Dennis L. Schossow
Aims Community College

Donald V. Nilson
Las Positas College

Lawrence Le Gree
Montcalm Community College

CHAPTER 1

# Shop Practices

Upon completion and review of this chapter, you should be able to:

- ❑ Describe three requirements for shop layout, and explain why these requirements are important.
- ❑ Observe all shop rules when working in the shop.
- ❑ Operate vehicles in the shop according to shop driving rules.
- ❑ Observe all shop housekeeping rules.
- ❑ Follow the necessary procedures to maintain satisfactory shop air quality.
- ❑ Fulfill employee obligations when working in the shop.
- ❑ Accept job responsibilities for each job completed in the shop.
- ❑ Describe the ASE technician testing and certification process, including the eight areas of certification.

## Shop Layout

There are many different types of shops in the automotive service industry including:

- new car dealer
- independent repair shop
- specialty shop
- service station
- fleet shop

**CAUTION:** Always know the location of all safety equipment in the shop, and be familiar with the operation of this equipment.

The layout in any shop is important to maintaining shop efficiency and contributing to safety. Shop layout includes bays for various types of repairs, space for equipment storage, and office locations. Safety equipment such as fire extinguishers, first-aid kits, and eyewash fountains must be in easily accessible locations. The location of each piece of safety equipment must be clearly marked. Areas such as the parts department and the parts cleaning area must be located so they are easily accessible from all areas of the shop. The service manager's office should also be centrally located. All shop personnel should familiarize themselves with the shop layout, especially the location of safety equipment. If you know the exact fire extinguisher locations, you may get an extinguisher into operation a few seconds faster. These few seconds could make the difference between a fire that is quickly extinguished, and one that gets out of control, causing extensive damage and personal injury! Most shops have specific bays for certain types of work, such as electrical repair, wheel alignment and tires, and machining (Figure 1-1).

The tools and equipment required for a certain type of work are stored in that specific bay. For example, the equipment for electrical and electronic service work is stored in the bay allotted to that type of repair. When certain bays are allotted to specific types of repair work, unnecessary equipment movement is eliminated. Each technician has his or her own tools on a portable roll cabinet, which is moved to the vehicle being repaired. Special tools are provided by the shop, and these tools may be located on tool boards attached to the wall. Other shops may have a tool room where special tools are located. Adequate work bench space must be provided in those bays where bench work is required.

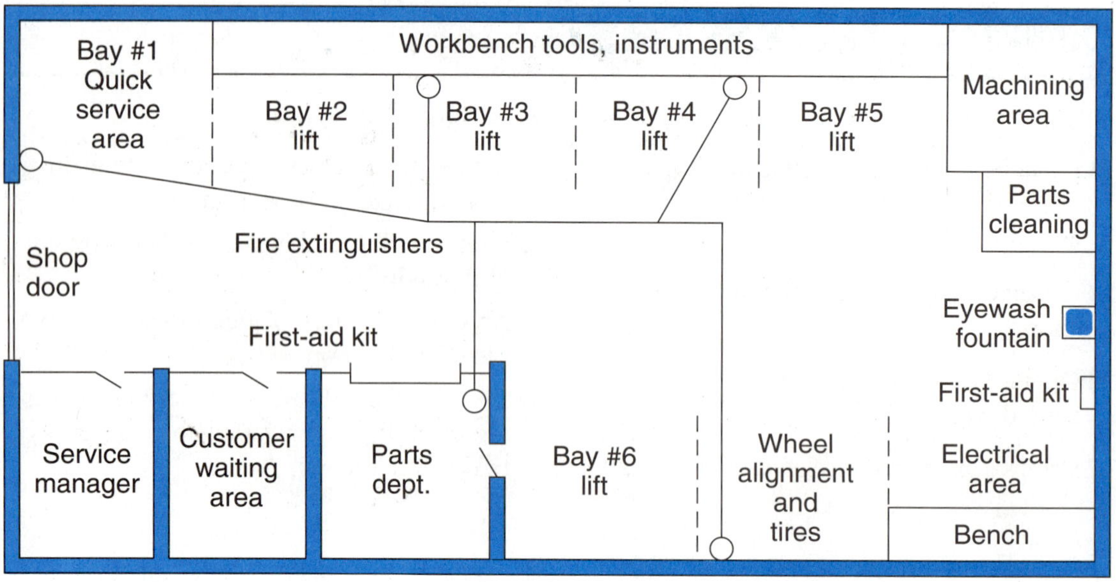


**Figure 1-1** Typical shop layout

## Shop Rules

**WARNING:** Shop rules, vehicle operation in the shop, and shop housekeeping are serious business. Each year a significant number of technicians are injured and vehicles are damaged because of shop rules being disregarded, careless vehicle operation, and sloppy housekeeping.

The application of some basic shop rules helps to prevent serious, expensive accidents. Failure to comply with shop rules may cause personal injury, or expensive damage to vehicles and shop facilities. It is the responsibility of the employer and all shop employees to make sure that shop rules are understood and followed until these rules become automatic habits. These basic shop rules should be followed:

1. Always wear safety glasses and other protective equipment that is required by a service procedure. For example, an asbestos parts washer must be used to avoid breathing asbestos dust into the lungs (Figure 1-2). *Asbestos dust is a known cause of lung cancer.* This dust is encountered in manual transmission clutch facings and brake linings.
2. Tie long hair securely behind the head, and do not wear loose, or torn, clothing.
3. Do not wear rings, watches, or loose, hanging jewelry. If jewelry such as a ring, metal watch band, or chain makes contact between an electrical terminal and ground, the jewelry becomes extremely hot, which results in severe burns.
4. Set the parking brake when working on a vehicle. If the vehicle has an automatic transmission, place the gear selector in park unless a service procedure requires another selector position. When the vehicle is equipped with a manual transmission, position the gear selector in neutral with the engine running, or reverse with the engine stopped.
5. Always connect a shop exhaust hose to the vehicle tail pipe, and be sure the shop exhaust fan is running. If it is absolutely necessary to operate a vehicle without a shop exhaust pipe connected to the tail pipe, open the large shop door to provide adequate ventilation. Carbon monoxide in the vehicle exhaust may cause severe headaches and other medical problems. High concentrations of carbon monoxide may result in death!
6. Keep hands, clothing, and wrenches away from rotating parts such as cooling fans. Remember that electric-drive fans may start turning at any time, even with the ignition off.

Carbon monoxide is a poisonous gas, and when breathed into the lungs it may cause headaches, nausea, ringing in the ears, tiredness, and heart flutter. In strong concentrations it causes death.

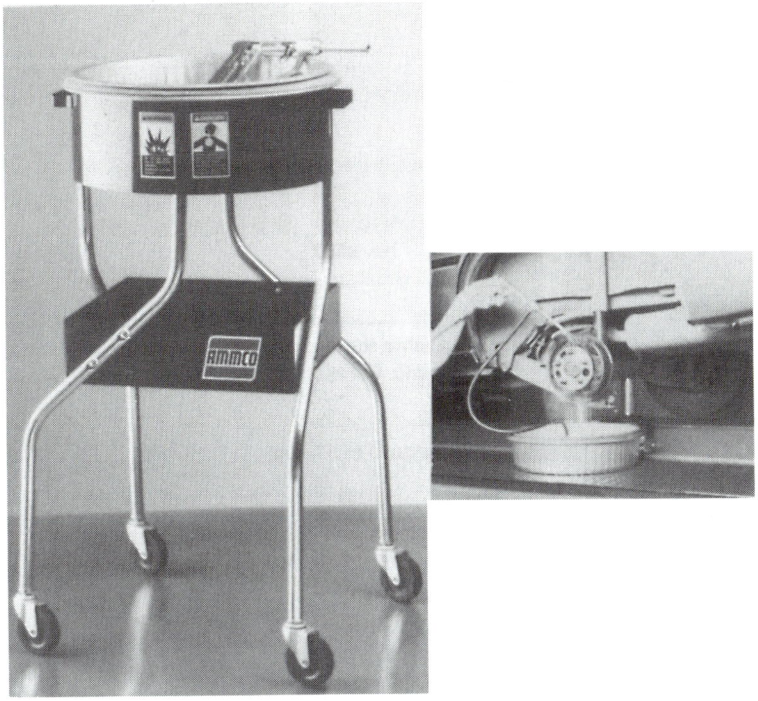

Figure 1-2 Brake assembly washer for asbestos dust (Courtesy of Hennessy Industries Inc.)

7. Always leave the ignition switch off unless a service procedure requires another switch position.
8. Do not smoke in the shop. If the shop has designated smoking areas, smoke only in these areas.
9. Store oily rags and other discarded combustibles in regulation, covered metal garbage containers.
10. Always use a wrench or socket that fits properly on the bolt. Do not substitute metric for English wrenches, or vice versa.
11. Keep tools in good condition. For example, do not use a punch or chisel with a mushroomed end. When struck with a hammer, a piece of the mushroomed metal could break off, resulting in severe eye injury or other injury.
12. Do not leave power tools running and unattended.
13. Prevent serious burns by avoiding contact with hot metal components such as exhaust manifolds, other exhaust system components, radiators, and some air conditioning hoses.
14. When lubricant such as engine oil is drained, always use caution because the oil could be hot enough to cause burns.
15. Before getting under a vehicle, be sure the vehicle is placed securely on safety stands.
16. Operate all shop equipment, including lifts, according to the equipment manufacturer's recommended procedure. Do not operate equipment unless you are familiar with the correct operating procedure.
17. Do not run or engage in horseplay in the shop.
18. Obey all state and federal fire, safety, and environmental regulations.
19. Do not stand in front of or behind vehicles.
20. Always place fender covers and a seat cover on a customer's vehicle before working on the car.
21. Inform the shop foreman of any safety dangers and suggestions for safety improvement.

## Vehicle Operation

When driving a customer's vehicle, certain precautions must be observed to prevent accidents and maintain good customer relations.

1. Prior to driving a vehicle always make sure the brakes are operating, and fasten the safety belt.
2. Check to be sure there is no person or objects under the car before you start the engine.
3. If the vehicle is parked on a lift, be sure the lift is fully down and the lift arms, or components, are not contacting the vehicle chassis.
4. Check to see if there are any objects directly in front of or behind the vehicle before driving away.
5. Always drive slowly in the shop, and watch carefully for personnel and other moving vehicles.
6. Make sure the shop door is up high enough so there is plenty of clearance between the top of the vehicle and the door.
7. Watch the shop door to be certain that it is not coming down as you attempt to drive under the door.
8. If a road test is necessary, obey all traffic laws, and never drive in a reckless manner.
9. Do not squeal tires when accelerating or turning corners.

If the customer observes that service personnel take good care of his or her car by driving carefully, and installing fender and seat covers, the service department image is greatly enhanced in the customer's eyes. These procedures impress upon the customer that shop personnel respect his or her car. Conversely, if grease spots are found on upholstery or fenders after service work is completed, the customer will probably think the shop is very careless, not only in car care, but also in service work quality.

## Housekeeping

**CUSTOMER CARE:** When a customer sees that you are concerned about his or her vehicle, and that you operate a shop with excellent housekeeping habits, the customer will be impressed, and will likely keep returning for service.

Excellent housekeeping involves general shop cleanliness, proper shop safety equipment in good working condition, and the proper maintenance of all shop equipment and tools.

Careful housekeeping habits prevent accidents and increase worker efficiency. Good housekeeping also helps to impress upon the customer that quality work is a priority in this shop. Follow these housekeeping rules:

1. Keep aisles and walkways clear of tools, equipment, and other items.
2. Be sure all sewer covers are securely in place.
3. Keep floor surfaces free of oil, grease, water, and loose material.
4. Place proper garbage containers in convenient locations, and empty the containers regularly.
5. Keep access to fire extinguishers unobstructed at all times, and check fire extinguishers for proper charge at regular intervals.

**SERVICE TIP:** When you are finished with a tool, never set it on the customer's car. After using a tool, the best place for it is in your tool box or on the work bench. Many tools have been lost by leaving them on customer's vehicles.

6. Keep tools clean and in good condition.
7. When not in use, store tools in their proper location.

8. Place oily rags and other combustibles in proper covered garbage containers.
9. Be sure that rotating components on equipment and machinery have guards, and schedule all shop equipment for regular service and adjustment.
10. Maintain benches and seats in a clean condition.
11. Keep parts and materials in their proper location.
12. When not in use, store creepers in a specific location. Do not leave them on the shop floor.
13. Keep the shop well lighted, and be sure all lights are in working order.
14. Replace frayed electrical cords on lights or equipment.
15. Clean walls and windows regularly.
16. Be sure stairs are clean, well lighted, and free of loose material.

If these housekeeping rules are followed, the shop will be a safer place to work, and customers will be impressed with the appearance of the premises.

## Air Quality

**CAUTION:** Never run the engine in a vehicle inside the shop without an exhaust hose connected to the tail pipe.

Vehicle exhaust contains small amounts of carbon monoxide, which is a poisonous gas. Strong concentrations of carbon monoxide may be fatal for human beings. All shop personnel are responsible for air quality in the shop. Shop management is responsible for an adequate exhaust system to remove exhaust fumes from the maximum number of vehicles that may be running in the shop at the same time. Technicians should never run a vehicle in the shop unless a shop exhaust hose is installed on the tail pipe of the vehicle. The exhaust fan must be switched on to remove exhaust fumes.

If shop heaters or furnaces have restricted chimneys, they release carbon monoxide emissions into the shop air. Therefore, chimneys should be checked periodically for restriction and proper ventilation.

Monitors are available to measure the level of carbon monoxide in the shop. Some of these monitors read the amount of carbon monoxide present in the shop air, and other monitors provide an audible alarm if the concentration of carbon monoxide exceeds the danger level.

Diesel exhaust contains some carbon monoxide, but particulates are also present in the exhaust from these engines. Particulates are basically small carbon particles, which can be harmful to the lungs.

The sulfuric acid solution in car batteries is a very corrosive, poisonous liquid. If a battery is charged with a fast charger at a high rate for a period of time, the battery becomes hot, and the sulfuric acid solution begins to boil. Under this condition, the battery may emit a strong sulfuric acid smell, and these fumes may be harmful to the lungs. If this condition occurs in the shop, the battery charger should be turned off, or the charger rate should be reduced considerably.

**WARNING:** When an automotive battery is charged, hydrogen gas and oxygen gas escape from the battery. If these gases are combined, they form water, but hydrogen gas by itself is very explosive. While a battery is charged, sparks, flames, and other sources of ignition must not be allowed near the battery.

**WARNING:** Breathing asbestos dust must be avoided because this dust is a known contributor to lung cancer.

Some automotive clutch facings and brake linings contain asbestos. Never use an air hose to blow dirt from these components, because this action disperses asbestos dust into the shop where

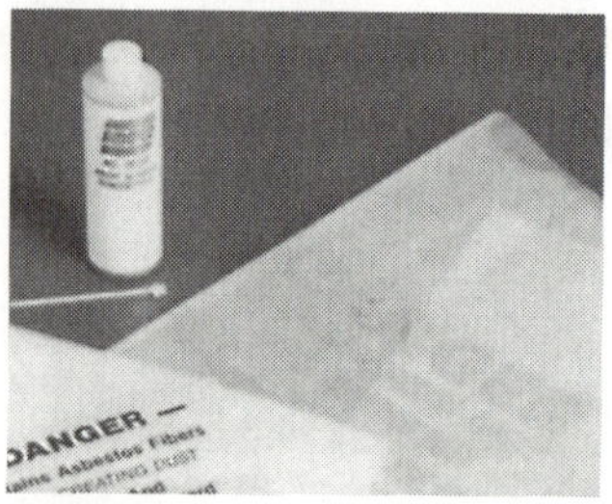


**Figure 1-3** Brake washer concentrate (Courtesy of Hennessy Industries Inc.)

it may be inhaled by technicians and other people in the shop. Parts washers approved by the Occupational Safety and Health Act (OSHA) must be used to clean the dust from these components (Figure 1-2). Brake washer concentrate is mixed with water in these parts washers (Figure 1-3). A catch basin with a removable liner is placed under the parts to be washed, and the washer sprays the cleaning solution on the parts. After the washing operation is completed, the liner containing the cleaning solution and asbestos dust is removed, sealed, and labeled for proper handling and disposal according to waste disposal laws.

Even though technicians take every precaution to maintain air quality in the shop, some undesirable gases may still get in the air. For example, exhaust manifolds may get oil on them during an engine overhaul. When the engine is started and these manifolds get hot, the oil burns off the manifolds and pollutes the shop air with oil smoke. Adequate shop ventilation must be provided to take care of this type of air contamination.

## Employer and Employee Obligations

When you begin employment, you enter into a business agreement with your employer. A business agreement involves an exchange of goods or services that have value. Although the automotive technician may not have a written agreement with his or her employer, the technician exchanges time, skills, and effort for money paid by the employer. Both the employee and the employer have obligations. The automotive technician's obligations include the following:

1. Productivity—As an automotive technician, you have a responsibility to your employer to make the best possible use of time on the job. Each job should be done in a reasonable length of time. Employees are paid for their skills, effort, and time.
2. Quality—Each repair job should be a quality job! Work should never be done in a careless manner. Nothing improves customer relations like quality workmanship.
3. Teamwork—The shop staff are a team and everyone, including technicians and management personnel, are team members. You should cooperate with and care about other team members. Each member of the team should strive for harmonious relations with fellow workers. Cooperative teamwork helps to improve shop efficiency, productivity, and customer relations. Customers may be "turned off" by bickering between shop personnel.
4. Honesty—Employers and customers expect, and deserve, honesty from automotive technicians. Honesty creates a feeling of trust between technicians, employers, and customers.
5. Loyalty—As an employee, you are obliged to act in the best interests of your employer, both on and off the job.
6. Attitude—Employees should maintain a positive attitude at all times. As in other professions, automotive technicians have days when it may be difficult to maintain a positive attitude. For example, there will be days when the technical problems on a certain vehicle are difficult to solve. However, developing a negative attitude certainly will not help the situation! A positive attitude has a positive effect on the job situation as well as on the customer and employer.

7. Responsibility—You are responsible for your conduct on the job and your work-related obligations. These obligations include always maintaining good workmanship and customer relations. Attention to details such as always placing fender and seat covers on customer vehicles prior to driving or working on the vehicle greatly improve customer relations.
8. Following directions—All of us like to do things "our way." Such action may not be in the best interests of the shop, and as an employee, you have an obligation to follow the supervisor's directions.
9. Punctuality and regular attendance—Employees have an obligation to be on time for work, and to be regular in attendance on the job. It is very difficult for a business to operate successfully if it cannot count on its employees to be on the job at the appointed time.
10. Regulations—Automotive technicians should be familiar with all state and federal regulations pertaining to their job situation, such as the Occupational Safety and Health Act (OSHA) and hazardous waste disposal laws. In Canada, employees should be familiar with work place hazardous materials information systems (WHMIS).

Employer- to-employee obligations should include:

1. Wages—The employer has a responsibility to inform the employee regarding the exact amount of financial remuneration to be received, and when it will be paid.
2. Fringe benefits—A detailed description of all fringe benefits should be provided by the employer. These benefits may include holiday pay, sickness and accident insurance, and pension plans.
3. Working conditions—A clean, safe work place must be provided by the employer. The shop must have adequate safety equipment and first-aid supplies. Employers must be certain that all shop personnel maintain the shop area and equipment to provide adequate safety and a healthy work place atmosphere.
4. Employee instruction—Employers must provide employees with clear job descriptions, and be sure that each worker is aware of his or her obligations.
5. Employee supervision—Employers should inform their workers regarding the responsibilities of their immediate supervisors and other management personnel.
6. Employee training—Employers must make sure that each employee is familiar with the safe operation of all the equipment that they are required to use in their job situation. Since automotive technology is changing rapidly, employers should provide regular update training for their technicians. Under the right-to-know laws, employers are required to inform all employees about hazardous materials in the shop. Employees should be familiar with work place hazardous materials information systems (WHMIS), which detail the labeling and handling of hazardous waste, and the health problems if exposed to hazardous waste.

## Job Responsibilities

An automotive technician has specific responsibilities regarding each job performed on a customer's vehicle. These job responsibilities include:

1. Do every job to the best of your ability. There is no place in the automotive service industry for careless workmanship! Automotive technicians and students must realize they have a very responsible job. During many repair jobs, you, as a student or technician working on a customer's vehicle, actually have the customer's life and the safety of his or her vehicle in your hands. For example, if you are doing a brake job and leave the wheel nuts loose on one wheel, that wheel may fall off the vehicle at high speed. This could result in serious personal injury for the customer and others, plus extensive

vehicle damage. If this type of disaster occurs, the individual who worked on the vehicle, and the shop may be involved in a very expensive legal action. As a student or technician working on customer vehicles, you are responsible for the safety of every vehicle that you work on! Even when careless work does not create a safety hazard, it leads to dissatisfied customers who often take their business to another shop, and nobody benefits when that happens.

2. Treat customers fairly and honestly on every repair job. Do not install parts that are unnecessary to complete the repair job.
3. Use published specifications. Do not guess at adjustments.
4. Follow the service procedures in the service manual provided by the vehicle manufacturer or an independent manual publisher.
5. When the repair job is completed, always be sure the customer's complaint has been corrected.
6. Do not be too concerned with work speed when you begin working as an automotive technician. Speed comes with experience.

## National Institute for Automotive Service Excellence (ASE) Certification

ASE has provided voluntary testing and certification of automotive technicians on a national basis for many years. The image of the automotive service industry has been enhanced by the ASE certification program. More than 265,000 ASE certified automotive technicians now work in a wide variety of automotive service shops. ASE provides certification in these eight areas of automotive repair:

1. engine repair
2. automatic transmissions/transaxles
3. manual drive train and axles
4. suspension and steering
5. brakes
6. electrical systems
7. heating and air conditioning
8. engine performance

A technician may take the ASE test and become certified in any, or all, of the eight areas. When a technician passes an ASE test in one of the eight areas, an Automotive Technician's shoulder patch is issued by ASE. If a technician passes all eight tests, he or she receives a Master Technician's shoulder insignia (Figure 1-4). Retesting at five-year intervals is required to remain certified.

**Figure 1-4** ASE certification shoulder insignia worn by automobile technicians and master technicians. (Courtesy of National Institute for Automotive Service Excellence [ASE])

**Table 1-1 SUSPENSION AND STEERING TEST SUMMARY**

| Content Area | Questions in Test | | Percentage of Test |
|---|---|---|---|
| A. Steering Systems Diagnosis and Repair | 9 | | 22.5% |
| 1. Steering Columns and Manual Steering Gears | | (2) | |
| 2. Power-Assisted Steering Units | | (4) | |
| 3. Steering Linkage | | (3) | |
| B. Suspension Systems Diagnosis and Repair | 13 | | 32.5% |
| 1. Front Suspensions | | (6) | |
| 2. Rear Suspensions | | (5) | |
| 3. Miscellaneous Service | | (2) | |
| C. Wheel Alignment Diagnosis, Adjustment, and Repair | 13 | | 32.5% |
| D. Wheel and Tire Diagnosis and Repair | 5 | | 12.5% |
| Total | 40 | | 100.0% |

(Courtesy of National Institute for Automotive Service Excellence [ASE])

The certification test in each of the eight areas contains 40 to 80 multiple-choice questions. The test questions are written by a panel of automotive service experts from various areas of automotive service such as automotive instructors, service managers, automotive manufacturer's representatives, test equipment representatives, and certified technicians. The test questions are pretested and checked for quality by a national sample of technicians. Most questions have the Technician A and Technician B format similar to the questions at the end of each chapter in this book. ASE regulations demand that each technician must have two years of working experience in the automotive service industry prior to taking a certification test or tests. However, relevant formal training may be substituted for one of the years of working experience. Contact ASE for details regarding this substitution. The contents of the Suspension and Steering test are listed in Table 1-1.

Shops that employ ASE certified technicians display an official ASE blue seal of excellence. This blue seal increases the customer's awareness of the shop's commitment to quality service, and the competency of certified technicians.

**Classroom Manual**
Chapter 1, page 1

## Shop Projects to Enhance the Theories in Chapter 1 of the Classroom Manual

1. Strike the tread of an inflated tire with a rubber hammer. What reaction did you get? Why did this reaction occur?
2. Push a tool box on casters sideways and straight ahead. What is the position of the casters in relation to the tool box direction? Why is this caster action important?
3. Place some cold water in a graduated heat-resistant container supported on a metal stand, and record the exact quantity of water. Heat the water with a propane torch until it boils, and record the exact quantity of water. What happened to the water when heated?
4. Obtain an empty, metal, 1 gallon container with a tight-fitting top. Heat the container until it is warm, and install the top securely. Allow the container to cool. Cooling time may be reduced by placing ice on the container. What happened to the container? Why did this happen?
5. Obtain a vacuum hand pump and a vacuum advance from a distributor. Gradually operate the pump and observe the vacuum diaphragm. How much vacuum is required to

begin moving the vacuum diaphragm? How much vacuum is required to move this diaphragm to the fully advanced position? What two forces move the diaphragm?

## Guidelines for Following Proper Shop Practices

1. Technicians must be familiar with shop layout, especially the location of safety equipment. This knowledge provides a safer, more efficient shop.
2. Shop rules must be observed by everyone in the shop to provide adequate shop safety, personal health protection, and vehicle protection.
3. The application of driving rules in the shop increases safety, protects customer vehicles and shop property, and improves the shop image in the eyes of the customer.
4. When good housekeeping habits are developed, shop safety is improved, worker efficiency is increased, and customers are impressed.
5. If some basic rules are followed to maintain shop air quality, the personal health of shop employees is improved.
6. When employers and employees accept and fulfill their obligations, personal relationships and general attitudes are greatly improved, shop productivity is increased, and customer relations are improved.
7. If a technician accepts certain job responsibilities, job quality improves and customer satisfaction increases.
8. ASE technician certification improves the quality of automotive repair and improves the image of the profession.

### CASE STUDY

A technician was removing and replacing the alternator on a General Motors car. After installing the replacement alternator and connecting the alternator battery wire, the technician proceeded to install the alternator belt. The rubber boot was still removed from the alternator battery terminal. While installing this belt, the technician's wrist watch expansion bracelet made electrical contact from the alternator battery terminal to ground on the alternator housing. Even though the alternator battery wire is protected with a fuse link that melted, a high current flowed through the wrist watch bracelet. This heated the bracelet to a very high temperature and severely burned the technician's arm.
This technician forgot two safety rules:

1. Never wear jewelry such as watches and rings while working in an automotive shop.
2. Before performing electrical work on a vehicle, disconnect the negative battery cable. If the vehicle is equipped with an air bag, wait 1 minute after this cable is disconnected.

### CASE STUDY

A technician was removing and replacing the starting motor with the solenoid mounted on top of the motor. The technician began removing the battery cable from the solenoid, and the ring on one of his fingers made contact between the end of the wrench and the engine block. The current flow through the ring was so high that the positive battery terminal melted out of the battery. The technician's finger was so badly burned that a surgeon had to cut the ring from his finger and repair the finger. This technician forgot to disconnect the negative battery cable before working on the vehicle!

## CASE STUDY

A technician had just replaced the engine in a Ford product, and she was performing final adjustments such as timing and air-fuel mixture. In this shop, the cars were parked in the work bays at an angle on both sides of the shop. With the engine running at fast idle, the automatic transmission suddenly slipped into reverse. The car went backwards across the shop and collided with a car in one of the electrical repair bays. Both vehicles were damaged to a considerable extent. Fortunately, no personnel were injured. This technician forgot to apply the parking brake while working on the vehicle!

## Terms to Know

Shop layout
Carbon monoxide
Asbestos parts washer
Diesel particulates
Sulfuric acid
National Institute for Automotive Service Excellence (ASE)
ASE technician certification
ASE blue seal of excellence

## ASE Style Review Questions

**1.** While discussing shop layout and safety:
*Technician A* says the location of each piece of safety equipment should be clearly marked.
*Technician B* says some shops have a special tool room where special tools are located.
Who is correct?
**A.** A only
**B.** B only
**C.** Both A and B
**D.** Neither A nor B

**2.** While discussing shop rules:
*Technician A* says breathing carbon monoxide may cause arthritis.
*Technician B* says breathing carbon monoxide may cause headaches.
Who is correct?
**A.** A only
**B.** B only
**C.** Both A and B
**D.** Neither A nor B

**3.** While discussing shop rules:
*Technician A* says breathing asbestos dust may cause heart defects.
*Technician B* says oily rags should be stored in uncovered garbage containers.
Who is correct?
**A.** A only
**B.** B only
**C.** Both A and B
**D.** Neither A nor B

**4.** While discussing shop rules:
*Technician A* says English tools may be substituted for metric tools.
*Technician B* says that foot injuries may be caused by loose sewer covers.
Who is correct?
**A.** A only
**B.** B only
**C.** Both A and B
**D.** Neither A nor B

5. While discussing air quality:
*Technician A* says a restricted chimney on a shop furnace may cause carbon monoxide gas in the shop.
*Technician B* says monitors are available to measure the level of carbon monoxide in the shop air.
Who is correct?
**A.** A only
**B.** B only
**C.** Both A and B
**D.** Neither A nor B

6. While discussing air quality:
*Technician A* says a battery gives off hydrogen gas during the charging process.
*Technician B* says a battery gives off oxygen gas during the charging process.
Who is correct?
**A.** A only
**B.** B only
**C.** Both A and B
**D.** Neither A nor B

7. While discussing air quality:
*Technician A* says diesel exhaust contains particulates.
*Technician B* says particulate emissions contain oxides of nitrogen.
Who is correct?
**A.** A only
**B.** B only
**C.** Both A and B
**D.** Neither A nor B

8. While discussing employer and employee responsibilities:
*Technician A* says employers are required to inform their employees about hazardous materials in the shop.
*Technician B* says that employers have no obligation to inform their employees about hazardous materials in the shop.
Who is correct?
**A.** A only
**B.** B only
**C.** Both A and B
**D.** Neither A nor B

9. While discussing ASE certification:
*Technician A* says a technician must pass four of the eight ASE certification tests to receive a master technician's shoulder patch.
*Technician B* says a technician must pass all eight ASE certification tests to receive a master technician's shoulder patch.
Who is correct?
**A.** A only
**B.** B only
**C.** Both A and B
**D.** Neither A nor B

10. While discussing ASE certification:
*Technician A* says retesting is required at five-year intervals to remain certified.
*Technician B* says a technician must have four years of automotive repair experience before writing an ASE certification test.
Who is correct?
**A.** A only
**B.** B only
**C.** Both A and B
**D.** Neither A nor B

# Tools and Safety Practices

CHAPTER 2

Upon completion and review of this chapter, you should be able to:

- ❑ Perform automotive measurements using United States customary (USC) and international system (SI) systems of weights and measures.
- ❑ Observe all personal safety precautions while working in the automotive shop.
- ❑ Demonstrate proper lifting procedures and precautions.
- ❑ Demonstrate proper vehicle lift operating and safety procedures.
- ❑ Observe all safety precautions when hydraulic tools are used in the automotive shop.
- ❑ Follow the recommended procedure while operating hydraulic tools such as presses, floor jacks, and vehicle lifts to perform automotive service tasks.
- ❑ Follow safety precautions regarding the use of power tools.
- ❑ Demonstrate proper safety precautions during the use of compressed air equipment.
- ❑ Follow safety precautions while using cleaning equipment in the automotive shop.

## Measuring Systems

Two systems of weights and measures are commonly used in the United States. One system of weights and measures is the United States customary system (USC). This system is commonly referred to as the English system. Well-known measurements for length in the USC system are the inch, foot, yard, and mile. In this system, the quart and gallon are common measurements for volume, and ounce, pound, and ton are measurements for weight. A second system of weights and measures is called the international system (SI). This system is often referred to as the metric system.

The United States customary (USC) system of weights and measures is commonly referred to as the English system.

The international system (SI) of weights and measures is called the metric system.

In the USC system, the basic linear measurement is the yard. The corresponding linear measurement in the metric system is the meter. Each unit of measurement in the metric system is related to the other metric units by a factor of 10. Thus, every metric unit can be multiplied or divided by 10 to obtain larger units (multiples), or smaller units (submultiples). For example, the meter may be divided by 10 to obtain centimeters (1/100 meter) or millimeters (1/1,000 meter).

The United States government passed the Metric Conversion Act in 1975 in an attempt to move American industry and the general public to accept and adopt the metric system. The automotive industry has adopted the metric system, and in recent years most bolts, nuts, and fittings on vehicles have been changed to metric. During the early 1980s some vehicles had a mix of English and metric bolts. Import vehicles have used the metric system for many years. Although the automotive industry has changed to the metric system, the general public in the United States has been slow to convert from the USC system to the metric system. One of the factors involved in this change is cost. What would it cost to change every highway distance and speed sign in the United States to read kilometers? Of course, the answer to that question is probably hundreds of millions or billions of dollars.

Service technicians must be able to work with both the USC and the metric system. One meter (m) in the metric system is equal to 39.37 inches (in) in the USC system. A metric tape measure may be graduated in millimeters, and 10 millimeters = 1 centimeter (Figure 2-1).

Some common equivalents between the metric and USC system are these:

1 meter (m) = 39.37 inches (in)
1 centimeter (cm) = 0.3937 inch
1 millimeter (mm) = 0.03937 inch
1 inch = 2.54 centimeters
1 inch = 25.4 millimeters

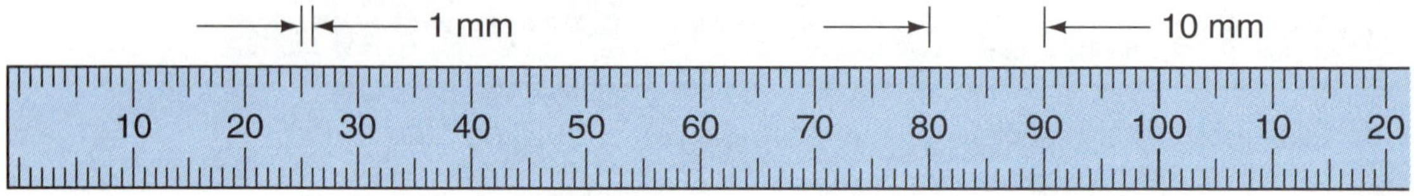


**Figure 2-1** Metric tape graduated in millimeters

In the USC system, phrases such as 1/8 of an inch are used for measurements. The metric system uses a set of prefixes. For example, in the word *kilometer*, the prefix kilo indicates 1,000, and this prefix indicates there are 1,000 meters in a kilometer. Common prefixes in the metric system are these:

| NAME | SYMBOL | MEANING |
|---|---|---|
| mega | M | one million |
| kilo | k | one thousand |
| hecto | h | one hundred |
| deca | da | ten |
| deci | d | one tenth of |
| centi | c | one hundreth of |
| milli | m | one thousandth of |
| micro | μ | one millionth of |

## Measurement of Mass

In the metric system, mass is measured in grams, kilograms, or tonnes. 1,000 grams (g) = 1 kilogram (kg). In the USC system, mass is measured in ounces, pounds, or tons. When converting pounds to kilograms, 1 pound = 0.453 kilograms.

## Measurement of Length

In the metric system, length is measured in millimeters, centimeters, meters, or kilometers. 10 millimeters (mm) = 1 centimeter (cm). In the USC system, length is measured in inches, feet, yards, or miles. When distance conversions are made between the two systems, some of the conversion factors are these:

1 inch = 25.4 millimeters
1 foot = 30.48 centimeters
1 yard = 0.91 meter
1 mile = 1.60 kilometers

## Measurement of Volume

In the metric system, volume is measured in milliliters, cubic centimeters, and liters. 1 cubic centimeter = 1 milliliter. If a cube has a length, depth, and height of 10 centimeters (cm), the volume of the cube is 10 cm x 10 cm x 10 cm = 1,000 $cm^3$ = 1 liter. When volume conversions are made between the two systems, 1 cubic inch = 16.38 cubic centimeters. If an engine has a displacement of 350 cubic inches, 350 x 16.38 = 5,733 cubic centimeters, and 5,733 ÷ 1,000 = 5.7 liters.

# Personal Safety

Personal safety is the responsibility of each technician in the shop. Always follow these safety practices:

1. Always use the correct tool for the job. If the wrong tool is used, it may slip and cause hand injury.

**Figure 2-2** Proper eye protection must be worn in the shop. (Courtesy of Mac Tools, Inc.)

2. Follow the car manufacturer's recommended service procedures.
3. Always wear eye protection such as safety glasses or a face shield (Figure 2-2).
4. Wear protective gloves when cleaning parts in hot or cold tanks, and when handling hot parts such as exhaust manifolds.
5. Do not smoke when working on a vehicle. A spark from a cigarette or lighter may ignite flammable materials in the work area.
6. When working on a running engine, keep hands and tools away from rotating parts. Remember that electric-drive fans may start turning at any time.
7. Do not wear loose clothing, and keep long hair tied behind your head. Loose clothing or long hair is easily entangled in rotating parts.
8. Wear safety shoes or boots.
9. Do not wear watches, jewelry, or rings when working on a vehicle. Severe burns occur when jewelry makes contact between an electrical terminal and ground. Rings may catch on an object, resulting in painful injury.
10. Always place a shop exhaust hose on the vehicle tail pipe if the engine is running, and be sure the exhaust fan is running. Carbon monoxide in the vehicle exhaust is harmful, or fatal, to the human body.
11. Be sure that the shop has adequate ventilation. Carbon monoxide is odorless, do not expect to be able to smell it.
12. Make sure the work area has adequate lighting.
13. Use trouble lights with steel or plastic cages around the bulb. If an unprotected bulb breaks, it may ignite flammable materials in the area.
14. When servicing a vehicle, always apply the parking brake, and place the transmission in park with an automatic transmission, or neutral with a manual transmission.
15. Avoid working on a vehicle parked on an incline.
16. Never work under a vehicle unless the vehicle chassis is supported securely on jack stands.
17. When one end of a vehicle is raised, place wheel chocks on both sides of the wheels remaining on the floor.
18. Be sure that you know the location of shop first-aid kits and eyewash fountains.
19. Familiarize yourself with the location of all shop fire extinguishers.
20. Do not use any type of open flame heater to heat the work area.
21. Collect oil, fuel, brake fluid, and other liquids in the proper safety containers.
22. Use only approved cleaning fluids and equipment. *Do not use gasoline to clean parts.*
23. Obey all state and federal safety, fire, and hazardous material regulations.
24. Always operate equipment according to the equipment manufacturer's recommended procedure.
25. Do not operate equipment unless you are familiar with the correct operating procedure.
26. Do not leave running equipment unattended.

27. Do not use electrical equipment, including trouble lights, with frayed cords.
28. Be sure the safety shields are in place on rotating equipment.
29. Before operating electric equipment, be sure the power cord has a ground connection.
30. When working in an area where extreme noise levels are encountered, wear ear plugs or covers.
31. Be sure all shop equipment is serviced regularly.

## Lifting and Carrying

Many automotive service jobs require heavy lifting. You should know your maximum weight-lifting ability, and do not attempt to lift more than this weight. If a heavy part exceeds your weight-lifting ability, have a co-worker help with the lifting job. Follow these steps when lifting or carrying an object:

1. If the object is to be carried, be sure your path is free from loose parts or tools.
2. Position your feet close to the object, and position your back reasonably straight for proper balance.
3. Keep your back and elbows as straight as possible. Continue to bend your knees until your hands reach the best lifting location on the object to be lifted.
4. Be certain the container is in good condition. If a container falls apart during the lifting operation, parts may drop out of the container and result in foot injury or part damage.
5. Maintain a firm grip on the object, and do not attempt to change your grip while lifting is in progress.
6. Straighten your legs to lift the object, and keep the object close to your body. Use leg muscles rather than back muscles.
7. If you have to change direction of travel, turn your whole body instead of twisting it.
8. Do not bend forward to place an object on a work bench or table. Position the object on the front surface of the work bench and slide it back. Do not pinch your fingers under the object while setting it on the front of the bench.
9. If the object must be placed on the floor or a low surface, bend your legs to lower the object. Do not bend your back forward, because this movement strains back muscles.
10. When a heavy object must be placed on the floor, locate suitable blocks under the object to prevent jamming your fingers.

## Hand-Tool Safety

Many shop accidents are caused by improper use and care of hand tools. These hand-tool safety steps must be followed:

1. Maintain tools in good condition and keep them clean. Worn tools may slip and result in hand injury. If a hammer is used with a loose head, the head may fly off and cause personal injury or vehicle damage. Your hand may slip off a greasy tool, and this action may cause some part of your body to hit the vehicle. For example, your head may hit the vehicle hood.
2. Using the wrong tool for the job may cause damage to the tool, fastener, or your hand, if the tool slips. If you use a screwdriver for a chisel or pry bar, the blade may shatter, causing serious personal injury.
3. Use sharp, pointed tools with caution. Always check your pockets before sitting on the vehicle seat. A screwdriver, punch, or chisel in the back pocket may put an expensive tear in the upholstery. Do not lean over fenders with sharp tools in your pockets.
4. Tool tips that are intended to be sharp should be kept in a sharp condition. Sharp tools, such as chisels, will do the job faster with less effort.

# Suspension and Steering Tools

**CUSTOMER CARE:** Some automotive service centers have a policy of performing some minor service as an indication of their appreciation to the customer. This service may include cleaning all the windows and/or vacuuming the floors before the car is returned to the customer.

Although this service involves more labor costs for the shop, it may actually improve profits over a period of time. When customers find their windows cleaned and/or the floors vacuumed, it impresses them with the quality of work you do, and the fact you care about their vehicle. They will likely return for service, and tell their friends about the quality of service your shop performs.

## Seal Drivers

**WARNING:** While using all suspension and steering tools, the vehicle manufacturer's recommended procedure in the service manual must be followed. Improper use of tools may lead to personal injury.

**CAUTION:** Do not hit the seal case with a hammer. This action will damage the seal.

**CAUTION:** Always be sure the seal goes into the housing squarely and evenly.

**SERVICE TIP:** When installing a seal, the garter spring must face toward the flow of lubricant.

Seal drivers are designed to fit squarely against the seal case, and inside the seal lip. A soft hammer is used to tap the seal driver and drive the seal straight into the housing. Some tool manufacturers market a seal driver kit with drivers to fit many common seals (Figure 2-3).

## Bearing Pullers

A variety of bearing pullers are available to pull different sizes of bearings in various locations (Figure 2-4). Some bearing pullers are slide-hammer-type pullers, while others are screw-type pullers.

## Axle Pullers

Axle pullers are used to pull rear axles in rear wheel drive vehicles. Most rear axle pullers are slide-hammer-type pullers (Figure 2-5).

## Stethoscope

A stethoscope is used to amplify sound and to diagnose the source of noises such as bearing noise. The stethoscope pickup is placed on the suspected noise source, and the ends of the two arms are placed in the technician's ears (Figure 2-6).

## Front Bearing Hub Tool

Front bearing hub tools are designed to remove and install front wheel bearings on front wheel drive cars (Figure 2-7). These bearing hub tools are usually designed for a specific make of vehicle.

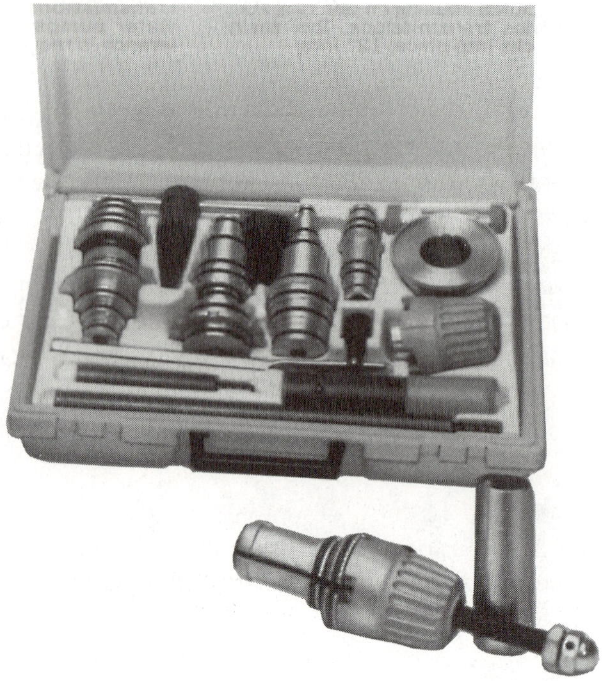

Figure 2-3 Seal drivers (Courtesy of Snap-On Tools Corporation)

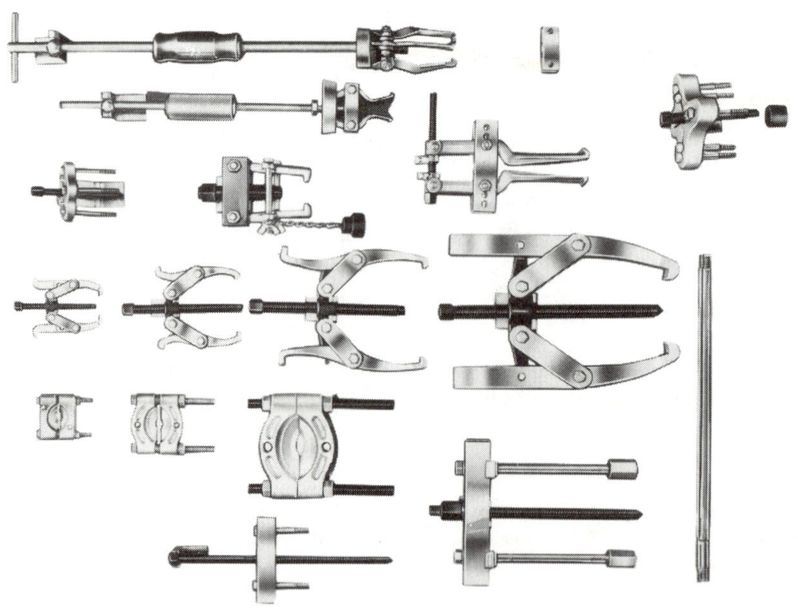

Figure 2-4 Bearing pullers (Courtesy of Owatonna Tool Company)

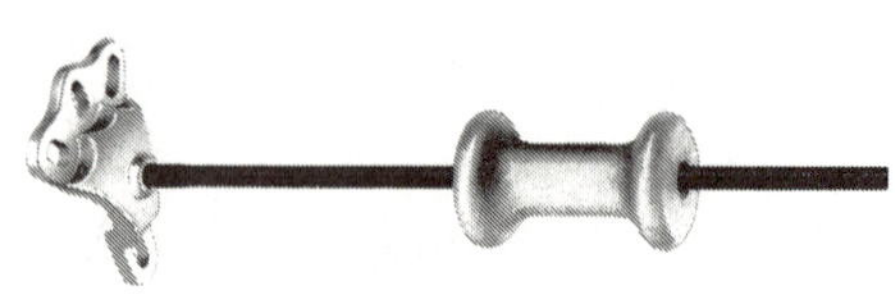

Figure 2-5 Rear axle puller (Courtesy of Mac Tools, Inc.)

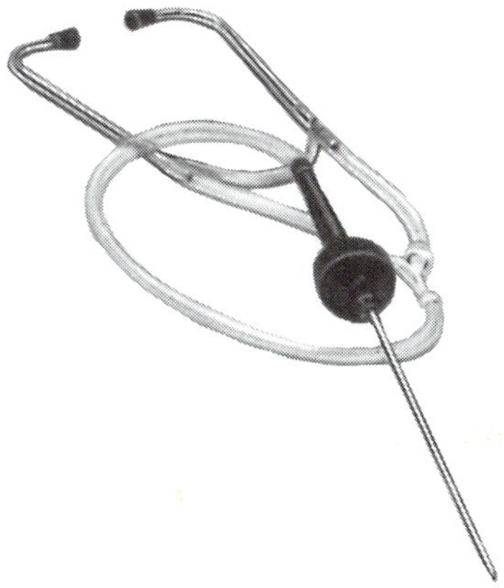

Figure 2-6 Stethoscope (Courtesy of Snap-On Tools Corporation)

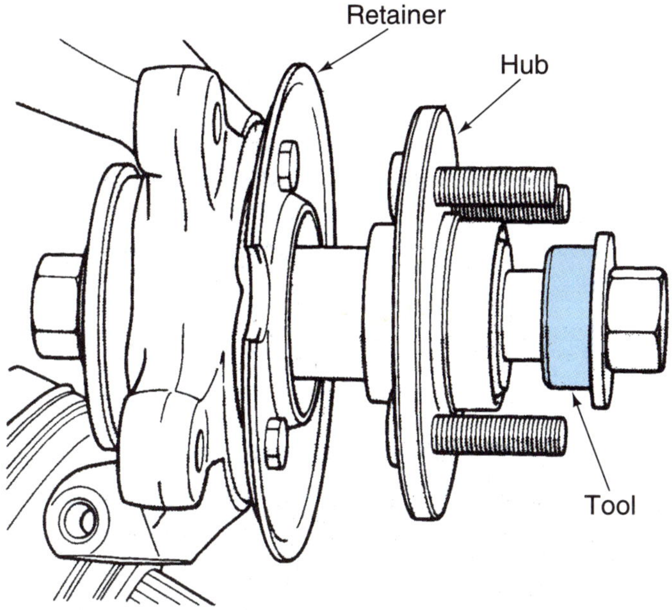


Figure 2-7 Front bearing hub tool (Courtesy of Chrysler Corporation)

## Dial Indicator

**SERVICE TIP:** Dial indicators must be kept clean and dry for accuracy and long life.

Dial indicators are used for measuring in many different locations. In the suspension and steering area, dial indicators are used for measurements such as tire runout and ball joint movement (Figure 2-8). Dial indicators have many different attaching devices to connect the indicator to the component to be measured.

Figure 2-8 Dial indicator (Courtesy of L.S. Starrett Company)

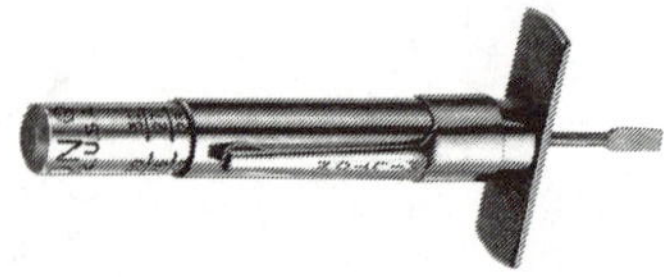

Figure 2-9 Tire tread depth gauge (Courtesy of Mac Tools, Inc.)

## Tire Tread Depth Gauge

A tire tread depth gauge measures the tread depth. This measurement should be taken at three or four locations around the tire circumference to obtain an average tread depth (Figure 2-9). This gauge is essential when making tire warranty adjustments.

## Tire Changer

**WARNING:** Never operate a tire changer until you are familiar with its operation. This action may lead to serious personal injury. Your instructor will explain and demonstrate the operation of this equipment. You should read the equipment operator's manual, and use this equipment under the instructor's supervision until you are familiar with it.

Tire changers are used to demount and mount tires (Figure 2-10). These changers may be used on most common tire sizes. There are a wide variety of tire changers available, and each one has somewhat different operating procedures. Always follow the procedure in the equipment operator's manual, and the directions provided by your instructor.

## Wheel Balancer

**WARNING:** Using a wheel balancer before you are familiar with its operation may result in serious personal injury and property damage.

Figure 2-10 Typical tire changer

**Figure 2-11** Typical electronic wheel balancer (Courtesy of Hunter Engineering Company)

Electronic wheel balancers are used in most automotive repair shops (Figure 2-11). Do not attempt to use this equipment until you have studied wheel balance theory. Your instructor will explain and demonstrate the use of this equipment before you attempt to use it. This equipment should be used under the supervision of your instructor until you are familiar with it.

## Machinist's Rule

A machinist's rule is used to perform many accurate measurements in the shop. Most machinist's rules are graduated in inches and millimeters (Figure 2-12).

## Ball Joint Removal and Pressing Tools

Ball joint removal and pressing tools are designed to remove and replace pressed-in ball joints on front suspension systems (Figure 2-13). The size of the removal and pressing tool must match the size of the ball joint.

## Tie Rod End and Ball Joint Puller

Some car manufacturers use the same puller to pull tie rod ends and remove ball joint studs from the steering knuckle (Figure 2-14).

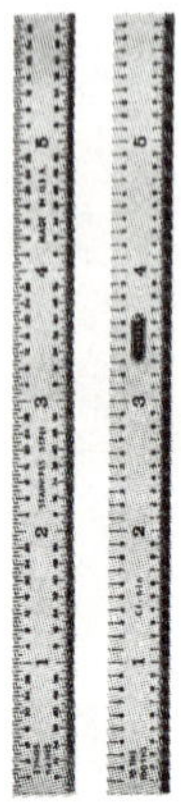

| Cat # | Length | Width | Graduations Inches Front | Back | Type | Wt. (lbs) |
|---|---|---|---|---|---|---|
| 9-GHCF645 | 6" | 1/2" | 10, 100 | 32, 64 | FLEX | 0.2 |
| 9-GHCF1265 | 12" | 1" | 10, 100 | 32, 64 | RIGID | 0.4 |
| 9-GHCF1865 | 18" | 1 1/8" | 10, 100 | 32, 64 | RIGID | 0.6 |
| 9-GHCF2465 | 21" | 1 1/8" | 10, 100 | 32, 64 | RIGID | 0.8 |
| 9-GHCF667ME | 6" | 3/4" | 32, 64 | mm,.5mm | RIGID | 0.2 |
| 9-GHCF1247M | 12' | 1/2" | 32, 64 | mm,.5mm | FLEX | 0.4 |
| **Cat #** | **Length** | **Width** | **Graduations Inches Front** | **Back** | **Type** | **Wt. (lbs)** |
| 9-GH3001 | 6" | 15/32" | 32, 64 | Dec. Eq. | FLEX | 0.2 |
| 9-GH300ME | 6" | 15/32" | mm, 64 | Metric Eq. | FLEX | 0.2 |
| 9-GH311ME | 6" | 3/4" | 16, 32 | – | FLEX | 0.2 |
| 9-GHCF616 | 6" | 15/32" | 10, 100 | 32, 64 | FLEX | 0.2 |
| 9-GH616 | 6" | 15/32" | 10, 100 | 32, 64 | FLEX | 0.2 |
| 9-GHCF676 | 6" | 3/4" | 8, 16 | 32, 64 | RIGID | 0.2 |
| 9-GHCF1227ME | 12" | 1" | 32, 64 | mm,.5mm | RIGID | 0.4 |
| 9-GH1210 | 12" | 1" | 16, 32 | 32, 64 | CENTER FINDING RIGID | 0.2 |

**Figure 2-12** Machinist's rule (Courtesy of Sears Industrial Tools)

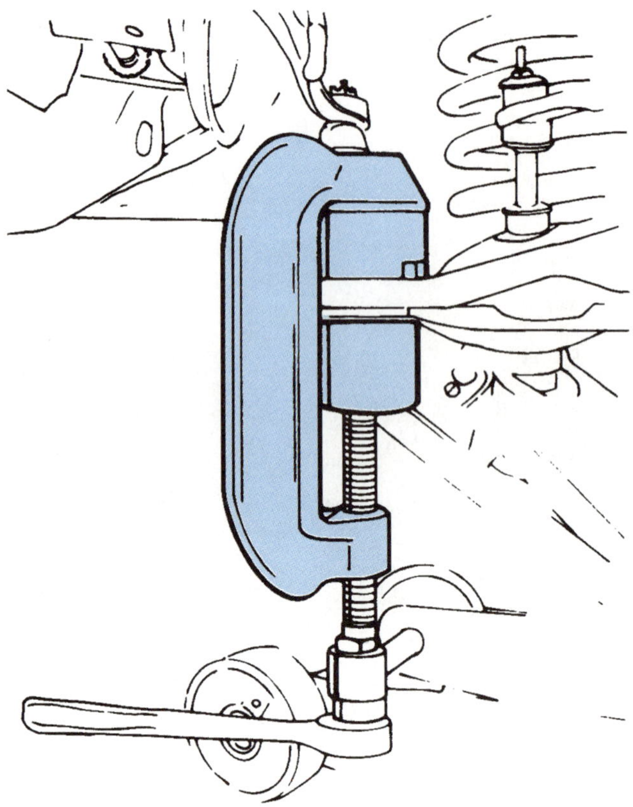

**Figure 2-13** Ball joint removal and pressing tools (Courtesy of Chevrolet Motor Division, General Motors Corporation)

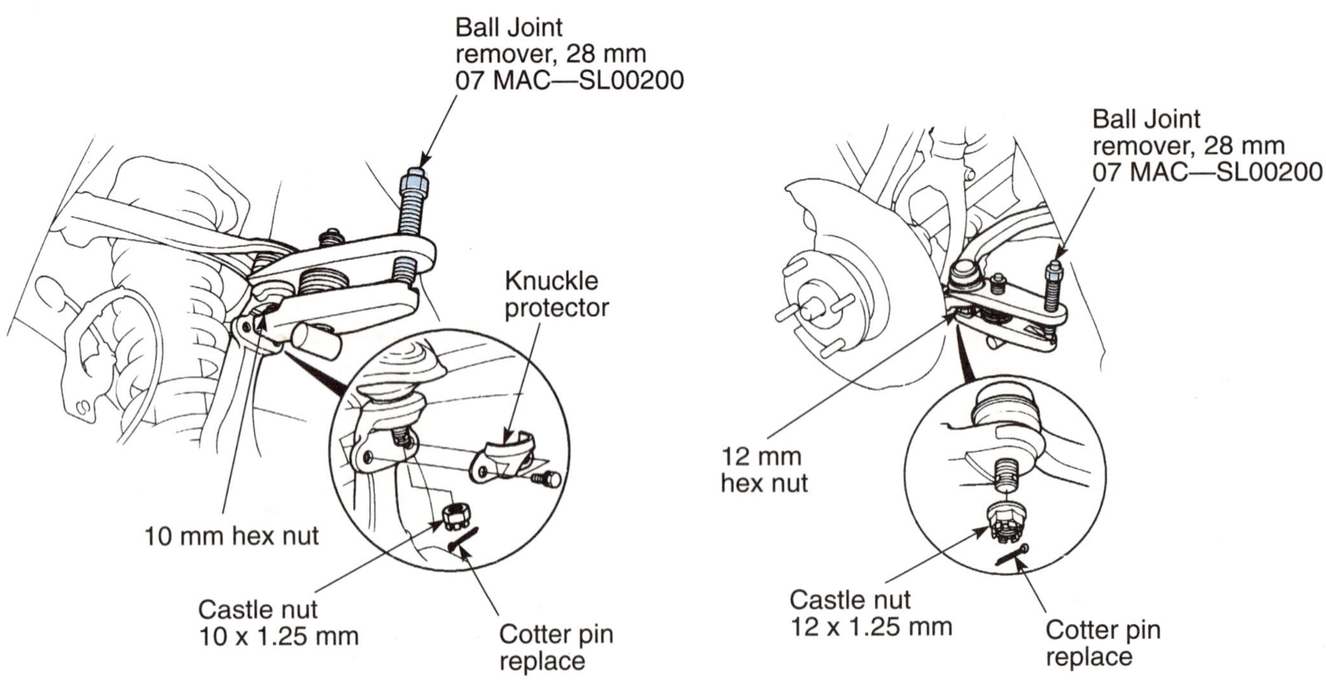


**Figure 2-14** Tie rod end and ball joint puller (Courtesy of American Honda Motor Co., Inc.)

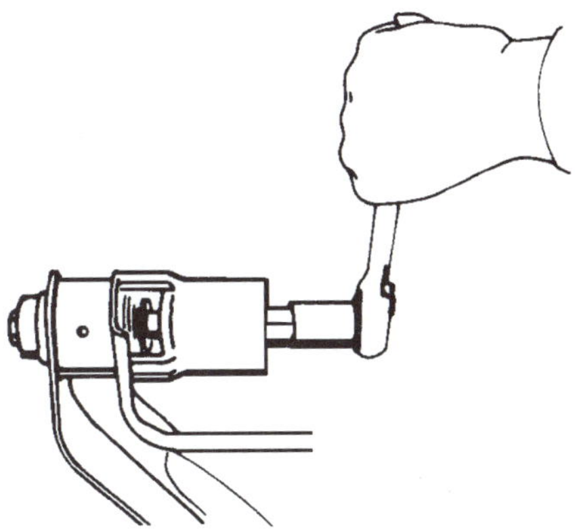

**Figure 2-15** Control arm bushing removal and replacement tools (Courtesy of Chevrolet Motor Division, General Motors Corporation)

## Control Arm Bushing Tools

A variety of control arm bushing removal and replacement tools are available to service control arm bushings. The bushing removal and installation tool must match the size of the control arm bushing (Figure 2-15).

## Coil Spring Compressor Tool

**WARNING:** There is a tremendous amount of energy in a compressed coil spring. Never disconnect any suspension component that will suddenly release this tension. This action may result in serious personal injury, and vehicle or property damage.

**WARNING:** The vehicle and equipment manufacturer's recommended procedures must be followed for each type of spring compressor tool.

There are many types of spring compressor tools available in the automotive service industry (Figure 2-16). These tools are designed to compress the coil spring and hold it in the compressed position while removing the strut from the coil spring, or performing other suspension work. Various types of spring compressor tools are required on different types of front suspension systems. While using spring compressor tools, the vehicle manufacturer's and equipment manufacturer's recommended procedures must be followed.

## Power Steering Pressure Gauge

**CAUTION:** The power steering pump delivers extremely high pressure during the pump pressure test. Always follow the recommended test procedure in the vehicle manufacturer's service manual to avoid personal injury during this test.

A power steering pump pressure gauge is used to test the power steering pump pressure (Figure 2-17). Since the power steering pump delivers extremely high pressure during this test, the recommended procedure in the vehicle manufacturer's service manual must be followed.

## Pitman Arm Puller

**CAUTION:** Never strike a puller with a hammer when it is installed and tightened. When this action is taken, personal injury may occur.

A pitman arm puller is a heavy-duty puller designed to remove the pitman arm from the pitman shaft (Figure 2-18).

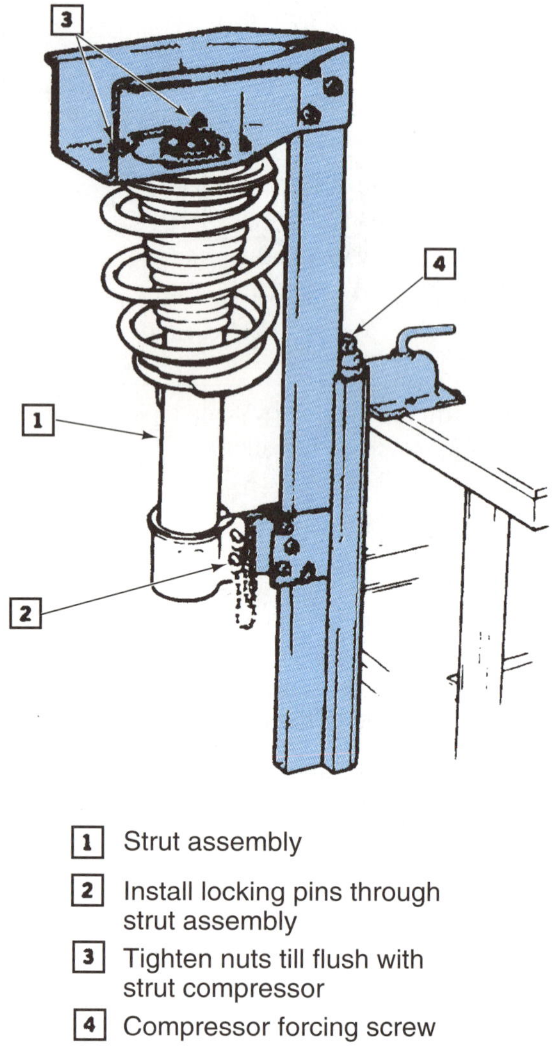


**Figure 2-16** Coil spring compressor tool (Courtesy of Oldsmobile Division, General Motors Corporation)

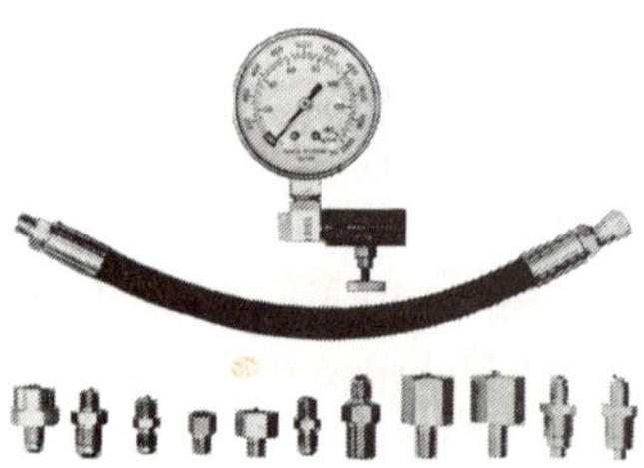

**Figure 2-17** Power steering pressure gauge (Courtesy of Mac Tools, Inc.)

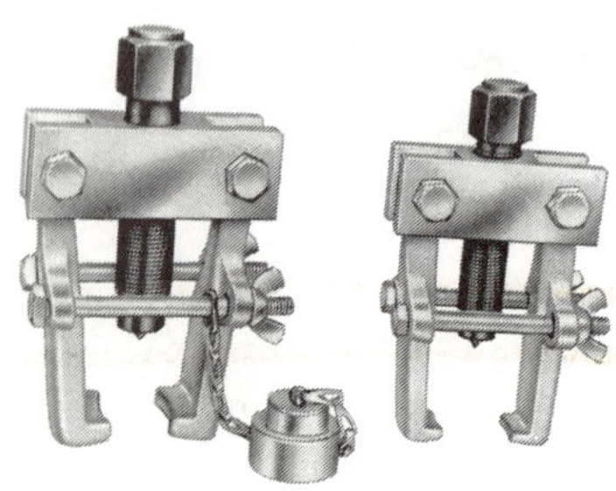

**Figure 2-18** Pitman arm puller (Courtesy of Mac Tools, Inc.)

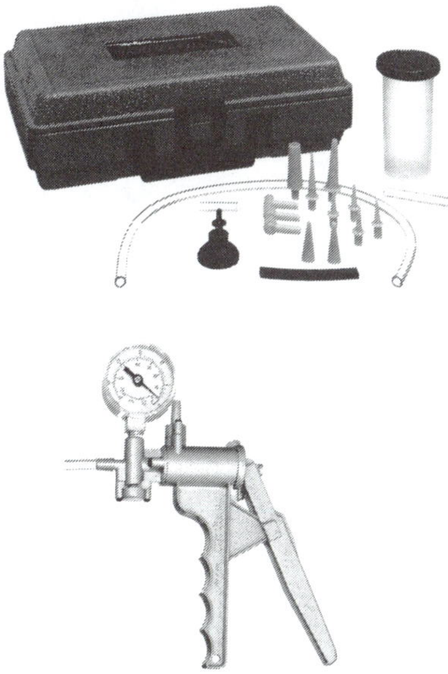

Figure 2-19 Vacuum hand pump (Courtesy of Mac Tools, Inc.)

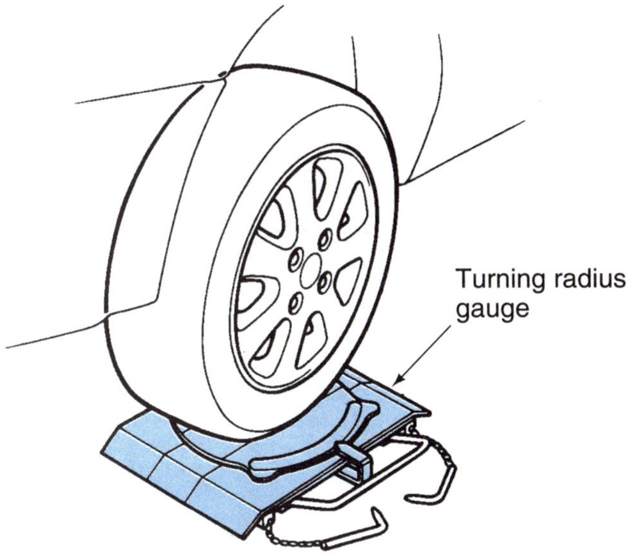


Figure 2-20 Turning radius gauge (Courtesy of American Honda Motor Co., Inc.)

## Vacuum Hand Pump

A vacuum hand pump creates vacuum to test vacuum-operated components and hoses (Figure 2-19).

## Turning Radius Gauge

Turning radius gauge turntables are placed under the front wheels during a wheel alignment. The top plate in the turning radius gauge rotates on the bottom plate to allow the front wheels to be turned during a wheel alignment. A degree scale and a pointer on the gauge indicate the number of degrees the front wheels are turned (Figure 2-20). If the car has four-wheel steering (4WS), the turning radius gauges are also placed under the rear wheels during a wheel alignment.

## Plumb Bob

A plumb bob is a metal weight with a tapered end (Figure 2-21). This weight is suspended on a string. Plumbers use a plumb bob to locate pipe openings directly below each other at the top and bottom of partitions. Some vehicle manufacturers recommend checking vehicle frame measurements with a plumb bob.

Figure 2-21 Plumb bob (Courtesy of Sears Industrial Tools)

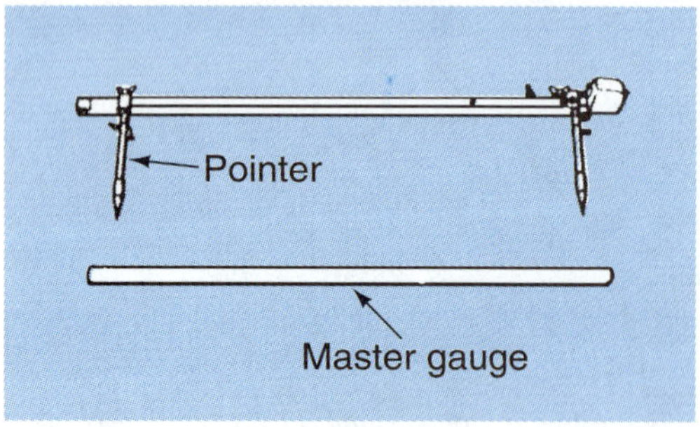


Figure 2-22 Tram gauge (Courtesy of Toyota Motor Corporation)

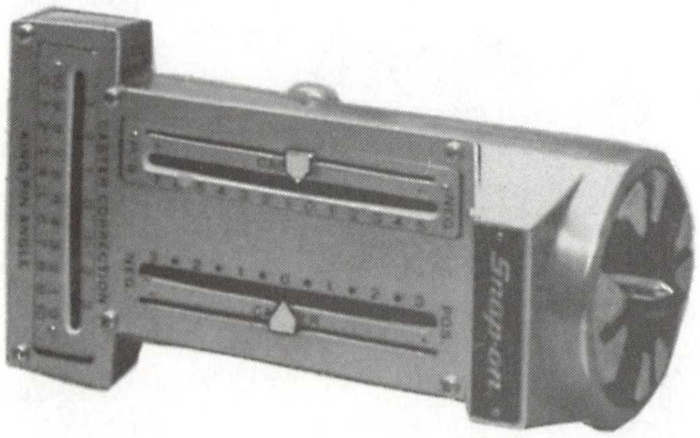

Figure 2-23 Magnetic wheel alignment gauge (Courtesy of Snap-On Tools Corporation)

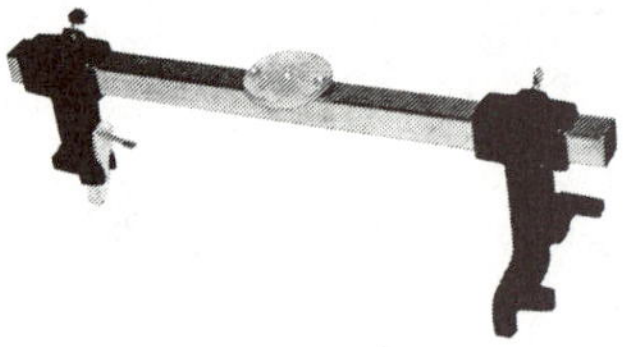

Figure 2-24 Rim clamps (Courtesy of Mac Tools, Inc.)

## Tram Gauge

A tram gauge is a long, straight graduated bar with an adjustable pointer at each end (Figure 2-22). The tram gauge is used for performing frame and body measurements.

## Magnetic Wheel Alignment Gauge

**SERVICE TIP:** Magnetic wheel alignment gauge mounting surfaces must be clean with no metal burrs.

Each magnetic alignment gauge contains a strong magnet, which holds the gauge securely on the front wheel hubs. The magnetic wheel alignment gauge is capable of measuring some of the front suspension alignment angles (Figure 2-23).

## Rim Clamps

When the wheel hub is inaccessible to the magnetic alignment gauge, an adjustable rim clamp may be attached to each front wheel, and the magnetic gauges may be attached to the rim clamp (Figure 2-24). Rim clamps are also used on computer wheel aligners.

## Brake Pedal Jack

A brake pedal jack must be installed between the front seat and the brake pedal to apply the brakes while checking some front wheel alignment angles (Figure 2-25).

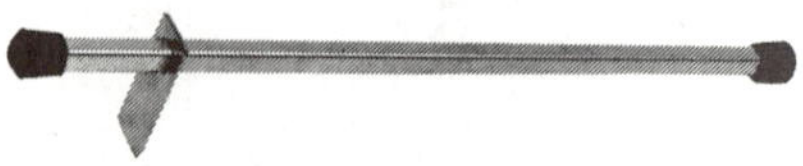

Figure 2-25 Brake pedal jack (Courtesy of Mac Tools, Inc.)

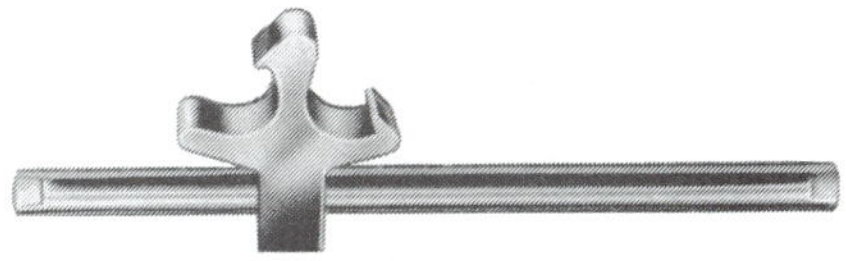

**Figure 2-26** Tie rod sleeve adjusting tool (Courtesy of Mac Tools, Inc.)

**Figure 2-27** Steering wheel locking tool (Courtesy of Mac Tools, Inc.)

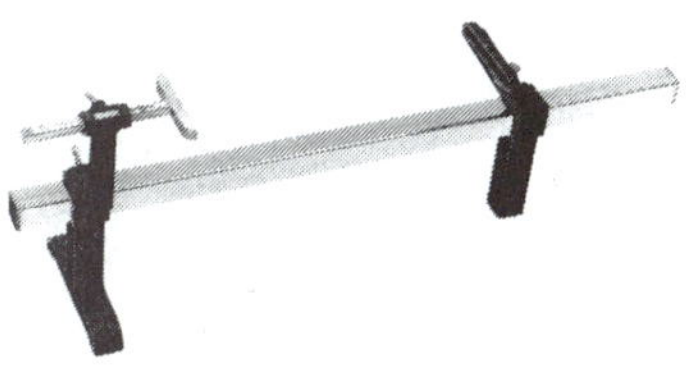

**Figure 2-28** Toe gauge (Courtesy of Mac Tools, Inc.)

## Tie Rod Sleeve Adjusting Tool

**CAUTION:** Do not use anything except a tie rod adjusting tool to adjust the tie rod sleeves. Tools such as a pipe wrench will damage the sleeves.

A tie rod sleeve adjusting tool is required to rotate the tie rod sleeves and perform some front wheel adjustments (Figure 2-26).

## Steering Wheel Locking Tool

A steering wheel locking tool is required to lock the steering wheel while performing some front suspension service (Figure 2-27).

## Toe Gauge

A toe gauge is a long, straight, graduated bar that may be used to measure front wheel toe (Figure 2-28).

## Track Gauge

A track and toe fiber-optic alignment system is available to measure front wheel toe, and determine if the rear wheels are tracking directly behind the front wheels. The front and rear fiber-optic gauges may be connected to the wheel hubs or to rim clamps attached to the wheel rims (Figure 2-29). A remote light source in the main control is sent through fiber-optic cables to the wheel gauges. A strong light beam between the front and rear wheel units informs the technician if the rear wheel tracking is correct.

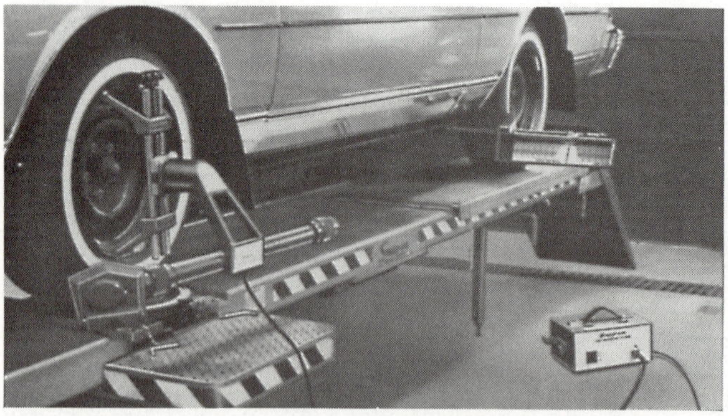

Figure 2-29 Fiber-optic track and toe gauge (Courtesy of Snap-On Tools Corporation)

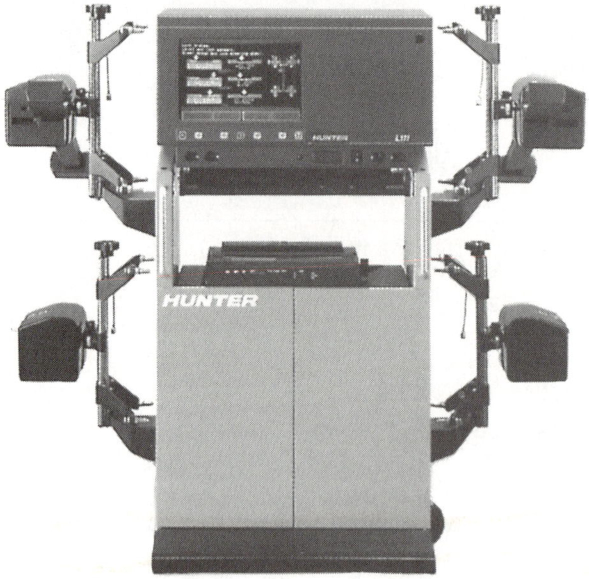


Figure 2-30 Computer wheel aligner (Courtesy of Hunter Engineering Company)

### Computer Wheel Aligner

Many automotive shops are equipped with computer wheel aligners (Figure 2-30). These wheel aligners perform all front and rear wheel alignment angles quickly and accurately.

### Scan tester

There are a variety of scan testers available for diagnosing automotive computer systems (Figure 2-31). These testers will obtain fault codes and perform other diagnostic functions on computer-controlled suspension systems.

## Hydraulic Pressing and Lifting Equipment

### Hydraulic Press

**WARNING:** When operating a hydraulic press, always be sure that the components being pressed are supported properly on the press bed with steel supports.

**Figure 2-31** Scan tester for diagnosing computer-controlled suspension systems (Courtesy of OTC Division, SPX Corporation)

**WARNING:** When using a hydraulic press, never operate the pump handle until the pressure gauge exceeds the maximum pressure rating of the press. If this pressure is exceeded, some part of the press may suddenly break and cause severe personal injury.

When two components have a tight precision fit between them, a hydraulic press is used to separate these components, or to press them together. The hydraulic press rests on the shop floor, and an adjustable steel beam bed is retained to the lower press frame with heavy steel pins. A hydraulic cylinder and ram is mounted on the top part of the press with the ram facing downward toward the press bed (Figure 2-32). The component being pressed is placed on the press bed with

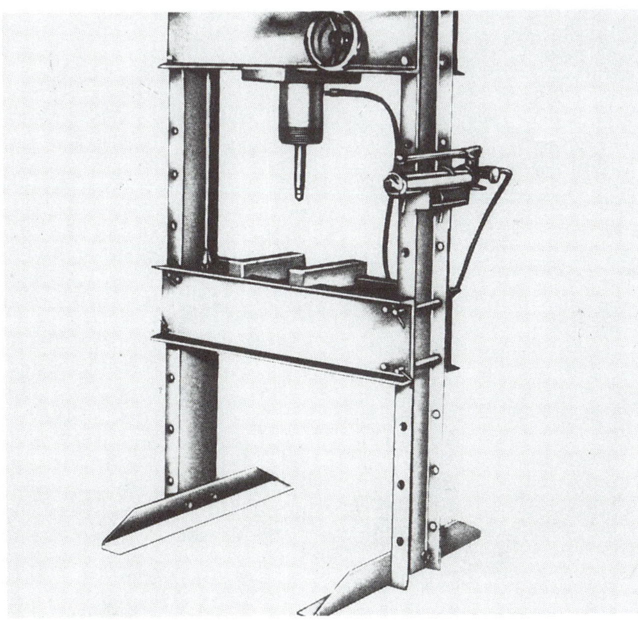

**Figure 2-32** Hydraulic press (Courtesy of Snap-on Tools Corporation)

appropriate steel supports. A hand-operated hydraulic pump is mounted on the side of the press. When the handle is pumped, hydraulic fluid is forced into the cylinder, and the ram is extended against the component on the press bed to complete the pressing operation. A pressure gauge on the press indicates the pressure applied from the hand pump to the cylinder. The press frame is designed for a certain maximum pressure, and this pressure must not be exceeded during hand pump operation.

## Floor Jack

**WARNING:** The maximum lifting capacity of the floor jack is usually written on a jack decal. Never lift a vehicle that exceeds the jack lifting capacity. This action may cause the jack to break or collapse, resulting in vehicle damage or personal injury.

A floor jack is a portable unit mounted on wheels. The lifting pad on the jack is placed under the chassis of the vehicle, and the jack handle is operated with a pumping action (Figure 2-33). This jack handle operation forces fluid into a hydraulic cylinder in the jack, and this cylinder extends to force the jack lift pad upward and lift the vehicle. Always be sure that the lift pad is positioned securely under one of the car manufacturer's recommended lift points. To release the hydraulic pressure and lower the vehicle, the handle or release lever must be turned slowly. Do not leave the jack handle where someone can trip over it.

## Lift

A lift may be referred to as a hoist.

**WARNING:** Always be sure that the lift arms are securely positioned under the car manufacturer's recommended lift points before raising a vehicle. These lift points are shown in the service manual.

**WARNING:** The maximum capacity of the vehicle lift is placed on an identification plate. Never lift a vehicle that is heavier than the maximum capacity of the lift.

A lift is used to raise a vehicle so the technician can work under the vehicle. The lift arms must be placed under the car manufacturer's recommended lift points prior to raising a vehicle. Twin posts

**Figure 2-33** Hydraulic floor jack (Courtesy of Lincoln St. Louis)

Figure 2-34 Lifts are used to raise a vehicle (Courtesy of Bear Automotive Equipment Service Company)

are used on some lifts, whereas other lifts have a single post (Figure 2-34). Some lifts have an electric motor that drives a hydraulic pump to create fluid pressure and force the lift upward. Other lifts use air pressure from the shop air supply to force the lift upward. If shop air pressure is used for this purpose, the air pressure is applied to fluid in the lift cylinder. A control lever or switch is placed near the lift. The control lever supplies shop air pressure to the lift cylinder, and the switch turns on the lift pump motor. Always be sure that the safety lock is engaged after the lift is raised. When the safety lock is released, a release lever is operated slowly to lower the vehicle.

## Lift Safety

**WARNING:** Do not raise a vehicle on a lift if the vehicle weight exceeds the maximum capacity of the lift.

**WARNING:** When a vehicle is raised on a lift, the vehicle must be raised high enough to allow engagement of the lift locking mechanism.

Special precautions and procedures must be followed when a vehicle is raised on a lift. Follow these steps for safety lift:

1. Always be sure the lift is completely lowered before driving a vehicle on or off the lift.
2. Do not hit or run over lift arms and adaptors when driving a vehicle on or off the lift. Have a co-worker guide you when driving a vehicle onto the lift. Do not stand in front of a lift with the car coming towards you.
3. Be sure the lift pads on the lift are contacting the car manufacturer's recommended lift points shown in the service manual. If the proper lift points are not used, components under the vehicle such as brake lines or body parts may be damaged. Failure to use the recommended lift points may cause the vehicle to slip off the lift, resulting in severe vehicle damage and personal injury.
4. Before a vehicle is raised or lowered, close the doors, hood, and trunk lid.

5. When a vehicle is lifted a short distance off the floor, stop the lift and check the contact between the hoist lift pads and the vehicle to be sure the lift pads are still on the recommended lift points.
6. When a vehicle is raised on a lift, be sure the safety mechanism is in place to prevent the lift from dropping accidentally.
7. Before lowering a vehicle on a lift, always make sure there are no objects, tools, or people under the vehicle.
8. Do not rock a vehicle on a lift during a service job.
9. When a vehicle is raised on a hoist, removal of some heavy components may cause vehicle imbalance on the lift. Since front wheel drive cars have the engine and transaxle located at the front of the vehicle, these cars have most of their weight on the front end. Removing a heavy rear end component on these cars may cause the back end of the car to rise off the lift. If this action is allowed to happen, the vehicle could fall off the lift!
10. Do not raise a vehicle on a lift with people in the vehicle.
11. When raising pickup trucks and vans on a lift remember these vehicles are higher than a passenger car. Be sure there is adequate clearance between the top of the vehicle and the shop ceiling, or components under the ceiling.
12. *Do not raise a four-wheel-drive vehicle with a frame contact lift, because this may damage axle joints.*
13. Do not operate a front wheel drive vehicle that is raised on a frame contact lift. This action may damage the front drive axles.

## Hydraulic Jack and Jack Stand Safety

**WARNING:** Always make sure the jack stand weight capacity rating exceeds the vehicle weight that is lowered onto the stands.

**WARNING:** Never lift a vehicle with a floor jack if the weight of the vehicle exceeds the rated capacity of the jack.

Accidents involving the use of floor jacks and jack stands may be avoided if these safety precautions are followed:

1. Never work under a vehicle unless jack stands are placed securely under the vehicle chassis, and the vehicle is resting on these stands (Figure 2-35).

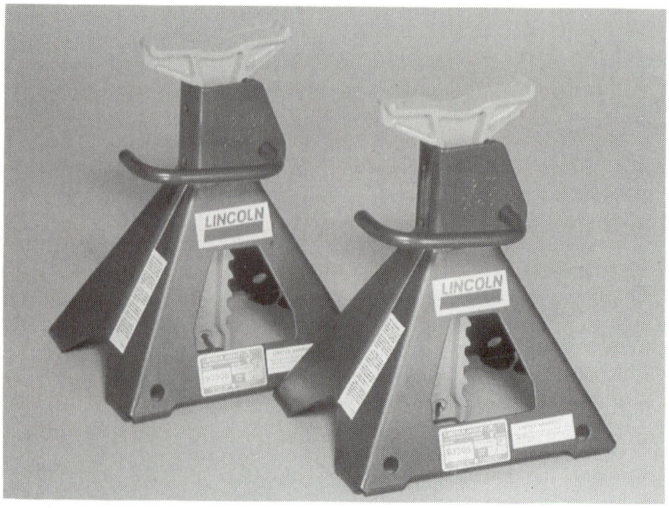


**Figure 2-35** Jack stands (Courtesy of Lincoln St. Louis)

2. Before lifting a vehicle with a floor jack, be sure that the jack lift pad is positioned securely under a recommended lift point on the vehicle. Lifting the front end of a vehicle with the jack placed under a radiator support may cause severe damage to the radiator and support.
3. Position the jack stands under a strong chassis member such as the frame or axle housing. The jack stands must contact the vehicle manufacturer's recommended lift points.
4. Since the floor jack is on wheels, the vehicle and jack tend to move as the vehicle is lowered from a floor jack onto jack stands. Always be sure the jack stands remain under the chassis member during this operation, and be sure the jack stands do not tip. All the jack stand legs must remain in contact with the shop floor.
5. When the vehicle is lowered from the floor jack onto jack stands, remove the floor jack from under the vehicle. Never leave a jack handle sticking out from under a vehicle. Someone may trip over the handle and injure themselves.

## Power Tool Safety

Power tools use electricity, shop air, or hydraulic pressure as a power source. Careless operation of power tools may cause personal injury and vehicle damage. Follow these steps for safe power tool operation:

1. Do not operate power tools with frayed power cords.
2. Be sure the power tool cord has a proper ground connection.
3. Do not stand on a wet floor while operating an electric power tool.
4. Always unplug an electric power tool before servicing the tool.
5. Do not leave a power tool running and unattended.
6. When using a power tool on small parts, do not hold the part in your hand. The part must be secured in a bench vise or with locking pliers.
7. Do not use a power tool on a job where the maximum capacity of the tool is exceeded.
8. Be sure that all power tools are in good condition, and always operate these tools according to the tool manufacturer's recommended procedure.
9. Make sure all protective shields and guards are in position.
10. Maintain proper body balance while using a power tool.
11. Always wear safety glasses or a face shield.
12. Wear ear protection.
13. Follow the equipment manufacturer's recommended maintenance schedule for all shop equipment.
14. Never operate a power tool unless you are familiar with the tool manufacturer's recommended operating procedure. Serious accidents occur from improper operating procedures.
15. Always make sure that the wheels on the electric grinder are securely attached and are in good condition.
16. Keep fingers and clothing away from grinding and buffing wheels. When grinding or buffing a small part, it should be held with a pair of locking pliers.
17. Always make sure the sanding or buffing disc is securely attached to the sander pad.
18. Use special heavy-duty sockets on impact wrenches. If ordinary sockets are used on an impact wrench, they may break and cause serious personal injury.
19. Never operate an air chisel unless the tool is securely connected to the chisel with the proper retaining device.
20. Never direct a blast of air from an air gun against any part of your body. If air penetrates the skin and enters the blood stream, it may cause very serious health problems and even death.

Tools operated by air pressure may be referred to as pneumatic tools.

# Compressed Air Equipment Safety

The shop air supply contains high-pressure air in the shop compressor and air lines. Serious injury or property damage may result from careless operation of compressed air equipment. When these steps are followed, safety is improved regarding compressed air equipment:

1. Wear safety glasses or a face shield for all shop tasks, including those tasks involving the use of compressed air equipment.
2. Wear ear protection when using compressed air equipment.
3. Always maintain air hoses and fittings in good condition. If an end suddenly blows off an air hose, the hose will whip around, and this may cause personal injury.
4. Do not direct compressed air against the skin. This air may penetrate the skin, especially through small cuts or scratches. If air penetrates the skin and enters the blood stream, it can be fatal or cause serious health complications. Use only air gun nozzles approved by the Occupational Safety and Health Act (OSHA).
5. Do not use an air gun to blow off clothing or hair.
6. Do not clean the work bench or floor with compressed air. This action may blow very small parts against your skin or into your eye. Small parts blown by compressed air may cause vehicle damage. For example, if the car in the next stall has the air cleaner removed, a small part may go into the carburetor or the throttle body. When the engine is started, this part will likely be pulled into the cylinder by engine vacuum, and the part will penetrate through the top of a piston.
7. Never spin bearings with compressed air, because the bearing will rotate at extremely high speed. Under this condition, the bearing may be damaged or it may disintegrate, causing personal injury.
8. Operate all pneumatic tools according to the tool manufacturer's recommended operating procedure.
9. Follow the equipment manufacturer's recommended maintenance schedule for all compressed air equipment.

**Classroom Manual**
Chapter 2, page 24

**Classroom Manual**
Chapter 2, page 26

# Cleaning Equipment Safety and Environmental Considerations

All technicians are required to clean parts during their normal work routines. Face shields and protective gloves must be worn while operating cleaning equipment. In most states, environmental regulations require that the runoff from steam cleaning must be contained in the steam cleaning system. This runoff cannot be dumped into the sewer system. Since it is expensive to contain this runoff in the steam cleaning system, the popularity of steam cleaning has decreased. The solution in hot and cold cleaning tanks may be caustic, and contact between this solution and skin or eyes must be avoided. Parts cleaning often creates a slippery floor, and care must be taken when walking in the parts cleaning area. The floor in this area should be cleaned frequently. When the cleaning solution in hot or cold cleaning tanks is replaced, environmental regulations require that the old solution be handled as hazardous waste. Use caution when placing aluminum or aluminum alloy parts in a cleaning solution. Some cleaning solutions will damage these components. Always follow the cleaning equipment manufacturer's recommendations.

**Classroom Manual**
Chapter 2, page 34

## Parts Washers with Electromechanical Agitation

Some parts washers provide electromechanical agitation of the parts to provide improved cleaning action (Figure 2-36). Hot tank parts washers may be heated with gas or electricity. Various water-

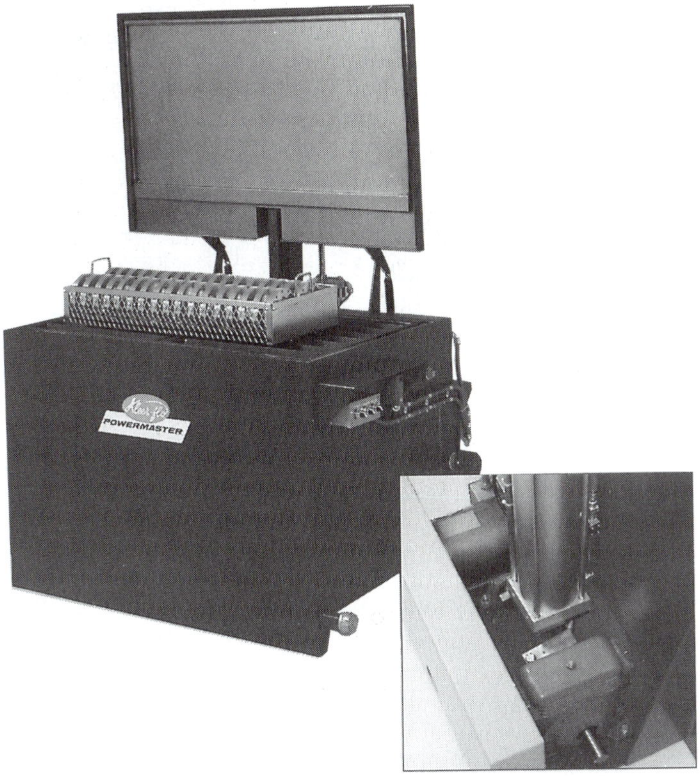


**Figure 2-36** Parts washer with electromechanical agitation (Courtesy of Kleer-flo Company)

based hot tank cleaning solutions are available depending on the type of metals being cleaned. For example, Kleer-flo Greasoff® number 1 powdered detergent is available for cleaning iron and steel. Nonheated electromechanical parts washers are also available, and these washers use cold tank cleaning solutions such as Kleer-flo Degreasol® formulas.

Many cleaning solutions, such as Kleer-flo Degreasol® 99R contain no ingredients listed as hazardous by the Environmental Protection Agency's RCRA Act. This cleaning solution is a blend of sulphur-free hydrocarbons, wetting agents, and detergents. Degreasol® 99R does not contain aromatic or chlorinated solvents, and conforms to California's Rule 66 for clean air. Always use the cleaning solution recommended by the equipment manufacturer.

## Cold Tank Parts Washer with Agitated Immersion Tank

Some parts washers have an agitated immersion chamber under the shelves that provides thorough parts cleaning. Folding workshelves provide a large upper cleaning area with a constant flow of solution from the dispensing hose (Figure 2-37). This cold tank parts washer operates on Kleer-flo Degreasol® 99R cleaning solution.

## Aqueous Parts Cleaning Tank

The aqueous parts cleaning tank uses a water-based environmentally friendly cleaning solution, such as Kleer-flo Greasoff® 2, rather than traditional solvents. The immersion tank is heated and agitated for effective parts cleaning (Figure 2-38). A sparger bar pumps a constant flow of cleaning solution across the surface to push floating oils away, and an integral skimmer removes these oils. This action prevents floating surface oils from redepositing on cleaned parts.

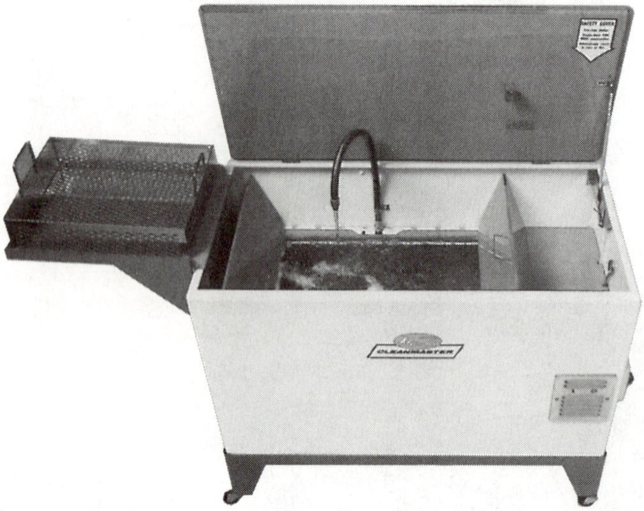

**Figure 2-37** Cold tank parts washer with agitated immersion tank (Courtesy of Kleer-flo Company)

**Figure 2-38** Aqueous parts cleaning tank (Courtesy of Kleer-flo Company)

# Safety Training Exercises

After the equipment operation and safety practices are explained by your instructor, complete these safety training exercises to demonstrate your understanding of equipment operation and safety procedures:

1. Find the lift points on a specific vehicle in the vehicle manufacturer's service manual. Under the supervision of your instructor, position this vehicle properly on a lift, then use the proper lift operating procedures to raise and lower the vehicle on the lift.
2. Find the lift points on a specific vehicle in the vehicle manufacturer's service manual. Under the supervision of your instructor, raise the front and rear suspension of this

vehicle with a floor jack, and lower the vehicle onto jack stands. Raise the vehicle with a floor jack, remove the jack stands, and lower the vehicle.

3. Find the lift points on a specific vehicle in the vehicle manufacturer's service manual. Under the supervision of your instructor, position this vehicle properly on a lift, and raise the vehicle. Follow proper service and safety procedures to remove and replace all four wheels. Remove two wheels with an electric impact wrench, and the other two wheels with an air impact wrench.
4. Draw a layout diagram of your automotive shop or shops, indicating the major service areas and service equipment, and clearly identifying the location of all safety equipment.

## Guidelines for Tools and Safety Practices

1. Two systems of measurement in common use are the United States customary (USC) and international system (SI).
2. The SI system may be called the metric system.
3. In the metric system, the units can be divided or multiplied by 10.
4. Personal safety is the responsibility of everyone in the shop.
5. When lifting heavy objects, always bend your knees rather than your back.
6. Many shop accidents are caused by the improper use of hand tools.
7. A seal driver should be used to install a seal rather than striking the seal case with a hammer.
8. Never operate any type of equipment unless you are familiar with the proper operating procedure.
9. A tremendous amount of energy is stored in a compressed coil spring. Always follow the tool and vehicle manufacturer's recommended procedure when using a spring compressing tool. Never disconnect any suspension component that would suddenly release the spring tension.
10. A power steering pump delivers extremely high pressure under certain test conditions. Always follow the vehicle manufacturer's recommended procedure during the pump pressure test.
11. Never exceed the rated capacity of a hydraulic press, hydraulic jack, vehicle lift, or jack stands.
12. When raising a vehicle with a lift or floor jack, always be sure the lifting equipment is contacting the vehicle on the manufacturer's recommended lift points.
13. After a vehicle is raised on a lift, be sure the lift locking mechanism is in place.
14. Never operate electric equipment with a frayed cord or without a ground wire.
15. Never direct a blast of compressed air against human flesh.
16. Always wear eye protection and protective gloves when cleaning parts in any type of cleaning solution.

## CASE STUDY

A technician proceeded to raise a vehicle on a lift to perform an oil and filter change, including a chassis lubrication, on a Grand Marquis. This lift was a twin-post type with separate front and rear lift posts. On this type of lift, the rear wheels must be positioned in depressions in the floor to position the rear axle above the rear lift arm. Then the front lift post and arms must be moved forward or rearward to position the front lift arm under the front suspension. The front lift arms must also be moved inward or outward so they are lifting on the vehicle manufacturer's specified lift points.

The technician carefully positioned the front lift post and arms properly, but he forgot to check the position of the rear tires in the floor depressions. The car was raised on the lift, and the technician proceeded with the service work. Suddenly there was a loud thump, and the rear of the car bounced up and down! The rear lift arms were positioned against the floor of the trunk rather than on the rear axle, and the lift arms punched through the floor of the trunk, barely missing the fuel tank. The technician was extremely fortunate the car did not fall off the lift, resulting in severe damage. If the rear lift arms had punctured the fuel tank, a disastrous fire might have occurred! Luckily, these things did not happen.

The technician learned a very important lesson about lift operation. Always follow all the recommended procedures in the lift operator's manual! The trunk floor was repaired at no cost to the customer, and fortunately the shop and the vehicle escaped without major damage.

## CASE STUDY

A technician was installing a rear axle bearing and seal on a Chevette, which is a small rear wheel drive car. The technician raised the vehicle on a lift to a comfortable working height, and proceeded to pull the rear axle with a slide hammer puller. This technician was very muscular, and weighed about 250 pounds. The rear axle was extremely tight, and the technician operated the slide hammer puller with all the strength he had. Each time he operated the slide hammer, the rear axle bounced sideways on the lift. Suddenly, the car moved sideways so much that the car fell off the lift and over on its side. Fortunately, the technician was able to jump out of the way of the falling vehicle, but the vehicle body suffered considerable damage.

The technician learned to avoid service operations on a lift that may cause the car to move on the lift. However, if such service operations must be performed, the car should be chained to the lift.

## Terms to Know

United States customary (USC)
International system (SI)
Seal drivers
Bearing pullers
Axle pullers
Stethoscope
Dial indicator
Tire tread depth gauge
Tire changer
Wheel balancer
Machinist's rule
Ball joint removal and pressing tools
Tie rod end and ball joint puller
Control arm bushing tools
Coil spring compressing tool
Power steering pressure gauge
Pitman arm puller
Vacuum hand pump
Turning radius gauge
Plumb bob
Tram gauge
Magnetic wheel alignment gauge
Rim clamps
Brake pedal jack
Tie rod sleeve adjusting tool
Steering wheel locking tool
Toe gauge
Track gauge
Computer wheel aligner
Scan tester
Hydraulic press
Floor jack
Jack stand
Vehicle lift
Pneumatic tools

## ASE Style Review Questions

1. While discussing systems of weights and measures:
   *Technician A* says the international system (SI) is called the metric system.
   *Technician B* says every unit in the metric system can be divided or multiplied by 10.
   Who is correct?
   **A.** A only
   **B.** B only
   **C.** Both A and B
   **D.** Neither A nor B

2. While discussing measurements in the metric system:
   *Technician A* says 1 liter is equal to 1,000 centimeters$^3$.
   *Technician B* says 1 mile is equal to 1.8 kilometers.
   Who is correct?
   **A.** A only
   **B.** B only
   **C.** Both A and B
   **D.** Neither A nor B

3. While discussing personal safety:
   *Technician A* says rings and jewelry may be worn in the automotive shop.
   *Technician B* says some electric-drive cooling fans may start turning at any time.
   Who is correct?
   **A.** A only
   **B.** B only
   **C.** Both A and B
   **D.** Neither A nor B

4. While discussing the proper way to lift heavy objects:
   *Technician A* says you should bend your back to pick up a heavy object.
   *Technician B* says you should bend your knees to pick up a heavy object.
   Who is correct?
   **A.** A only
   **B.** B only
   **C.** Both A and B
   **D.** Neither A nor B

5. While discussing steering and suspension tools:
   *Technician A* says there is no energy in a compressed coil spring.
   *Technician B* says that a coil spring compressor tool must be used to compress the coil spring before removing any suspension component that would release the spring tension.
   Who is correct?
   **A.** A only
   **B.** B only
   **C.** Both A and B
   **D.** Neither A nor B

6. While discussing hand tool safety:
   *Technician A* says a screwdriver may be used as a chisel.
   *Technician B* says a screwdriver may be used as a pry bar.
   Who is correct?
   **A.** A only
   **B.** B only
   **C.** Both A and B
   **D.** Neither A nor B

7. While discussing power tool safety:
   *Technician A* says an electric power tool cord does not require a ground.
   *Technician B* says frayed electric cords should be replaced.
   Who is correct?
   **A.** A only
   **B.** B only
   **C.** Both A and B
   **D.** Neither A nor B

8. While discussing shop cleaning equipment safety:
   *Technician A* says hot tanks may contain caustic solutions.
   *Technician B* says some metals such as aluminum may dissolve in hot tanks.
   Who is correct?
   **A.** A only
   **B.** B only
   **C.** Both A and B
   **D.** Neither A nor B

9. While discussing hydraulic equipment:
*Technician A* says jack stands have a maximum weight capacity.
*Technician B* says before raising a vehicle on a lift, the doors, trunk, and hood should be closed.
Who is correct?
**A.** A only
**B.** B only
**C.** Both A and B
**D.** Neither A nor B

10. While discussing vehicle lifts:
*Technician A* says a four-wheel drive vehicle may be lifted with a frame contact lift.
*Technician B* says a vehicle should not be raised on a lift with people in the vehicle.
Who is correct?
**A.** A only
**B.** B only
**C.** Both A and B
**D.** Neither A nor B

CHAPTER 3

# Wheel Bearing and Seal Service

Upon completion and review of this chapter, you should be able to:

- ❑ Diagnose bearing defects.
- ❑ Clean and repack wheel bearings.
- ❑ Reassemble and adjust wheel bearings.
- ❑ Remove and replace wheel bearing seals.
- ❑ Diagnose wheel bearings on the vehicle.
- ❑ Diagnose wheel bearing hub units.
- ❑ Remove and replace front drive axles.
- ❑ Remove and replace wheel bearing hub units.
- ❑ Remove and replace rear axle bearings on rear wheel drive cars.

**Basic Tools**

Basic technician's tool set

Service manual

Inch-pound torque wrench

Foot-pound torque wrench

Fine-toothed round and flat files

Wheel bearing grease

Differential lubricant

## Diagnosis of Bearing Defects

Bearings are designed to provide long life, but there are many causes of premature bearing failure. If a bearing fails, the technician must decide if the bearing failure was caused by normal wear, or if the bearing failed prematurely. For example, if a front wheel bearing fails on a car that is one year old with an original odometer reading of 15,000 miles (24,000 kilometers), experience tells us the bearing failure is premature because front wheel bearings normally last for a much longer mileage period. Always listen to the customer's complaints, and obtain as much information as possible from the customer. Ask the customer specific questions about abnormal or unusual vehicle noises and operation. If a bearing fails prematurely, there must be some cause for the failure. The causes of premature bearing failure are these:

1. lack of lubrication
2. improper type of lubrication
3. incorrect end-play adjustment (where applicable)
4. misalignment of related components such as shafts or housings
5. excessive bearing load
6. improper installation or service procedures
7. excessive heat
8. dirt or contamination

Bearing galling refers to metal smears on the ends of the rollers.

Bearing abrasive step wear is a fine circular wear pattern on the ends of the rollers.

Etching appears as a loss of material on the bearing rollers and races. Bearing surfaces are gray or grayish black.

Bearing indentations are surface depressions on the rollers and races.

When a bearing fails prematurely, the technician must correct the cause of this failure to prevent the new bearing from failing. The types of bearing failures and the necessary corrective service procedures are provided in Figures 3-1 and 3-2. Fatigue spalling appears as flaking of surface metal on bearing rollers and races. Bearing brinelling shows up as straight-line indentations on the races and rollers. Bearing smears appear as metal loss in a circular, blotched pattern around the bearing races and rollers. Bearing frettage shows up as a fine, corrosive wear pattern around the bearing races and rollers. This wear pattern is circular on the races.

The first indication of bearing failure is usually a howling noise while the bearing is rotating. The howling noise will likely vary depending on the bearing load. A front wheel bearing usually provides a more noticeable howl when the vehicle is turning a corner, because this places additional thrust load on the bearing. A defective rear axle bearing usually provides a howling noise that is more noticeable at lower speeds. The howling noise is more noticeable when driving on a narrow street with buildings on each side, because the noise vibrates off the nearby buildings. A rear axle bearing noise is present during acceleration and deceleration, because the vehicle weight places a load on the bearing regardless of the operating condition. The rear axle bearing noise may be somewhat more noticeable during deceleration because there is less engine noise at that time.

**Classroom Manual**
Chapter 3, page 41

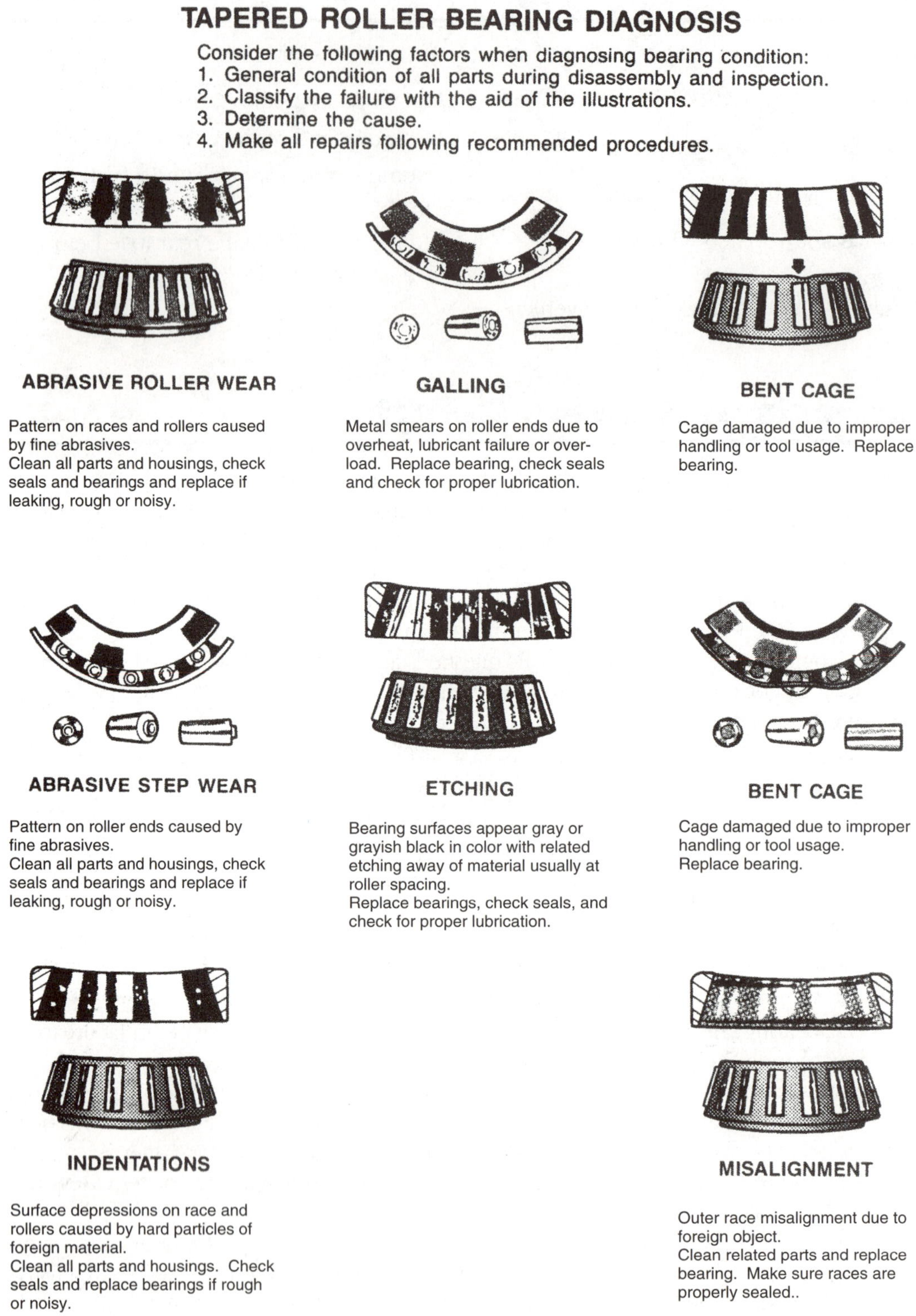


**Figure 3-1** Bearing failures and corrective procedures (Courtesy of Chevrolet Motor Division, General Motors Corporation)

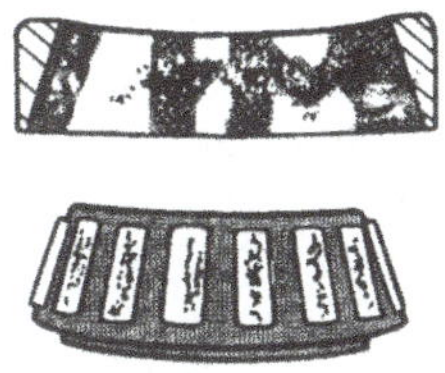

**FATIGUE SPALLING**

Flaking of surface metal resulting from fatigue.
Replace bearing, clean all related parts.

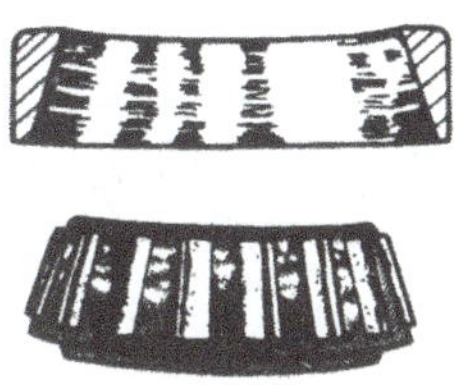

**CAGE WEAR**

Wear around outside diameter of cage and roller pockets caused by abrasive material and inefficient lubrication.
Clean related parts and housings. Check seals and replace bearings

**SMEARS**

Smearing of metal due to slippage. Slippage can be caused by poor fits, lubrication, overheating, overloads or handling damage.
Replace bearings, clean related parts and check for proper fit and lubrication.

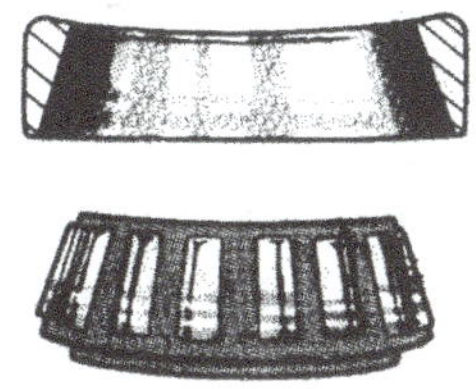

**STAIN DISCOLORATION**

Discoloration can range from light brown to black caused by incorrect lubricant or moisture.
Re-use bearings if stains can be removed by light polishing or if no evidence of overheating is observed. Check seals and related parts for damage.

**HEAT DISCOLORATION**

Heat discoloration can range from faint yellow or dark blue resulting from overload or incorrect lubricant. Excessive heat can cause softening of races or rollers. To check for loss of temper on races or rollers a simple file test may be made. A file drawn over a tempered part will grab and cut metal, whereas, a file drawn over a hard part will glide readily with no metal cutting.
Replace bearings if overheating damage is indicated.
Check seals and other parts.

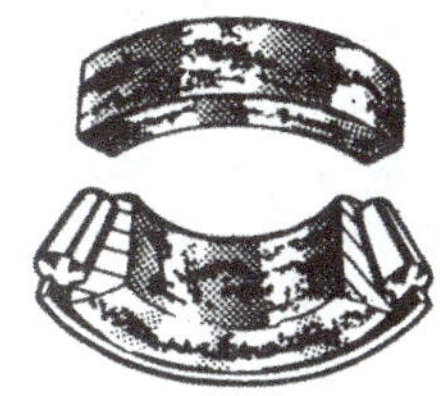

**FRETTAGE**

tiCorrosion set up by small relative movement of parts with no lubrication. Replace bearing. Clean related parts. Check seals and check for proper lubrication.

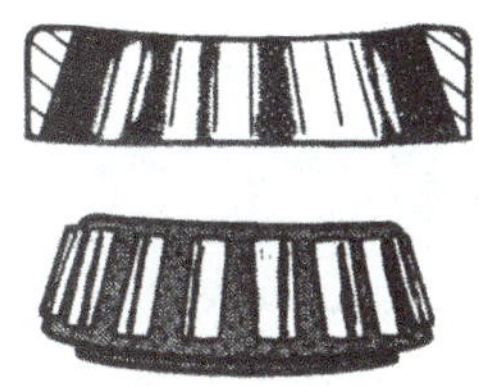

**BRINELLING**

Surface indentations in raceway caused by rollers either under impact loading or vibration while the bearing is not rotating.
Replace bvearing if rough or noisy

**CRACKED INNER RACE**

Race cracked due to improper fit, cocking, or poor bearing seats.
Replace bearing and correct bearing seats.

**Figure 3-2** Bearing failures and corrective procedures continued (Courtesy of Chevrolet Motor Division, General Motors Corporation )

## Service and Adjustment of Tapered-Roller-Bearing-Type Wheel Bearings

### Cleaning Bearings

**CAUTION:** Always wear safety goggles when working in the shop.

**WARNING:** Do not aim compressed air directly against any part of your body. Compressed air can penetrate human flesh and enter the blood stream with very serious consequences.

**WARNING:** Do not spin the bearing at high speed with compressed air. Bearing damage or disintegration may result. Bearing disintegration may cause serious personal injury.

**WARNING:** Never strike a bearing with a ball peen hammer. This action will damage the bearing, and the bearing may shatter, causing severe personal injury.

**CAUTION:** Never wash a sealed bearing or a bearing that is shielded on both sides because solvent may enter the bearing and destroy the lubricant in the bearing, resulting in very short bearing life.

Two separate tapered roller bearings are used in the front wheel hubs of many rear wheel drive cars. The rear wheel hubs on some front wheel drive cars have the same type of bearings. Similar service and adjustment procedures apply to these tapered roller bearings.

These bearings should be cleaned, inspected, and packed with wheel bearing grease at the vehicle manufacturer's recommended service intervals. Pry the grease seal out of the inner hub opening with a large flat-blade screwdriver, and discard the seal. This seal should always be replaced when the bearings are serviced. Do not attempt to wash sealed bearings or bearings that are shielded on both sides. If a bearing is sealed on one side, it may be washed in solvent and repacked with grease.

Bearings may be placed in a tray and lowered into a container of clean solvent. A brush may be used to remove old grease from the bearing (Figure 3-3). The bearings may be dried with com-

**Figure 3-3** Cleaning bearing with solvent (Courtesy of the Timken Company)

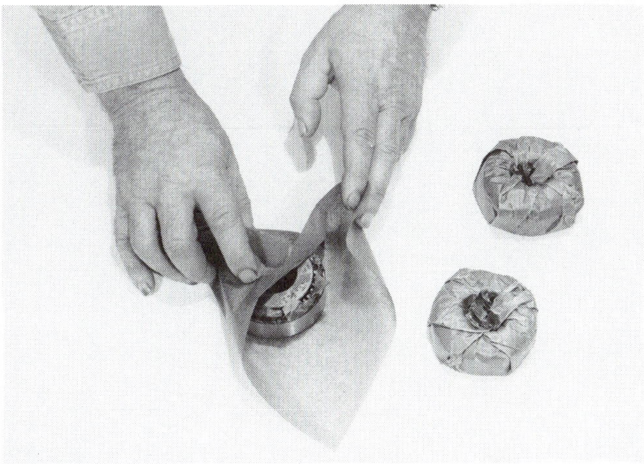

**Figure 3-4** Wrapping a bearing in waterproof paper (Courtesy of the Timken Company)

pressed air after the cleaning operation. Be sure the shop air supply is free from moisture, which causes rust formation in the bearing. After all the old grease has been cleaned from the bearing, rinse the bearing in clean solvent and dry it thoroughly with compressed air.

When bearing cleaning is completed, bearings should be inspected for the defects illustrated in Figures 3-1 and 3-2. If any of these conditions are present on the bearing, replacement is necessary. Tapered roller bearings and their matching outer races must be replaced as a set. If the bearing installation is not done immediately, cover the bearings with a protective lubricant and wrap them in waterproof paper (Figure 3-4). Be sure to identify the bearings, or lay them in order, so you reinstall them in their original location. Do not clean bearings or races with paper towels. If you are using a shop towel for this purpose, be sure it is lint-free. Lint from shop towels or paper towels may contaminate the bearing.

Bearing races and the inner part of the wheel hub should be thoroughly cleaned with solvent and dried with compressed air. Inspect the seal mounting area in the hub for metal burrs. Remove any burrs with a fine round file. Bearing races must be replaced if any of the defects described in Figures 3-1 and 3-2 are found. The proper bearing race driving tool must be used to remove the bearing races (Figure 3-5). If a driver is not available for the bearing races, a long punch and hammer may be used to drive the races from the hub. When a hammer and punch are used for this purpose, be careful not to damage the hub inner surface with the punch.

**Special Tools**

Bearing driver

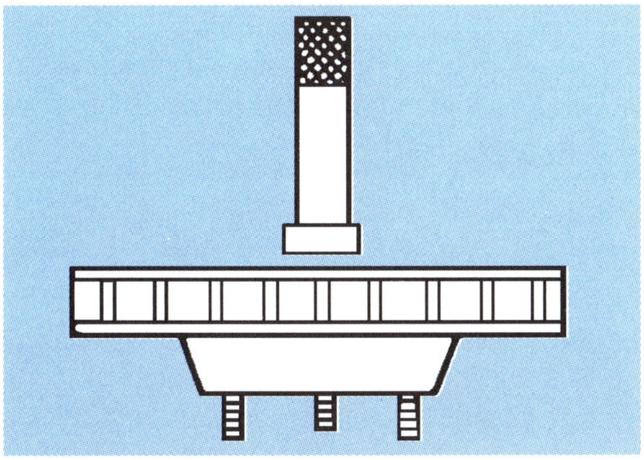

**Figure 3-5** Bearing race removal (Courtesy of Chevrolet Motor Division, General Motors Corporation)

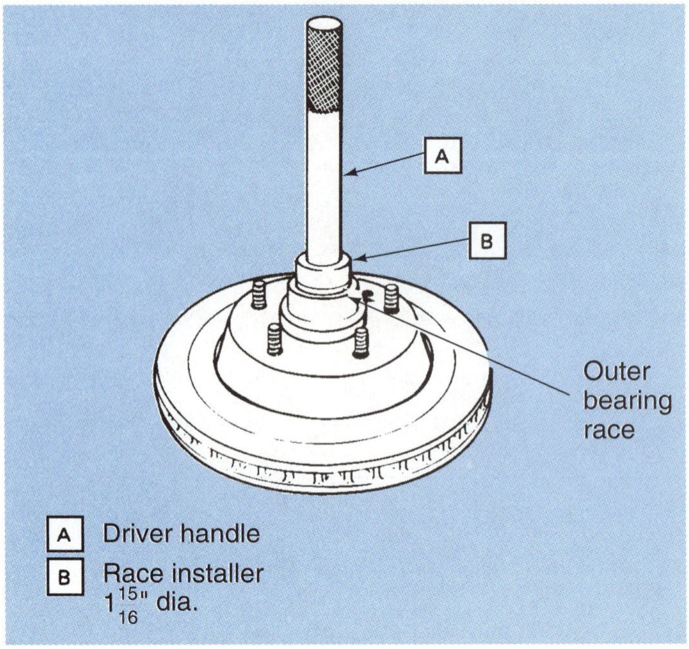


**Figure 3-6** Bearing race installation (Courtesy of Chevrolet Motor Division, General Motors Corporation)

The new bearing races should be installed in the hub with the correct bearing race driving tool (Figure 3-6). When bearings and races are replaced, be sure they are the same as the original bearings. The part numbers should be the same on the old bearings and the replacement bearings.

Inspect the bearing and seal mounting surfaces on the spindle. Small metal burrs may be removed from the spindle with a fine-toothed file. If the spindle is severely scored in the bearing or seal mounting areas, spindle replacement is necessary.

## Bearing Lubrication and Assembly

**CAUTION:** Cleanliness is very important during wheel bearing service. Always maintain cleanliness of hands, tools, work area, and all related bearing components. One small piece of dirt in a bearing will cause bearing failure.

**CAUTION:** Always keep grease containers covered when not in use. Uncovered grease containers are easily contaminated with dirt and moisture.

After the bearings and races have been cleaned and inspected, the bearings should be packed with grease. Always use the vehicle manufacturer's specified wheel bearing grease. Place a lump of grease in the palm of one hand and grasp the bearing in the other hand. Force the widest edge of the bearing into the lump of grease, and squeeze the grease into the bearing. Continue this process until grease is forced into the bearing around the entire bearing circumference. Place a coating of grease around the outside of the rollers, and apply a light coating of grease to the races. A bearing packing tool may be used to force grease into the bearings rather than the hand method (Figure 3-7).

**SERVICE TIP:** When a lip seal is installed, the garter spring should always face toward the flow of lubricant.

Figure 3-7 Mechanical wheel bearing packer (Courtesy of the Timken Company)

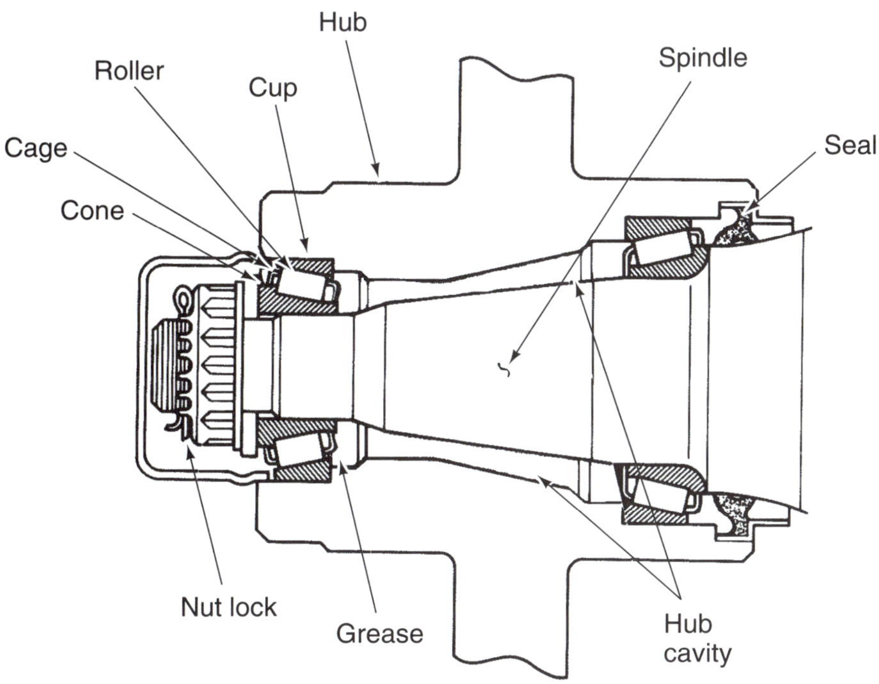


Figure 3-8 Wheel bearing lubrication (Courtesy of Chrysler Corporation)

Place some grease in the wheel hub cavity and position the inner bearing in the hub (Figure 3-8). Check the fit of the new bearing seal on the spindle and in the hub. The seal lip must fit snugly on the spindle, and the seal case must fit properly in the hub opening. The part number on the old seal and the replacement seal should be the same. Be sure the seal is installed in the proper

**Special Tools**

Figure 3-9 Seal installation (Courtesy of Pontiac Motor Division, General Motors Corporation)

direction with the garter spring and higher part of the lip toward the lubricant in the hub. The new inner bearing seal must be installed in the hub with a suitable seal driver (Figure 3-9).

Place a light coating of wheel bearing grease on the spindle and slide the hub assembly onto the spindle. Install the outer wheel bearing and be sure there is adequate lubrication on the bearing and race. Be sure the washer and nut are clean and install these components on the spindle (Figure 3-10). Tighten the nut until it is finger tight.

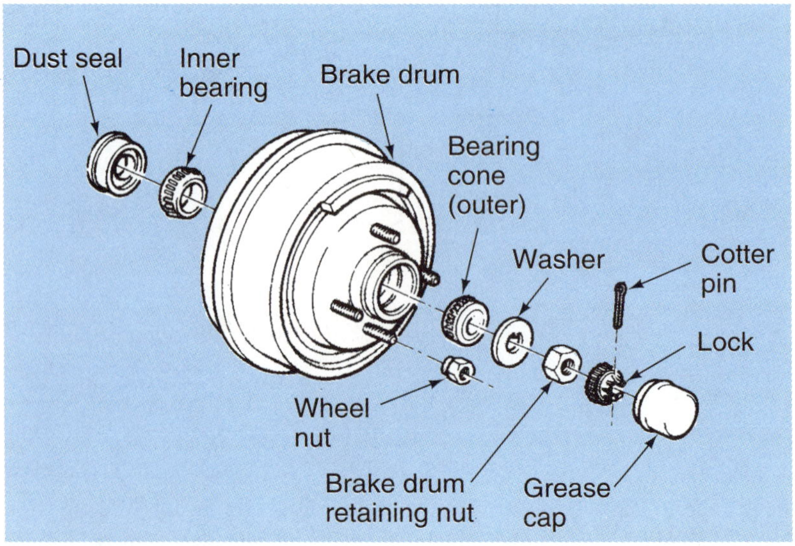


Figure 3-10 Installation of wheel bearings and related components (Courtesy of Chrysler Corporation)

## Photo Sequence 1

## Typical Procedure for Adjusting Rear Wheel Bearings on a Front Wheel Drive Car

**P1-1** Always make sure the car is positioned safely on a lift before working on the vehicle.

**P1-2** Remove the dust cap from the wheel hub.

**P1-3** Remove the cotter pin and nut retainer from the bearing adjusting nut.

**P1-4** Tighten the bearing adjusting nut to 17 to 25 ft lbs.

**P1-5** Loosen the bearing adjusting nut one-half turn.

**P1-6** Tighten the bearing adjusting nut to 10 to 15 ft lbs.

**P1-7** Position the adjusting nut retainer over the adjusting nut so the slots are aligned with the holes in the nut and spindle.

**P1-8** Install a new cotter pin and bend the ends around the retainer flange.

**P1-9** Install the dust cap and be sure the hub rotates freely.

### Wheel Bearing Adjustment with Two Separate Tapered Roller Bearings in the Wheel Hub

**CUSTOMER CARE:** Never sell a customer automotive service that is not required on his or her car. Selling preventative maintenance is a sound business practice, and may save a customer some future problems. An example of preventative maintenance is selling a cooling system flush when the cooling system is not leaking but the manufacturer's recommended service interval has elapsed. If customers find out they were sold some unnecessary service, they will probably never return to the shop. They will likely tell their friends about their experience, and that kind of advertising the shop can do without.

Loose front wheel bearing adjustment results in lateral front wheel movement and reduced directional stability. If the wheel bearing adjusting nut is tightened excessively, the bearings may overheat, resulting in premature bearing failure. The bearing adjustment procedure may vary depending on the make of vehicle. Always follow the procedure in the vehicle manufacturer's service manual. A typical bearing adjustment procedure follows:

1. With the hub and bearings assembled on the spindle, tighten the adjusting nut to 17 to 25 ft-lbs (23 to 34 Nm) while the hub is rotated.
2. Loosen the adjusting nut 1/2 turn and retighten it to 10 to 15 in-lbs (1.0 to 1.7 Nm). This specification varies depending on the make of vehicle. Always use the manufacturer's specifications.
3. Position the adjusting nut retainer over the nut so the retainer slots are aligned with the cotter key hole in the spindle.
4. Install a new cotter key, and bend the ends around the retainer flange.
5. Install the grease cap, and make sure the hub and drum rotate freely.

Photo Sequence 1 shows a typical procedure for adjusting rear wheel bearings on a front wheel drive car.

## Wheel Hub Unit Diagnosis

**Special Tools**

Dial indicator

When wheel bearings and hubs are an integral assembly, the bearing end play should be measured with a dial indicator stem mounted against the hub. If the end play exceeds 0.005 in (0.127 mm) as the hub is moved in and out, the hub and bearing assembly should be replaced. This specification is typical, but the vehicle manufacturer's specifications must be used. Hub and bearing replacement is also necessary if the bearing is rough or noisy. Integral-type bearing and hub assemblies are used on the front and rear wheels on some front wheel drive cars.

When the front wheel bearings are mounted in the steering knuckle, the wheel bearings may be checked with the vehicle raised on the hoist and a dial indicator positioned against the outer wheel rim lip as shown (Figure 3-11).

When the wheel is moved in and out, the maximum bearing movement on the dial indicator should be as follows:

0.020 in (0.508 mm) for 13-in (33-cm) wheels

0.023 in (0.584 mm) for 14-in (35.5-cm) wheels

0.025 in (0.635 mm) for 15-in (38-cm) wheels

If the bearing movement is excessive, check the hub nut torque before replacing the bearing. When this torque is correct and bearing movement is excessive, the bearing should be replaced.

When a wheel is removed to service the wheel bearings, proper balance must be maintained between the wheel and tire and the hub. Therefore, the tire, wheel, and hub stud should be chalk marked prior to removal as shown (Figure 3-12).

Figure 3-11 Wheel bearing diagnosis on vehicle (Courtesy of Chrysler Corporation)

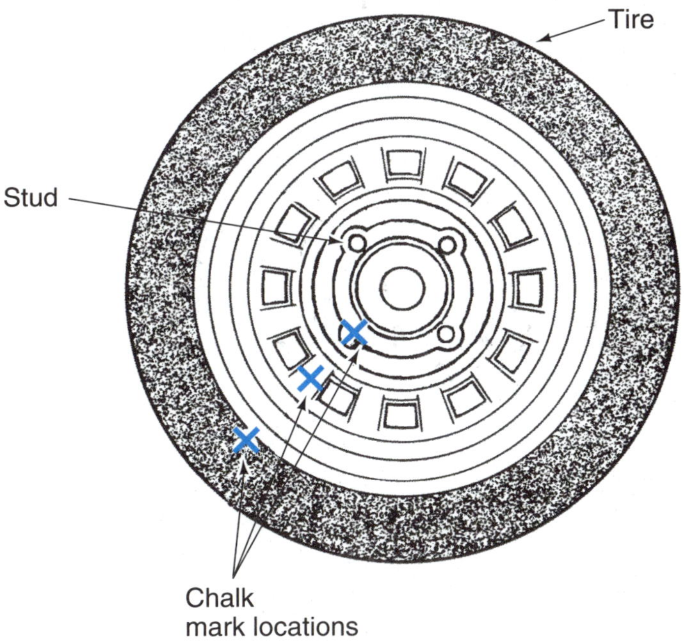


Figure 3-12 Chalk marking on wheel, tire, and stud (Courtesy of Chrysler Corporation)

## Front Drive Axle Diagnosis

On many front wheel drive vehicles, the front drive axles must be removed before the wheel hub unit or steering knuckle and bearing can be detached. Therefore, we will discuss front drive axle diagnosis and removal. Since drive axle noises may be confused with front wheel bearing noise, a brief discussion of drive axle noises and problems may be helpful. A defective inner drive axle joint usually causes a vibration when the vehicle is decelerating at 35 to 45 miles per hour (mph), or 56 to 72 kilometers per hour (kph). When an outer drive axle joint is worn, a clicking noise is heard during a hard turn below 20 mph (32 kph). To determine which drive axle has the defective joint, lift the vehicle on a hoist, and allow the front wheels to drop down. This action will position the axle joints at a different angle than when the car is driven on the road. Lift the lower control arms one at a time with a floor jack, and place the transmission drive to simulate the driving conditions that provided the vibration or noise. If the vibration or noise occurs with one lower control arm lifted, that side has the defective drive axle joint.

## Drive Axle Removal

Many drive axles have a circlip on the inner joint extension that holds the inner joint into the differential side gear. Drive axle systems vary depending on the vehicle. Follow the drive axle removal procedure in the vehicle manufacturer's service manual. A general front drive axle removal procedure follows:

1. Loosen the front wheel nuts and hub nuts.
2. Lift the vehicle on a hoist and be sure the hoist safety mechanism is in place, then remove the front wheels and tires.
3. Remove the brake calipers and rotors. Connect a piece of wire from the calipers to a suspension or chassis component. Do not allow the calipers to hang on the end of the brake line.
4. Install protective drive axle boots if these are supplied by the car manufacturer.
5. Remove the ball joint to steering knuckle clamp bolt (Figure 3-13).
6. Pry the ball joint stud from the steering knuckle.
7. Pull the inner axle from the transaxle (Figure 3-14), and do not allow the axle to drop down at a severe angle.
8. Remove the hub nut and washer, and separate the outer axle joint from the wheel hub. Some outer axle joint splines, such as those used on early model Ford Escorts or Lynx, are slightly spiraled. On this type of outer axle joint, a special puller is required to separate the axle joint from the wheel (Figure 3-15).
9. Remove the drive axle from the chassis.

**Special Tools**

Front wheel bearing hub pulling and installing tools

Reverse the drive axle removal procedure for front drive axle installation.

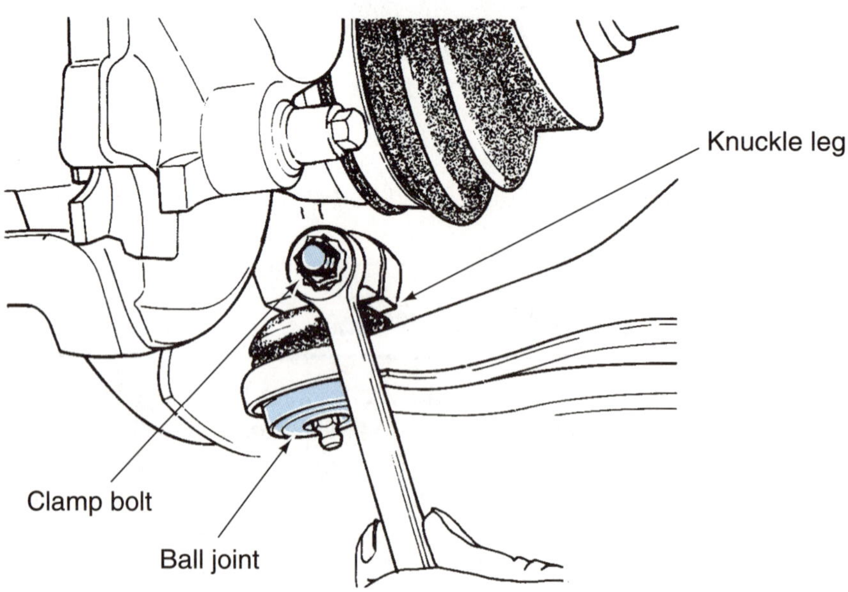


**Figure 3-13** Removal of ball joint to steering knuckle clamp bolt (Courtesy of Chrysler Corporation)

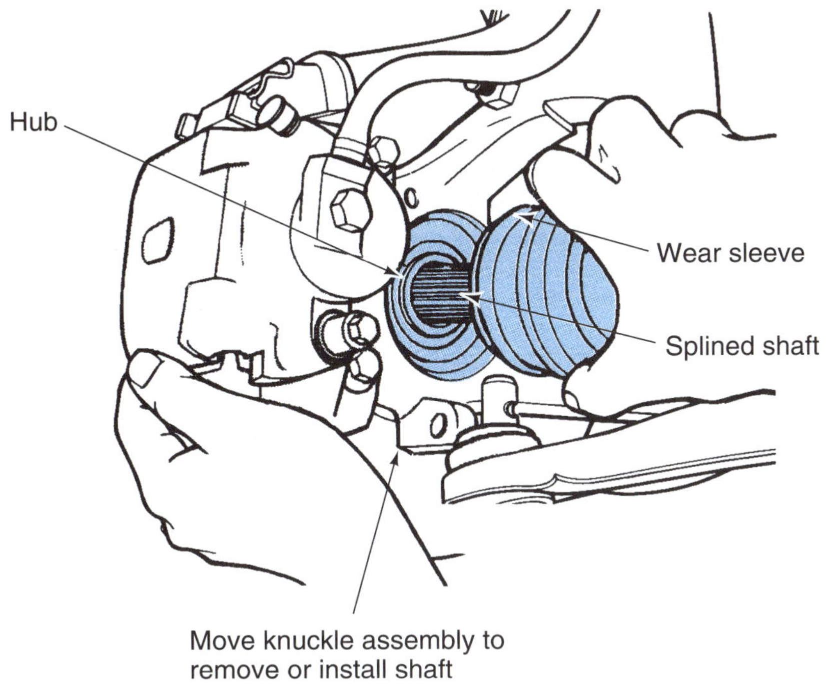


**Figure 3-14** Removal of inner axle joint from transaxle (Courtesy of Chrysler Corporation)

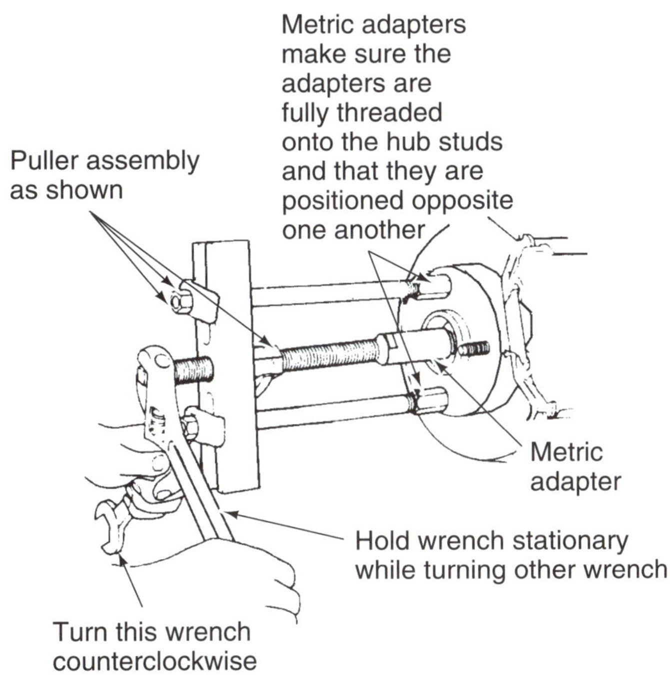


**Figure 3-15** Removal of outer axle joint from front wheel hub (Courtesy of Ford Motor Company)

# Special Procedures for Drive Axle Removal

Some car manufacturers recommend removal of the strut from the steering knuckle, rather than removal of the ball joint, when the drive axles are removed. This type of suspension system has a threaded nut on the ball joint stem to hold the ball joint into the steering knuckle. If an eccentric

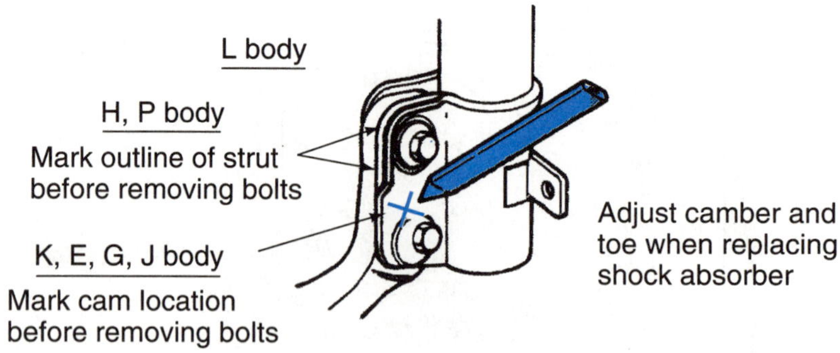


**Figure 3-16** Marking eccentric strut bolt before removal (Courtesy of Chrysler Corporation)

camber adjustment bolt is positioned in the strut, the bolt head position should be marked in relation to the strut before the bolt is removed (Figure 3-16).

On these suspension systems, the brake calipers and brake line clamps should be removed before the drive axles. Some car manufacturers supply a slide-hammer-type puller to remove the inner axle joints from the transaxle.

On Chrysler manual and automatic transaxles, the speedometer gear must be removed from the right differential extension housing before the right drive axle is removed (Figure 3-17).

Early model Chrysler transaxles had circlip retainers, which held the inner drive axle joints into the differential side gears. The differential cover had to be removed and these circlips had to

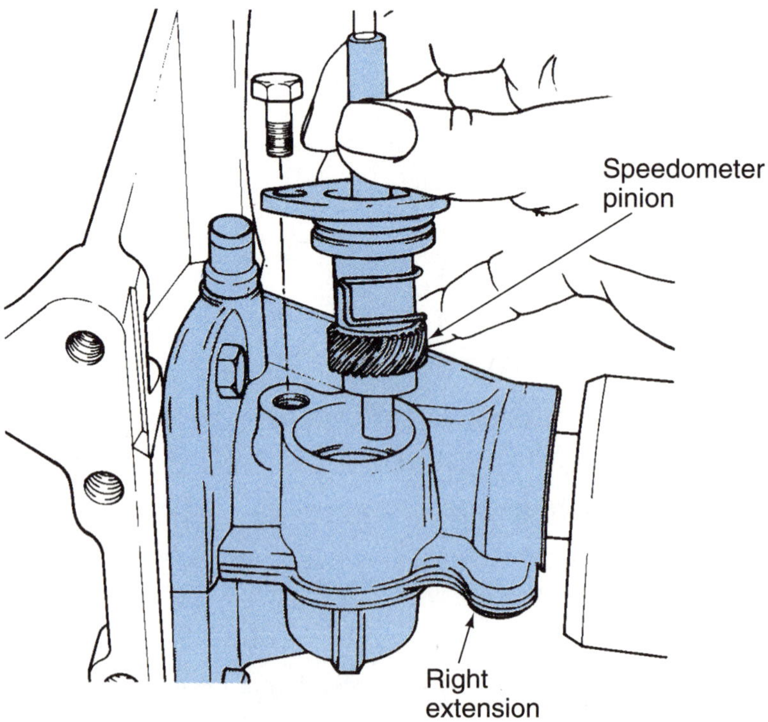


**Figure 3-17** The speedometer drive gear must be removed before the right axle is removed. (Courtesy of Chrysler Corporation)

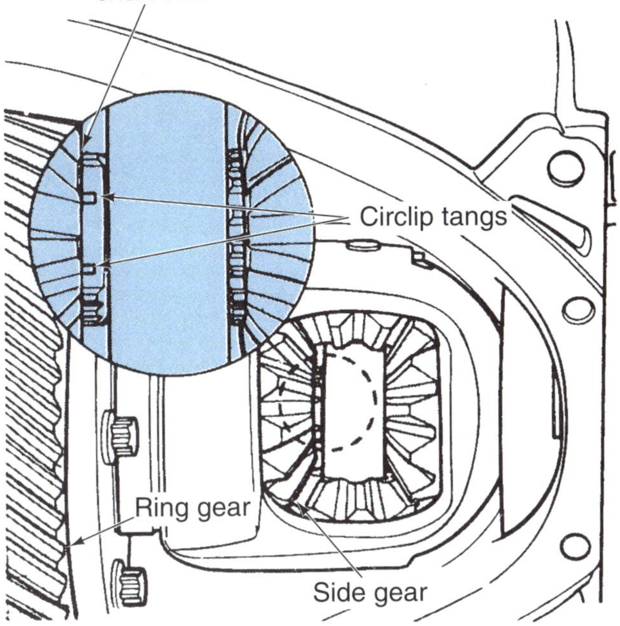


**Figure 3-18** Compressible circlips in early model Chrysler transaxles (Courtesy of Chrysler Corporation)

be compressed with needle nose pliers before the drive axles could be removed from the differential side gears. The drive axles had to be rotated until a flat area on the axle and the circlip ends were visible (Figure 3-18).

The inner joint housings did not have tripod springs on early models. Later-model Chrysler drive axles may be pulled from the differential side gears without collapsing the circlips. To determine which type of circlip is used, grasp the inner joint housing and try to pull it out of the transaxle. If the inner joint housing moves outward and springs back into the transaxle, the joint has a later model circlip. When the inner joint is solid in the transaxle, the older model circlip, which requires compressing before removal, is used.

On Ford automatic transaxles (ATX), the left inner axle joint is inset into the transaxle housing. In this location, it is impossible to pry the axle joint from the transaxle. Therefore, the right axle joint must be removed first. A special tool is then used to drive the left inner axle joint from the differential side gear (Figure 3-19).

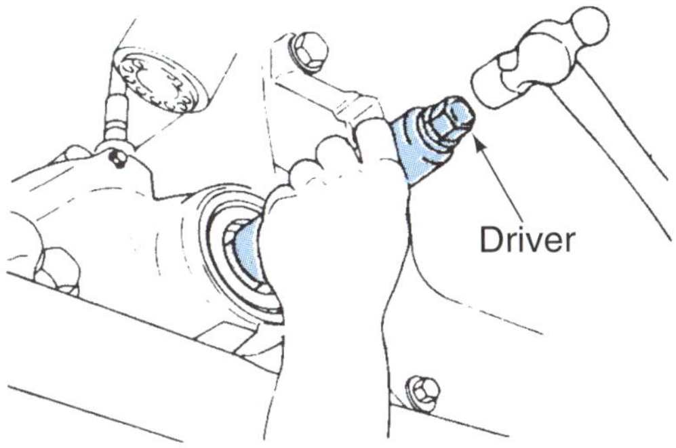


**Figure 3-19** Removal of left inner drive axle joint on Ford automatic transaxles (Courtesy of Ford Motor Company)

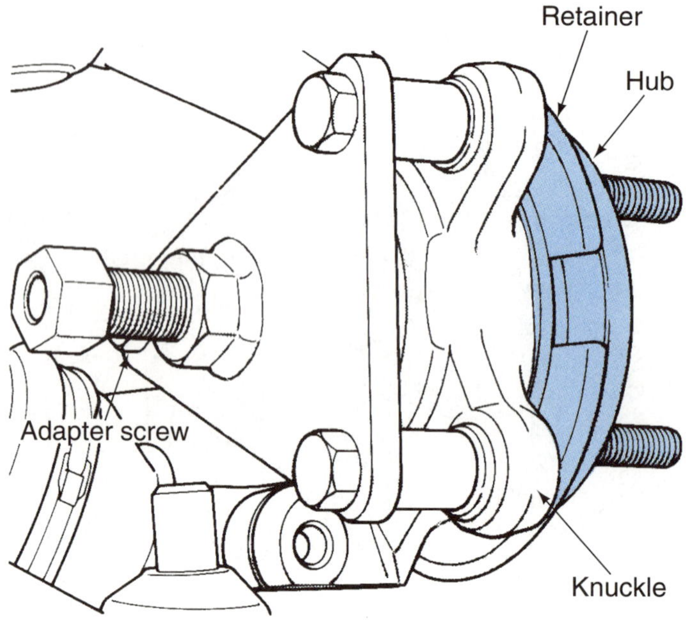


**Figure 3-20** Wheel hub removal from bearing (Courtesy of Chrysler Corporation)

If both left and right drive axles are removed at the same time on Ford transaxles, the differential side gears may become dislocated. When this occurs, the differential must be removed to realign the side gears. The special tool for driving out the left inner axle joint may be left in place to support the side gears on ATX models. On manual transaxles (MTX), shipping plugs T81P-1177-B should be installed in each side gear when the drive axles are removed.

## Front Wheel Bearing Hub Unit Removal and Replacement

The front wheel bearing removal and replacement procedure varies depending on the vehicle and the type of front wheel bearing. Always follow the front wheel bearing removal and replacement procedure in the manufacturer's service manual. This procedure applies to front wheel bearing units that are pressed into the steering knuckle. When front wheel bearing replacement is necessary, the steering knuckle must be removed, and the wheel hub must be pressed from the bearing with a special tool (Figure 3-20).

A special puller is used to remove and replace the wheel bearing in the knuckle (Figure 3-21 and Figure 3-22).

The wheel hub must be pulled into the wheel bearing with a special tool (Figure 3-23). The proper driving tool is used to install the seal behind the bearing in the knuckle (Figure 3-24).

When two separate roller bearings are mounted in the steering knuckle, the bearing races must be driven from the knuckle with a hammer and punch. These bearings must be lubricated with wheel bearing grease prior to installation as described previously in this chapter. When the wheel bearings are removed, all wheel bearing seals must be replaced as described previously. A staked-type hub nut must be replaced if it is removed.

On these front wheel drive cars, the hub nut torque applies the correct adjustment on the front wheel bearings. Therefore, this torque is extremely important. With the brakes applied, the

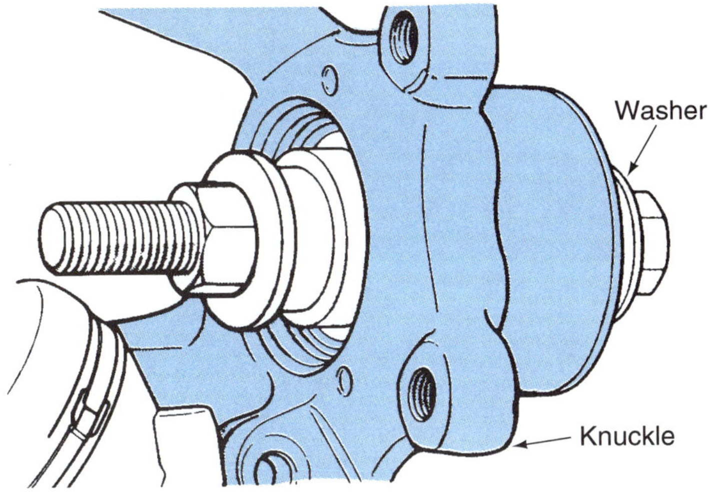


**Figure 3-21** Wheel bearing removal from knuckle (Courtesy of Chrysler Corporation)

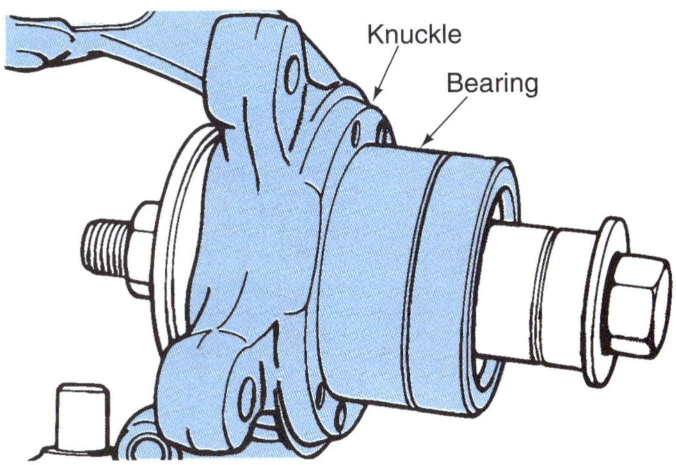


**Figure 3-22** Wheel bearing installation in knuckle (Courtesy of Chrysler Corporation)

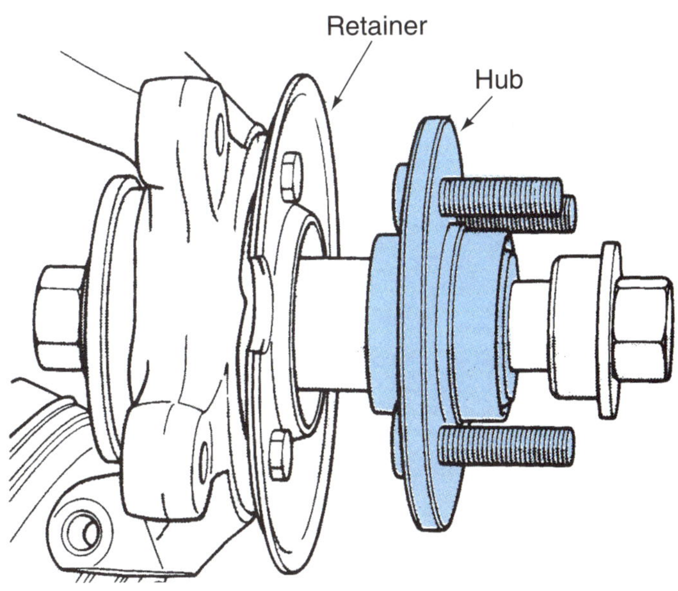


**Figure 3-23** Wheel hub installation in wheel bearing (Courtesy of Chrysler Corporation)

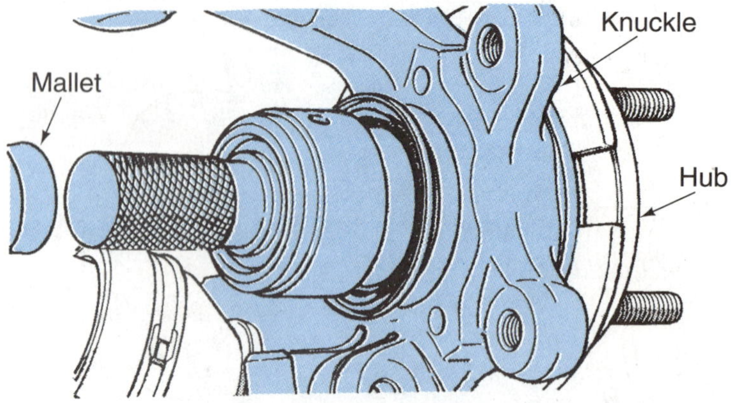


**Figure 3-24** Seal installation in steering knuckle (Courtesy of Chrysler Corporation)

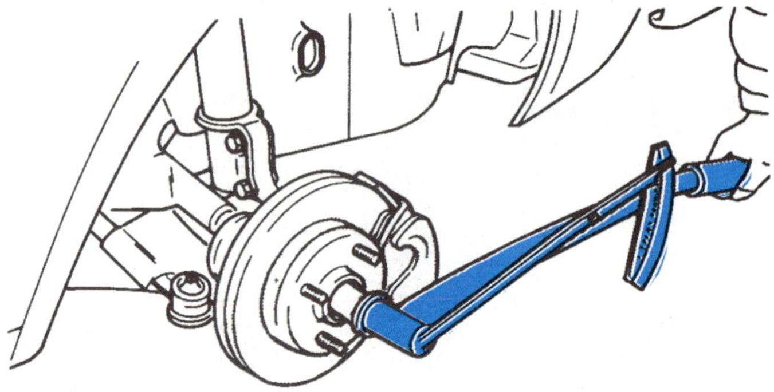

**Figure 3-25** Hub nut torquing (Courtesy of Chrysler Corporation)

hub nut should be tightened to the specified torque (Figure 3-25). When the hub nut is torqued to specifications, the nut lock and cotter pin should be installed (Figure 3-26).

**CAUTION:** Never reuse a cotter pin.

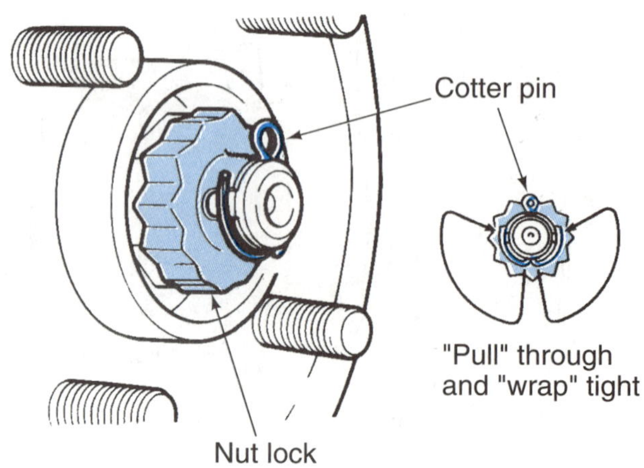


**Figure 3-26** Nut lock and cotter pin installation (Courtesy of Chrysler Corporation)

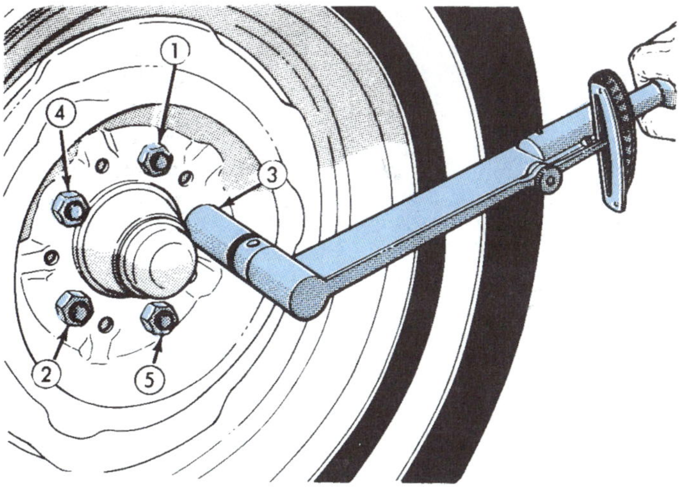


Figure 3-27 Wheel nut tightening sequence (Courtesy of Chrysler Corporation)

After the wheel is installed, the wheel nuts should be tightened in sequence to the specified torque (Figure 3-27).

On cars with the front wheel bearings mounted in the steering knuckles, never move a car unless the front hub nuts are torqued to specifications. Lack of bearing preload could damage the bearings if the hub nuts are not tightened to specifications. If the car must be moved when the drive axles are removed, place a large bolt and nut with suitable washers through the front wheel bearing and tighten the nut to specifications.

## Rear Axle Bearing and Seal Service, Rear Wheel Drive Cars

**CAUTION:** Use extreme caution when diagnosing problems with a vehicle raised on a hoist and the engine running with the transmission in drive. Keep away from rotating wheels, drive shafts, or drive axles.

**Special Tools**

Technician's stethoscope

Rear axle bearing noise may be diagnosed with the vehicle raised on a hoist. Be sure the hoist safety mechanism is engaged after the vehicle is raised on the hoist. With the engine running and the transmission in drive, operate the vehicle at moderate speed (35 to 45 mph) (56 to 72 kph) and listen with a stethoscope placed on the rear axle housing directly over the axle bearings. If grinding or clicking noises are heard, bearing replacement is necessary.

A stethoscope is a tool that amplifies sound. The pickup on the stethoscope rests against the suspected component, and extensions from the tool are placed in your ears. A doctor uses a stethoscope to check heart beats, or to listen to your lungs as you breathe.

Many axle shafts in rear wheel drive cars have a roller bearing and seal at the outer end (Figure 3-28). These axle shafts are often retained in the differential with "C" locks that must be removed before the axles.

The rear axle bearing removal and replacement procedure varies depending on the vehicle make and model year. Always follow the rear axle bearing removal and replacement procedure in the manufacturer's service manual. A typical rear axle shaft removal and replacement procedure on a rear wheel drive car with "C" lock axle retainers follows:

1. Loosen the rear wheel nuts and chalk mark the rear wheel position in relation to the rear axle.
2. Raise the vehicle on a hoist and make sure the hoist safety mechanism is in place.
3. Remove the rear wheels and brake drums, or calipers and rotors.

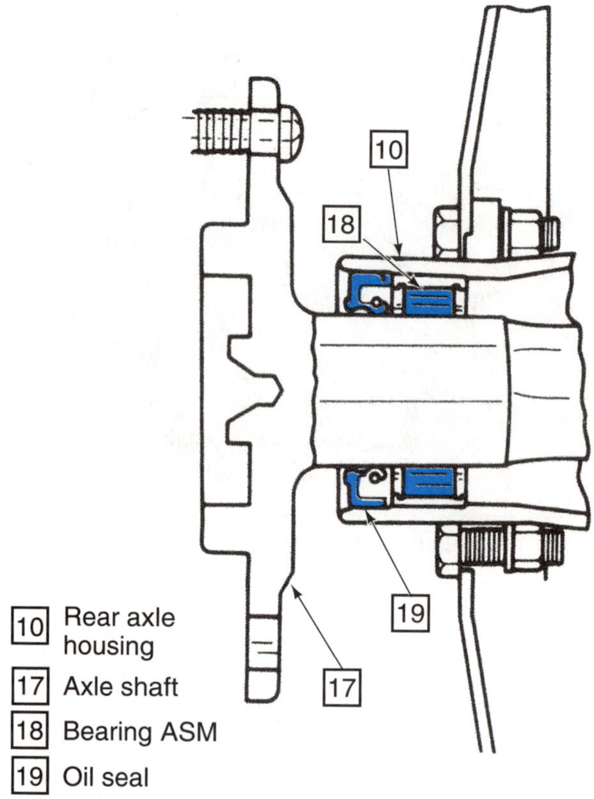


**Figure 3-28** Rear axle roller bearing and seal, rear wheel drive car (Courtesy of Chevrolet Motor Division, General Motors Corporation)

4. Place a drain pan under the differential and remove the differential cover. Discard the old lubricant.
5. Remove the differential lock bolt, pin, pinion gears, and shaft (Figure 3-29).
6. Push the axle shaft inward and remove the axle "C" lock.
7. Pull the axle from the differential housing.

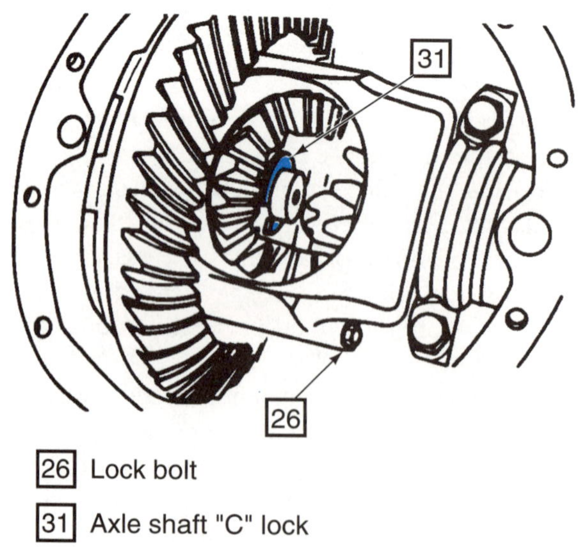


**Figure 3-29** Rear axle "C" lock, lock bolt, and pinion gears (Courtesy of Chevrolet Motor Division, General Motors Corporation)

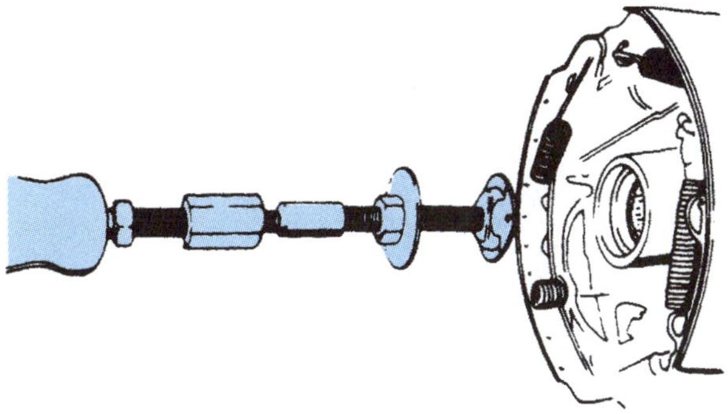

**Figure 3-30** Rear axle bearing puller (Courtesy of Chevrolet Motor Division, General Motors Corporation)

Reverse the axle removal procedure to reinstall the axle. Always use a new differential cover gasket, and fill the differential to the bottom of the filler plug opening with the manufacturer's recommended lubricant. Be sure all fasteners, including the wheel nuts, are tightened to the specified torque.

A typical axle bearing and seal removal procedure follows:

**Special Tools**

Seal puller

Bearing puller

1. Remove the axle seal with a seal puller.
2. Use the proper bearing puller to remove the axle bearing (Figure 3-30).
3. Clean the axle housing seal and bearing mounting area with solvent and a brush. Clean this area with compressed air.
4. Check the seal and bearing mounting area in the housing for metal burrs and scratches. Remove any burrs or irregularities with a fine-toothed round file.
5. Wash the axle shaft with solvent and blow it dry with compressed air.
6. Check the bearing contact area on the axle for roughness, pits, and scratches. If any of these conditions are present, axle replacement is necessary.
7. Be sure the new bearing fits properly on the axle and in the housing. Install the new bearing with the proper bearing driver (Figure 3-31). The bearing driver must apply pressure to the outer race that is pressed into the housing.

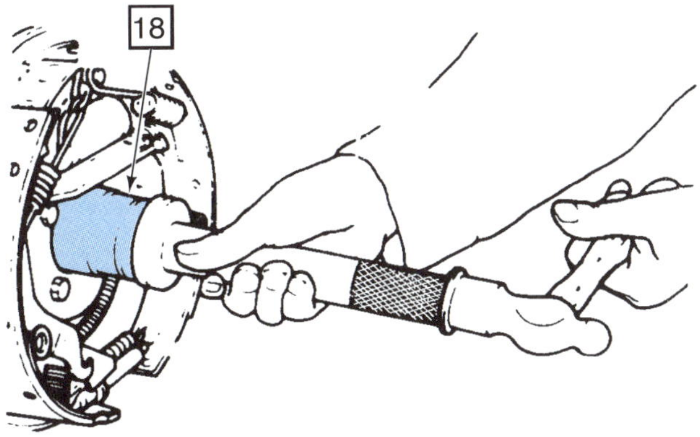


**Figure 3-31** Rear axle bearing driver (Courtesy of Chevrolet Motor Division, General Motors Corporation)

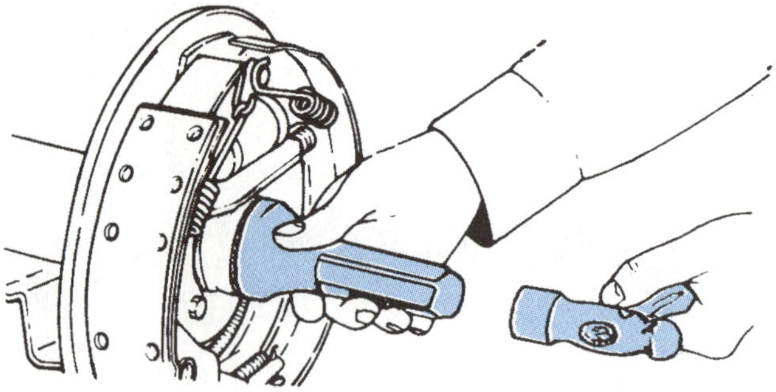

**Figure 3-32** Installing rear axle seal (Courtesy of Chevrolet Motor Division, General Motors Corporation)

8. Be sure the new seal fits properly on the axle shaft and in the housing. Make sure the garter spring on the seal faces toward the differential. Use the proper seal driver to install the new seal in the housing (Figure 3-32).
9. Lubricate the bearing, seal, and bearing surface on the axle with the manufacturer's specified differential lubricant.
10. Reverse the rear axle removal procedure to reinstall the rear axle.
11. Be sure all fasteners are tightened to the specified torque.

**WARNING:** Never use an acetylene torch to heat axle bearings or adaptor rings during the removal and replacement procedure. The heat may cause fatigue in the steel axle and the axle may break suddenly, causing the rear wheel to fall off. This action will likely result in severe vehicle damage and personal injury.

**Special Tools**

Axle puller

**Classroom Manual**
Chapter 3, page 54

Some rear axles have a sealed bearing that is pressed onto the axle shaft and held in place with an adaptor ring. These rear axles usually do not have "C" locks in the differential. A retainer plate is mounted on the axle between the bearing and the outer end of the axle. This plate is bolted to the outer end of the differential housing. After the axle retainer plate bolts are removed, a slide-hammer-type puller is attached to the axle studs to remove this type of axle. When this type of axle bearing is removed, the adapter ring must be split with a hammer and chisel while the axle is held in a vise. Do not heat the adapter ring or the bearing with an acetylene torch during the removal or installation process. After the adapter ring is removed, the bearing must be pressed from the axle shaft, and the bearing must not be reused. A new bearing and adapter ring must be pressed onto the axle shaft. The bearing removal and replacement procedure is shown (Figure 3-33).

# Guidelines for Servicing Wheel Bearings and Seals

1. When a wheel bearing fails prematurely, determine the cause of the failure and correct this cause to prevent a second bearing failure.
2. Before a tire and wheel assembly is removed from a car, chalk mark the wheel and tire assembly in relation to the hub or axle to maintain wheel balance when the wheel is reinstalled.
3. Replace tapered roller bearings and races as a matched set.
4. When cleaning bearings, never spin the bearings with compressed air.

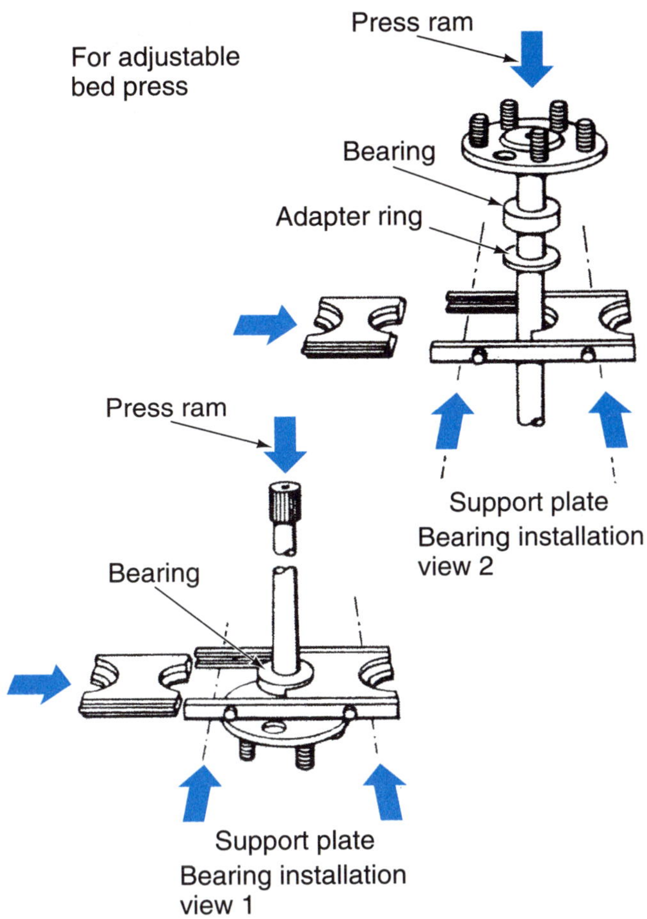


**Figure 3-33** Axle bearing and adapter ring removal and replacement. (Courtesy of Ford Motor Company)

5. Do not wash sealed bearings or bearings that are shielded on both sides.
6. When drying bearings with compressed air, be sure the shop air supply is free from moisture.
7. Do not wipe bearings with paper towels or shop towels with lint on them.
8. Always inspect bearing and seal mounting areas for metal burrs and scratches. Burrs may be removed with a fine-toothed file.
9. While servicing wheel bearings, always keep hands, tools, and work area clean.
10. When two separate tapered roller bearings are mounted in the wheel hub, follow the car manufacturer's recommended bearing adjustment procedure.
11. Check wheel bearing hub units for end play with a dial indicator.
12. On many front wheel drive vehicles the front drive axle must be removed to service the wheel bearing hub unit.
13. When front wheel bearings are mounted in the steering knuckle, the front hub nuts must be tightened to the specified torque before the vehicle is moved.
14. Do not use an acetylene torch to heat rear axle bearings and adaptors that are pressed onto the axle shaft.
15. On some rear wheel drive cars, the rear axle "C" locks inside the differential must be removed prior to rear axle removal.

## CASE STUDY

A customer complains of a bearing noise in the right front wheel of his full-size Ford wagon. The customer says the right front outer wheel bearing has been replaced twice in the last year, and this is the third failure. The technician asks the customer about the mileage intervals between bearing replacements, and the customer indicates that the wheel bearing has lasted about 8,000 miles (12,800 kilometers) each time it has been replaced. The technician finds out from the customer that no other work was done on the car each time the bearing was replaced.

When the technician removes the right front wheel and hub, the outer bearing rollers and races are badly scored. After cleaning both bearings, races, and hub, a closer examination of the outer bearing race shows an uneven wear pattern, which indicates misalignment. The technician removes the outer bearing race and finds a small metal burr behind the bearing race. This burr caused race misalignment and excessive wear on the race and rollers. The burr is removed with a fine-toothed file. The inner bearing and race have indentation wear because metal particles from the outer bearing contaminated the lubricant in the hub. The inner bearing race is removed, and the hub and spindle are thoroughly cleaned. Both bearings and the inner seal are replaced and the bearings and hub are repacked with grease. After reinstalling the hub, the bearings are carefully adjusted to the manufacturer's specifications, and the wheel nuts are tightened to the specified torque. A road test indicates that the bearing noise has been eliminated.

One small metal burr caused this customer a considerable amount of unnecessary expense. This experience proves that a technician's diagnostic capability is extremely important!

# Terms to Know

Bearing galling
Bearing abrasive step wear
Bearing etching
Bearing fatigue spalling
Bearing brinelling
Bearing abrasive roller wear
Bearing smears
Bearing frettage
Bearing heat discoloration
Bearing stain discoloration
Circlip
"C" locks
Stethoscope

# ASE Style Review Questions

1. While discussing defective bearings:
*Technician A* says brinelling appears as indentations across the bearing races.
*Technician B* says brinelling occurs while the bearing is rotating.
Who is correct?
**A.** A only
**B.** B only
**C.** Both A and B
**D.** Neither A nor B

2. While discussing bearing defects:
*Technician A* says misalignment wear on a front wheel bearing could be caused by a metal burr behind one of the bearing races.
*Technician B* says misalignment wear on a front wheel bearing could be caused by a bent front spindle.
Who is correct?
**A.** A only
**B.** B only
**C.** Both A and B
**D.** Neither A nor B

3. While discussing wheel bearing service:
*Technician A* says sealed bearings should be washed in solvent.
*Technician B* says shielded bearings should be washed in solvent.
Who is correct?
**A.** A only
**B.** B only
**C.** Both A and B
**D.** Neither A nor B

4. While discussing seal service:
*Technician A* says a seal should be installed with the garter spring facing toward the flow of lubricant.
*Technician B* says a seal should be installed with the garter spring facing in the same direction as the flow of lubricant.
Who is correct?
**A.** A only
**B.** B only
**C.** Both A and B
**D.** Neither A nor B

5. A front wheel bearing in which the hub and bearing unit is bolted to the steering knuckle has 0.010 in (0.254 mm) of hub end play.
*Technician A* says the bearing assembly is satisfactory.
*Technician B* says the hub nut should be retorqued to correct the end play.
Who is correct?
**A.** A only
**B.** B only
**C.** Both A and B
**D.** Neither A nor B

6. While discussing front wheel bearing service on a vehicle with the bearing hub unit pressed into the steering knuckle:
*Technician A* says the front wheel bearings may be damaged if this front wheel drive vehicle is moved without the hub nuts torqued to specifications.
*Technician B* says moving the vehicle under this condition does not affect the front wheel bearings.
Who is correct?
**A.** A only
**B.** B only
**C.** Both A and B
**D.** Neither A nor B

7. While discussing front wheel bearing service on a front wheel drive car with two separate tapered roller bearings mounted in the front steering knuckles:
*Technician A* says the hub nut torque supplies the correct bearing adjustment.
*Technician B* says that the brake should be applied while the front hub nuts are torqued.
Who is correct?
**A.** A only
**B.** B only
**C.** Both A and B
**D.** Neither A nor B

8. While discussing rear axle bearing service on a rear wheel drive car with press-on-type rear axle bearings:
*Technician A* says these bearings must not be reused after they have been pressed from the axle shaft.
*Technician B* says the bearing adapter ring should be removed by splitting the ring with a hammer and chisel.
Who is correct?
**A.** A only
**B.** B only
**C.** Both A and B
**D.** Neither A nor B

9. While discussing wheel bearing adjustment in a rear wheel drive car with two separate tapered roller bearings in each front wheel hub:
*Technician A* says after these bearings are cleaned and repacked with grease, the bearing adjusting nut should be tightened to 17 to 25 ft-lbs (23 to 34 Nm), loosened 1/2 turn, and retightened to 10 to 15 in-lbs (1.0 to 1.7 Nm).
*Technician B* says this type of bearing should be spun with compressed air during the cleaning process.
Who is correct?
**A.** A only
**B.** B only
**C.** Both A and B
**D.** Neither A nor B

10. A front wheel drive vehicle has 14-in (35.5-cm) rims and front wheel bearings mounted in the steering knuckles. The vehicle is lifted on a hoist and a dial indicator is positioned against the outer rim lip. The total inward and outward rim movement is 0.035 in (0.889 mm).
*Technician A* says that the wheel bearing may require replacement.
*Technician B* says that the hub nut torque should be checked prior to bearing replacement.
Who is correct?
**A.** A only
**B.** B only
**C.** Both A and B
**D.** Neither A nor B

## Table 3-1 ASE Task

Diagnose and service front and/or rear wheel bearings.

| Problem Area | Symptoms | Possible Causes | Classroom Manual | Shop Manual |
|---|---|---|---|---|
| **NOISES** | Growling noise at low speed, more noticeable while cornering | Defective front wheel bearings | 41 | 41 |
| | Growling noise at low speed | Defective rear wheel bearings | 54 | 41 |
| **STEERING QUALITY** | Reduced directional stability | Loose front wheel bearings | 47 | 49 |

CHAPTER 4

# Wheel and Tire Servicing and Balancing

Upon completion and review of this chapter, you should be able to:

- ❑ Diagnose tire thump and vibration problems.
- ❑ Diagnose steering pull problems related to tire condition.
- ❑ Rotate tires according to the vehicle manufacturer's recommended procedure.
- ❑ Remove and replace wheel and tire assemblies.
- ❑ Demount, inspect, repair, and remount tires.
- ❑ Inspect wheel rims.
- ❑ Measure tire and wheel radial and lateral runout.
- ❑ Diagnose problems caused by excessive radial or lateral wheel and tire runout.
- ❑ Measure tire tread wear.
- ❑ Perform off-car static wheel balance procedures.
- ❑ Perform off-car dynamic wheel balance procedures.
- ❑ Diagnose tire wear problems caused by tire and wheel imbalance.
- ❑ Perform on-car balance procedures.

**Basic Tools**

Basic technician's tool set

Service manual

Tire repair kit

Tread depth gauge

## Tire Noises and Steering Problems

### Diagnosis of Tire Noises

**SERVICE TIP:** Tire noise varies with road surface conditions, whereas differential noise is not affected when various road surfaces are encountered.

Uneven tread surfaces may cause tire noises that seem to originate elsewhere in the vehicle. These noises may be confused with differential noise. Differential noise usually varies with acceleration and deceleration, while tire noise remains more constant in relation to these forces. Tire noise is most pronounced on smooth asphalt road surfaces at speeds of 15 to 45 mph (24 to 72 kph).

Tire thump may be defined as a pounding noise caused by tire and wheel rotation.

Tire vibration is a fast shaking of the tire, which is transferred to the chassis and passenger compartment.

### Tire Thump and Vibration

When tire thump and vibration are present, check these items:

1. cupped tire treads
2. excessive tire radial runout
3. manufacturing defects such as heavy spots, weak spots, or tread chunking
4. incorrect wheel balance.

### Steering Pull

A vehicle should maintain the straight-ahead forward direction on smooth, straight road surfaces without excessive steering wheel correction by the driver. If the steering gradually pulls to one side on a smooth, straight road surface, a tire, steering, or suspension defect is present. Tires of different types, sizes, designs, or inflation pressures on opposite sides of a vehicle cause steering pull. Sometimes a tire manufacturing defect occurs in which the belts are wound off-center on the tire. This condition is referred to as conicity. A cone-shaped object rolls in the direction of its smaller diameter. Similarly, a tire with conicity tends to lead, or pull, to one side, which causes the vehicle to follow the action of the tire (Figure 4-1).

Steering pull is the tendency of the steering to gradually pull to the right or left when the vehicle is driven straight ahead on a reasonably smooth, straight road.

Tire conicity occurs when the tire belt is wound off-center in the manufacturing process creating a cone-shaped belt, which results in steering pull.

**SERVICE TIP:** Tire conicity is not visible. It can only be diagnosed by changing the tire and wheel position.

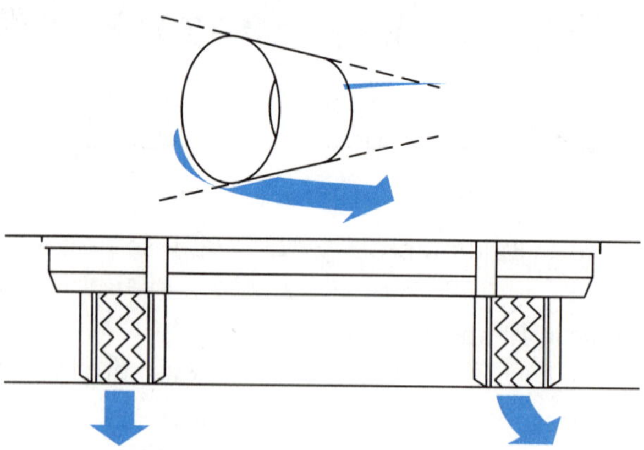

Figure 4-1 Tire conicity (Courtesy of Chrysler Corporation)

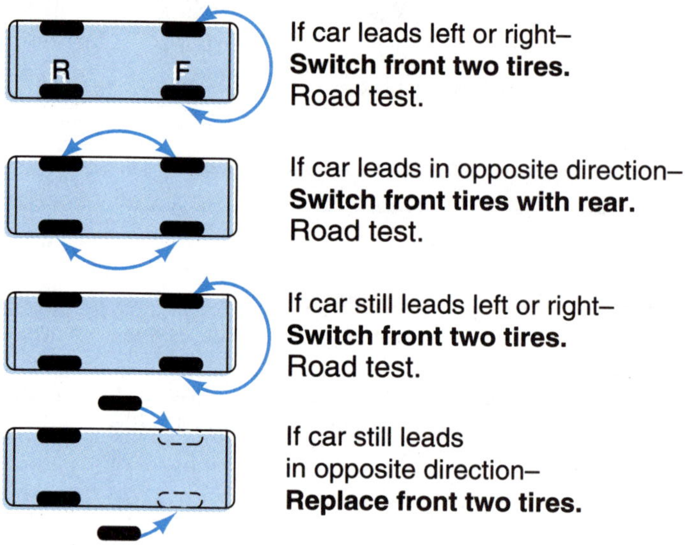


Figure 4-2 Tire conicity diagnosis (Courtesy of Chrysler Corporation)

Since tire conicity cannot be diagnosed by a visual inspection, it must be diagnosed by switching the two front tires, and reversing the front and rear tires (Figure 4-2). Incorrect front suspension alignment angles also cause steering pull.

## Tire Rotation

Driving habits determine tire life to a large extent. Severe brake applications, high-speed driving, turning at high speeds, rapid acceleration and deceleration, and striking curbs are just a few driving habits that shorten tire life. Most car manufacturers recommend tire rotation at specified intervals to obtain maximum tire life. The exact tire rotation procedure depends on the model year, the type of tires, and whether the vehicle has a conventional spare or a compact spare (Figure 4-3). Tire rotation procedures do not include the compact spare. The vehicle manufacturer provides tire rotation information in the owner's manual and service manual. Vehicle manufacturers usually recommend different tire rotation procedures for bias ply tires than for radial tires (Figure 4-4).

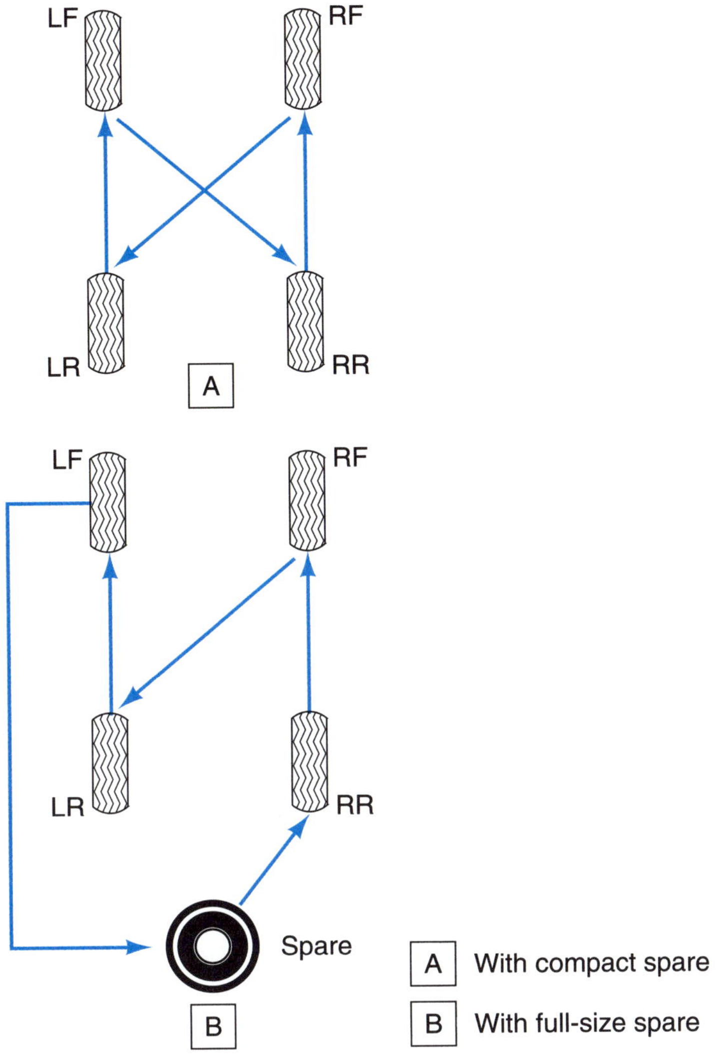


**Figure 4-3** Radial tire rotation procedure (Courtesy of Chevrolet Motor Division, General Motors Corporation)

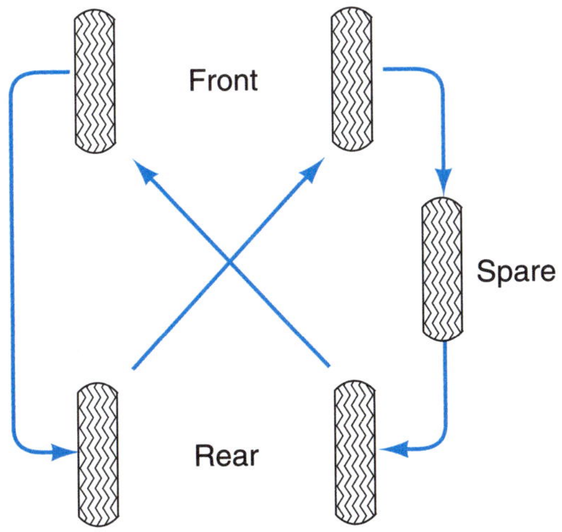


**Figure 4-4** Bias ply tire rotation procedure (Courtesy of Pontiac Motor Division, General Motors Corporation)

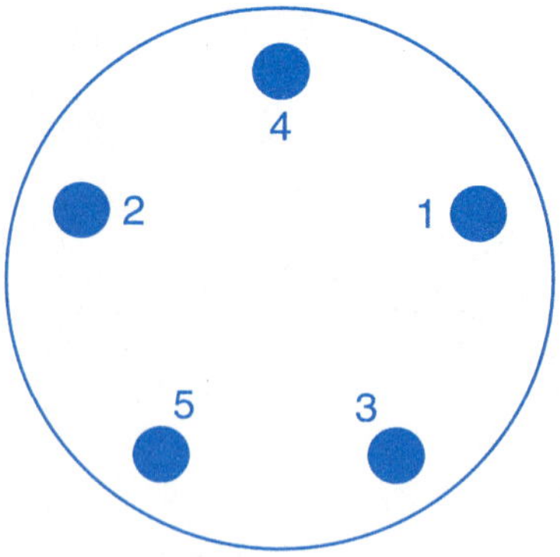


**Figure 4-5** Wheel nut tightening sequence (Courtesy of Chevrolet Motor Division, General Motors Corporation)

When tires and wheels are installed on a vehicle, it is very important that the wheel nuts are torqued to manufacturer's specifications in the proper sequence (Figure 4-5). Do not use an impact wrench when tightening wheel nuts to the specified torque.

# Wheel and Tire Service

## Wheel and Tire Removal

**CUSTOMER CARE:** During many automotive service operations such as tire and wheel service, the technician literally has the customer's life in his or her hands! Always perform tire and wheel service carefully and thoroughly. Always watch for unsafe tire or wheel conditions, and report these problems to the customer. When you prove to the customer that you are concerned about vehicle safety, you will probably have a steady customer.

**SERVICE TIP:** Some wheel covers have fake plastic lug nuts that must be removed to access the lug nuts. Be careful not to break the fake lug nuts.

When it is necessary to remove a wheel and tire assembly, follow these steps:

1. Remove the wheel cover. If the vehicle is equipped with antitheft locking wheel covers, the lock bolt for each wheel cover is located behind the ornament in the center of the wheel cover. A special key wrench is supplied to the owner for ornament and lock bolt removal. If the customer's key wrench has been lost, a master key is available from the vehicle dealer.
2. Loosen the wheel lug nuts about one-half turn, but do not remove the wheel nuts. Some vehicles are equipped with antitheft wheel nuts. A special lug nut key is supplied to the vehicle owner. This lug nut key has a hex nut on the outer end and a special internal projection that fits in the wheel nut opening. Install the lug nut key on the lug nuts and connect the lug nut wrench on the key hex nut to loosen the lug nuts.

**CAUTION:** Before the vehicle is raised on a hoist, be sure that the hoist is lifting on the car manufacturer's recommended lift points.

**CAUTION:** If the vehicle is lifted with a floor jack, place safety stands under the suspension or frame and lower the vehicle onto the safety stands. Then remove the floor jack from under the vehicle.

3. Raise the vehicle on a hoist or with a floor jack to a convenient working level.
4. Chalk mark the tire, wheel, and one of the lug nuts, so the wheel and tire can be reinstalled in the same position.

**WARNING:** If heat is used to loosen a rusted wheel, the wheel and/or wheel bearings may be damaged.

5. Remove the lug nuts and the wheel and tire assembly. If the wheel is rusted and will not come off, hit the inside of the wheel with a large rubber mallet. Do not hit the wheel with a steel hammer, because this action could damage the wheel. Do not heat the wheel.

## Tire and Wheel Service Precautions

There are many different types of tire changing equipment in the automotive service industry. However, specific precautions apply to the use of any tire changing equipment. These precautions include the following:

1. Before you operate any tire changing equipment, always be absolutely certain that you are familiar with the operation of the equipment.
2. When operating tire changing equipment, always follow the equipment manufacturer's recommended procedure.
3. Always deflate a tire completely before attempting to demount the tire.
4. Clean the bead seats on the wheel rim before mounting the tire on the wheel rim.
5. Lubricate the outer surface of the tire beads with rubber lubricant before mounting the tire on the wheel rim.
6. When the tire is mounted on the wheel rim, be sure the tire is positioned evenly on the wheel rim.
7. While inflating a tire, do not stand directly over the tire. An air hose extension allows the technician to stand back from the tire during the inflation process.
8. Do not overinflate tires.
9. When mounting tires on cast aluminum alloy wheel rims or cast magnesium alloy wheel rims, always use the tire changing equipment manufacturer's recommended tools and procedures.

### Tire Demounting

**WARNING:** If hand tools or tire irons are used to demount tires, tire bead and wheel rim damage may occur.

Always use a tire changer to demount tires (Figure 4-6). Do not use hand tools or tire irons for this purpose. A typical tire demounting procedure follows:

**Special Tools**

Tire changer

1. Remove the valve core and be sure the tire is completely deflated.
2. Place the wheel and tire on the tire changer with the narrow bead ledge facing upward.
3. Follow the operating procedure recommended by the manufacturer of the tire changer to force the tire bead inward and separate it from the rim on both sides.
4. Push one edge of the top bead into the drop center of the rim.
5. Place the tire changer's bar or lever between the bead and the rim on the opposite side of the rim from where the bead is in the drop center.
6. Operate the tire changer to rotate the bar or lever and move the bead over the top of the rim.
7. Repeat steps 4, 5, and 6 to move the lower bead over the top of the rim.

**Figure 4-6** Tire changer

## Tire Inspection and Repair

To find a leak in a tire and wheel, inflate the tire to the pressure marked on the sidewall, and then submerge the tire and wheel in a tank of water. An alternate method of leak detection is to sponge soapy water on the tire and wheel. Bubbles will appear wherever the leak is located in the tire or wheel. Mark the leak location in the tire or wheel rim with a crayon, and mark the tire at the valve stem location so the tire can be reinstalled in the same position on the wheel to maintain proper balance.

A puncture is the most common cause of a tire leak, and many punctures can be repaired satisfactorily. Do not attempt to repair punctures over 1/4 inch in diameter. Punctures in the sidewalls or on the tire shoulders should not be repaired. The repairable area in belted bias ply tires is approximately the width of the belts (Figure 4-7). The belts in radial tires are wider than those in bias ply tires. The repairable area in radial tires is also the width of the belts. Since compact spare tires have thin treads, do not attempt to repair these tires.

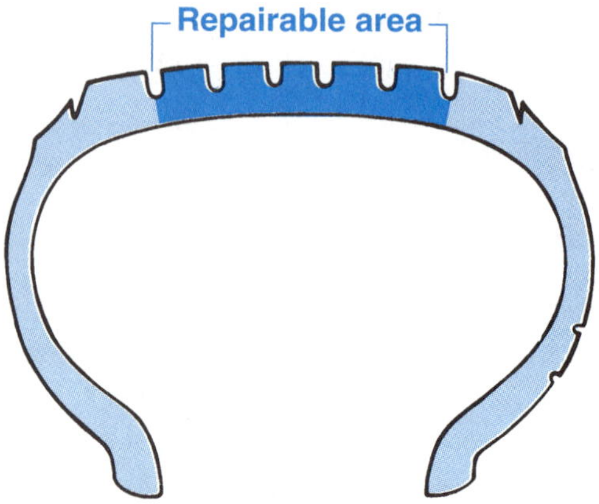


**Figure 4-7** Repairable areas on bias ply and belted bias ply tires (Courtesy of Cadillac Motor Car Division, General Motors Corporation)

Inspect the tire and do not repair a tire with any of these defects, signs of damage, or excessive wear:

1. tires with the wear indicators showing
2. tires worn until the fabric or belts are exposed
3. bulges or blisters
4. ply separation
5. broken or cracked beads
6. cuts or cracks anywhere in the tire

Since most vehicles are equipped with tubeless tires, we will discuss this type of tire repair. If the cause of the puncture, such as a nail, is still in the tire, remove it from the tire. Most punctures can be repaired from inside the tire with a service plug or vulcanized patch service kit. The instructions of the tire service kit manufacturer should be followed, but we will discuss three common tire repair procedures.

## Plug Installation Procedure

1. Buff the area around the puncture with a wire brush or wire buffing wheel.
2. Select a plug slightly larger than the puncture opening, and insert the plug in the eye of the insertion tool.
3. Wet the plug and the insertion tool with vulcanizing fluid.
4. While holding and stretching the plug, pull the plug into the puncture from the inside of the tire (Figure 4-8). The head of the plug should contact the inside of the tire. If the plug pulls through the tire, repeat the procedure.
5. Cut the plug off 1/32 inch from the tread surface. Do not stretch the plug while cutting.

## Cold Patch Installation Procedure

**CAUTION:** Radial tire patches should have arrows that must be positioned parallel to the radial plies.

1. Buff the area around the puncture with a wire brush or buffing wheel.
2. Apply vulcanizing fluid to the buffed area and allow it to dry until it is tacky.
3. Peel the backing from the patch, and apply the patch over the puncture. Center the patch over the puncture.
4. Run a stitching tool back and forth over the patch to improve bonding.

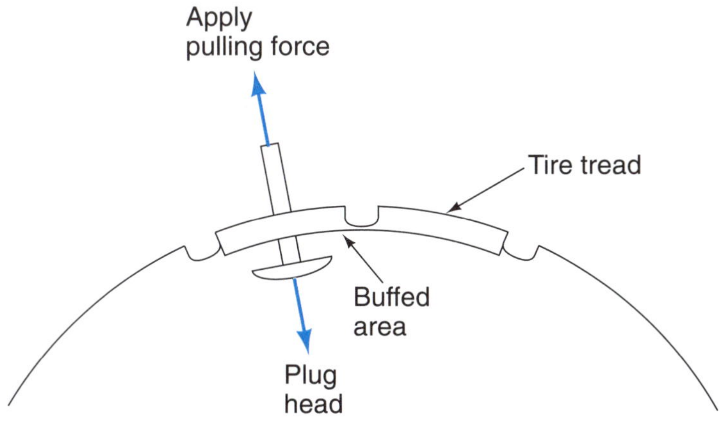


**Figure 4-8** Plug installation procedure

### Hot Patch Installation Procedure

1. Buff the area around the puncture with a wire brush or buffing wheel.
2. Apply vulcanizing fluid to the buffed area, if required.
3. Peel the backing from the patch and install the patch so it is centered over the puncture on the inside of the tire. Many hot patches are heated with an electric heating element clamped over the patch. This element should be clamped in place for the amount of time recommended by the equipment or patch manufacturer.
4. After the heating element is removed, allow the patch to cool for a few minutes and be sure the patch is properly bonded to the tire.

## Wheel Rim Service

**WARNING:** The use of abrasive cleaners, alkaline-base detergents, or caustic agents on aluminum or magnesium wheel rims may cause discoloration or damage to the protective coating.

Steel rims should be spray cleaned with a water hose. Aluminum or magnesium wheel rims should be cleaned with a mild soap and water solution, and rinsed with clean water. The use of abrasive cleaners, alkaline-base detergents, or caustic agents may damage aluminum or magnesium wheel rims. Clean the rim bead seats on these wheel rims thoroughly with the mild soap and water solution. The rim bead seats on steel wheel rims should be cleaned with a wire brush or coarse steel wool.

**WARNING:** Steel wheel rims must not be welded, heated, or peened with a ball peen hammer. These procedures may weaken the rim and create a safety hazard.

**CAUTION:** Installing an inner tube to correct leaks in a tubeless tire or wheel rim is not an approved procedure.

Steel wheel rims should be inspected for excessive rust and corrosion, cracks, loose rivets or welds, bent or damaged bead seats, and elongated lug nut holes. Aluminum or magnesium wheel rims should be inspected for damaged bead seats, elongated lug nut holes, cracks, and porosity. If any of these conditions are present on either type of wheel rim, replace the wheel rim.

Many shops always replace the tire valve assembly when a tire is repaired or replaced. This policy helps to prevent future problems with tire valve leaks. The inner end of the valve may be cut off with a pair of diagonal pliers, and then the outer end may be pulled from the rim. Coat the new valve with rubber tire lubricant and pull it into the rim opening with a special puller screwed onto the valve threads.

### Wheel Rim Leak Repair

A wheel rim leak may be repaired if the leak is not caused by excessive rust on a steel rim, and the rim is in satisfactory condition. Follow these steps for wheel rim leak repair:

1. Use #80-grit sandpaper to thoroughly clean the area around the leak on the tire side of the rim.
2. Use a shop towel to remove any grit from the leak area.
3. Be sure the wheel rim is at room temperature, and apply a heavy coating of silicone rubber sealer over the leak area.
4. Spread the sealer over the entire sanded area with a putty knife.
5. Allow the sealer to cure for 6 hours before remounting the tire.

# Tire Remounting Procedure

1. Be sure the wheel rim bead seats are thoroughly cleaned.
2. Coat the tire beads and the wheel rim bead seats with rubber tire lubricant.
3. Secure the wheel rim on the tire changer with the narrow bead ledge facing upward, and place the tire on top of the wheel rim with the bead on the lower side of the tire in the drop center of the wheel rim.
4. Use the tire changer bar or lever under the tire bead to install the tire bead over the wheel rim. Always operate the tire changer with the manufacturer's recommended procedure.
5. Repeat the procedure in steps 3 and 4 to install the upper bead over the wheel rim.
6. Rotate the tire on the wheel rim until the crayon mark is aligned with the valve stem. This mark was placed on the tire prior to demounting.

**WARNING:** When a bead expander is installed around the tire, never exceed 10 psi (69 kPa) pressure in the tire. A higher pressure may cause the expander to break and fly off the tire, causing serious personal injury or property damage.

**WARNING:** When a bead expander is not used, never exceed 40 psi (276 kPa) tire pressure to move the tire beads out tightly against the wheel rim. A higher pressure may blow the tire bead against the rim with excessive force, and this action could burst the rim or tire, resulting in serious personal injury or property damage.

**WARNING:** While inflating a tire, do no stand directly over a tire. In this position, serious injury could occur if the tire or wheel rim flies apart.

7. Follow the recommended procedure supplied by the manufacturer of the tire changer to inflate the tire. This procedure may involve the use of a bead expander installed around the center of the tire tread to expand the tire beads against the wheel rim. If a bead expander is used, inflate the tire to 10 psi (69 kPa) to move the beads out tightly against the wheel rim. Never exceed this pressure with a bead expander installed on the tire. Always observe the circular marking around the tire bead as the tire is inflated. This mark should be centered around the wheel rim (Figure 4-9). Always observe both beads while a tire is inflated. If the circular mark around the tire bead is not centered on the rim, deflate the tire and center it on the wheel rim. If a bead expander is not used, never exceed 40 psi (276 kPa) to try and move tire beads out tightly against the rim. When either tire bead will not move out tightly against the wheel rim with 40 psi (276 kPa) tire pressure, deflate the tire and center it on the wheel rim again.

Photo Sequence 2 shows a typical procedure for demounting and mounting a tire on a wheel assembly.

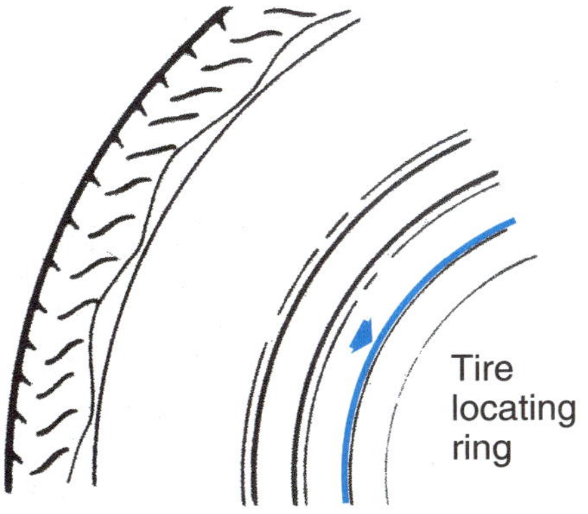


**Figure 4-9** The circular ring around the tire bead must be centered on the wheel rim. (Courtesy of Oldsmobile Division, General Motors Corporation)

# Photo Sequence 2
## Typical Procedure for Demounting and Mounting a Tire on a Wheel Assembly

**P2-1** A typical tire changer.

**P2-2** Demounting the tire from the wheel begins with releasing the air, removing the air valve core, and unseating the tire from the rim. The machine does the unseating. The technician merely guides the operating lever.

**P2-3** Once both sides of the tire are unseated, place the assembly onto the machine.

**P2-4** Depress the pedal that clamps the wheel to the tire machine.

**P2-5** Lower the machine's arm into position on the tire and wheel assembly.

**P2-6** Insert the tire iron between the tire and wheel. Depress the pedal that causes the wheel to rotate. This will free the tire from the wheel.

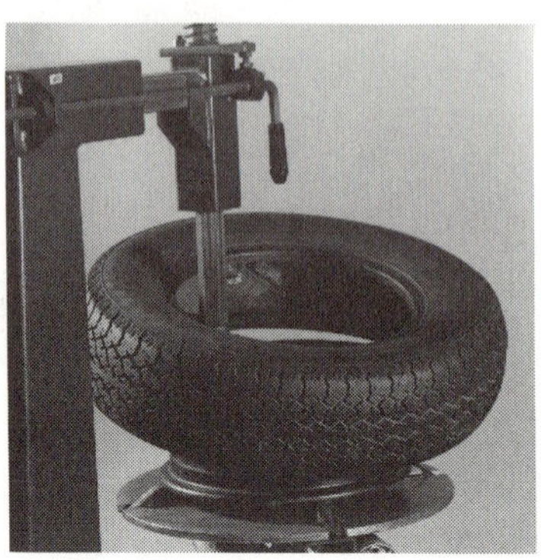

**P2-7** After the tire is totally free from the rim, remove the tire.

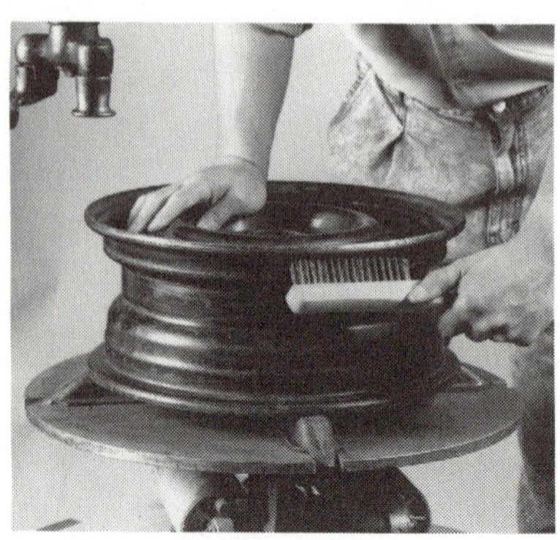

**P2-8** To prepare the wheel for the mounting of a new tire, use a wire brush to remove all of the dirt and rust from the sealing surface.

**P2-9** Rubber compound should be liberally applied to the bead area of the new tire.

# Typical Procedure for Demounting and Mounting a Tire on a Wheel Assembly (continued)

**P2-10** Place the tire onto the wheel assembly and lower the arm onto the assembly. As the wheel rotates, the tire will be forced over the rim.

**P2-11** After the tire is completely over the rim, install the air ring over the tire. Activate this to seat the tire against the wheel.

**P2-12** Reinstall the air valve core and inflate the tire to the recommended pressure.

# Tire and Wheel Runout Measurement

Ideally, a tire and wheel assembly should be perfectly round. However, this condition is rarely achieved. A tire and wheel assembly that is out-of-round is said to have radial runout. If the radial runout exceeds manufacturer's specifications, a vibration may occur, because the radial runout causes the spindle to move up and down (Figure 4-10). A defective tire with a variation in stiffness may also cause this up-and-down spindle action.

Radial tire runout refers to excessive variations in the tread circumference.

A dial indicator gauge may be positioned against the center of the tire tread as the tire is rotated slowly to measure radial runout (Figure 4-11). Radial runout of more than 0.060 in (1.5 mm) will cause vehicle shake. If the radial runout is between 0.045 in to 0.060 in (1.1 mm to 1.5 mm), vehicle shake may occur. These are typical radial runout specifications. Always consult the vehicle manufacturer's specifications. Mark the highest point of radial runout on the tire with a crayon, and mark the valve stem position on the tire.

**Special Tools**

Tire runout gauge

If the radial tire runout is excessive, demount the tire and check the runout of the wheel rim with a dial indicator positioned against the lip of the rim while the rim is rotated (Figure 4-12). Use a crayon to mark the highest point of radial runout on the wheel rim. Radial wheel runout should not exceed 0.035 in (0.9 mm), whereas the maximum lateral wheel runout is 0.045 in (1.1 mm). If the highest point of wheel radial runout coincides with the chalk mark from the highest point of maximum tire radial runout, the tire may be rotated 180° on the wheel to reduce radial runout. Tires or wheels with excessive runout are usually replaced.

Lateral tire runout may be defined as excessive variations in the sidewalls of the tire.

Lateral tire runout may be measured with a dial indicator located against the sidewall of the tire. Excessive lateral runout causes the tire to waddle as it turns, and this waddling sensation may be transmitted to the passenger compartment (Figure 4-13). A chassis waddling action may also be caused by a defective tire in which the belt is not straight. If the lateral runout exceeds 0.080 in (2.0 mm), wheel shake problems will occur on the vehicle. Chalk mark the tire and wheel at the highest point of radial runout. When the tire runout is excessive, the tire should be removed from the wheel, and the wheel lateral runout should be measured with a dial indicator positioned against the edge of the wheel as the wheel is rotated. Wheels or tires with excessive lateral runout should be replaced.

**Classroom Manual**
Chapter 4, page 59

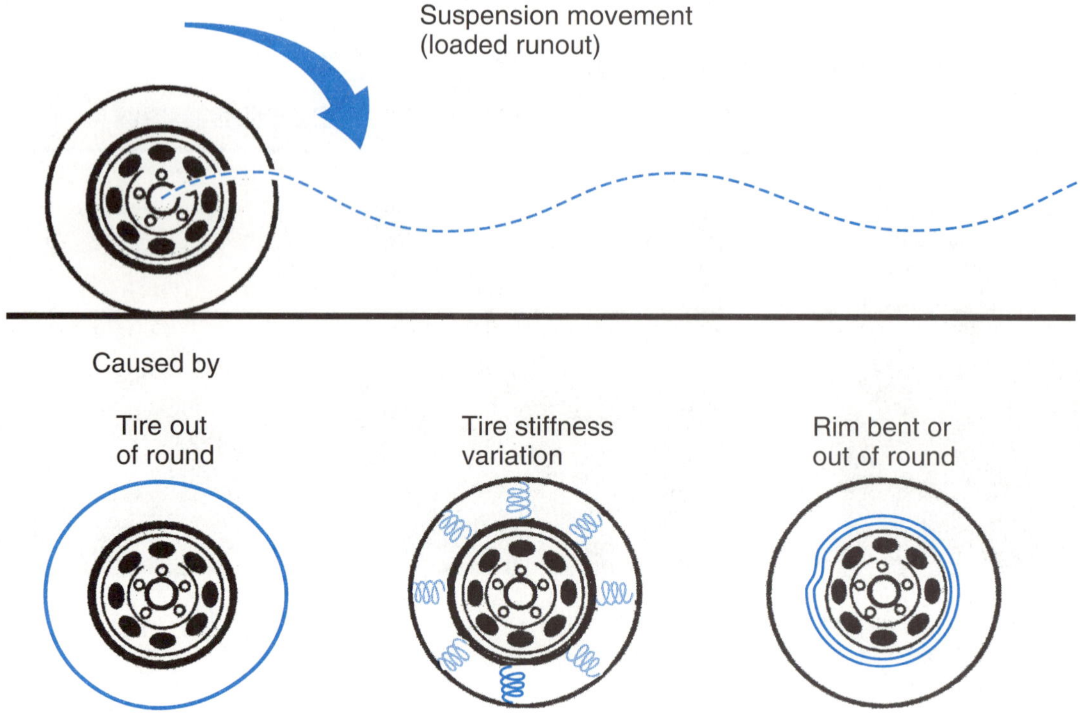


Figure 4-10 Vertical tire and wheel vibrations caused by radial tire or wheel runout, or variation in tire stiffness (Courtesy of Oldsmobile Division, General Motors Corporation)

Figure 4-11 Measuring tire radial runout (Courtesy of Chrysler Corporation)

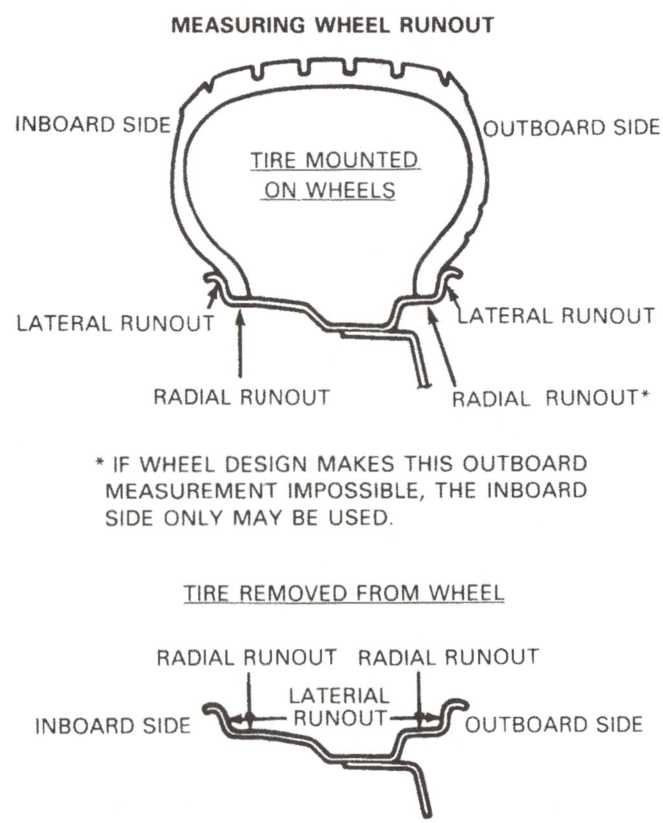


Figure 4-12 Measuring wheel radial runout (Courtesy of Oldsmobile Division, General Motors Corporation)

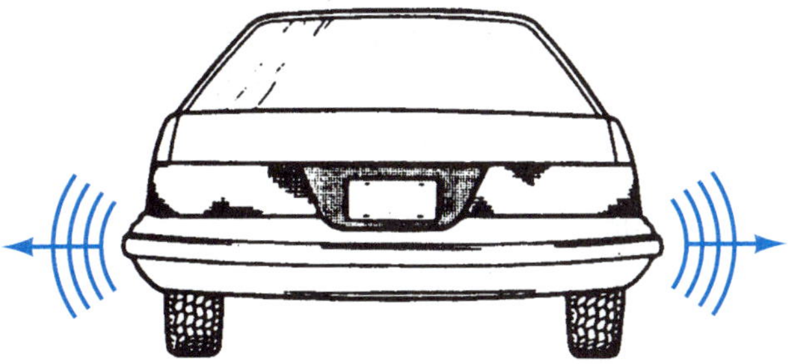


**Figure 4-13** Chassis waddling action caused by lateral tire or wheel runout, or by a defective tire with a belt that is not straight (Courtesy of Oldsmobile Division, General Motors Corporation)

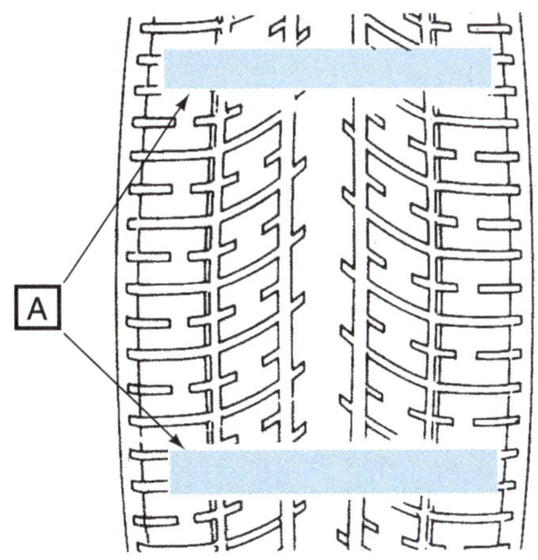


**Figure 4-14** Tire tread wear indicators (Courtesy of Oldsmobile Division, General Motors Corporation)

# Tread Wear Measurement

On most tires, the tread wear indicators appear as wide bands across the tread when tread depth is worn to 1/16 in (1.6 mm). Most tire manufacturers recommend tire replacement when the wear indicators appear across two or more tread grooves at three locations around the tire (Figure 4-14). If tires do not have wear indicators, a tread depth gauge may be used to measure the tread depth (Figure 4-15). The tread depth gauge reads in 32nds of an inch. Tires with 2/32 inch of tread depth, or less, should be replaced.

**Figure 4-15** Tread depth gauge (Courtesy of Snap-On Tools Corporation)

## Preliminary Wheel Balancing Checks

These preliminary checks should be completed before a wheel and tire are balanced:

1. Check for objects in tire tread.
2. Check for objects inside tire.
3. Inspect the tread and sidewall.
4. Check inflation pressure.
5. Measure tire and wheel runout.
6. Check wheel bearing adjustment.
7. Check for mud collected on the inside of the wheel.

▲ **WARNING:** On many wheel balancers, the tire and wheel are spun at high speed during the dynamic balance procedure. Be sure that all wheel weights are attached securely, and check for other loose objects on the tire and wheel, such as stones in the tread. If loose objects are detached from the tire or wheel at high speed, they may cause serious personal injury or property damage.

▲ **WARNING:** On the type of wheel balancer that spins the tire and wheel at high speed during the dynamic balance procedure, always attach the tire and wheel assembly securely to the balancer. Follow the equipment manufacturer's recommended wheel-mounting procedure. If the tire and wheel assembly ever becomes loose on the balancer at high speed, serious personal injury or property damage may result.

■ **CAUTION:** Prior to spinning a tire and wheel at high speed on a wheel balancer, always lower the protection shield over the tire. This shield provides protection in case anything flies off the tire or wheel.

All of the items on the preliminary check list influence wheel balance or safety. Therefore, it is extremely important that the preliminary checks are completed. Since a tire and wheel assembly is rotated at high speed during the dynamic balance procedure, it is very important that objects such as stones are removed from the treads. Centrifugal force may dislodge objects from the treads and cause serious personal injury. For this reason, it is also extremely important that the old wheel weights are removed from the wheel prior to balancing, and the new weights are attached securely to the wheel during the balance procedure. Follow these preliminary steps before attempting to balance a wheel and tire:

1. When most types of wheel balancers are used, the wheel and tire must be removed from the vehicle and installed on the balancer. All mud, dust, and debris must be washed from the wheel after it is removed.
2. Objects inside a tire, such as balls of rubber, make balance impossible. When the wheel and tire assembly is mounted on the balancer, be absolutely sure that the wheel is securely tightened on the balancer. As the tire is rotated slowly, listen for objects rolling inside the tire. If such objects are present, they must be removed prior to wheel balancing.
3. The tire should be inspected for tread and sidewall defects before the balance procedure. These defects create safety hazards, and they may influence wheel balance. For example, tread chunking makes the wheel balancing difficult.

## Tire Inflation Pressure

A tire depends on correct inflation pressure to maintain its correct shape and support the vehicle weight. Excessive inflation pressure causes these problems:

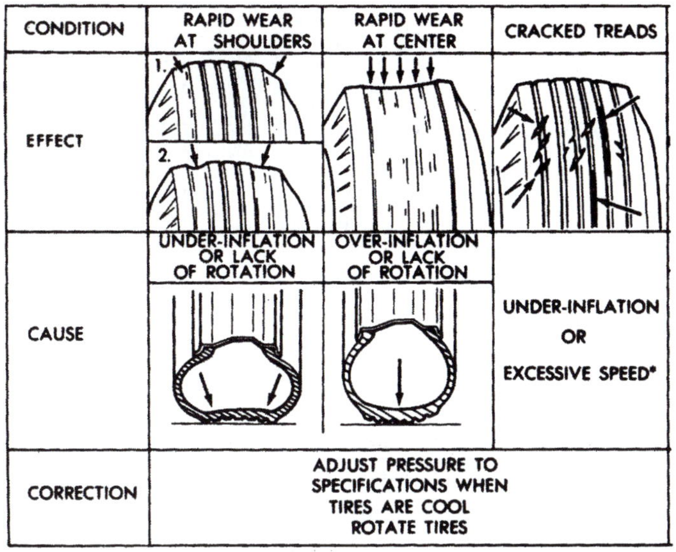


**Figure 4-16** Tire tread wear caused by underinflation and overinflation (Courtesy of Chrysler Corporation)

1. excessive center tread wear (Figure 4-16)
2. hard ride
3. damage to tire carcass

When tires are underinflated, these problems will be evident:

1. excessive wear on each side of the tread (Figure 4-16)
2. hard steering
3. wheel damage
4. excessive heat buildup in the tire, and possible severe tire damage with resultant hazardous driving

The tires should be inflated to the car manufacturer's recommended pressure prior to the balance procedure. Loose wheel bearings allow lateral wheel shaking, and simulate an imbalanced wheel condition when the vehicle is in motion. Therefore, wheel bearing adjustments should be checked when wheel balance conditions are diagnosed.

## Static Wheel Balance Procedure

Many different types of wheel balancers are available in the automotive service industry. The exact wheel balance procedure may vary depending on the type of wheel balancer. One of the simplest types of wheel balancers for static balancing is the bubble balancer (Figure 4-17).

Static wheel balance refers to proper balance of the tire when it is at rest.

**Special Tools**

Off-car wheel balancer

**Figure 4-17** Bubble-type wheel balancer (Courtesy of Mac Tools Inc.)

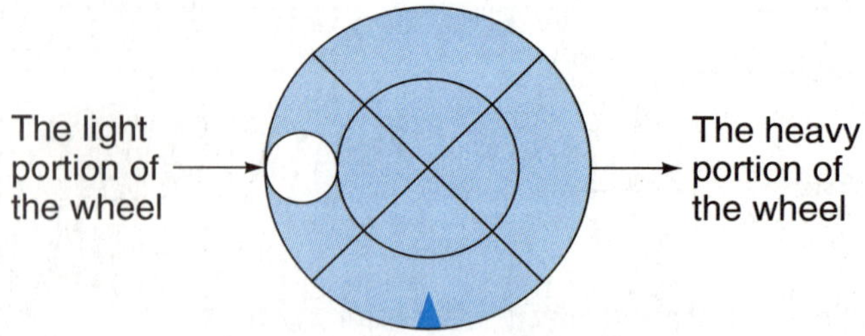


**Figure 4-18** Bubble and triangular mark on a bubble balancer (Courtesy of the Bada Company)

The bubble balancer will only perform a static balance procedure. This type of balancer does not have dynamic balance capabilities. Therefore, the required static weight should be divided, and half of this weight should be placed on each side of the wheel directly opposite each other. When this type of wheel balancer is used to static balance a wheel, use this procedure.

1. Complete the preliminary checks outlined previously.
2. Adjust the legs on the balancer to center the bubble.
3. Position the wheel securely on the balancer, and align the triangular mark with the valve stem on the wheel.
4. Observe the spirit level on the balancer to determine the heavy spot and the light spot on the wheel. The light spot will be directly opposite the heavy spot (Figure 4-18).
5. Place enough weight at the wheel light spot until the spirit level indicates a balanced wheel.
6. Chalk mark from the weight position on the sidewall radially around the tire to the inside bead.
7. Remove the wheel from the balancer and divide the weight by 2.
8. Install the required weights securely on the rim at the chalk-marked positions. The combined weight of the two wheel weights must equal the original weight installed in step 5 (Figure 4-19).
9. Reinstall the wheel on the balancer and check for proper balance.

Many wheel weights have a spring clamp, which holds the weight on the edge of the rim. A special pair of wheel weight pliers is used to tap the wheel weight onto the rim and remove the weight from the rim (Figure 4-20). Magnesium or aluminum wheels may require the use of stick-on wheel weights. This type of weight must also be used on some Cadillac wheels on which the wheel covers interfere with the conventional weights. Regardless of the wheel weight type, they must be attached securely to the wheel.

On some types of wheel balancers during the static balance procedure, the wheel is allowed to rotate by gravity. A heavy spot rotates the tire until this spot is at the bottom. The necessary static

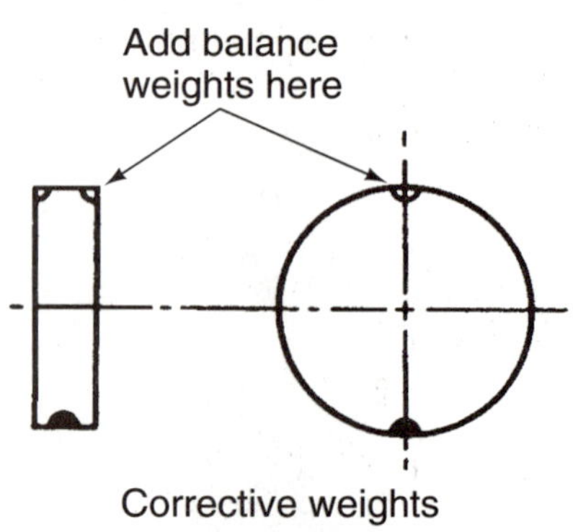


**Figure 4-19** Dynamic wheel balance procedure (Courtesy of Oldsmobile Division, General Motors Corporation)

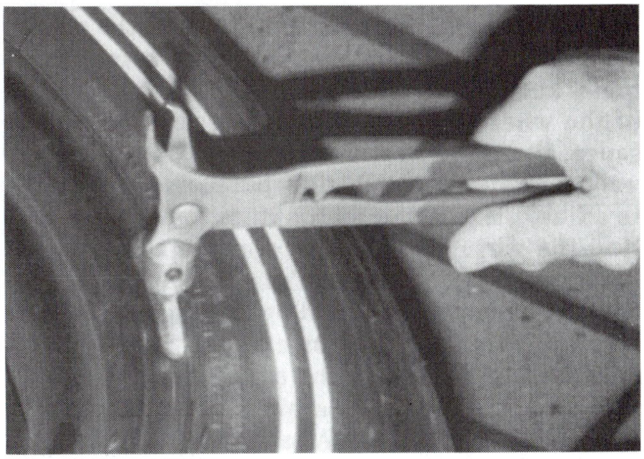

Figure 4-20 Wheel weights are removed and installed with special wheel weight pliers.

balance weights are then added at the top of the wheel 180° from the heavy spot. When the wheel and tire assembly is balanced statically, gravity does not rotate the wheel from the at-rest position. Rotate the tire by hand and check the static balance at 120° intervals.

The electronic wheel balancer is the most common type in use at the present time (Figure 4-21). There are many different types of electronic wheel balancers in the automotive service industry. With an electronic balancer, the operator must enter the wheel diameter, width, and offset. The balancer must have this information to perform its calculations.

## Dynamic Wheel Balance Procedure

Dynamic imbalance is corrected with lead weights. Many electronic wheel balancers have an electric drive motor that spins the tire at high speed. On some electronic wheel balancers, the tire is spun by hand during the balance procedure. The electronic-type balancer performs static and dynamic balance calculations simultaneously, and indicates the correct weight size and location to the operator. The tire and wheel assembly must be fastened securely on the balancer with the correct size adapters. Off-car electronic high-speed wheel balancers have a safety hood that must be

Dynamic wheel balance refers to proper balance of the tire during tire and wheel rotation.

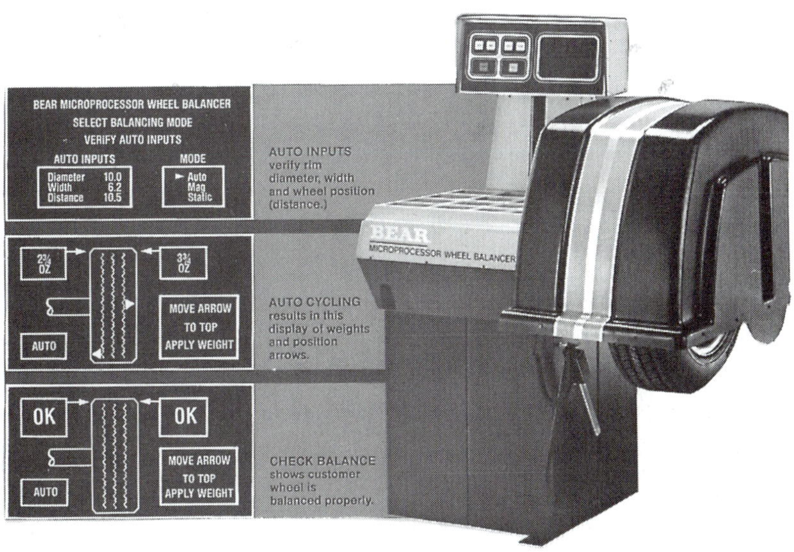


Figure 4-21 Electronic wheel balancer (Courtesy of Bear Automotive Service Equipment Company)

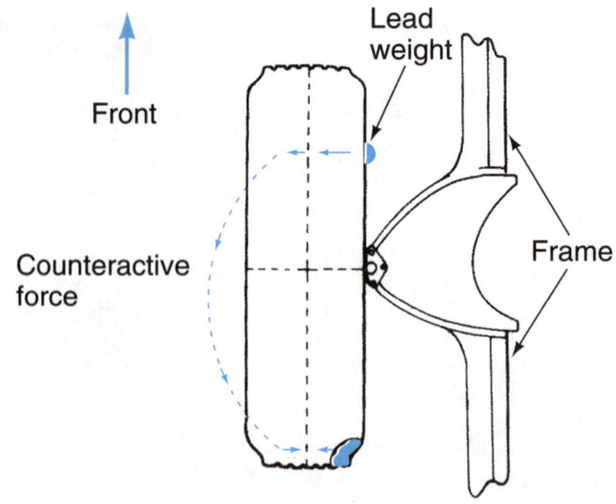


Figure 4-22 Dynamic wheel balance with heavy spot on the inside edge of the tread

positioned over the tire prior to dynamic balancing. Hand-spun electronic balancers do not require a safety hood. The preliminary checks mentioned previously in this chapter also apply to dynamic balancing. Wheel balancers must always be operated according to the manufacturer's instructions.

When the heavy spot is located on the inside edge of the tread, the correct size weight installed on the inside of the wheel 180° from the heavy spot provides proper dynamic balance (Figure 4-22). When the heavy spot is on the outside edge of the tread, the correct size wheel weight installed at location C 180° from the heavy spot on the outside of the wheel provides proper dynamic balance (Figure 4-23).

Some dynamic wheel balance procedures require the installation of the correct size wheel weight 180° from the heavy spot, and an equal weight on the opposite side of the wheel 180° from the first weight. This second weight maintains static balance (Figure 4-24). Static or dynamic wheel imbalance causes cupped tire wear and bald spots on the tire tread (Figure 4-25).

## On-Car Wheel Balancing

**Special Tools**
On-car wheel balancer

On-car wheel balancers contain a drum driven by an electric motor. This drum is positioned against the tire on the vehicle, which allows the electric motor to rotate the wheel. Many on-car balancers have a strobe light with a meter, and an electronic vibration sensor. Off-car wheel balancing and

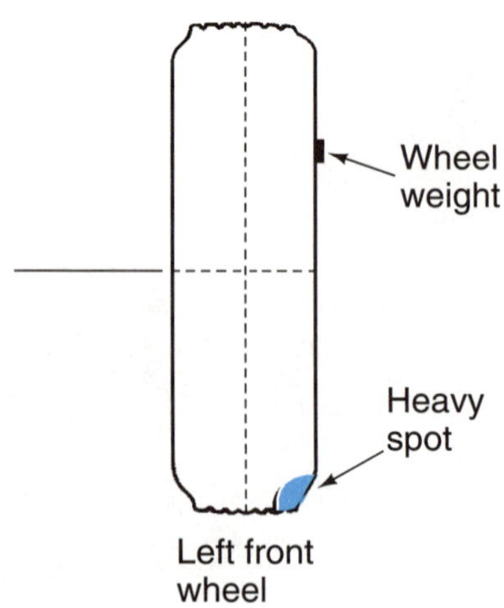


Figure 4-23 Dynamic wheel balance with heavy spot on the outside edge of the tread

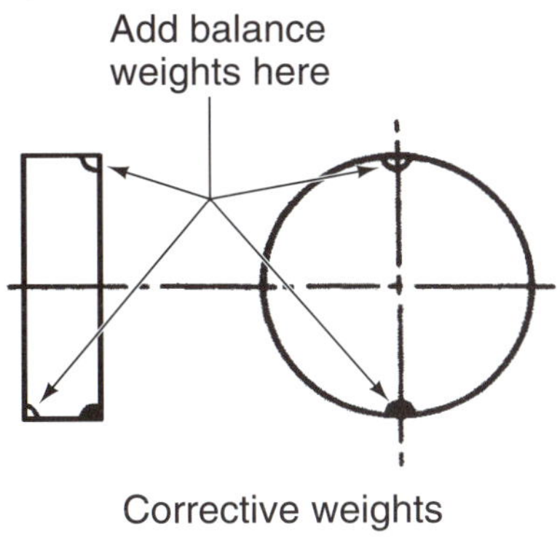


**Figure 4-24** Dynamic wheel balance procedure (Courtesy of Oldsmobile Division, General Motors Corporation)

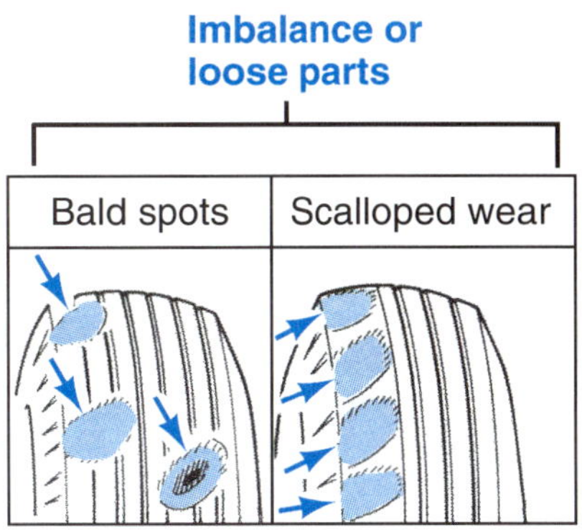


**Figure 4-25** Cupped tire wear and bald spots on tire tread caused by static or dynamic wheel imbalance (Courtesy of Chrysler Corporation)

on-car wheel balancing are a complementary combination for fine-tuning wheel balance. For example, a wheel vibration problem may still exist after an off-car balance procedure. If this problem occurs, the on-car balancer may be used to correct the problem. The on-car balance procedure corrects imbalance problems in all rotating components, including brake drums or rotors. When a front wheel on a rear wheel drive car is balanced, use this procedure:

1. Perform the preliminary checks listed previously in this chapter.
2. Raise the wheel being balanced 5 in (12 cm) off the floor.
3. Install the electronic vibration sensor between the lower control arm and the floor.
4. Chalk mark a reference mark on the outer sidewall of the tire.
5. Spin the wheel just fast enough to produce vibration on the front bumper.
6. When vibration causes strobe light flashing, move the balancer drum away from the tire and allow the tire to spin freely.
7. Shine the strobe light around the tire sidewall and note the chalk-mark position.
8. Note the pointer position on the meter.
9. Use the balancer's brake plate to slow and stop the wheel.
10. Rotate the wheel until the chalk mark is in the exact position where it appeared under the strobe light. The heavy spot is now at the bottom of the wheel and the balancing weight should be attached 180° from the heavy spot. Install the amount of weight indicated on the meter.
11. Spin the wheel again (Figure 4-26). If the wheel balance is satisfactory, the meter pointer will read in the balanced position. When the pointer does not indicate a bal-

**Figure 4-26** On-car wheel balancer (Courtesy of Hunter Engineering Company)

anced wheel, shine the strobe light on the tire sidewall. If the installed wheel weight is at the 12 o'clock position, additional weight is required, whereas the 6 o'clock weight position indicates excessive weight. A 3 o'clock or 9 o'clock weight position may be corrected by moving the weight 1 in (2.5 cm) toward the 12 o'clock position.

Additional on-car wheel balancer precautions are these:

1. Do not spin the wheel at excessive speeds.
2. Do not spin the front wheels on a front wheel drive vehicle with the floor jack under the chassis and the suspension dropped downward. Under this condition, severe angles exist in the front drive axle joints, and these joints may be damaged if the wheels are rotated with the balancer. Place the floor jack under the lower control arm to raise the wheel.

### Rear Wheel Balancing

*If an on-car balancer is used on the rear wheels of a rear wheel drive car, the technician must determine if the vehicle has a conventional or a limited slip differential.* With the transmission in park, or in gear with a manual transmission, rotate one rear wheel by hand. If the vehicle has a limited slip differential, the rear wheels will not rotate, whereas a free-turning wheel indicates a conventional differential.

**CAUTION:** When a rear wheel is rotated with an on-car balancer, do not allow the speed indicated on the speedometer to exceed 35 mph (56 kph).

When the vehicle has a conventional differential, use the same front wheel balance procedure on the rear wheels. Raise only the wheel to be balanced from the floor, and do not allow the speed indicated on the speedometer to exceed 35 mph (56 kph). When the other rear tire is resting on the floor, the speed of the wheel being balanced is 70 mph (112 kph). If the vehicle has a limited slip differential, proceed as follows for on-car rear wheel balancing:

1. Raise the vehicle and place safety stands under the frame to support the vehicle weight.
2. Lift the rear axle housing with a floor jack to reduce the universal joint angles, but do not take the vehicle weight off the safety stands.
3. Remove the wheel that is not being balanced. Install and torque the wheel nuts on the brake drum to hold the drum in place.
4. Balance the opposite wheel using the same procedure as for front wheels.
5. Leave the balanced wheel on the vehicle and install and balance the other wheel. Be sure to torque the wheel nuts properly. Do not allow rear wheel speed to exceed 55 mph (90 kph) on the speedometer.

**Classroom Manual**
Chapter 4, page 76

## Guidelines for Servicing Wheels and Tires

1. Tire noises vary with road conditions, but differential noise is not affected by road conditions.
2. Steering pull may be caused by defects in the suspension or steering systems, and tire conicity.
3. Tire conicity occurs when the belt in a tire is wound off-center during the manufacturing process.
4. Tire conicity cannot be diagnosed by a visual inspection.
5. The tire rotation procedure varies depending on the model year, type of tires, and whether the vehicle has a compact spare or a conventional spare.
6. Hand tools and tire irons should not be used to demount tires.
7. Tire punctures over 1/4 inch in diameter should not be repaired.
8. Compact spare tires should not be repaired.
9. Do not use caustic agents, alkaline-base detergents, or abrasive cleaners to clean aluminum or magnesium wheels.

10. When a bead expander is used to mount a tire, do not increase tire pressure above 10 psi (69 kPa).
11. When a bead expander is not used to mount a tire, do not increase tire pressure above 40 psi (276 kPa) to move the tire beads out against the wheel rim.
12. While mounting a tire, be sure that the tire beads and wheel rim bead seats are coated with rubber tire lubricant.
13. Radial tire and wheel runout causes tire thump.
14. Lateral tire and wheel runout causes tire and vehicle waddling.
15. Balls of rubber or other objects inside a tire make proper balance impossible.
16. Aluminum or magnesium wheels require the use of stick-on wheel weights.
17. Electronic wheel balancers perform the necessary wheel weight calculations and indicate the exact locations where these weights should be installed.

## CASE STUDY

A customer complains about severe front end waddling on a 1993 Cadillac Brougham equipped with steel-belted radial tires. The technician questions the customer further regarding this problem and learns that the problem only occurs after the car has sat over night, and it only lasts for approximately six blocks when the customer starts driving the car.

The technician carefully checks all steering and suspension components, wheel bearing adjustment, and tire runout. No problems are found during these checks, and the technician asks the customer to leave the car in the shop over night. The technician installs the front tires on the rear of the car and rotates the rear tires to the front, and the car is left in the shop over night. A road test the next morning indicates rear-end waddling, which proves the problem must be caused by a tire problem such as an improperly installed steel belt, since the tires were the only thing changed on the car. The technician installs and balances two new tires on the rear wheel rims, and installs these wheels and tires on the front of the car. The front tires and wheels were moved to the rear wheels. The customer reports later that the problem has been completely corrected.

## Terms to Know

Tire thump
Tire vibration
Radial runout
Lateral runout
Steering pull
Tire conicity
Tire rotation
Antitheft wheel covers
Antitheft lug nuts
Tread wear indicators
Heavy spots
Ply separation
Static imbalance
Dynamic imbalance

## ASE Style Review Questions

1. While discussing a tire thumping problem:
*Technician A* says this problem may be caused by cupped tire treads.
*Technician B* says a heavy spot in the tire may cause this complaint.
Who is correct?
**A.** A only
**B.** B only
**C.** Both A and B
**D.** Neither A nor B

2. While discussing a vehicle that pulls to one side:
*Technician A* says that excessive radial runout on the right front tire may cause this problem.
*Technician B* says that tire conicity may be the cause of this complaint.
Who is correct?
**A.** A only
**B.** B only
**C.** Both A and B
**D.** Neither A nor B

3. While discussing the results of tire underinflation:
*Technician A* says underinflation causes excessive wear on the outer edges of the tire tread.
*Technician B* says that underinflation causes excessive wear on the center of the tire tread.
Who is correct?

**A.** A only  **C.** Both A and B
**B.** B only  **D.** Neither A nor B

4. While discussing tire noise:
*Technician A* says that tire noise varies with road surface conditions.
*Technician B* says that tire noise remains constant when the vehicle is accelerated and decelerated.
Who is correct?

**A.** A only  **C.** Both A and B
**B.** B only  **D.** Neither A nor B

5. A tire has 0.060 in (1.52 mm) of radial runout:
*Technician A* says the runout is excessive and must be corrected.
*Technician B* says the tire is satisfactory.
Who is correct?

**A.** A only  **C.** Both A and B
**B.** B only  **D.** Neither A nor B

6. A tire has 0.030 in (0.762 mm) lateral runout:
*Technician A* says that the runout is satisfactory.
*Technician B* says that the runout is excessive and must be corrected.
Who is correct?

**A.** A only  **C.** Both A and B
**B.** B only  **D.** Neither A nor B

7. While discussing wheel balance:
*Technician A* says that static imbalance causes wheel tramp.
*Technician B* says that static imbalance causes rapid wear on suspension and steering components.
Who is correct?

**A.** A only  **C.** Both A and B
**B.** B only  **D.** Neither A nor B

8. While discussing tire wear:
*Technician A* says that static imbalance causes feathered tread wear.
*Technician B* says that dynamic imbalance causes cupped wear and bald spots on the tire tread.
Who is correct?

**A.** A only  **C.** Both A and B
**B.** B only  **D.** Neither A nor B

9. While discussing on-car wheel balancing:
*Technician A* says that during an on-car wheel balancing procedure on a rear wheel of a rear wheel drive car, the speed indicated on the speedometer should not exceed 65 mph (105 kph).
*Technician B* says that the speed indicated on the speedometer on this car must not exceed 35 mph (56 kph).
Who is correct?

**A.** A only  **C.** Both A and B
**B.** B only  **D.** Neither A nor B

10. While discussing wheel balance:
*Technician A* says that balls of rubber inside a tire make wheel balancing impossible.
*Technician B* says that wheel weights may be positioned on the rim flanges to correct imbalance caused by a ball of rubber inside the tire.
Who is correct?

**A.** A only  **C.** Both A and B
**B.** B only  **D.** Neither A nor B

## Table 4-1 ASE TASK

Diagnose tire wear patterns; determine needed repairs.

| Problem Area | Symptoms | Possible Causes | Classroom Manual | Shop Manual |
|---|---|---|---|---|
| **TIRE TREAD WEAR** | Wear on center or edges of tread | Overinflation or underinflation | 74 | 81 |
| | Cupped wear on tread | Improper wheel balance | 77 | 85 |

## Table 4-2 ASE TASK

Inspect tires; check and adjust air pressure.

| Problem Area | Symptoms | Possible Causes | Classroom Manual | Shop Manual |
|---|---|---|---|---|
| NOISE | Tire thumping | Tire defects, weak spots, bulges | 77 | 67 |

## Table 4-3 ASE TASK

Diagnose wheel/tire vibration, shimmy, and noise problems; determine needed repairs.

| Problem Area | Symptoms | Possible Causes | Classroom Manual | Shop Manual |
|---|---|---|---|---|
| RIDE QUALITY | Vibration above a certain speed | Improper static wheel balance | 76 | 83 |
| | Shimmy above a certain speed | 1. Improper dynamic wheel balance | 78 | 83 |
| | | 2. Excessive lateral runout | 63 | 77 |
| TIRE NOISE | Thumping | 1. Excessive tire radial runout | 63 | 77 |
| | | 2. Cupped tire treads | 77 | 85 |
| | | 3. Improper wheel balance | 77 | 81 |
| | | 4. Defects, heavy spots, bulges | 63 | 73 |

## Table 4-4 ASE TASK

Rotate tires/wheels according to manufacturer's recommendations.

| Problem Area | Symptoms | Possible Causes | Classroom Manual | Shop Manual |
|---|---|---|---|---|
| TIRE TREAD WEAR | Premature tread wear | Lack of tire rotation | 67 | 69 |

## Table 4-5 ASE TASK

Measure wheel, tire, axle, and hub runout; determine needed repairs.

| Problem Area | Symptoms | Possible Causes | Classroom Manual | Shop Manual |
|---|---|---|---|---|
| NOISE | Tire thumping | Excessive radial wheel or tire runout | 77 | 67 |
| VIBRATION | Chassis waddle | Excessive lateral wheel or tire runout | 63 | 77 |

## Table 4-6 ASE TASK

Diagnose tire pull (lead) problems; determine corrective actions.

| Problem Area | Symptoms | Possible Causes | Classroom Manual | Shop Manual |
|---|---|---|---|---|
| DIRECTIONAL CONTROL | Steering pull | Tire conicity | 63 | 67 |

## Table 4-7 ASE TASK

Balance wheel and tire assembly (static and/or dynamic).

| Problem Area | Symptoms | Possible Causes | Classroom Manual | Shop Manual |
|---|---|---|---|---|
| RIDE QUALITY | Chassis vibration above a certain speed | 1. Improper static wheel balance | 76 | 81 |
| | Chassis shimmy above a certain speed | 2. Improper dynamic wheel balance | 78 | 83 |

CHAPTER 5

# Shock Absorber and Strut Diagnosis and Service

Upon completion and review of this chapter, you should be able to:

- ❑ Perform a visual shock absorber inspection.
- ❑ Perform a shock absorber bounce test, and determine shock absorber condition.
- ❑ Determine shock absorber condition from a manual shock absorber test.
- ❑ Remove and replace shock absorbers.
- ❑ Diagnose shock absorber and strut noise complaints.
- ❑ Remove and replace front and rear struts.
- ❑ Remove struts from coil springs.
- ❑ Install coil springs on struts.
- ❑ Follow the vehicle manufacturer's recommended strut disposal procedure.
- ❑ Perform off-car strut cartridge replacement procedures.
- ❑ Perform on-car strut cartridge replacement procedures.

## Shock Absorber Visual Inspection

**Basic Tools**

Basic technician's tool set

Service manual

Hydraulic floor jack

Jack stands

Center punch

**SERVICE TIP:** During a visual shock absorber inspection, check the rebound, or strikeout, bumpers on the control arms, or chassis. If the rebound bumpers are severely worn, the shock absorbers may be worn out.

### Bolt Mounting

Shock absorbers should be inspected for loose mounting bolts and worn mounting bushings. If these components are loose, rattling noise is evident, and replacement of the bushings and bolts is necessary.

### Bushing Condition

In some shock absorbers, the bushing is permanently mounted in the shock, and the complete unit must be replaced if the bushing is worn. When the mounting bushings are worn, the shock absorber will not provide proper spring control.

### Oil Leakage

Shock absorbers and struts should be inspected for oil leakage. A slight oil film on the lower oil chamber is acceptable. Any indication of oil dripping is not acceptable, and unit replacement is necessary (Figure 5-1).

Struts and shock absorbers should be inspected visually for a bent condition and severe dents or punctures. When any of these conditions are present, unit replacement is required (Figure 5-2).

A strut is similar internally to a shock absorber, but a strut also supports the steering knuckle. In many applications, the coil spring is mounted on the strut.

## Shock Absorber or Strut Bounce Test

**CUSTOMER CARE:** Always be willing to spend a few minutes explaining problems, including safety concerns, regarding the customer's vehicle. When customers understand why certain repairs are necessary, they feel better about spending the money. For example, if you explain that worn-out shock absorbers cause excessive chassis oscillations, and may result in loss of steering control on irregular road surfaces, the customer is more receptive to spending the money for shock absorber replacement.

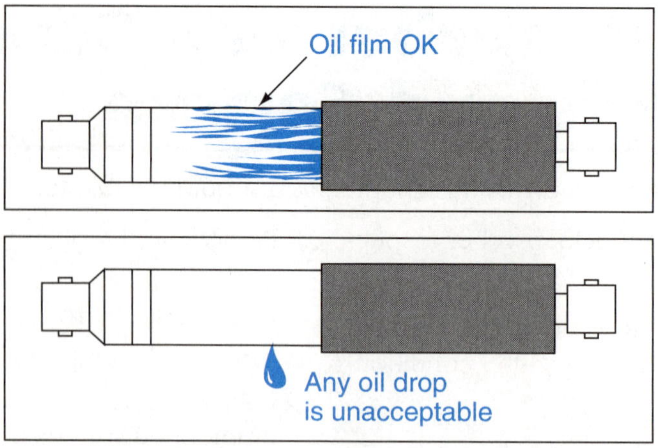


**Figure 5-1** Shock absorber and strut oil leak diagnosis (Courtesy of Delco Products Division of General Motors Corporation)

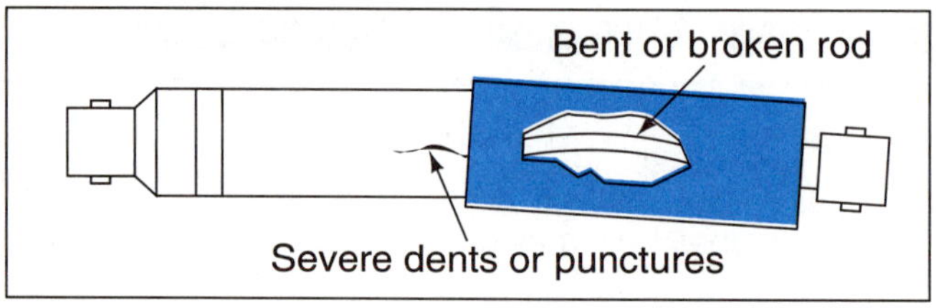


**Figure 5-2** Damaged shock absorber inspection (Courtesy of Delco Products Division of General Motors Corporation)

When the bounce test is performed, the bumper is pushed downward with considerable weight applied on each corner of the vehicle. The bumper is released after this action, and one free upward bounce should stop the vertical chassis movement if the shock absorber or strut provides proper spring control. Shock absorber replacement is required if more than one free upward bounce occurs.

## Shock Absorber Manual Test

**CAUTION:** Gas-filled shock absorbers will extend when disconnected.

Upward wheel movement is referred to as wheel jounce, and downward wheel movement is called rebound.

A manual test may be performed on shock absorbers. When this test is performed, disconnect the lower end of the shock and move the shock up and down as rapidly as possible. A satisfactory shock absorber should offer a strong, steady resistance to movement on the entire compression and rebound strokes. The amount of resistance may be different on the compression stroke compared to the rebound stroke. If a loss of resistance is experienced during either stroke, shock replacement is essential.

Some defective shock absorbers or struts may have internal clunking, clicking, and squawking noises, or binding conditions. When these shock absorber noises or conditions are experienced, shock absorber or strut replacement is necessary.

## Air Shock Absorber Diagnosis and Replacement

Air shock absorbers are similar to conventional shocks, except they have air pressure applied to them to control chassis curb height. Some air shock absorbers must be pressurized with the shop air hose. This type of unit contains a valve for inflation purposes. Other air shock absorbers are inflated by an on-board compressor with interconnecting plastic lines between the compressor and the shocks. Shock absorber lines must be inspected for breaks, cracks, and sharp bends. If any of these defects are present, the line must be replaced. The shock absorber lines must be secured to the chassis, and they must not rub against other components.

When air shock absorbers slowly lose their air pressure and reduce the curb riding height, shock replacement is required. Before removing an air shock, relieve the air pressure in the shock.

## Shock Absorber Replacement

**WARNING:** Never apply heat to the lower shock absorber or strut chamber with an acetylene torch. Excessive heat may cause a shock absorber or strut explosion, which could result in personal injury.

When shock absorber replacement is necessary, follow this procedure:

1. Before replacing rear shock absorbers, lift the vehicle on a hoist and support the rear axle on jack stands so the shock absorbers are not fully extended.
2. When a front shock absorber is changed, lift the front end on the vehicle with a floor jack, and then place jack stands under the lower control arms. Lower the vehicle onto the jack stands and remove the floor jack.
3. Disconnect the upper shock mounting nut and grommet.
4. Remove the lower shock mounting nut or bolts, and remove the shock absorber.
5. Reverse steps 1 through 4 to install the new shock absorber and grommets.
6. With the full vehicle weight supported on the suspension, tighten the shock mounting nuts to the specified torque.

**Classroom Manual**
Chapter 5, page 85

## Diagnosis of Front Spring and Strut Noise

Strut chatter may be heard when the steering wheel is turned with the vehicle not moving, or moving at low speed. To verify the location of this chattering noise, place one hand on a front coil spring while someone turns the steering wheel. If strut chatter is present, the spring binds and releases as it turns. This condition is caused by the upper spring seat binding against the strut bearing mount. A revised spring seat is available to correct this problem on some models.

A noise that occurs on sharp turns or during front suspension jounce may be caused by one of the following problems:

1. interference between the upper strut rebound stop and the upper mount or strut tower
2. interference between the coil spring and tower (Figure 5-3)
3. interference between the coil spring and the upper mount (Figure 5-4)

On some models, these coil-spring interference problems may be corrected by installing upper coil spring spacers on top of the coil spring. Spring removal from the strut is required to install these spacers.

Strut chatter is heard as a clicking noise when the front wheels are turned.

A strut tower is a raised, circular, reinforced area inboard of the front fenders, that supports the upper end of the coil spring and strut assembly.

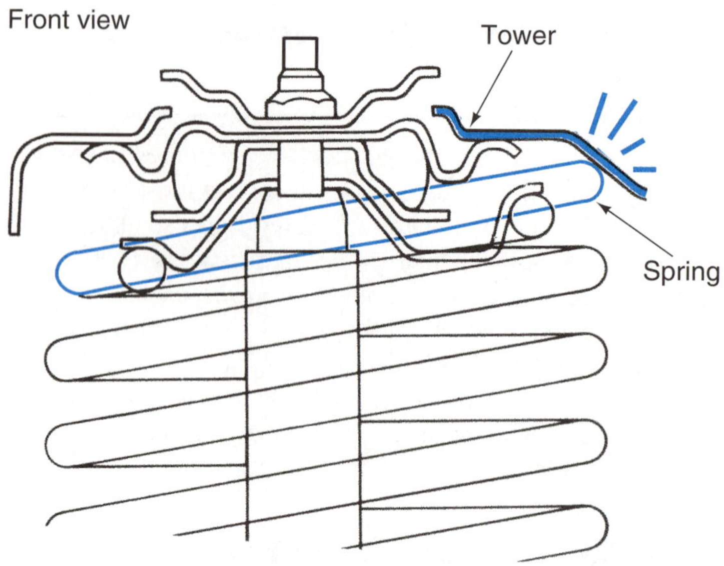


**Figure 5-3** Coil spring to upper tower interference (Courtesy of Chrysler Corporation)

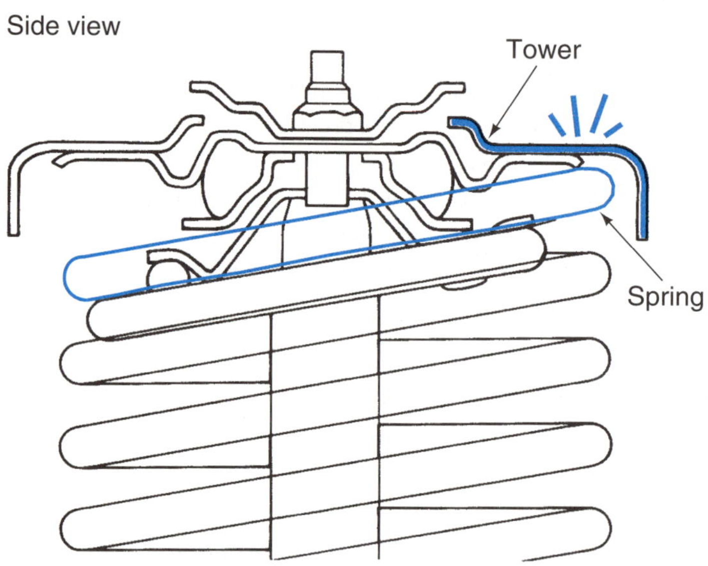


**Figure 5-4** Coil spring to upper mount interference (Courtesy of Chrysler Corporation)

An eccentric camber bolt has an oblong head, which provides inward or outward steering knuckle movement in relation to the strut as this bolt is rotated.

If the lateral movement of the strut rod and nut above the strut tower exceeds 3/16 in (4.76 mm), the upper strut bearing and mount assembly should be replaced.

Worn spring insulators or broken coil springs cause a rattling noise on road irregularities. Broken coil springs result in reduced curb riding height and harsh riding.

## Strut Removal and Replacement

Before a front strut and spring assembly is removed, the strut must be removed from the steering knuckle, and top strut mount bolts must be removed from the strut tower. If an eccentric camber

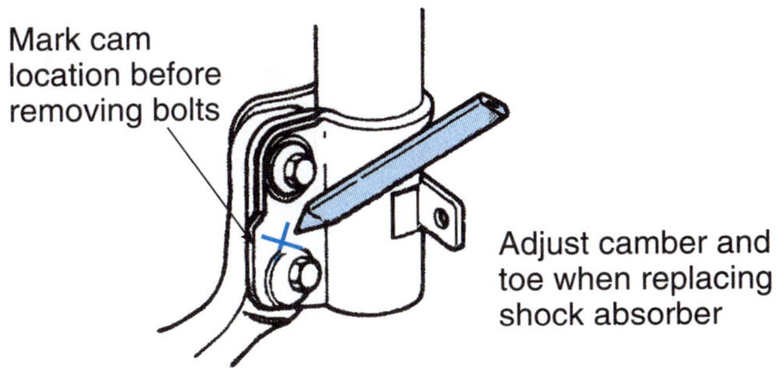


**Figure 5-5** Camber bolt marking for strut removal (Courtesy of Chrysler Corporation)

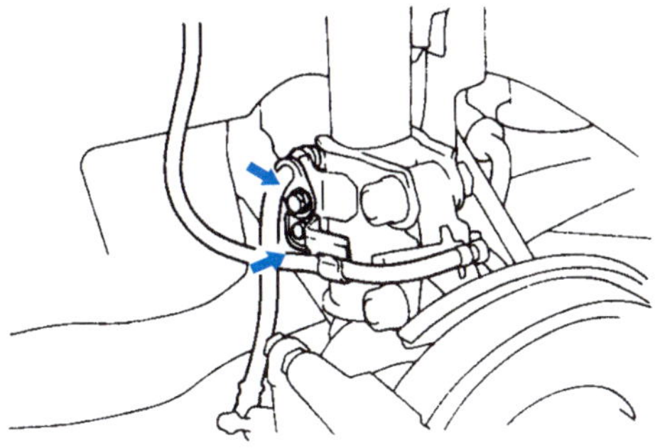

**Figure 5-6** Brake line and ABS wheel speed sensor wire removal from strut (Courtesy of Toyota Motor Corporation)

bolt is used to attach the strut to the knuckle, always mark the bolt head in relation to the strut and reinstall the bolt in the same position (Figure 5-5).

Always follow the vehicle manufacturer's recommended procedure in the service manual for removal of the strut and spring assembly. A typical procedure for strut and spring assembly removal follows:

1. Raise the vehicle on a hoist or floor jack. If a floor jack is used to raise the vehicle, lower the vehicle onto jack stands placed under the chassis so the lower control arms and front wheels drop downward. Remove the floor jack from under the vehicle.
2. Remove the brake line and antilock brake system (ABS) wheel speed sensor wire from clamps on the strut (Figure 5-6). In some cases, these clamps may have to be removed from the strut.
3. Remove the strut to steering knuckle retaining bolts, and remove the strut from the knuckle (Figure 5-7).
4. Remove the upper strut mounting bolts on top of the strut tower, and remove the strut and spring assembly (Figure 5-8).

## Removal of Strut from Coil Spring

**WARNING:** Always use a coil spring compressing tool according to the tool or vehicle manufacturer's recommended service procedure. Be sure the tool is properly installed on the spring. If a coil spring slips off the tool when the spring is compressed, severe personal injury or property damage may occur.

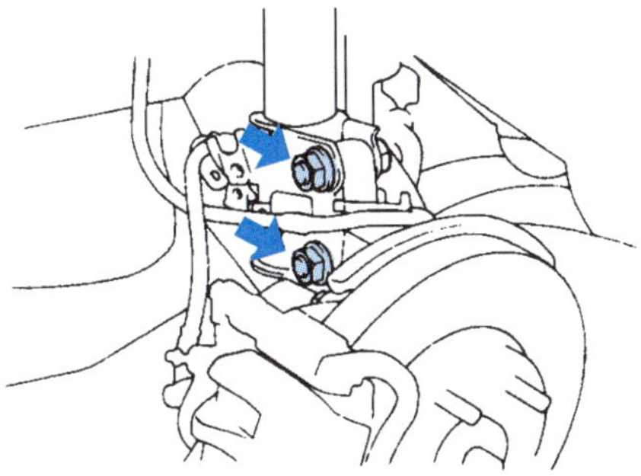

**Figure 5-7** Removing strut to steering knuckle retaining bolts (Courtesy of Toyota Motor Corporation)

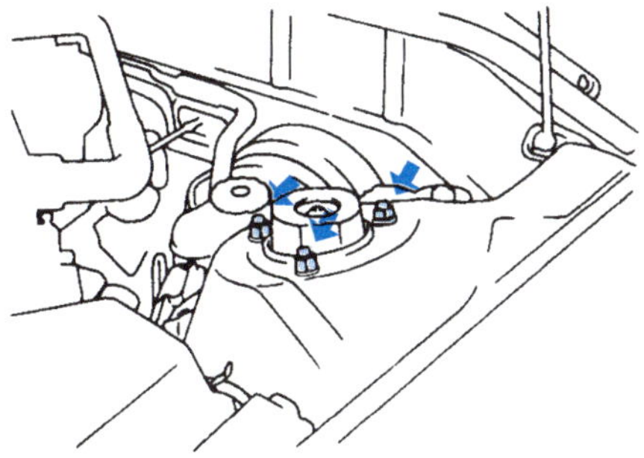

**Figure 5-8** Removing upper strut mounting bolts on top of strut tower (Courtesy of Toyota Motor Corporation)

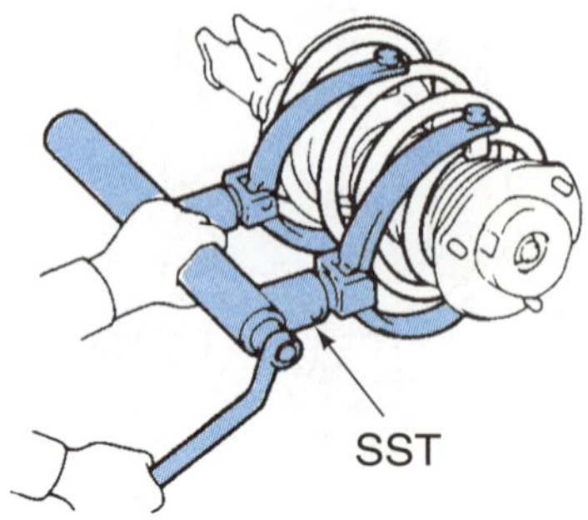


**Figure 5-9** The spring compressing tool is tightened until all the spring tension is removed from the upper strut mount. (Courtesy of Toyota Motor Corporation)

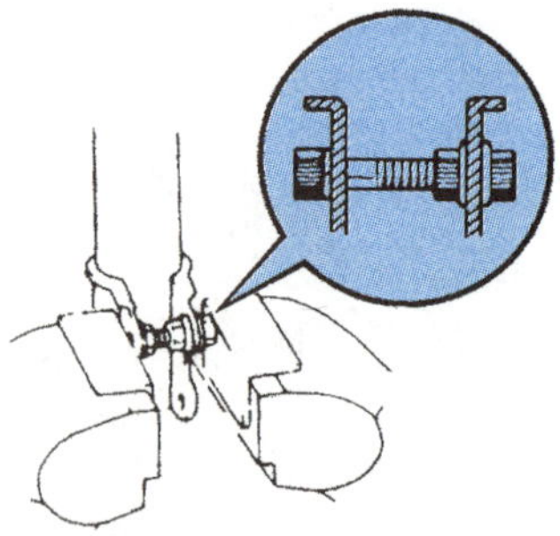

**Figure 5-10** Bolt and two nuts installed in upper strut to knuckle mounting hole to hold strut and spring assembly in a vise (Courtesy of Toyota Motor Corporation)

**WARNING:** Never loosen the upper strut mount retaining nut on the end of the strut rod unless the spring is compressed enough to remove all spring tension from the upper strut mount. If this nut is loosened with spring tension on the upper mount, this mount becomes a very dangerous projectile that may cause serious personal injury or property damage.

**CAUTION:** Never clamp the lower strut or shock absorber chamber in a vise with excessive force. This action may distort the lower chamber and affect piston movement in the strut or shock absorber.

The coil spring must be compressed with a special tool before the strut can be removed. All the tension must be removed from the upper spring seat before the upper strut piston rod nut is loosened. Many different spring-compressing tools are available, and they must always be used according to the manufacturer's recommended procedure. *If the coil spring has an enamel-type coating, tape the spring where the compressing tool contacts the spring.* The spring may break prematurely if this coating is chipped. A typical procedure for removing a strut from a coil spring follows:

**Special Tools**

Coil spring compressing tool

1. Install the spring compressing tool on the coil spring according to the tool or vehicle manufacturer's recommended procedure.
2. Turn the nut on top of the compressing tool until all the spring tension is removed from the upper strut mount (Figure 5-9).
3. Install a bolt and two nuts in the upper strut to knuckle mounting bolt holes. Install a nut on each side of the strut flange (Figure 5-10). Clamp this bolt securely in a vise to hold the strut and coil assembly and the compressing tool.
4. Use the bar on the spring compressing tool to hold the strut and spring assembly from turning, and loosen the nut on the upper strut mount (Figure 5-11). Be sure all the spring tension is removed from the upper strut mount before loosening this nut.

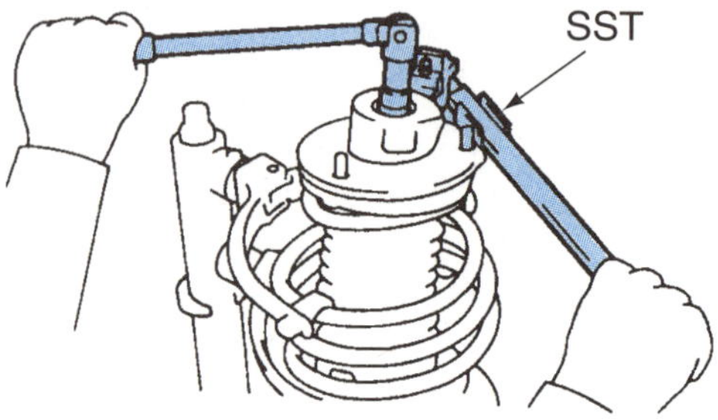


**Figure 5-11** Removal of nut from strut piston rod (Courtesy of Toyota Motor Corporation)

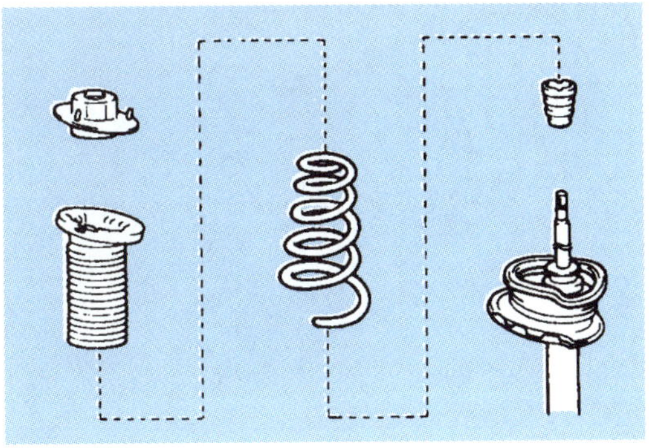

Figure 5-12 Removal of upper strut mount, upper insulator, spring, spring bumper, and lower insulator (Courtesy of Toyota Motor Corporation)

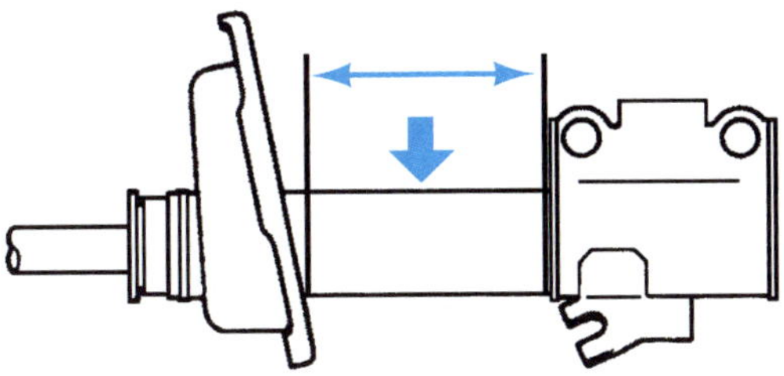

Figure 5-13 Strut drilling location (Courtesy of Toyota Motor Corporation)

5. Remove the upper strut mount, upper insulator, coil spring, spring bumper, and lower insulator (Figure 5-12).

## Strut Disposal Procedure

**WARNING:** Always follow the vehicle manufacturer's recommended procedure for strut disposal. Do not throw gas-filled struts or shock absorbers in a fire of any kind, or a dumpster. If the vehicle manufacturer recommends drilling the strut to release the gas charge, drill the strut at the manufacturer's recommended location.

The following is a typical strut drilling procedure that is performed prior to strut disposal:

1. Fully extend the strut rod.
2. Center punch the strut at the manufacturer's recommended drilling location (Figure 5-13).
3. Drill a small hole at the center-punched position.

## Installation of Coil Spring on Strut

A typical procedure for installing a coil spring on a strut follows:

1. Install the lower insulator on the lower strut spring seat, and be sure the insulator is properly seated (Figure 5-14).
2. Install the spring bumper on the strut rod (Figure 5-15).

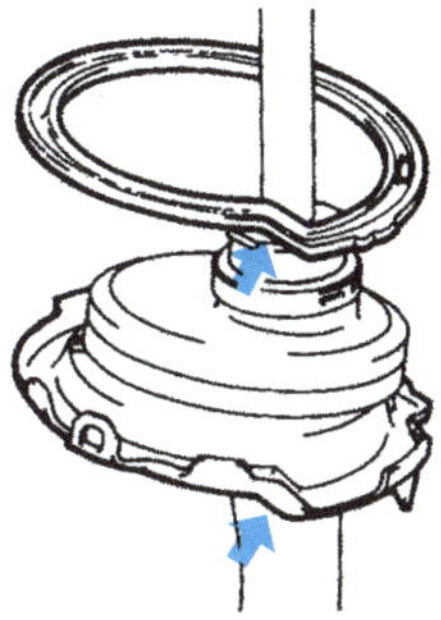

Figure 5-14 Insulator installation on lower spring seat (Courtesy of Toyota Motor Corporation)

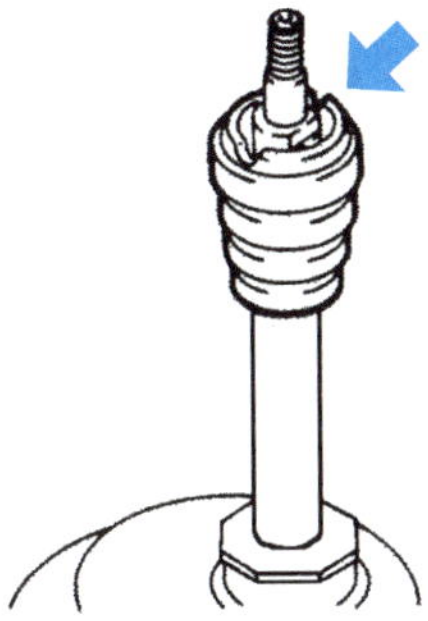

Figure 5-15 Spring bumper installation on strut piston rod (Courtesy of Toyota Motor Corporation)

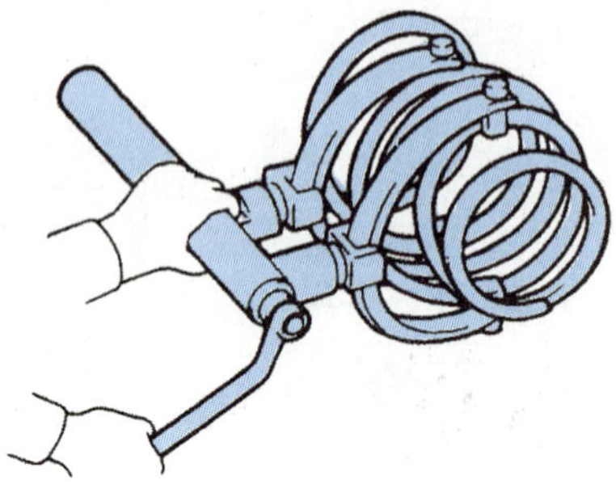

Figure 5-16 Installing compressed spring on strut with compressing tool (Courtesy of Toyota Motor Corporation)

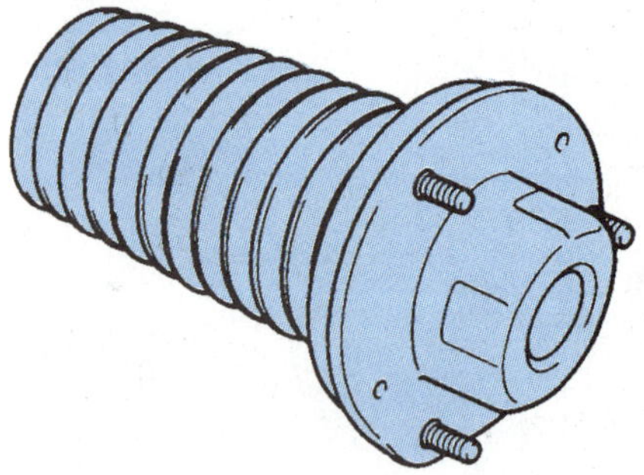

Figure 5-17 Upper insulator and upper strut mount installed on coil spring (Courtesy of Toyota Motor Corporation)

3. With the coil spring compressed in the spring compressing tool, install the spring on the strut (Figure 5-16). Be sure the spring is properly seated on the lower insulator spring seat.
4. Be sure the strut piston rod is fully extended and install the upper insulator on top of the coil spring.
5. Install the upper strut mount on the upper insulator (Figure 5-17).
6. Be sure the spring, upper insulator, and upper strut mount are properly positioned and seated on the coil spring and strut piston rod (Figure 5-18).
7. Install a bolt in the upper strut to knuckle retaining bolt, and clamp this bolt in a vise to hold the strut, spring, and compressing tool as in the disassembly procedure.
8. Use the compressing tool bar to hold the strut and spring from turning, and tighten the strut piston rod nut to the specified torque (Figure 5-19).
9. Rotate the upper strut mount until the lowest bolt in this mount is aligned with the tab on the lower spring seat (Figure 5-20).
10. Gradually loosen the nut on the compressing tool until all the spring tension is released from the tool, and remove the tool from the spring.

## Installation of Spring and Strut Assembly in Vehicle

A typical installation procedure for a spring and strut assembly follows:

1. Install the strut and spring assembly with the upper strut mounting bolts extending through the bolt holes in the strut tower. Tighten the nuts on the upper strut mounting bolts to the specified torque (Figure 5-21).

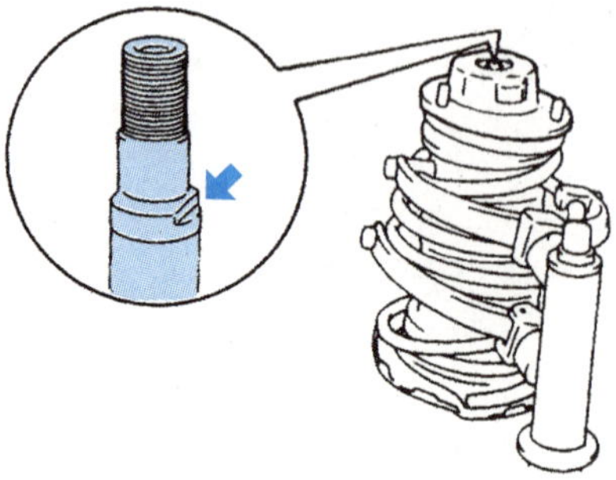

Figure 5-18 Upper strut mount properly positioned on strut piston rod (Courtesy of Toyota Motor Corporation)

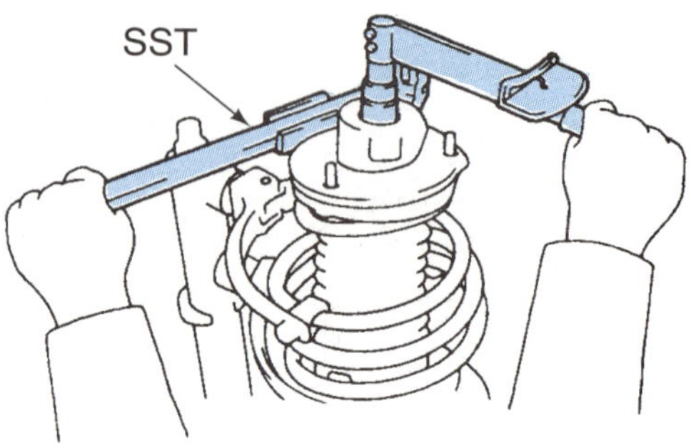


Figure 5-19 Tightening nut on strut piston rod (Courtesy of Toyota Motor Corporation)

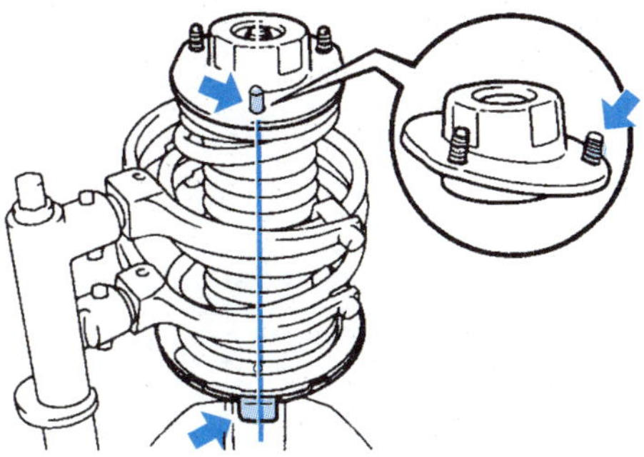

Figure 5-20 Aligning lowest bolt on upper strut mount with tab on lower spring seat (Courtesy of Toyota Motor Corporation)

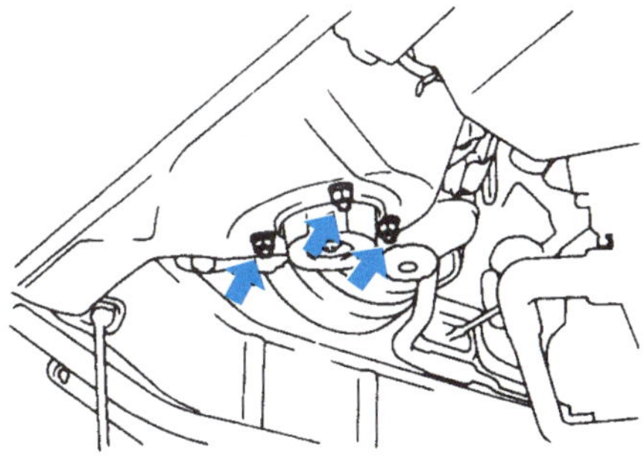

Figure 5-21 Nuts installed on upper strut mounting bolts (Courtesy of Toyota Motor Corporation)

2. Install the lower end of the strut in the steering knuckle to the proper depth, and tighten the strut to knuckle retaining bolts to the specified torque (Figure 5-22). If one of the strut to knuckle bolts is an eccentric camber bolt, be sure the eccentric is aligned with the mark placed on the strut during the removal procedure.
3. Install the brake hose in the clamp on the strut. Place the ABS wheel-speed sensor wire in the strut clamp if the vehicle is equipped with ABS (Figure 5-23).

Photo Sequence 3 shows a typical procedure for removing and replacing a MacPherson strut.

## Rear Strut Replacement

The rear strut replacement procedure varies depending on the type of rear suspension. Always follow the vehicle manufacturer's recommended procedure outlined in the appropriate service manual. A typical rear strut replacement procedure follows:

1. Lift the vehicle with a floor jack and lower the vehicle onto safety stands placed under the chassis to support the vehicle weight.
2. Place the floor jack under the lower control arm and operate the jack to support some of the spring tension.
3. Remove the tire and wheel assembly.
4. Remove the strut to spindle bolts.
5. Pull upward on the strut to remove the strut from the spindle. If necessary, lower the floor jack slightly to remove the strut from the spindle.
6. Disconnect the upper strut mount from the chassis, and remove the strut.
7. When the new strut and/or mount is installed, reverse steps 1 through 6. Tighten all the bolts to the specified torque.

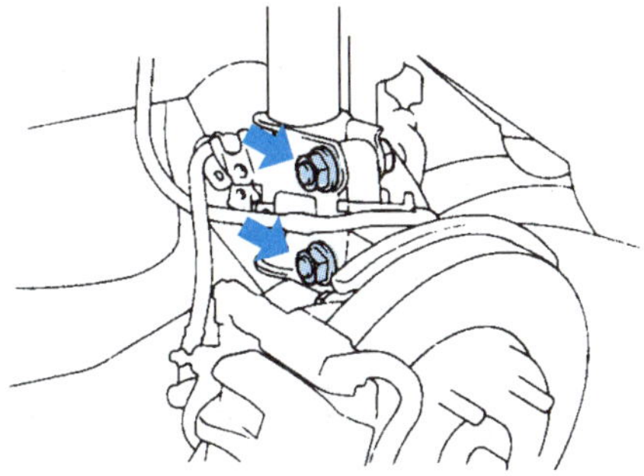

Figure 5-22 Lower end of strut installed in steering knuckle (Courtesy of Toyota Motor Corporation)

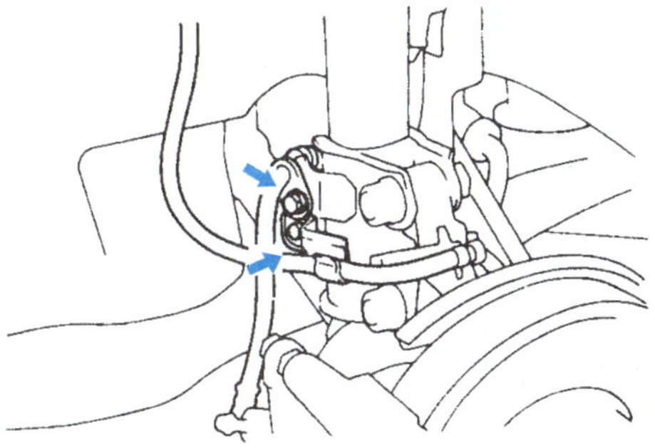

Figure 5-23 Brake hose and ABS wire installed in strut clamps (Courtesy of Toyota Motor Corporation)

# Photo Sequence 3
## Typical Procedure for Removing and Replacing a MacPherson Strut

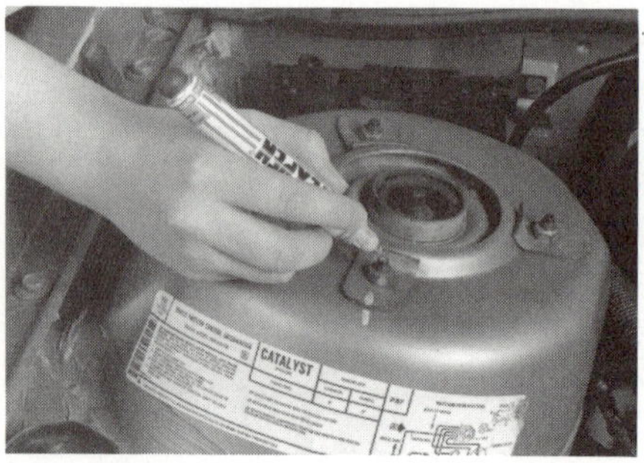


**P3-1** The top of the strut assembly is mounted directly to the chassis of the car. Prior to loosening the strut to the chassis bolts, scribe alignment marks on the strut bolts and the chassis.

**P3-2** With the top strut bolts or nuts removed, raise the car to a working height It is important that the car be supported on its frame and not on its suspension components.

**P3-3** Remove the wheel assembly. The strut is accessible from the wheel well after the wheel is removed.

**P-3-4** Remove the bolt that fastens the brake line or hose to the strut assembly.

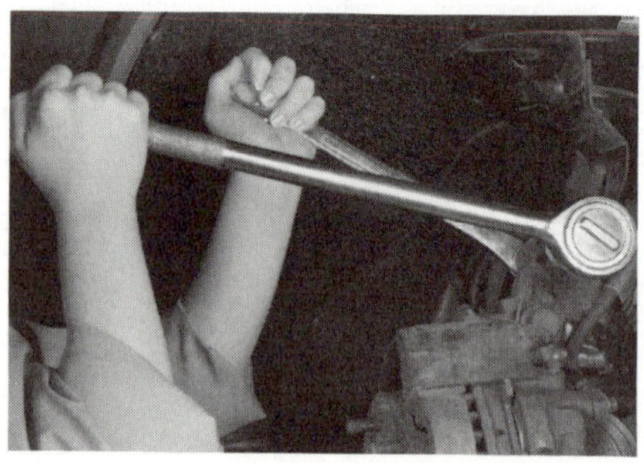

**P3-5** Remove the strut to steering knuckle bolts.

**P3-6** Support the steering knuckle with wire and remove the strut assembly from the car.

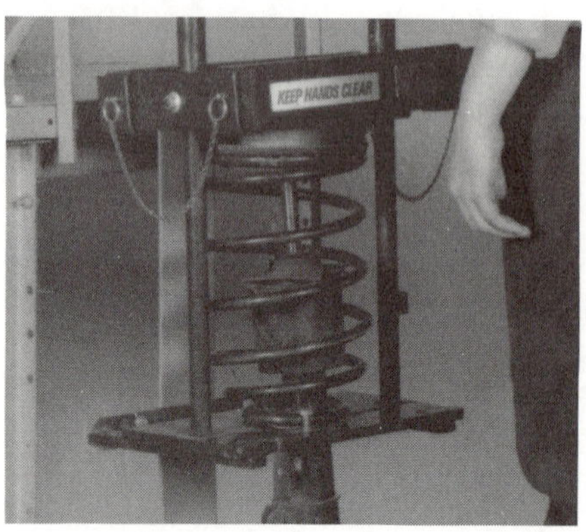


**P3-7** Install the strut assembly into the proper type spring compressor. Then compress the spring until it is possible to safely loosen the retaining bolts.

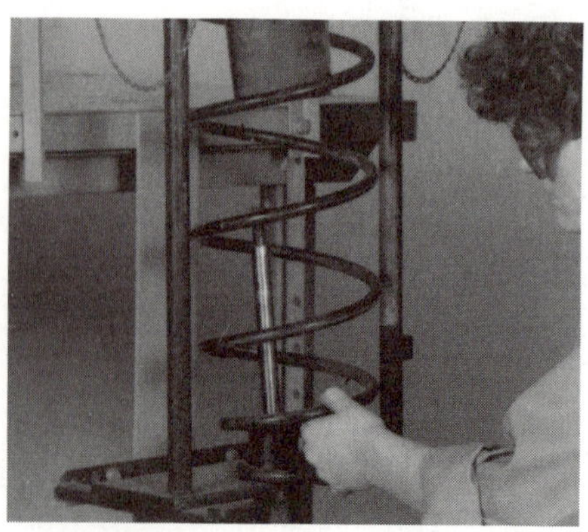

**P3-8** Remove the old strut assembly from the spring and install the new strut. Compress the spring to allow for reassembly and tighten the retaining bolts.

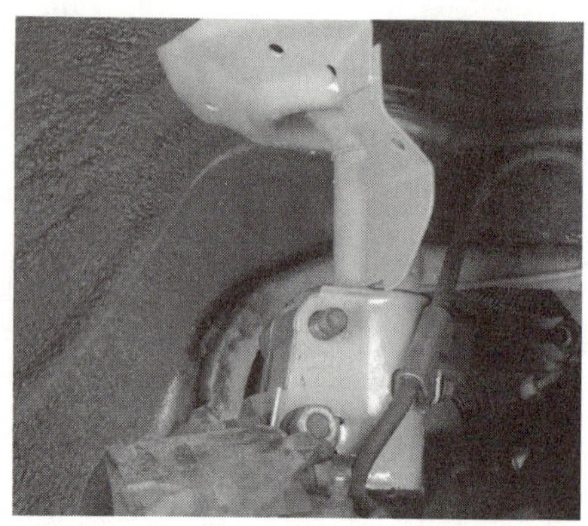

**P3-9** Reinstall the strut assembly into the car. Make sure all bolts are properly tightened and in the correct locations.

The rear coil springs may be removed from the rear struts using the same basic procedure for spring removal from the front struts.

## Installing Strut Cartridge, Off-Car

**SERVICE TIP:** Check the cost of the strut cartridges versus the price of new struts. Give customers the best value for their repair dollar!

Many struts are a sealed unit, and thus rebuilding is impossible. However, some manufacturers supply a replacement cartridge that may be installed in the strut housing after the strut has been removed from the vehicle. Always follow the strut cartridge manufacturer's recommended installation procedure. The following is a typical off-car strut cartridge installation procedure:

1. Install a bolt and two nuts in the upper strut to knuckle mounting bolt hole. Place a nut on the inside and outside of the strut flange.
2. Clamp this bolt in a vise to hold the strut.
3. Locate the line groove near the top of the strut body, and use a pipe cutter installed in this groove to cut the top of the strut body.
4. After the cutting procedure remove the strut piston assembly from the strut (Figure 5-24).

A strut cartridge contains the inner working part of the strut, which may be installed in the outer housing of the old strut.

**Special Tools**

Pipe cutter

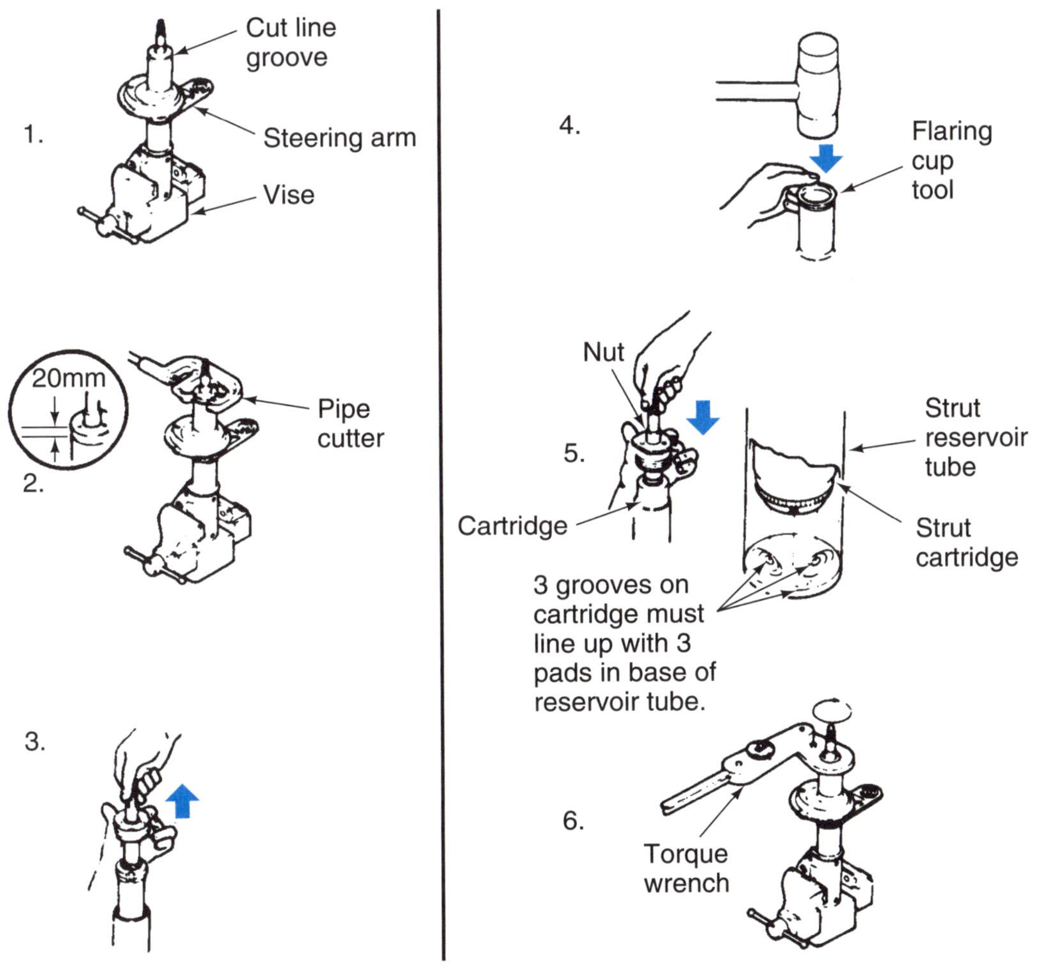


**Figure 5-24** Installation of strut cartridge (Courtesy of Pontiac Motor Division, General Motors Corporation)

5. Remove the strut from the vise and dump the oil from the strut.
6. Place the special tool supplied by the vehicle manufacturer or cartridge manufacturer on top of the strut body. Strike the tool with a plastic hammer until the tool shoulder contacts the top of the strut body. This action removes burrs from the strut body, and places a slight flare on the body.
7. Remove the tool from the strut body.
8. Place the new cartridge in the strut body, and turn the cartridge until it settles into indentations in the bottom of the strut body.
9. Place the new nut over the cartridge.
10. Using a special tool supplied by the vehicle or cartridge manufacturer, tighten the nut to the specified torque.
11. Move the strut piston rod in and out several times to check for proper strut operation.

## Installing Strut Cartridge, On-Car

**WARNING:** If a vehicle is hoisted, or lifted, in any way during an on-car strut cartridge replacement, the coil spring may fly off the strut, causing vehicle damage and personal injury.

On some vehicles, the front strut cartridge may be removed and replaced with the strut installed in the vehicle. Always consult the vehicle manufacturer's service manual for the proper strut service procedure. A typical on-car strut cartridge installation procedure follows:

1. Remove the nuts from the upper strut mount, and remove the strut mount cover (Figure 5-25).
2. Use the special tool supplied by the vehicle manufacturer and a number 50 torx bit to remove the nut from the strut piston rod (Figure 5-26). Never lift or hoist the vehicle once this nut is removed.
3. Remove the bushing from the upper strut mount (Figure 5-27).
4. Thread the special tool supplied by the vehicle manufacturer onto the top of the strut piston rod, and push this rod downward into the strut (Figure 5-28). Remove this tool and then remove the jounce bumper retainer (Figure 5-29).

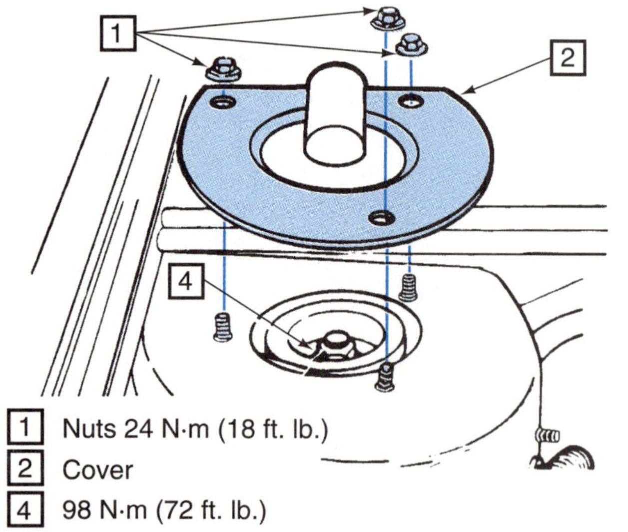


**Figure 5-25** Removing upper strut mount nuts and cover (Courtesy of Delco Products Division of General Motors Corporation)

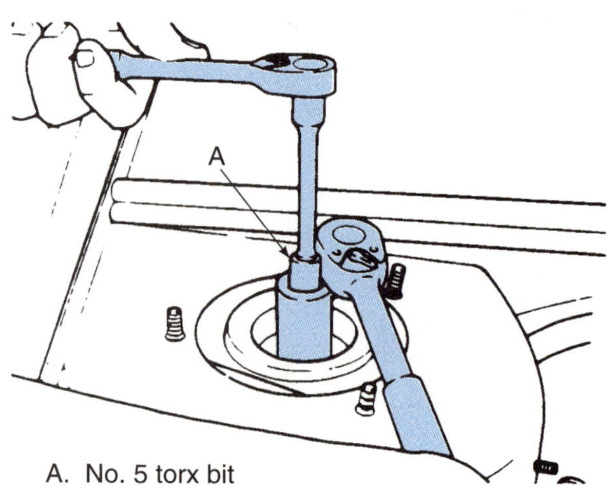


**Figure 5-26** Removing nut from the strut piston rod (Courtesy of Delco Products Division of General Motors Corporation)

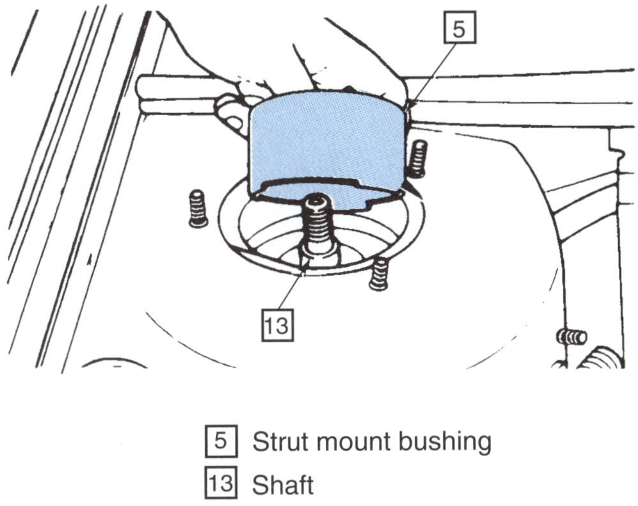


**Figure 5-27** Removing upper strut mount bushing (Courtesy of Delco Products Division of General Motors Corporation)

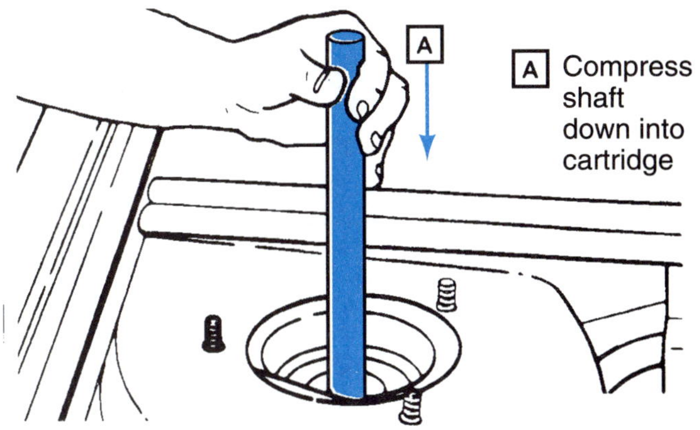


**Figure 5-28** Pushing strut piston rod downward (Courtesy of Delco Products Division of General Motors Corporation)

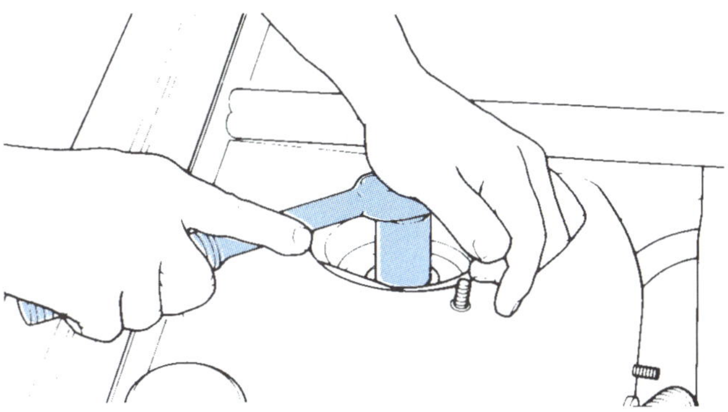

**Figure 5-29** Removing jounce bumper retainer (Courtesy of Delco Products Division of General Motors Corporation)

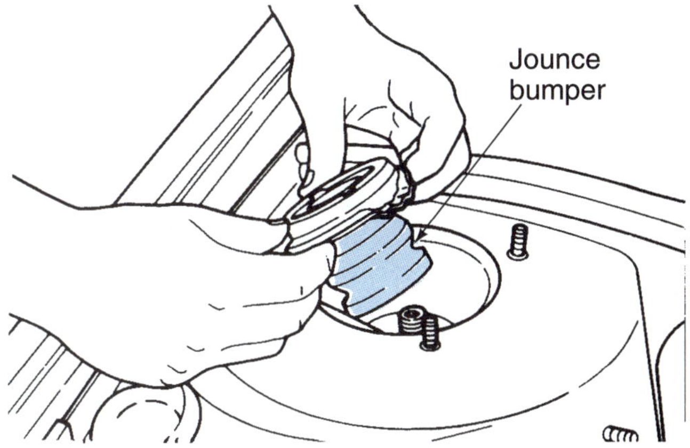


**Figure 5-30** Removing jounce bumper (Courtesy of Delco Products Division of General Motors Corporation)

5. Remove jounce bumper (Figure 5-30).
6. Thread the special tool onto the strut piston rod, and re-extend this rod (Figure 5-31). Use the special tool supplied by the vehicle manufacturer to remove the closure nut on top of the strut (Figure 5-32).
7. Grasp the top of the strut piston rod and remove the strut valve mechanism (Figure 5-33).
8. Remove the oil from the strut tube with a hand-operated suction pump.
9. Install the new strut cartridge in the strut, and tighten the closure nut to the specified torque (Figure 5-34). Reverse steps 1 through 6 to complete the strut cartridge replacement. Place a light coating of the vehicle manufacturer's recommended engine oil on the upper strut mount bushing prior to bushing installation.

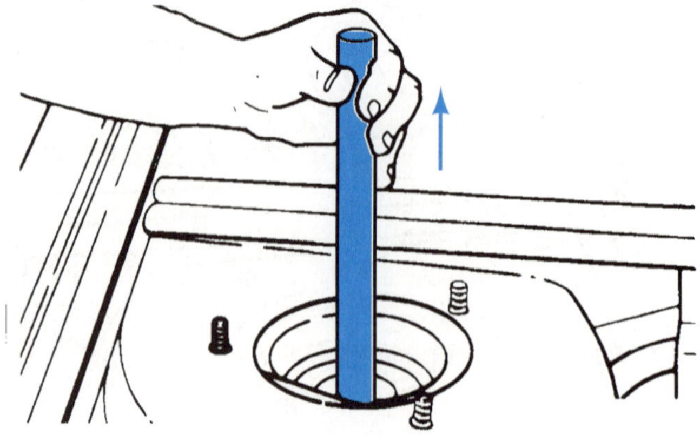

Figure 5-31 Re-extending strut piston rod (Courtesy of Delco Products Division of General Motors Corporation)

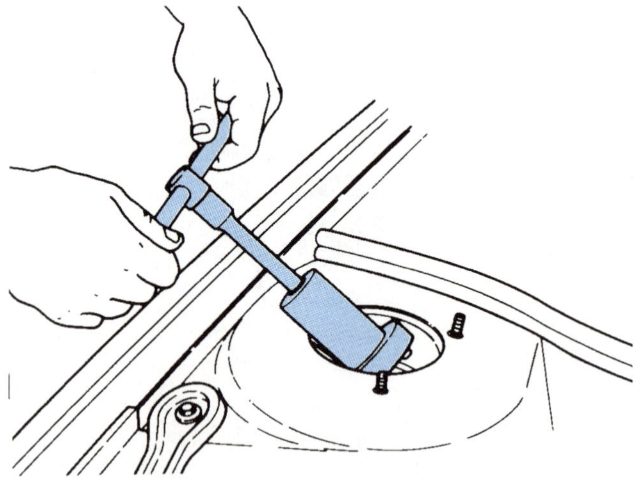

Figure 5-32 Removing strut closure nut (Courtesy of Delco Products Division of General Motors Corporation)

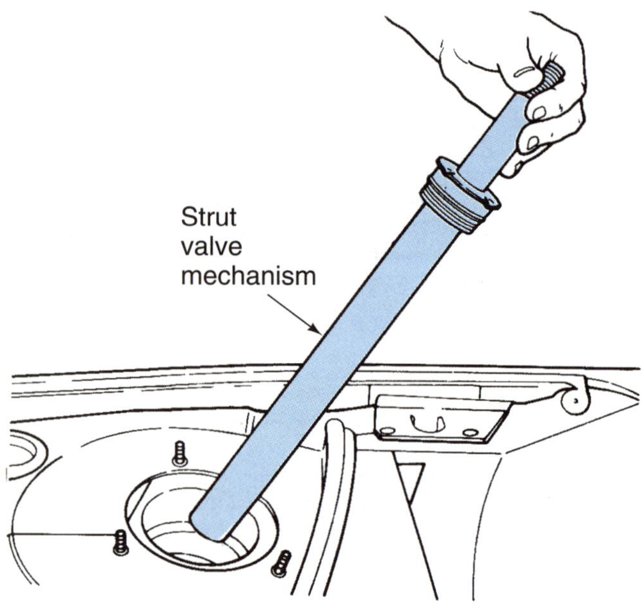


Figure 5-33 Removing strut valve mechanism (Courtesy of Delco Products Division of General Motors Corporation)

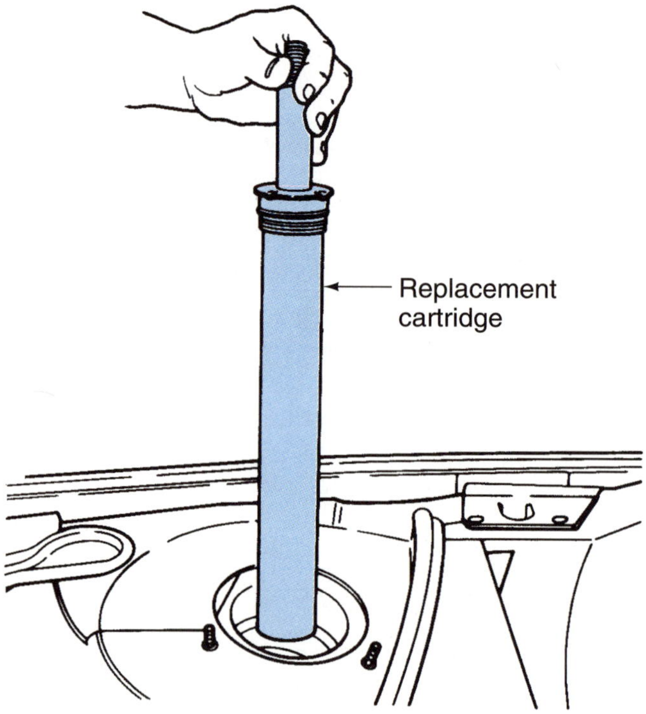


Figure 5-34 Installing new strut cartridge (Courtesy of Delco Products Division of General Motors Corporation)

# Guidelines for Servicing Shock Absorbers and Struts

1. Shock absorber condition may be determined by performing a visual inspection and a bounce test, or manual test.
2. A slight oil film on the lower shock absorber chamber is acceptable, but any sign of excessive oil leaks from a shock absorber or strut indicates that replacement is necessary.

3. Shock absorbers or struts that are bent, dented, or punctured should be replaced.
4. During the bounce test, if the vehicle bumper does more than one free upward bounce, the shock absorber or strut is defective.
5. During a manual test, the shock absorber should offer a strong, steady resistance to movement on the entire compression and rebound strokes. The amount of resistance may be greater on the rebound stroke than on the compression stroke, depending on shock absorber ratio. If there is a reduction, or loss, of resistance on either stroke, shock absorber replacement is necessary.
6. Front strut chatter may be caused by upper strut mount binding.
7. Front strut noise on sharp turns or during suspension jounce may be caused by interference between the coil spring and the strut tower, or between the coil spring and the upper mount.
8. If one of the front strut to steering knuckle bolts has an eccentric cam, the cam position should be marked on the strut prior to bolt removal.
9. When a coil spring compressing tool is used, it is extremely important to follow the tool manufacturer's and vehicle manufacturer's recommended spring compressing procedure. Never loosen the strut piston rod nut until the spring is compressed so all the tension is removed from the upper strut mount.
10. If the vehicle manufacturer recommends drilling a gas-filled strut or shock absorber prior to strut disposal, always drill the hole at the manufacturer's specified location.
11. When a strut and coil spring are reassembled, always make sure the spring and insulators are properly seated on the spring seats.
12. Before assembling the strut and coil spring in the vehicle, always make sure the upper strut mount is properly aligned with the lower spring seat.
13. Off-car strut cartridge replacements are possible on some vehicles, whereas on other cars on-car strut cartridge replacement is possible. On some struts, cartridge replacement is not possible, and struts must be replaced if they are defective.

## CASE STUDY

A customer complained about a squeaking noise in the rear suspension of a 1993 Dodge Aries. The customer said the noise occurred during normal driving at lower speeds. The technician lifted the car on a hoist and made a check of all rear suspension bushings in shock absorbers, track bar, and trailing arms. The spring insulators and all suspension bolts were checked visually. All of these items were in good condition, and there was no evidence of a squeaking noise as the chassis was bounced gently. Since the exact source of chassis noise can sometimes be difficult to locate, the technician performed a visual check of bushings, insulators, and fasteners in the front suspension. No problems were found in the front suspension.

One of the first requirements for successful automotive diagnosis is to obtain as much information as possible from the customer. The customer with the squeaking rear suspension in a Dodge Aries lived in a part of the country where cold temperatures occur in the winter. This complaint occurred in January. The technician questioned the customer further about the conditions when the squeaking suspension noise was heard, and the customer revealed that the noise occurred when the temperature was severely cold. The customer also indicated that the noise disappeared at warmer temperatures.

The technician informed the customer that the only way to find the exact cause of this annoying squeak was to leave the car on the lot at the shop all night, and check it first thing in the morning when it was colder. The customer complied with this suggestion, and the technician drove the car into the shop immediately the next morning. The squeaking noise occurred as the car was driven across the parking lot. The technician lifted the car on a hoist and listened with a stethoscope at each rear suspension bushing as a co-worker gently bounced the rear suspension. No squeaking noise was heard at any of the rear suspension bushings. However, when the stethoscope pickup was placed on the left rear shock absorber, the squeaking noise was loud and clear. The shock absorber was quickly removed, and the squeaking noise was gone when the rear chassis was bounced gently. Replacement of the left rear shock absorber corrected this complaint, and made a customer happy.

## Terms to Know

Shock absorber or strut bounce test
Shock absorber manual test
Strut chatter
Strut tower
Eccentric camber bolt
Coil spring compressing tool
Spring insulators
Strut cartridge

## ASE Style Review Questions

1. A slight oil film appears on the lower shock absorber oil chamber, and the shock absorber performs satisfactorily in a bounce test:
*Technician A* says the shock absorber is satisfactory.
*Technician B* says the shock absorber must be replaced.
Who is correct?
**A.** A only
**B.** B only
**C.** Both A and B
**D.** Neither A nor B

2. While discussing a shock absorber and strut bounce test:
*Technician A* says the shock absorber is satisfactory if the bumper makes two free upward bounces.
*Technician B* says the shock absorber is satisfactory if the bumper makes three free upward bounces.
Who is correct?
**A.** A only
**B.** B only
**C.** Both A and B
**D.** Neither A nor B

3. While discussing a shock absorber manual test:
*Technician A* says the shock absorber's resistance to movement should be jerky and erratic.
*Technician B* says the shock absorber's resistance to movement may be greater on the rebound stroke compared to the compression stroke.
Who is correct?
**A.** A only
**B.** B only
**C.** Both A and B
**D.** Neither A nor B

4. When the front wheels are turned on a vehicle equipped with front struts, the left front coil spring provides a chattering action and noise:
*Technician A* says the strut has internal defects and strut replacement is necessary.
*Technician B* says the upper strut bearing and mount is defective.
Who is correct?
**A.** A only
**B.** B only
**C.** Both A and B
**D.** Neither A nor B

5. While discussing enamel-coated coil springs:
*Technician A* says if the enamel-type coating on a coil spring is chipped, the spring may break prematurely.
*Technician B* says coil springs should be taped in the compressing tool contact areas to prevent chipping of the enamel-type coating.
Who is correct?
**A.** A only
**B.** B only
**C.** Both A and B
**D.** Neither A nor B

6. While discussing gas-filled shock absorbers:
*Technician A* says gas-filled shock absorbers may be heated with an acetylene torch.
*Technician B* says gas-filled shock absorbers should be disposed of in an incinerator.
Who is correct?
**A.** A only
**B.** B only
**C.** Both A and B
**D.** Neither A nor B

7. While discussing gas-filled shock absorbers:
*Technician A* says if a gas-filled shock absorber is compressed to its shortest length, it should remain compressed.
*Technician B* says if a gas-filled shock absorber is compressed to its shortest length, it should re-extend.
Who is correct?
**A.** A only
**B.** B only
**C.** Both A and B
**D.** Neither A nor B

8. While discussing gas-filled shock absorber or strut disposal:
*Technician A* says a gas-filled shock absorber may be drilled at any location to relieve the gas charge prior to disposal.
*Technician B* says a gas-filled strut must be drilled at the location specified by the vehicle manufacturer prior to disposal.
Who is correct?
**A.** A only
**B.** B only
**C.** Both A and B
**D.** Neither A nor B

9. While discussing strut and coil spring assembly:
*Technician A* says the upper insulator should be installed between the coil spring and the upper strut mount.
*Technician B* says the upper insulator should be installed between the upper strut mount and the strut tower.
Who is correct?
**A.** A only
**B.** B only
**C.** Both A and B
**D.** Neither A nor B

10. While discussing strut cartridge replacement:
*Technician A* says during an on-car front strut cartridge replacement, the front of the vehicle may be lifted with a floor jack.
*Technician B* says when a strut cartridge is replaced for the first time, the oil must be removed from the strut with a hand-operated suction pump.
Who is correct?
**A.** A only
**B.** B only
**C.** Both A and B
**D.** Neither A nor B

## Table 5-1 ASE TASK

Inspect and replace shock absorbers.

| Problem Area | Symptoms | Possible Causes | Classroom Manual | Shop Manual |
|---|---|---|---|---|
| RIDE QUALITY | Excessive chassis oscillations | Worn out shock absorbers | 85 | 92 |
| NOISE | Rattling on road irregularities | Worn shock absorber bushings, grommets | 90 | 91 |

## Table 5-2 ASE TASK

Inspect and replace air shock absorbers, lines, and fittings.

| Problem Area | Symptoms | Possible Causes | Classroom Manual | Shop Manual |
|---|---|---|---|---|
| CURB RIDING HEIGHT | Low curb riding height | Worn-out air shocks, lines, fittings | 96 | 93 |

## Table 5-3 ASE TASK

Inspect and replace front suspension system coil springs and spring insulators (silencers).

| Problem Area | Symptoms | Possible Causes | Classroom Manual | Shop Manual |
|---|---|---|---|---|
| NOISE | Rattling on road irregularities | 1. Worn spring insulators<br>2. Broken coil spring | 91<br>91 | 95<br>97 |
| RIDE HARSHNESS | Low curb riding height | Weak coil springs | 91 | 97 |

## Table 5-4 ASE TASK

Inspect and replace MacPherson strut upper bearing and mount.

| Problem Area | Symptoms | Possible Causes | Classroom Manual | Shop Manual |
|---|---|---|---|---|
| NOISE | Chatter while cornering | Worn upper strut mount | 91 | 93 |

## Table 5-5 ASE TASK

Inspect and replace MacPherson strut cartridge or assembly.

| Problem Area | Symptoms | Possible Causes | Classroom Manual | Shop Manual |
|---|---|---|---|---|
| RIDE QUALITY | Excessive chassis oscillations | Worn-out strut cartridge | 91 | 101 |

CHAPTER 6

# Front Suspension System Service

Upon completion and review of this chapter, you should be able to:

- ❑ Measure curb riding height.
- ❑ Diagnose and correct curb riding height problems.
- ❑ Adjust torsion bars.
- ❑ Diagnose front suspension noise and body sway.
- ❑ Check ball joint condition, and remove and replace ball joints.
- ❑ Remove and replace steering knuckles, and check knuckle condition.
- ❑ Remove and replace lower control arms, and check control arm and bushing condition.
- ❑ Remove and replace coil springs, and check spring and insulator condition.
- ❑ Remove and replace upper control arms, and check control arm and bushing condition.
- ❑ Remove and replace control arm bushings.
- ❑ Inspect and replace rebound bumpers.
- ❑ Diagnose, remove, and replace stabilizer bars.
- ❑ Diagnose, remove, and replace strut rods.
- ❑ Diagnose, remove, and replace leaf springs.
- ❑ Replace torsion bars, and check torsion bar condition.

## Curb Riding Height Measurement

Regular inspection and proper maintenance of suspension systems is extremely important to maintaining vehicle safety. The curb riding height is determined mainly by spring condition. Other suspension components, such as control arm bushings, affect curb riding height if they are worn. Since incorrect curb riding height affects most of the other suspension angles, this measurement is critical.

Reduced curb riding height on the front suspension may cause decreased directional stability. If the curb riding height is reduced on one side of the front suspension, the steering may pull to one side. Reduced rear suspension height increases steering effort and causes rapid steering wheel return after turning a corner. Harsh riding occurs when the curb riding height is less than specified. The curb riding height must be measured at the vehicle manufacturer's specified location, which varies depending on the type of suspension system.

When the vehicle is on a level floor or an alignment rack, measure the curb riding height from the floor to the center of the lower control arm mounting bolt on both sides of the front suspension (Figure 6-1). On the rear suspension system, measure the curb riding height from the floor to the center of the strut rod mounting bolt (Figure 6-2).

If the curb riding height is less than specified, the control arms and bushings should be inspected and replaced as necessary. When the control arms and bushings are in normal condition, the reduced curb riding height may be caused by sagged springs that require replacement.

**Basic Tools**

Basic technician's tool set

Service manual

Floor jack

Jack stands

Pry bar

3/8 electric drill and drill bits

Directional stability refers to the tendency of the vehicle steering to remain in the straight-ahead position when driven straight ahead on a reasonably smooth, level road surface.

## Front Suspension Diagnosis and Service

### Torsion Bar Adjustment

**CUSTOMER CARE:** Each time undercar service is performed, make a quick inspection of suspension and steering components. Advise the customer regarding any necessary repairs. This procedure often obtains additional work for the shop, and the customer will be impressed that you are interested in his or her car and personal safety.

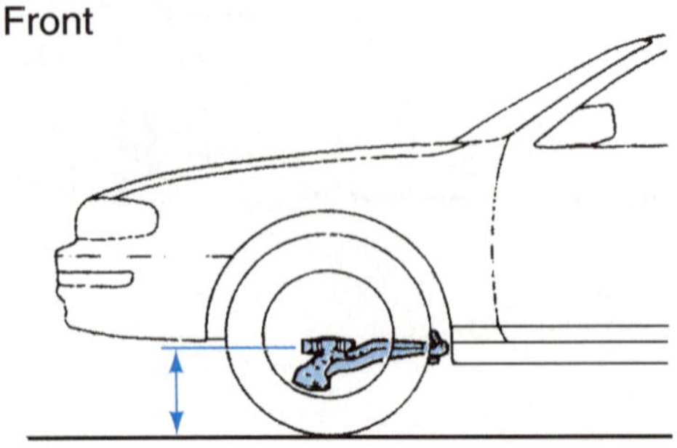


**Figure 6-1** Curb riding height measurement, front suspension (Courtesy of Toyota Motor Corporation)

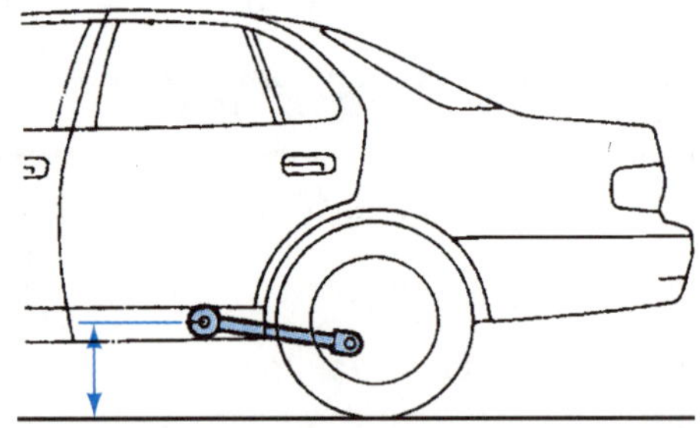


**Figure 6-2** Curb riding height measurement, rear suspension (Courtesy of Toyota Motor Corporation)

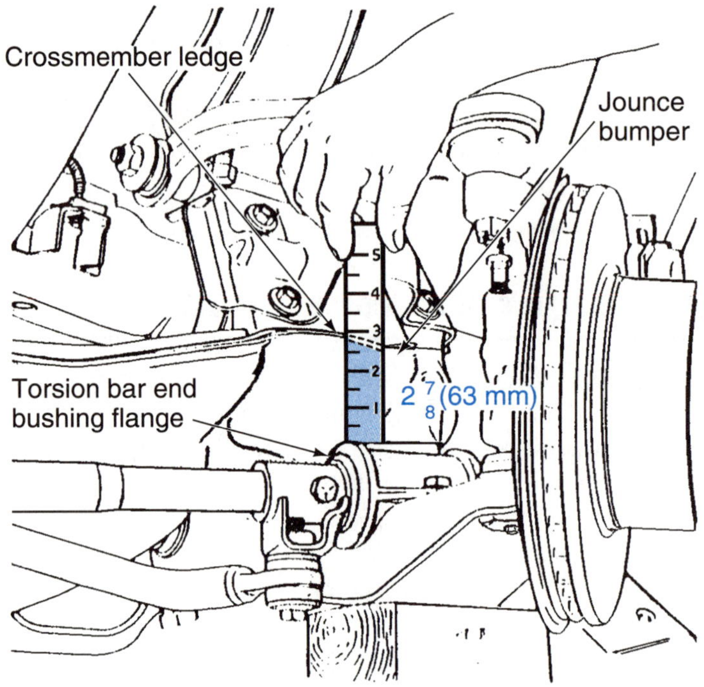


**Figure 6-3** Curb riding height measurement, torsion bar suspension (Courtesy of Chrysler Corporation)

**Special Tools**

Machinist's rule

On torsion bar front suspension systems, the torsion bars may be adjusted to correct the curb riding height. The curb riding height must be measured at the location specified by the vehicle manufacturer. On some torsion bar front suspension systems, the curb riding height is measured between the torsion bar bushing flange and the crossmember ledge (Figure 6-3).

If the curb riding height is not correct on a torsion bar front suspension, the torsion bar anchor adjusting bolts must be rotated until the curb riding height equals the vehicle manufacturer's specifications (Figure 6-4).

## Checking Ball Joints

**SERVICE TIP:** Regular chassis lubrication at the vehicle manufacturer's recommended service interval is one of the keys to long ball joint life. Always advise the customer of this fact.

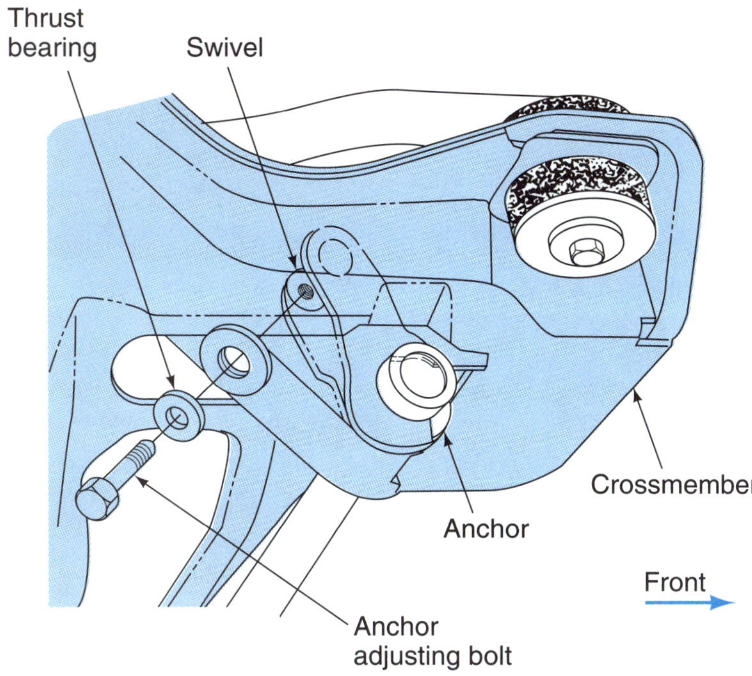


**Figure 6-4** Curb riding height adjustment on torsion bar front suspension (Courtesy of Chrysler Corporation)

**Wear Indicators.** Some ball joints have a grease fitting installed in a floating retainer. The grease fitting and retainer may be used as a ball joint wear indicator. With the vehicle weight resting on the wheels, grasp the grease fitting and check for movement (Figure 6-5).

Some car manufacturers recommend ball joint replacement if any grease fitting movement is present. In some other ball joints, the grease fitting retainer extends a short distance through the ball joint surface (Figure 6-6). On this type of joint, replacement is necessary if the grease fitting shoulder is flush with, or inside, the ball joint cover.

## Ball Joint Unloading

On many suspension systems, ball joint looseness is not apparent until the weight has been removed from the joint. When the coil spring is positioned between the lower control arm and the

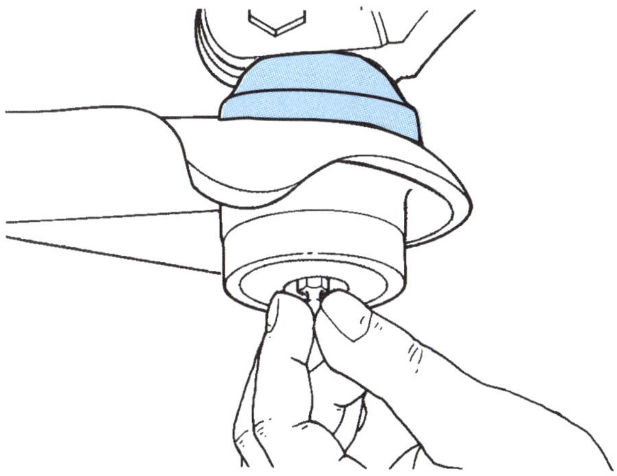

**Figure 6-5** Ball joint grease fitting wear indicator (Courtesy of Chrysler Corporation)

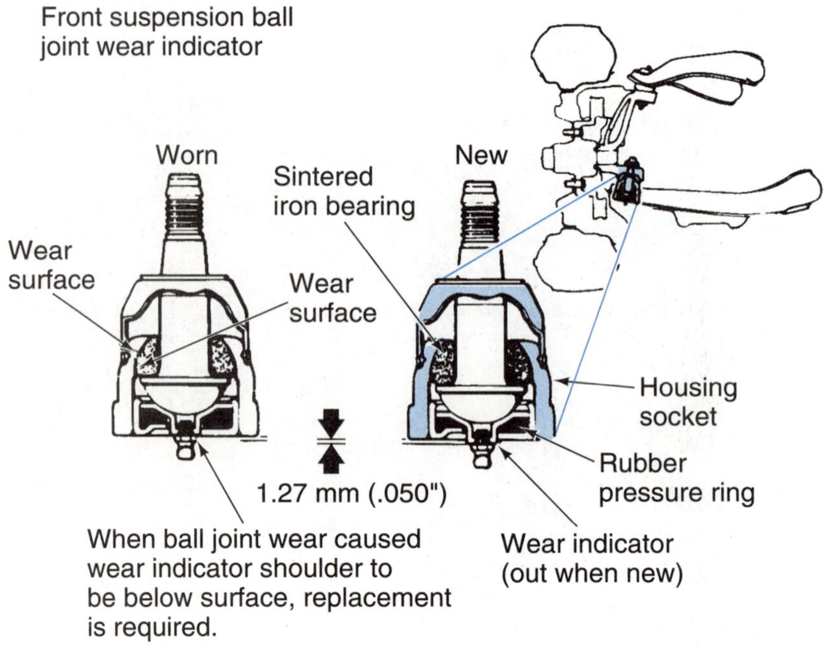


**Figure 6-6** Ball joint wear indicator with grease fitting extending from ball joint surface (Courtesy of Chevrolet Motor Division, General Motors Corporation)

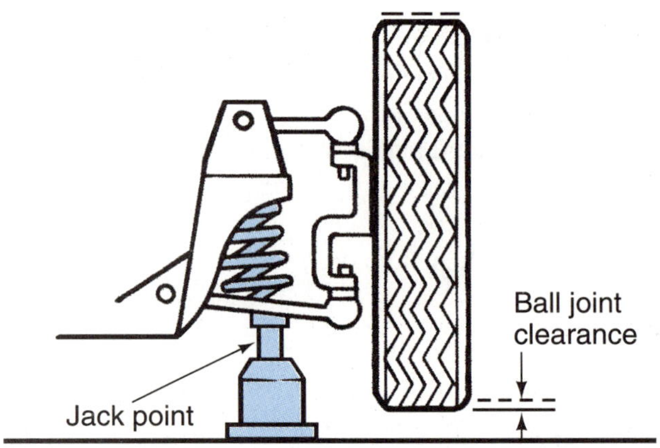


**Figure 6-7** Floor jack position to check ball joint wear with spring between the lower control arm and the chassis (Courtesy of Sealed Power Corporation)

chassis, place a floor jack near the outer end of the lower control arm and raise the tire off the floor (Figure 6-7). Be sure the rebound bumper is not in contact with the control arm or frame.

When the coil springs are positioned between the upper control arm and the chassis, place a steel block between the upper control arm and the frame. With this block in place, raise the tire off the floor with a floor jack under the front crossmember (Figure 6-8).

## Ball Joint Axial Measurement

**Special Tools**

Dial indicator for ball joint measurement

The vehicle manufacturer may provide ball joint axial (vertical) and radial (horizontal) tolerances. A dial indicator is one of the most accurate ball joint measuring devices (Figure 6-9). Always install the dial indicator at the vehicle manufacturer's recommended location for ball joint measurement.

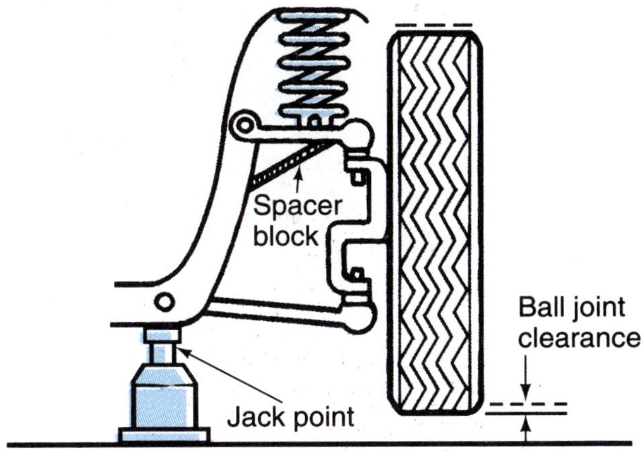


**Figure 6-8** Floor jack position to check ball joint wear with spring between the upper control arm and the chassis (Courtesy of Sealed Power Corporation)

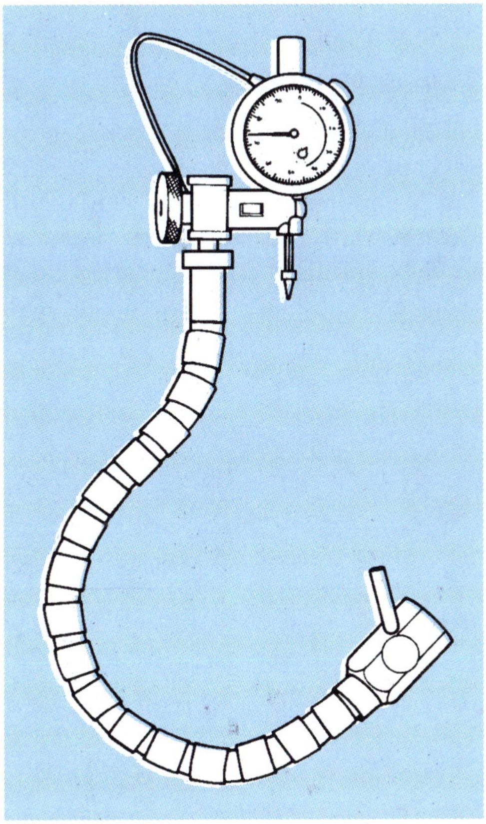

**Figure 6-9** Dial indicator designed for ball joint measurement (Courtesy of Moog Automotive)

Clean the top end of the lower ball joint stud, and position the dial indicator stem against the top end of this stud (Figure 6-10). Depress the dial indicator plunger approximately 0.250 in.

Lift upward with a pry bar under the tire and observe the dial indicator reading. If the vertical ball joint movement exceeds the manufacturer's specifications, ball joint replacement is required.

**Figure 6-10** Dial indicator installed for axial ball joint measurement (Courtesy of Moog Automotive)

**Figure 6-11** Dial indicator installed to measure radial ball joint movement (Courtesy of Moog Automotive)

## Ball Joint Radial Measurement

Worn ball joints cause improper camber and caster angles, which result in reduced directional stability and tire tread wear. Connect the dial indicator to the lower control arm of the ball joint being checked and position the dial indicator stem against the edge of the wheel rim (Figure 6-11).

Be sure the front wheel bearings are adjusted properly prior to the ball joint radial measurement. While an assistant grasps the top and bottom of the raised tire and attempts to move the tire and wheel horizontally in and out, observe the reading on the dial indicator (Figure 6-12).

The lower ball joint on a MacPherson strut front suspension should be checked for radial movement with a dial indicator when the tire is lifted off the floor (Figure 6-13). Since the spring load is carried by the upper and lower spring seats when the tire is lifted off the floor, it is not necessary to

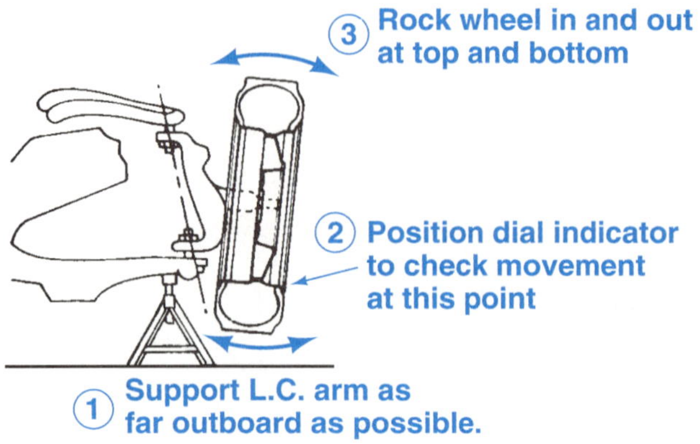


**Figure 6-12** Checking radial ball joint movement (Courtesy of Chevrolet Motor Division, General Motors Corporation

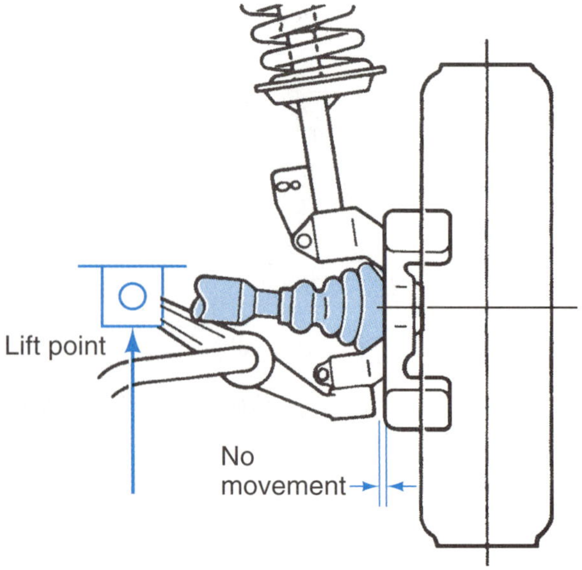


**Figure 6-13** Ball joint wear check on MacPherson strut front suspension (Courtesy of Ford Motor Company)

unload this type of ball joint. Photo Sequence 4 shows a typical procedure for measuring the lower ball joint radial movement on a MacPherson strut front suspension.

## Checking Ball Joints, Twin I-Beam Axles

Follow these steps to check the ball joints on a twin I-beam axle:

1. Lift the front end of the vehicle with a floor jack and place jack stands near the outer end of the I-beams.
2. Lower the vehicle onto the jack stands.
3. While an assistant grasps the wheel at the bottom and moves the wheel in and out, watch for movement between the lower part of the axle jaw and the spindle lower arm. If this movement exceeds 1/32 inch, replace the lower ball joint.
4. As an assistant grasps the top of the wheel and moves the wheel in and out, watch for movement between the upper axle jaw and the spindle upper arm. If this movement exceeds 1/32 inch, replace the upper ball joint.

**Classroom Manual**
Chapter 6, page 108

## Front Steering Knuckle Diagnosis and Service, Front Wheel Drive Car

**Front Steering Knuckle Diagnosis.** Steering looseness may be caused by a worn tie-rod end opening or ball joint opening in the steering knuckle. If these knuckle openings are worn, reduced directional stability may be experienced. Since a bent steering knuckle affects front suspension alignment angles, this problem may cause reduced directional stability and tire tread wear. Many steering arm and knuckle assemblies must be replaced as a unit. If the steering arms are bent, knuckle replacement is required.

**Steering Knuckle Removal and Replacement.** The steering knuckle replacement procedure varies depending on the type of front suspension. Always follow the vehicle manufacturer's steering knuckle replacement procedure in the service manual. Follow these steps to replace the steering knuckle on a front wheel drive vehicle with a MacPherson strut front suspension system.

1. Remove the wheel cover and loosen the front wheel nuts and the drive axle nut.
2. Lift the vehicle chassis on a hoist and allow the front suspension to drop downward. Remove the front wheel, brake caliper, brake rotor, and drive axle nut. Tie the brake caliper to a suspension component. Do not allow the caliper to hang on the end of the brake hose.
3. Remove the inner end of the drive axle from the transaxle with a pulling or prying action. On some Chrysler products, the differential cover must be removed and axle circlips compressed before the drive axle is removed from the transaxle.
4. Remove the outer end of the drive axle from the steering knuckle. On some early model Ford Escorts and Lynx, a puller is required for this operation.
5. Be sure the vehicle weight is supported on the hoist with the front suspension dropped downward. Disconnect the lower ball joint nuts from the mounting bolts in the lower control arm (Figure 6-14).

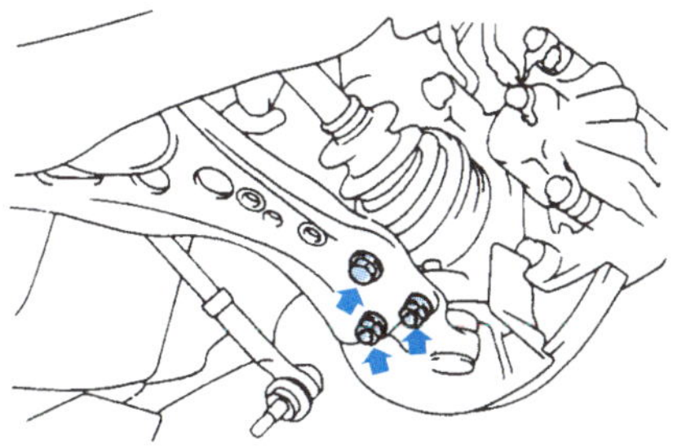

**Figure 6-14** Removal of nuts from lower ball joint mounting bolts under the lower control arm (Courtesy of Toyota Motor Corporation)

## Photo Sequence 4
# Typical Procedure for Measuring the Lower Ball Joint Radial Movement on a MacPherson Strut Front Suspension

**P4-1** Raise the front suspension with a floor jack and place jack stands under the chasis at the vehicle manufacturer's recommended lift points.

**P4-2** Grasp the front tire at the top and bottom and rock the tire inward and outward while a co-worker visually checks for movement in the front wheel bearing. If there is movement in the front wheel bearing, adjust or replace the bearing.

**P4-3** Position a dial indicator against the inner edge of the rim at the bottom. Preload and zero the dial indicator.

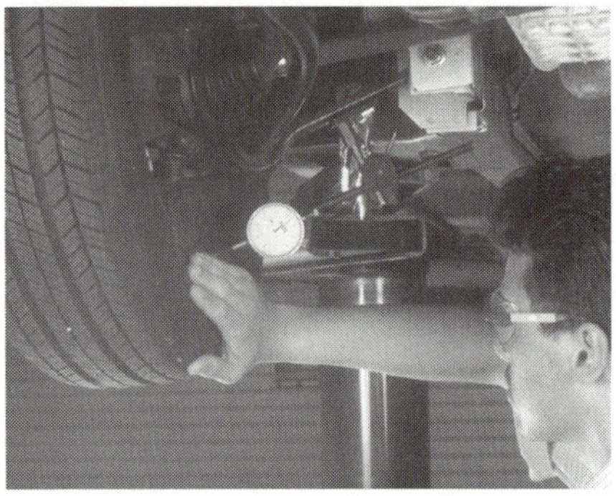

**P4-4** Grasp the bottom of the tire and pull outward.

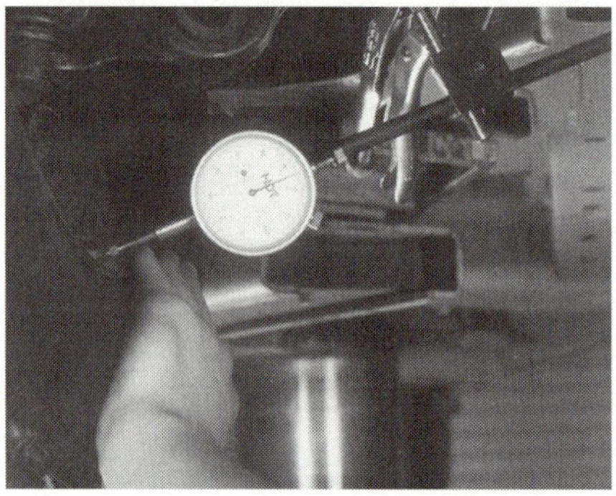

**P4-5** With the tire held outward, read the dial indicator.

**P4-6** Push the bottom of the tire inward and be sure the dial indicator reading is zero. Adjust the dial indicator as required.

**P4-7** Grasp the bottom of the tire and pull outward.

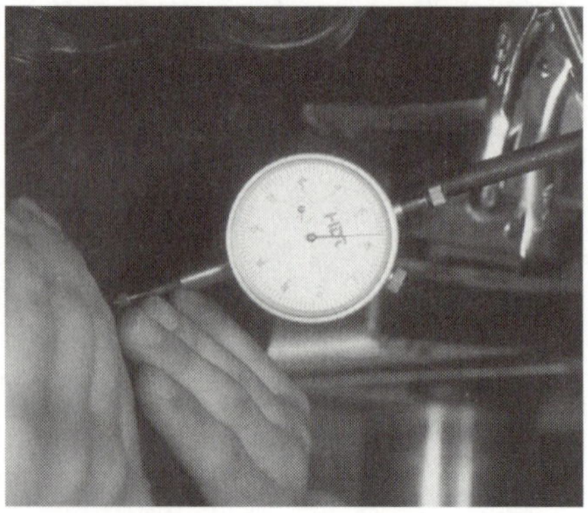

**P4-8** With the tire held in this position, read the dial indicator.

**P4-9** If the dial indicator reading is more than specified, replace the lower ball joint.

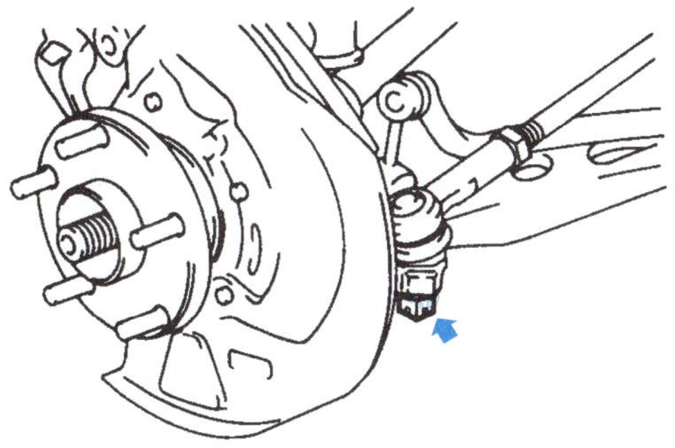

**Figure 6-15** Removing cotter key and nut from outer tie-rod end (Courtesy of Toyota Motor Corporation)

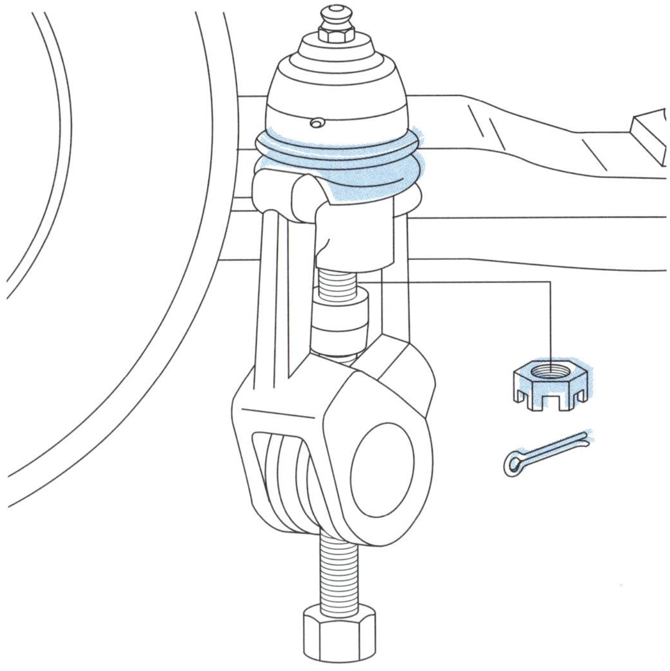

**Figure 6-16** Removing outer tie-rod end from steering knuckle (Courtesy of Chrysler Corporation)

6. Remove the cotter key from the outer tie-rod nut (Figure 6-15). Remove the outer tie-rod nut, and then use a puller to disconnect the tie-rod end from the steering knuckle (Figure 6-16).
7. If an eccentric cam is used on one of the strut-to-knuckle bolts, mark the cam and bolt position in relation to the strut, and remove the strut-to-knuckle bolts.
8. Remove the knuckle from the strut, and lift the knuckle out of the chassis (Figure 6-17).
9. Pry the dust deflector from the steering knuckle with a large flat-blade screwdriver (Figure 6-18).
10. Use a puller to remove the ball joint from the steering knuckle (Figure 6-19). Check the ball joint and tie-rod end openings in the knuckle for wear and out-of-round. Replace the knuckle if these openings are worn or out-of-round.

**Special Tools**

Ball joint puller

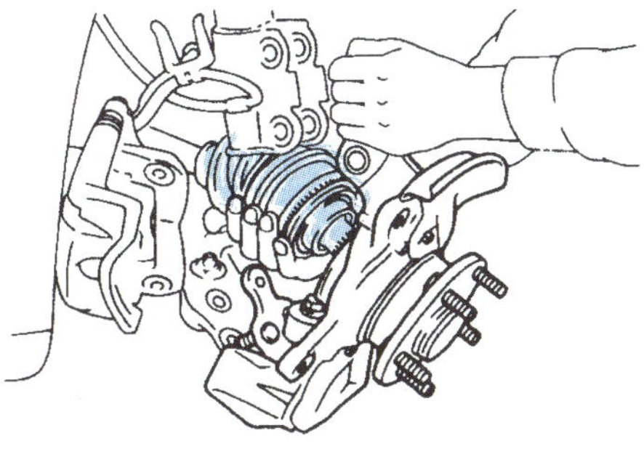

**Figure 6-17** Removing the steering knuckle from the lower control arm (Courtesy of Toyota Motor Corporation)

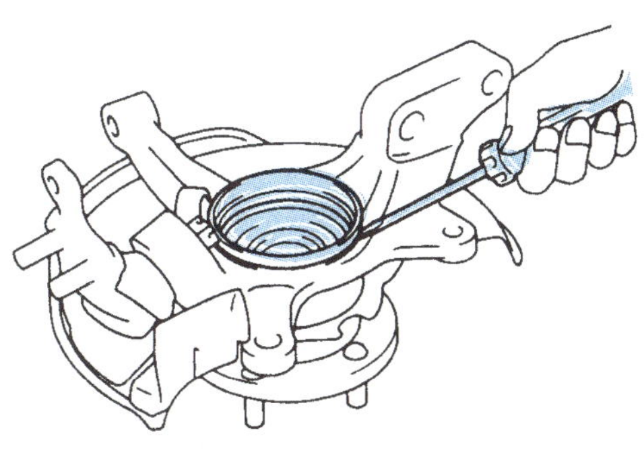

**Figure 6-18** Removing dust deflector from steering knuckle (Courtesy of Toyota MotorCorporation)

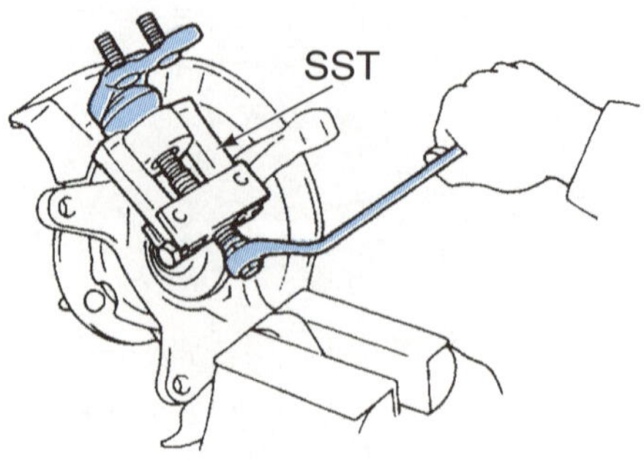


**Figure 6-19** Removing ball joint from steering knuckle (Courtesy of Toyota Motor Corporation)

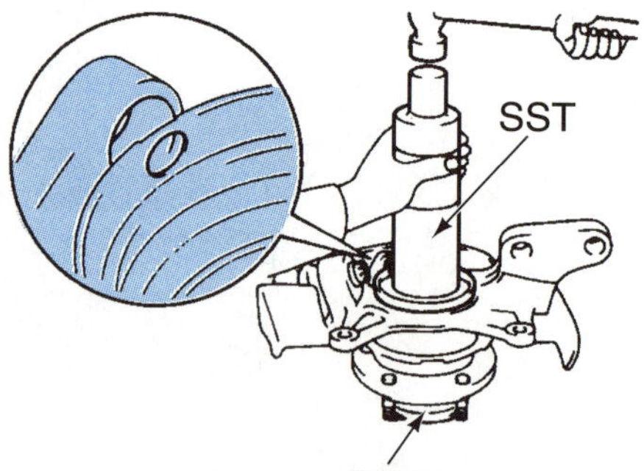


**Figure 6-20** Installing dust deflector in steering knuckle (Courtesy of Toyota Motor Corporation)

11. Use the proper driving tool to reinstall the dust deflector in the steering knuckle (Figure 6-20).
12. Reverse steps 1 through 10 to reinstall the knuckle. Service or replace the wheel bearing as required. (Refer to Chapter 3 for wheel bearing service.) Torque all nuts to specifications and install cotter keys as required.

## Front Steering Knuckle Replacement Rear Wheel Drive Cars

Proceed as follows for steering knuckle replacement on long and short arm front suspension systems with the coil spring positioned between the lower control arm and the frame.

1. Remove the wheel cover and loosen the wheel nuts.
2. Lift the vehicle with a floor jack and place jack stands under the chassis so the front suspension drops downward. Lower the vehicle onto the jack stands, and remove the floor jack.
3. Remove the front wheel, brake caliper, and the brake rotor and hub with the wheel bearings. Attach the brake caliper to a chassis component with a piece of wire. Do not allow the caliper to hang on the end of the brake hose.
4. Remove the outer tie-rod cotter key and nut, and remove the outer tie-rod end from the steering arm.
5. Remove the cotter keys from the upper and lower ball joint nuts and loosen, but do not remove, the nuts.

**CAUTION:** During the knuckle replacement procedure, the floor jack supports the spring tension when the ball joints are disconnected from the knuckle. Do not lower the floor jack. Chain the coil spring to the lower control arm as a safety precaution.

6. Support the lower control arm with a floor jack, and raise the floor jack until the spring tension is supported by the jack.
7. Remove both ball joint nuts, and remove the ball joint studs from the steering knuckle. A threaded spreader tool may be used to push the upper and lower ball joints from the knuckle (Figures 6-21 and 6-22).

**Special Tools**

Ball joint spreader tool

8. Pull upward on the upper control arm to remove the upper ball joint stud from the knuckle. Remove the knuckle from the lower ball joint stud. Check the ball joint and tie rod end openings in the knuckle for wear and out-of-round. Replace the knuckle if these conditions are present.
9. Reverse steps 1 through 8 to install the steering knuckle. Tighten the ball joint nuts and the outer tie-rod nut to the specified torque, and install the cotter keys in these nuts. Adjust the front wheel bearings, and tighten the wheel nuts to the specified torque.

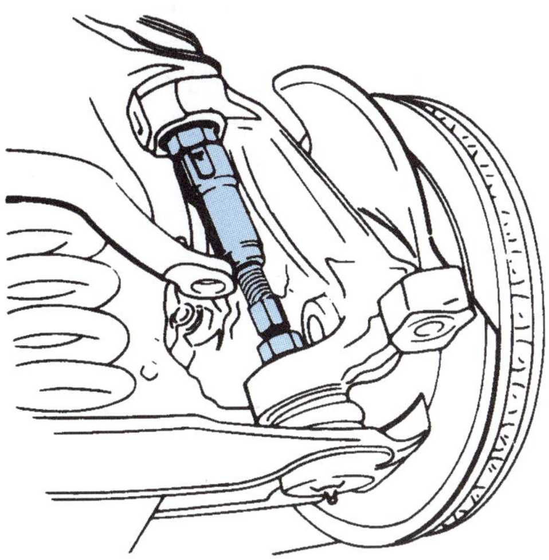

Figure 6-21 Removing the lower ball joint from the steering knuckle (Courtesy of Chevrolet Motor Division, General Motors Corporation)

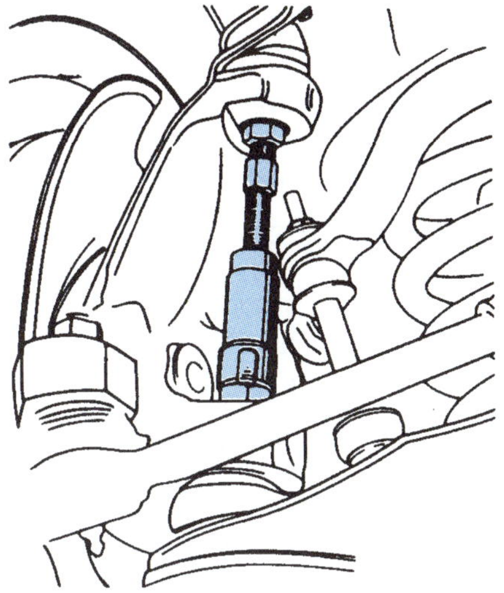

Figure 6-22 Removing the upper ball joint from the steering knuckle (Courtesy of Chevrolet Motor Division, General Motors Corporation)

## Ball Joint Replacement

**Results of Ball Joint Wear.** Worn ball joints affect steering angles and cause reduced directional stability and excessive tire tread wear.

**Ball Joint Removal and Replacement.** Ball joints may be threaded, pressed, bolted, or riveted into the control arms. The ball joint replacement procedure varies depending on the type of suspension and the method of ball joint attachment. Always follow the vehicle manufacturer's recommended ball joint replacement procedure in the service manual. Following are typical steps that apply to all three methods of ball joint attachment.

1. Remove the wheel cover and loosen the wheel nuts.
2. Lift the vehicle with a floor jack and place jack stands under the chassis, so the front suspension is allowed to drop downward. Lower the vehicle onto the jack stands, and remove the floor jack.
3. Remove the wheel, and place a floor jack under the outer end of the lower control arm. Operate the floor jack and raise the lower control arm until the ball joints are unloaded. Remove other components such as the brake caliper and rotor, and drive axle, as required to gain access to the ball joints.
4. Remove the cotter key in the ball joint, or joints, that requires replacement, and loosen, but do not remove, the ball joint stud nuts.
5. Loosen the ball joint stud tapers in the steering knuckle. A threaded expansion tool is available for this purpose.
6. Remove the ball joint nut, and lift the knuckle off the ball joint stud. Block or tie up the knuckle and hub assembly to access the ball joint.
7. If the ball joint is riveted to the control arm, drill and punch out the rivets, and bolt the new ball joint to the control arm (Figure 6-23).
8. If the ball joint is pressed into the lower control arm, remove the ball joint dust boot, and use a pressing tool to remove and replace the ball joint (Figures 6-24 and 6-25).
9. If the ball joint housing is threaded into the control arm, use the proper size socket to remove and install the ball joint. The replacement ball joint must be torqued to the manufacturer's specifications. If a minimum of 125 foot-pounds of torque cannot be obtained, the control arm threads are damaged and control arm replacement is necessary.

**Special Tools**

Ball joint pressing tools

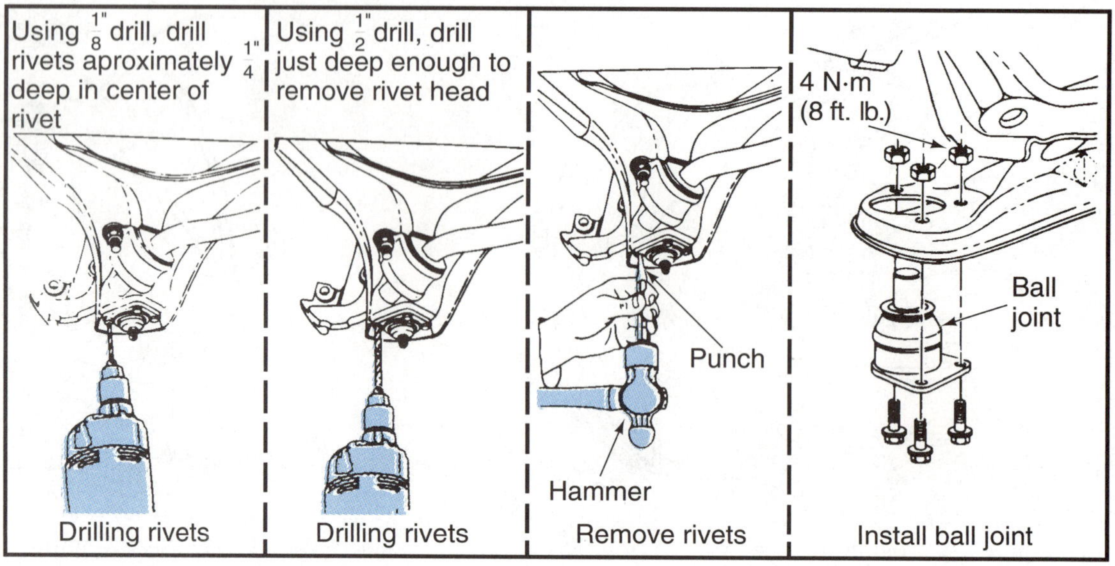


**Figure 6-23** Replacement of riveted ball joint (Courtesy of Pontiac Motor Division, General Motors Corporation)

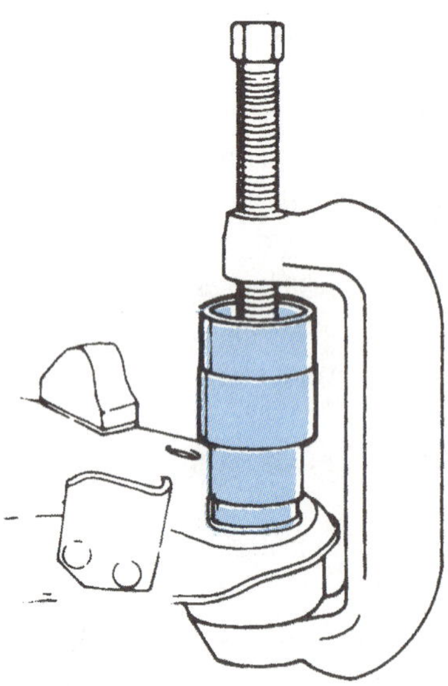

**Figure 6-24** Removing pressed-in ball joint from lower control arm (Courtesy of Chevrolet Motor Division, General Motors Corporation)

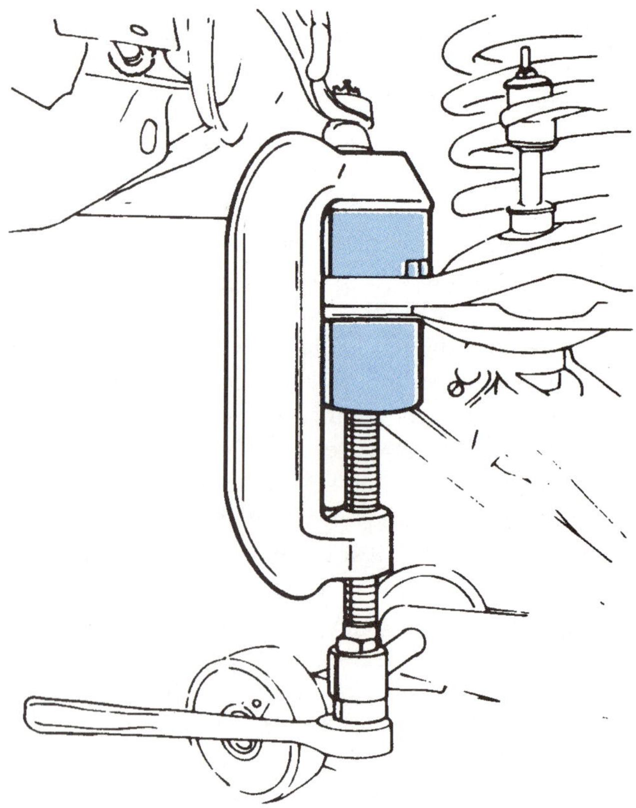

**Figure 6-25** Installing pressed-in ball joint in lower control arm (Courtesy of Chevrolet Motor Division, General Motors Corporation)

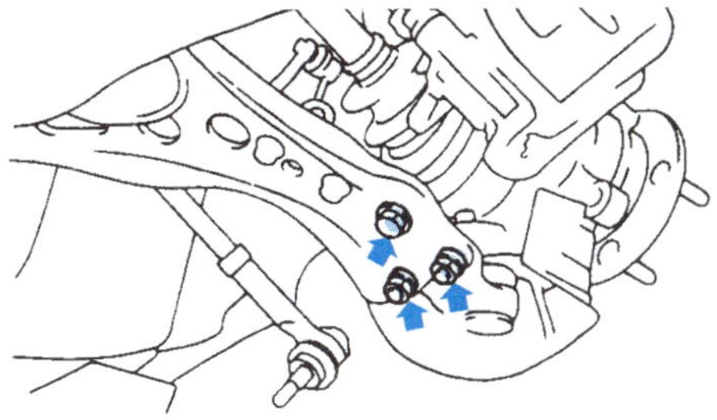

Figure 6-26 Installing ball joint retaining bolt and nuts in lower control arm (Courtesy of Toyota Motor Corporation)

10. If the ball joint is bolted to the lower control arm, install the new ball joint and tighten the bolt and nuts to the specified torque (Figure 6-26).
11. Clean and inspect the ball joint stud tapered opening in the steering knuckle. If this opening is out-of-round or damaged, the knuckle must be replaced.
12. Check the fit of the ball joint stud in the steering knuckle opening. This stud should fit snugly in the opening and only the threads on the stud should extend through the knuckle. If the ball joint stud fits loosely in the knuckle tapered opening, either this opening is worn or the wrong ball joint has been supplied.

**CAUTION:** Never back off a ball joint stud nut to align the cotter key openings in the nut and stud. Always tighten the nut to the next hole to install the cotter key.

13. Install the ball joint stud in the steering knuckle opening, making sure the stud is straight and centered. Install the stud nut and tighten this nut to the specified torque. Install a new cotter key through the stud and nut. Do not loosen the nut to align the nut and stud openings.
14. Reassemble the components that were removed in step 3. Make sure the wheel nuts are tightened to the specified torque.
15. After ball joint replacement, the front suspension alignment should be checked.

# Control Arm Diagnosis and Service

## Control Arm Diagnosis and Replacement

Upper and lower control arms should be inspected for cracks, bent conditions, and worn bushings. If the control arm bushings are worn, steering is erratic, especially on irregular road surfaces. Worn control arm bushings may cause a rattling noise while driving on irregular road surfaces. Dry or worn control arm bushings may cause a squeaking noise on irregular road surfaces. Caster and camber angles on the front suspension are altered by worn upper and lower control arm bushings. Incorrect caster or camber angles may cause the vehicle to pull to one side. Tire wear may be excessive when the camber angle is not within specifications.

Caster angle on the front suspension is the tilt of a line that intersects the center of the lower ball joint and the center of the upper strut mount in relation to a vertical line through the center of the wheel and spindle viewed from the side of the vehicle.

## Lower Control Arm Replacement, MacPherson Strut Suspension

The upper or lower control arm removal and replacement procedure varies depending on the type of suspension. Always follow the recommended procedure in the vehicle manufacturer's service manual. Following is a lower control arm replacement procedure for a MacPherson strut front suspension:

1. Remove the wheel cover and loosen the wheel nuts and drive axle nut.
2. Lift the front of the vehicle with a floor jack, and place jack stands under the chassis. Lower the vehicle onto the jack stands and allow the front suspension to drop downward.

Camber angle on the front, or rear, suspension is the tilt of a line inward, or outward, in relation to the vertical line through the center of the wheel and tire viewed from the front.

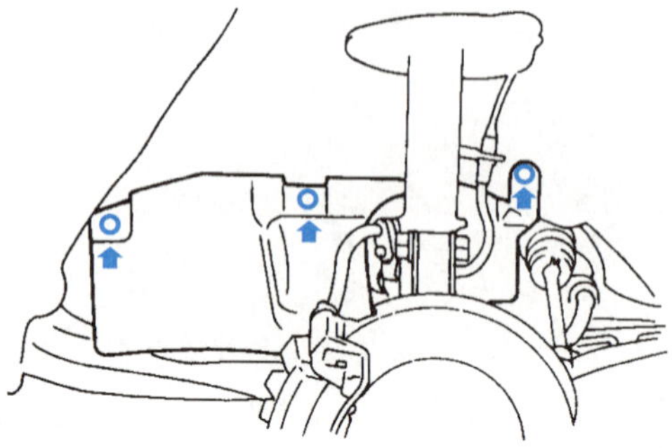

Figure 6-27 Removal of front fender apron seal (Courtesy of Toyota Motor Corporation)

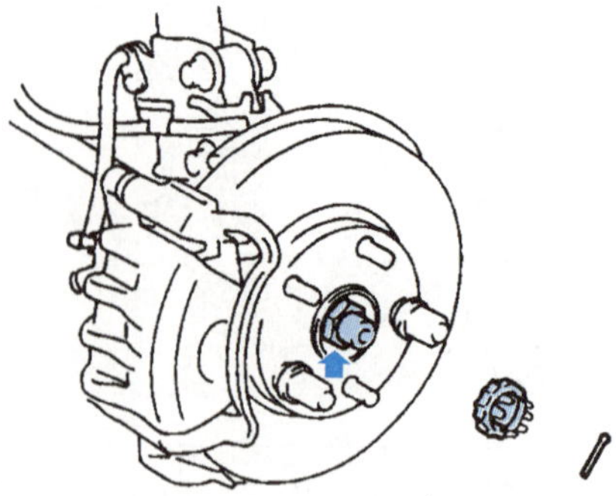

Figure 6-28 Removal of drive axle cotter key, lock cap, and nut (Courtesy of Toyota Motor Corporation)

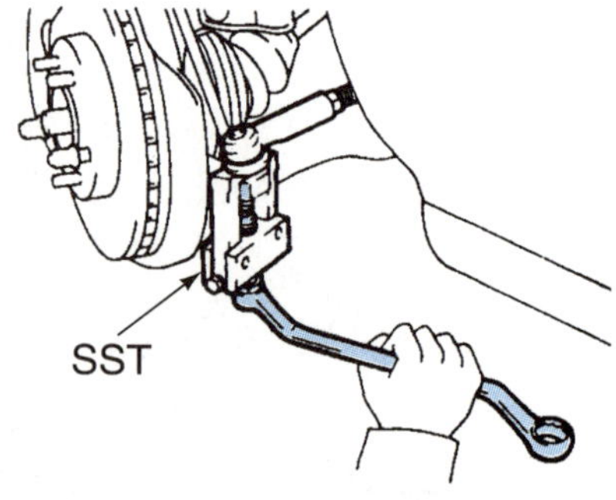


Figure 6-29 Removing outer tie-rod end (Courtesy of Toyota Motor Corporation)

**Special Tools**

Tie-rod end puller

3. Remove the front wheel and the front fender apron seal (Figure 6-27). Remove the drive axle nut (Figure 6-28).
4. Remove the cotter key and nut from the outer tie-rod end, and use a puller to remove this tie-rod end from the steering arm (Figure 6-29).
5. Remove the stabilizer brackets from the lower control arm (Figure 6-30).
6. Remove the lower ball joint nuts and bolt from the lower control arm (Figure 6-31).
7. Remove the drive axle from the axle hub (Figure 6-32), and secure the drive axle to a suspension component with a piece of wire (Figure 6-33). Do not allow the drive axle to hang downward.

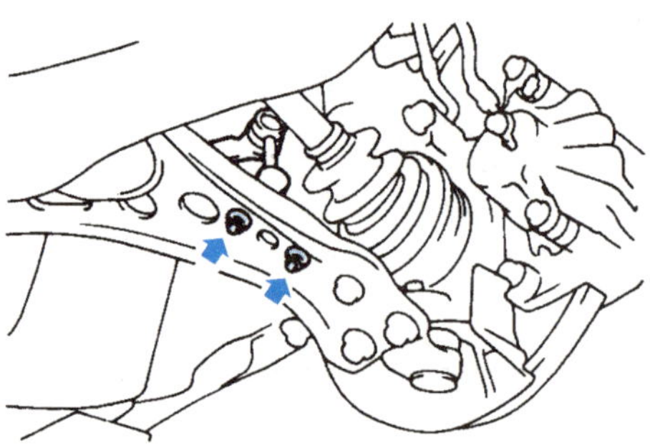

Figure 6-30 Removing stabilizer brackets from lower control arm (Courtesy of Toyota Motor Corporation)

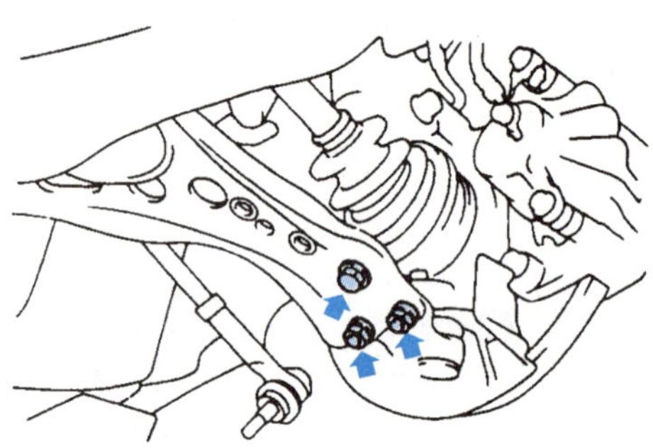

Figure 6-31 Removing lower ball joint bolt and nuts from lower control arm (Courtesy of Toyota Motor Corporation)

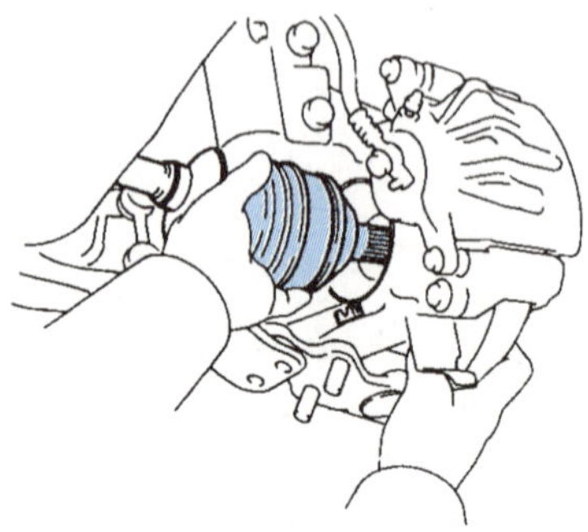

Figure 6-32 Removing drive axle from axle hub (Courtesy of Toyota Motor Corporation)

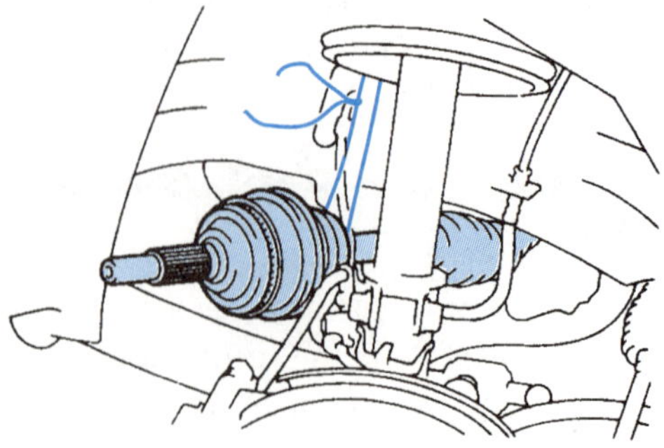

Figure 6-33 Securing drive axle with a piece of wire (Courtesy of Toyota Motor Corporation)

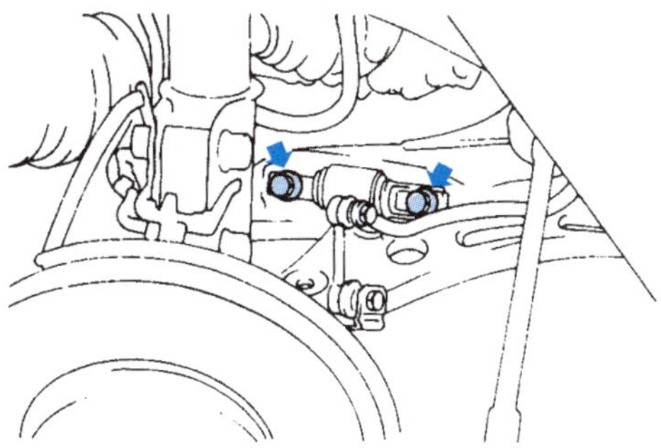

**Figure 6-34** Removing bolts at the front side of the lower control arm (Courtesy of Toyota Motor Corporation)

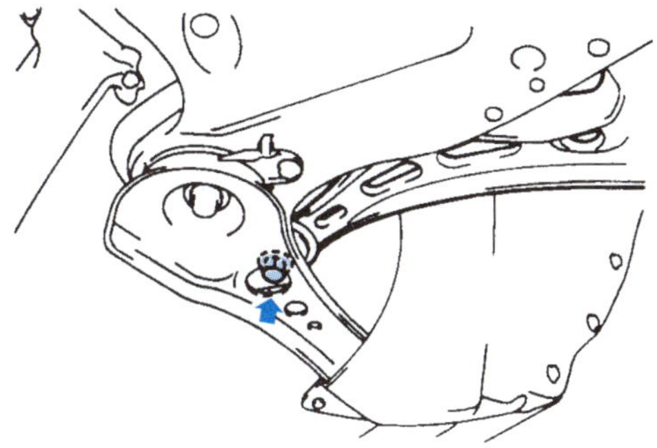

**Figure 6-35** Removing bolt and nut from rear side of the lower control arm (Courtesy of Toyota Motor Corporation)

8. Remove the two attaching bolts at the front side of the lower control arm (Figure 6-34).
9. Remove the bolt and nut on the rear side of the lower control arm (Figure 6-35), and remove the lower arm bushing stopper from the lower arm shaft.

Reverse steps 1 through 9 to install the replacement control arm. Tighten all bolts and nuts to the specified torque, and install cotter keys as required.

## Lower Control Arm and Spring Replacement, Long and Short Arm Suspension

**WARNING:** During control arm and spring replacement, the coil spring tension is supported by a compressing tool or floor jack. Always follow the vehicle manufacturer's recommended control arm and spring replacement procedures very carefully. Serious personal injury or property damage may occur if the spring tension is released suddenly.

Broken coil springs may cause a rattling noise while driving on irregular road surfaces. Weak or broken coil springs also reduce curb riding height. A rattling noise while driving on road irregularities may also be caused by worn or broken spring insulators. Since weak or broken coil springs affect front suspension alignment angles, this problem may cause reduced directional stability, excessive tire wear, and harsh riding.

Follow this procedure when a lower control arm and/or spring is replaced on a vehicle with a long and short arm suspension system with the coil springs positioned between the lower control arm and the frame:

1. Lift the vehicle on a hoist until the tires are a short distance off the floor, and allow the front suspension to drop downward. An alternate method is to lift the vehicle with a floor jack, and then support the chassis securely on jack stands so the front suspension drops downward.
2. Disconnect the lower end of the shock absorber. On some applications, the shock absorber must be removed.
3. Disconnect the stabilizer bar from the lower control arm.
4. Install a spring compressor and turn the spring compressor bolt until the spring is compressed (Figure 6-36). Make sure all the spring tension is supported by the compressing tool.
5. Place a floor jack under the lower control arm, and raise the jack until the control arm is raised and the rebound bumper is not making contact.
6. Remove the lower ball joint cotter key and nut, and use a threaded expansion tool to loosen the lower ball joint stud.

**Special Tools**

Spring compressing tool, long and short arm suspension

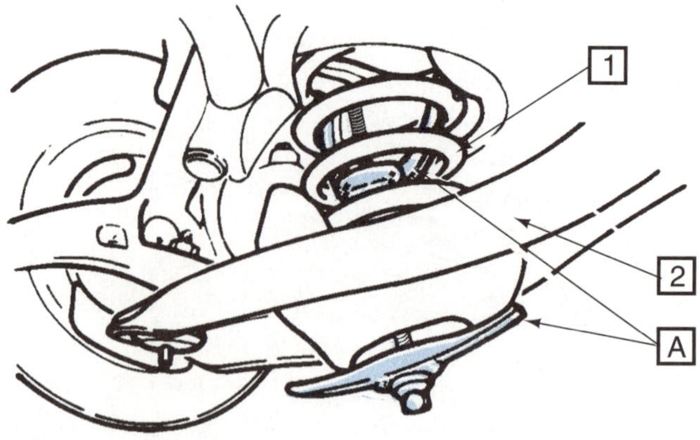


**Figure 6-36** Spring compressing tool installed on spring in long and short arm suspension system (Courtesy of Chevrolet Motor Division, General Motors Corporation)

7. Lower the floor jack very slowly to lower the control arm and coil spring.
8. Disconnect the lower control arm inner mounting bolts, and remove the lower control arm.
9. Rotate the compressing tool bolt to release the spring tension, and remove the spring from the control arm.
10. Inspect the lower control arm for a bent condition or cracks. If either of these conditions is present, replace the control arm. Visually inspect all control arm bushings. Loose or worn bushings must be replaced. Visually inspect upper and lower spring insulators for cracks and wear, and inspect the spring seat areas in the chassis and lower control arm. Worn or cracked spring insulators must be replaced.
11. Reverse steps 1 through 9 to install the lower control arm. Be sure the coil spring and insulators are properly seated in the lower control arm and in the upper spring seat.

## Upper Control Arm Removal and Replacement, Long and Short Arm Suspension System

Proceed as follows when replacing an upper control arm on a long and short arm suspension system with the front coil springs located between the lower control arm and the frame.

1. Remove the wheel cover and loosen the wheel nuts.
2. Lift the vehicle with a floor jack and install jack stands under the chassis. Lower the vehicle onto the jack stands so the front suspension drops downward.
3. Remove the front wheel and tire.
4. Place a floor jack under the lower control arm and raise this arm. Make sure the rebound bumper is not making contact.
5. Remove the cotter key and loosen, but do not remove, the upper ball joint nut.
6. Use a threaded expansion tool to loosen the ball joint stud in the control arm. Make sure the floor jack is supporting the spring tension.

**CAUTION:** Since the floor jack is supporting the spring tension, do not lower the floor jack until the ball joint is reconnected and the ball joint nut is tightened to the specified torque.

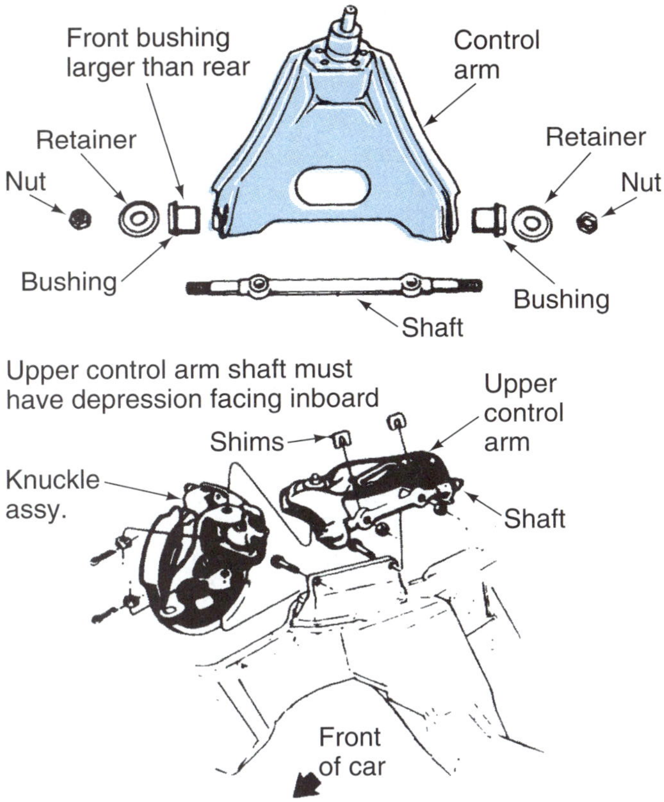


**Figure 6-37** Upper control arm components (Courtesy of Chevrolet Motor Division, General Motors Corporation)

7. Remove the ball joint nut and the inner control arm mounting bolts. If shims are located on the inner control arm mounting bolts, note the shim position.
8. Remove the control arm from the chassis.
9. Visually inspect the upper control arm for a bent or cracked condition. Inspect the control arm bushing openings for wear, and replace the control arm if it is bent, cracked, or worn in the bushing areas. Inspect the control arm shaft and bushings for wear, and replace all worn parts (Figure 6-37).
10. Install the control arm and place the original number of shims on the inner mounting bolts. Tighten these bolts to the manufacturer's specifications.
11. Install the ball joint stud in the steering knuckle, and tighten the stud nut to the specified torque.
12. Install the upper ball joint cotter key and the front wheel. Tighten the wheel nuts to the specified torque.

The front suspension alignment angles should be checked after upper control arm replacement.

## Rebound Bumpers

**SERVICE TIP:** Excessively worn rebound bumpers indicate worn-out shock absorbers, reduced curb riding height, or driving continually on irregular road surfaces.

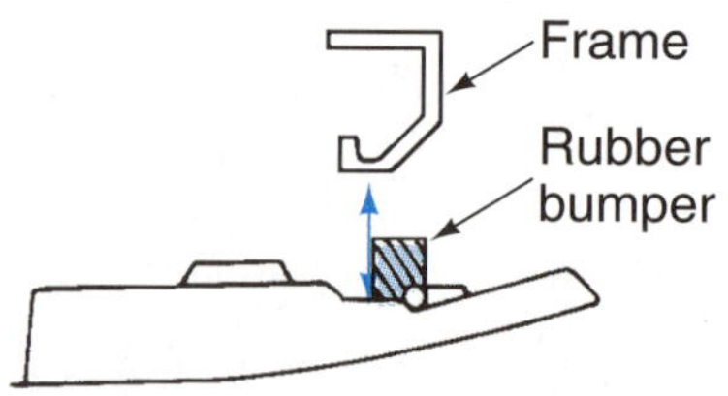


**Figure 6-38** Rebound bumper (Courtesy of Sealed Power Corporation)

Rebound bumpers may be called strikeout bumpers.

Rebound bumpers are usually bolted to the lower control arm or to the chassis. Inspect the rebound bumpers for cracks, wear, and flattened conditions (Figure 6-38). Damaged rebound bumpers may be caused by sagged springs and insufficient curb riding height, or worn-out shock absorbers and struts. If the rebound bumpers must be replaced, remove the mounting bolts and the bumper. Install the new bumper and tighten the mounting bolts to the specified torque.

## Front Lower Control Arm Bushing Removal and Replacement

**Special Tools**

Control arm bushing removal, replacement, and flaring tools

All suspension bushings should be checked periodically for wear, looseness, and deterioration. These bushings are important to providing quiet suspension operation and to preventing the transmission of suspension vibration to the chassis and passenger compartment. When a bushing is contained in a steel sleeve, press only on the outer sleeve during the removal and replacement procedure. The control arm bushing removal and replacement procedure varies depending on the type of suspension system. Always follow the vehicle manufacturer's recommended procedure in the service manual. Special bushing removal and replacement tools are required for control arm bushing replacement. A front control arm bushing removal tool and spacer is used to remove the control arm bushing (Figure 6-39). The spacer is installed between the bushing support lugs to prevent distorting these lugs during the bushing removal process.

A different tool is required to install the new front control arm bushing (Figure 6-40). The same spacer that is used during the bushing removal is installed between the bushing lugs during bushing installation.

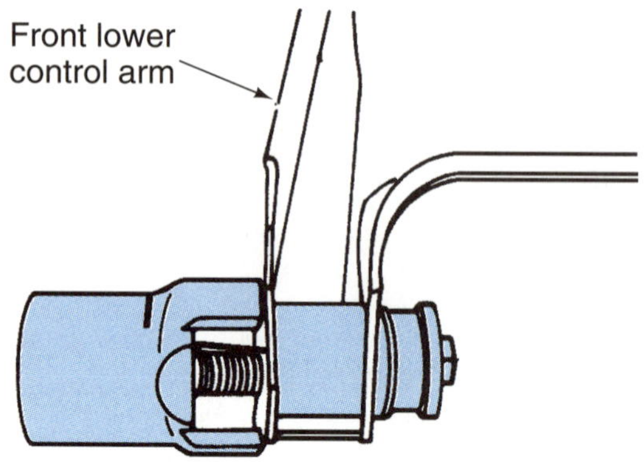


**Figure 6-39** Removal of front control arm bushing (Courtesy of Chevrolet Motor Division, General Motors Corporation)

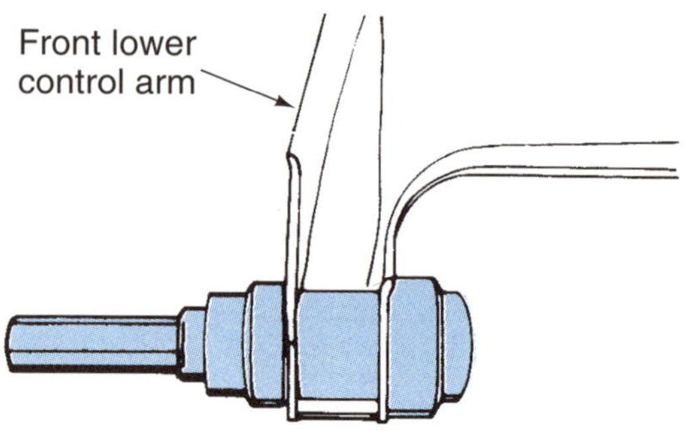


**Figure 6-40** Installation of front control arm bushing (Courtesy of Chevrolet Motor Division, General Motors Corporation)

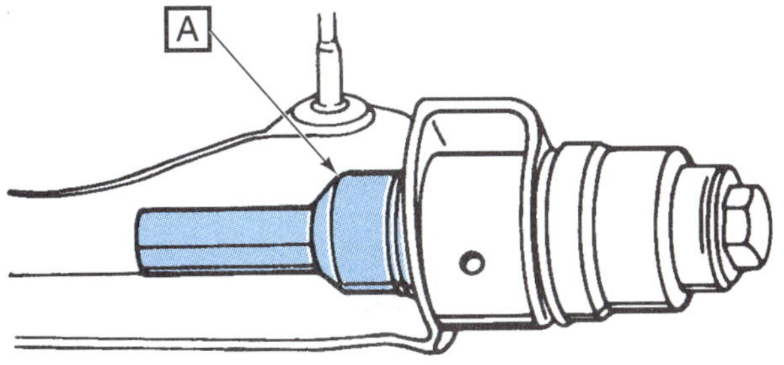


Figure 6-41 Flaring the front control arm bushing (Courtesy of Chevrolet Motor Division, General Motors Corporation)

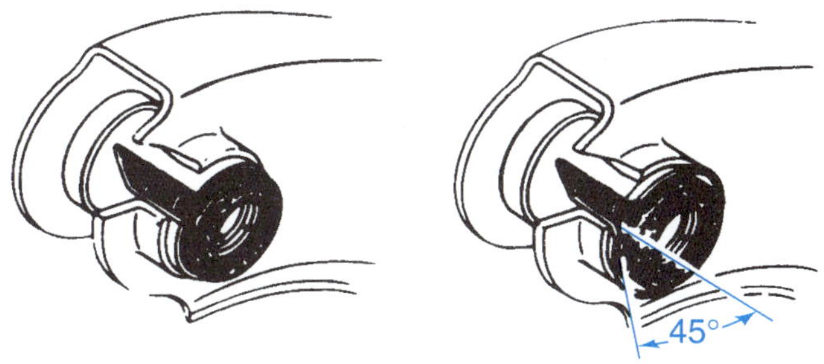


Figure 6-42 Front control arm bushing after flaring (Courtesy of Chevrolet Motor Division, General Motors Corporation)

After the front control arm bushing is installed, a bushing flaring tool is installed to flare the bushing (Figure 6-41). Notice that a spacer is installed between the bushing lugs during the flaring process. Flaring retains the bushing securely (Figure 6-42).

The rear control arm bushing is replaced with the same procedure as the front control arm bushing. Some of the same special tools are used for rear control arm bushing replacement (Figures 6-43 and 6-44).

## Stabilizer Bar Diagnosis and Replacement

A stabilizer bar may be referred to as a sway bar.

Worn stabilizer bar mounting bushings, grommets, or mounting bolts cause a rattling noise as the vehicle is driven on irregular road surfaces. A weak stabilizer bar or worn bushings and grommets cause harsh riding and excessive body sway while driving on irregular road surfaces. Worn or very dry stabilizer bar bushings may cause a squeaking noise on irregular road surfaces. All stabilizer bar components should be visually inspected for wear. Stabilizer bar removal and replacement procedures vary depending on the vehicle. Always follow the vehicle manufacturer's recommended procedure in the service manual. Following is a typical stabilizer bar removal and replacement procedure:

1. Lift the vehicle on a hoist and allow both sides of the front suspension to drop downward as the vehicle chassis is supported on the hoist.

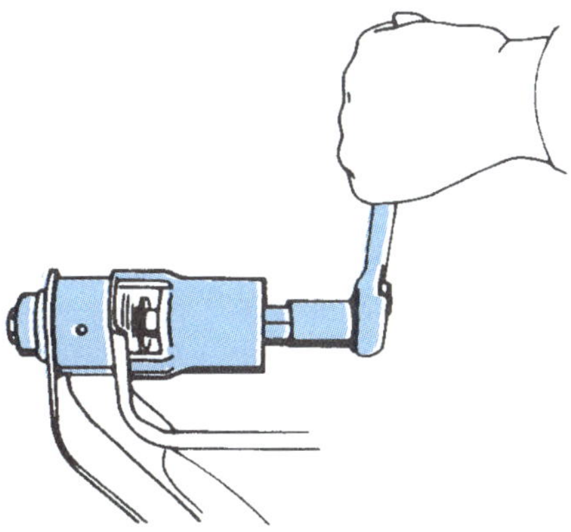

Figure 6-43 Rear control arm bushing removal (Courtesy of Chevrolet Motor Division, General Motors Corporation)

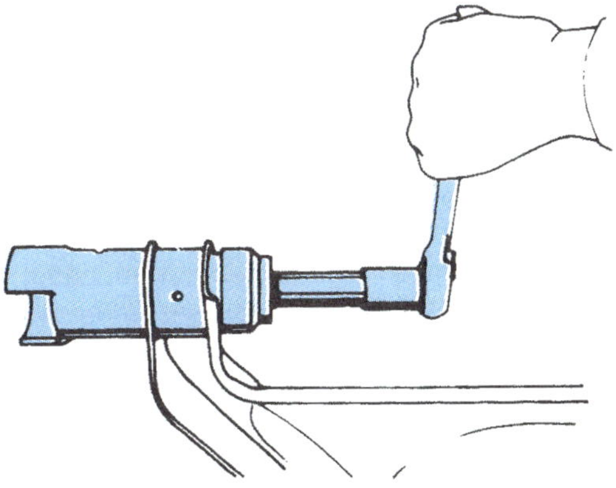

Figure 6-44 Rear control arm bushing installation (Courtesy of Chevrolet Motor Division, General Motors Corporation)

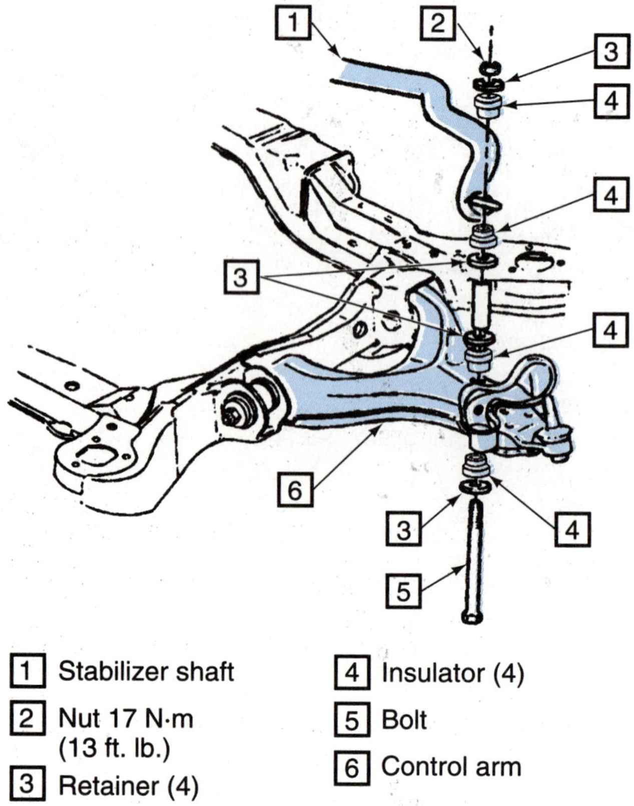


**Figure 6-45** Stabilizer bar components (Courtesy of Oldsmobile Division, General Motors Corporation)

2. Remove the mounting bolts at the outer ends of the stabilizer bar and remove the bushings, grommets, brackets, or spacers (Figure 6-45).
3. Remove the mounting bolts in the center area of the stabilizer bar.
4. Remove the stabilizer bar from the chassis.
5. Visually inspect all stabilizer bar components such as bushings, bolts, and spacer sleeves. Replace the stabilizer bar, grommets, bushings, brackets, or spacers as required. Split bushings may be removed over the stabilizer bar. Bushings that are not split must be pulled from the bar.
6. Reverse steps 2 through 4 to install the stabilizer bar. Make sure all stabilizer bar components are installed in the original position, and tighten all fasteners to the specified torque.

Some vehicle manufacturers specify that stabilizer bars must have equal distances between the outer bar ends and the lower control arms. Always refer to the manufacturer's recommended measurement procedure. If this measurement is required, adjust the nut on the outer stabilizer bar mounting bolt until equal distances are obtained between the outer bar ends and the lower control arms. Worn grommets cause these distances to be unequal.

## Strut Rod Diagnosis and Replacement

Some front suspension systems have a strut rod connected from the lower control arm to the frame. Rubber grommets isolate the rod from the chassis components. Worn strut rod grommets may allow the lower control arm to move rearward or forward. This movement changes the caster angle, which may affect steering quality and cause the vehicle to pull to one side. A bent strut rod also causes steering pull. Worn strut rod grommets or loose mounting bolts cause a rattling noise

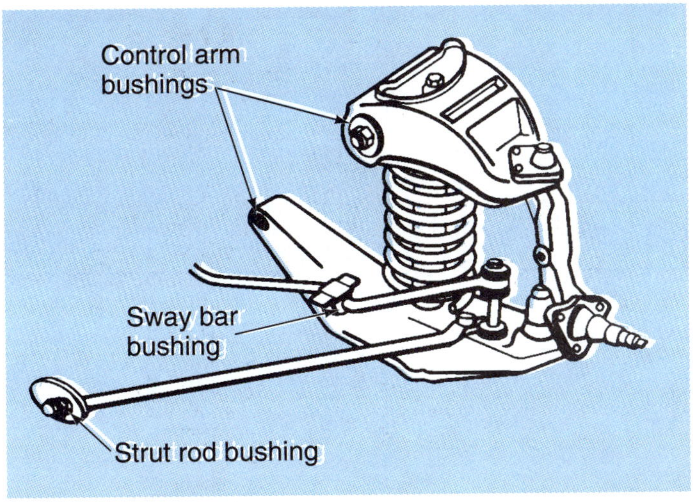


**Figure 6-46** Strut rod removal and replacement (Courtesy of Sealed Power Corporation)

while driving on irregular road surfaces. Inspect the strut rod grommets visually for wear and deterioration. With the vehicle lifted on a hoist, grasp the strut rod firmly and apply vertical and horizontal force to check the rod and grommets for movement. Worn grommets must be replaced. If a strut rod is bent, replace the rod.

Follow these steps for strut rod replacement:

1. Lift the vehicle on a hoist.
2. Remove the strut rod nut from the front end of the rod.
3. Remove the strut rod bolts from the lower control arm.
4. Pull the strut rod rearward to remove the rod (Figure 6-46).
5. Remove the bushings from the opening in the chassis.
6. Visually inspect the strut rod, bushings, washers, and retaining bolts. Replace all worn parts. Reverse steps 1 through 5 to reinstall the strut rod. Tighten the strut rod nut and bolts to the specified torque.

Since strut rod and bushing conditions affect front suspension alignment, check front suspension alignment after strut rod service.

## Front Leaf Spring Inspection and Replacement

Front leaf springs are used in some truck suspension systems. These springs are mounted longitudinally with a spring on each side of the suspension. Many leaf springs have plastic silencers between the spring leafs. If these silencers are worn out, creaking and squawking noises will be heard when the vehicle is driven over road irregularities at low speeds.

When the silencers require checking or replacement, lift the vehicle with a floor jack and support the frame on jack stands so the suspension moves downward. With the vehicle weight no longer applied to the springs, the spring leafs may be pried apart with a pry bar to remove and replace the silencers. On some leaf springs, the clamps must be removed to replace the silencers.

Leaf springs should be inspected for a sagged condition, which causes the curb riding height to be less than specified. Leaf springs should also be visually inspected for broken leafs, broken center bolts, and worn shackles or bushings. Weak or broken leaf springs affect front suspension alignment angles and cause excessive tire tread wear, reduced directional stability, and harsh riding. A rattling noise while driving on irregular road surfaces may be caused by worn shackles or bushings. A broken center bolt usually allows one side of the front suspension to move rearward in relation to the other side. This action may cause steering pull to one side. Worn shackles and

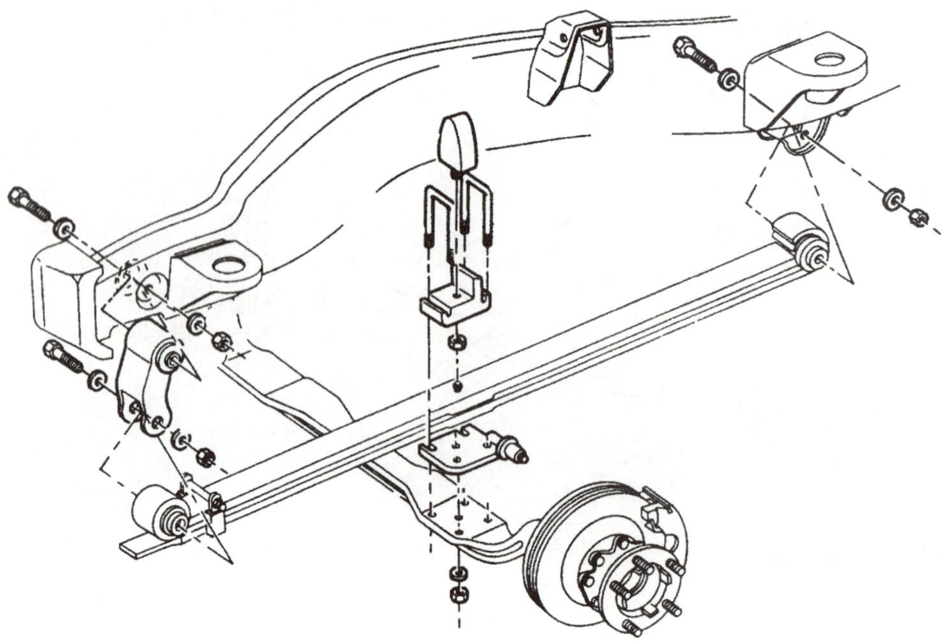

**Figure 6-47** Front leaf spring (Courtesy of Chevrolet Motor Division, General Motors Corporation)

bushings lower curb riding height and cause a rattling noise on road irregularities. Following is a typical front leaf spring replacement procedure:

1. Lift the front end of the vehicle with a floor jack under the front axle and place jack stands under the frame. Lower the vehicle weight onto the jack stands, and leave the floor jack under the front axle to support some of the front suspension weight.
2. Remove the nuts from the spring U-bolts, and remove the U-bolts and lower spring plate. If the spring plate is attached to the shock absorber, this plate may be left on the shock absorber and moved out of the way. If shims are positioned between the spring plate and the front axle, be sure to note the position and the number of shims. These shims must be reinstalled in their original position because they set the front wheel caster angle.
3. Be sure the floor jack is lowered enough to relieve the vehicle weight from the springs.
4. Remove the front shackle assembly.
5. Remove the rear spring mounting bolt, and remove the spring from the chassis (Figure 6-47).
6. Check all the spring hangers, bolts, bushings, and shackle plates for wear, and replace as required. Springs with broken or sagged leafs are usually sent to a spring rebuilding shop for repair.
7. Check the spring center bolt to be sure it is not broken.
8. Reverse steps 1 through 5 to install the front springs. Tighten all bolts to the manufacturer's specifications.

## Removing and Replacing Transversely Mounted Torsion Bars

Worn torsion bar components such as pivot cushion bushings and control arm bushings result in harsh riding and suspension noise when driving on road irregularities. A worn torsion bar hex and anchor may cause a rattling noise while driving on irregular road surfaces. Weak torsion bars or those with worn bushings or anchors cause reduced curb riding height, which may result in

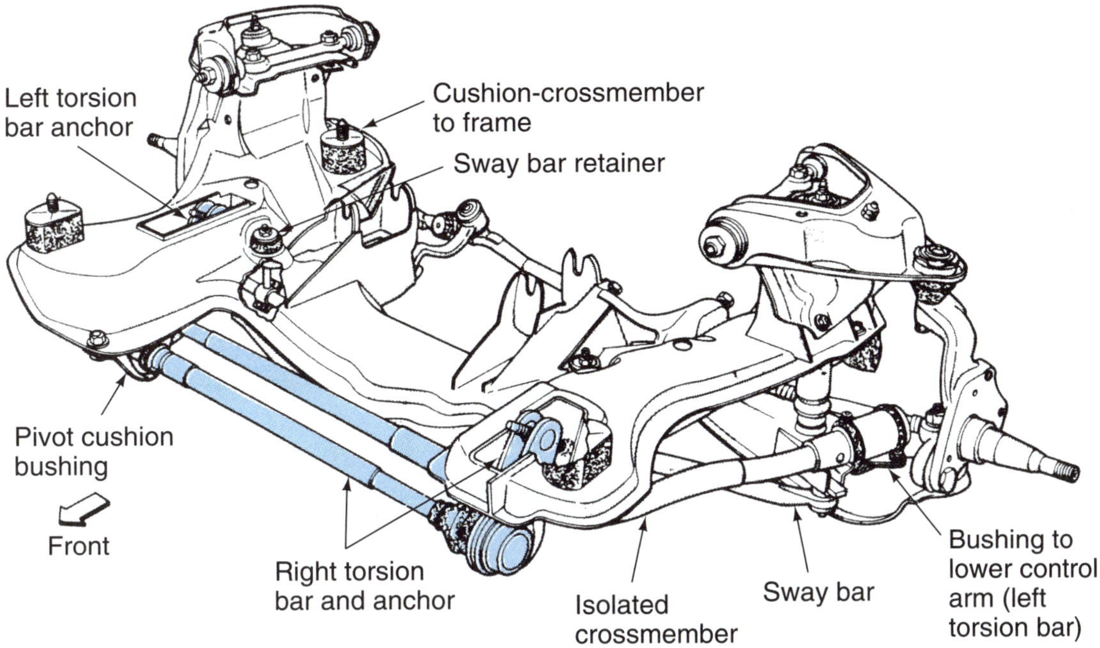


**Figure 6-48** Torsion bar anchor adjusting bolts (Courtesy of Chrysler Corporation)

reduced directional stability and excessive tire tread wear. Always follow the vehicle manufacturer's recommended procedure in the service manual for torsion bar removal and replacement. Following is a typical torsion bar removal and replacement procedure:

1. Lift the vehicle with a floor jack and place jack stands under the chassis. Lower the vehicle onto the jack stands and allow the front suspension to drop downward.
2. Loosen the torsion bar anchor adjusting bolts on both torsion bars (Figure 6-48).
3. Remove the anchor adjusting bolt on the torsion bar you wish to remove.
4. Install a floor jack under each lower control arm and raise these arms until the specified vehicle height is obtained between the chassis and the torsion bar bushing flange (Figure 6-49).
5. Remove the stabilizer bar attaching bolt, grommets, washer, and sleeve from the lower control arm.
6. Remove the two retaining bolts between the torsion bar bushing and the lower control arm.
7. Remove the two retaining bolts in the pivot cushion bushing.
8. Remove the torsion bar and anchor assembly from the crossmember.
9. Remove the anchor assembly from the end of the torsion bar.
10. Visually inspect the torsion bar bushing and pivot cushion bushing for cuts, tears, wear, and looseness. Inspect the anchor assembly for wear in the hex opening, and inspect the hex on the end of the torsion bar for wear (Figure 6-50). Make sure the torsion bar seal is in satisfactory condition. Replace all worn parts.
11. Place chassis lubricant on the torsion bar hex, and the hex opening in the anchor. Assemble the anchor on the end of the torsion bar so the ears are nearly straight upward with the torsion bar in a horizontal position. Reverse steps 1 through 8 to reinstall the torsion bar. Tighten all bolts to the specified torque. Adjust the torsion bar anchor bolts to obtain the specified curb riding height.

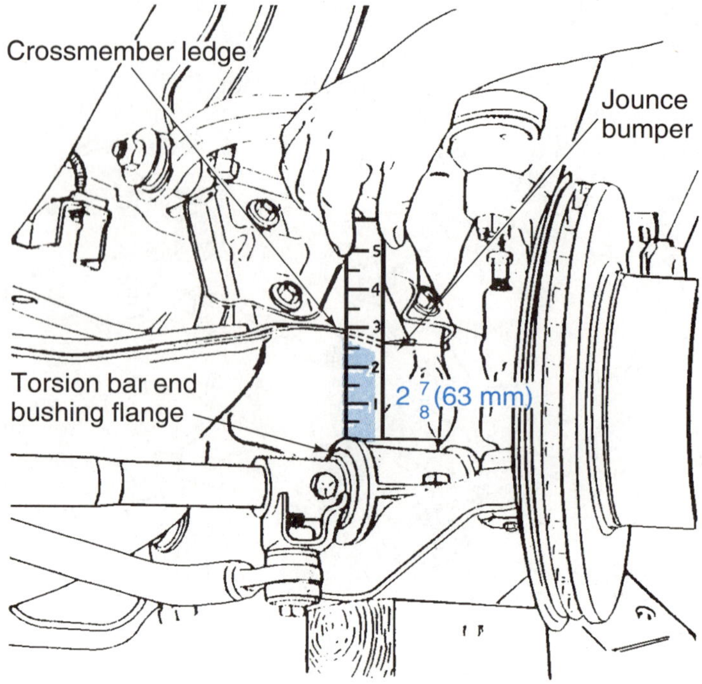


**Figure 6-49** Specified curb riding height between chassis and torsion bar bushing flange (Courtesy of Chrysler Corporation)

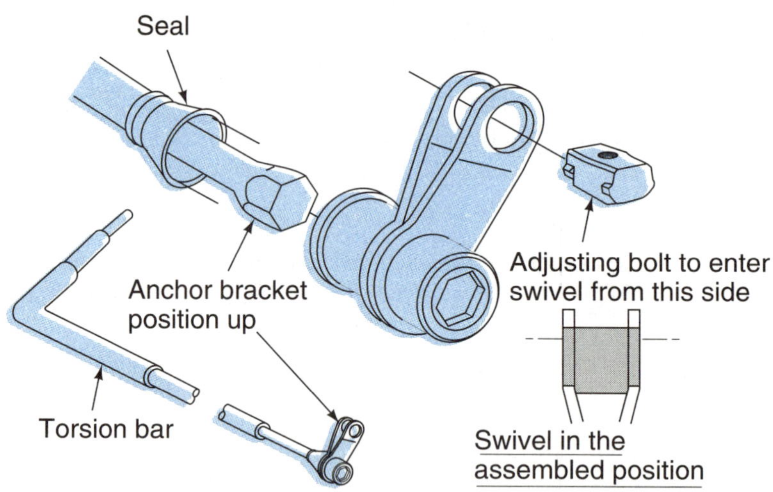


**Figure 6-50** Torsion bar and anchor (Courtesy of Chrysler Corporation)

# Guidelines for Servicing Front Suspension Systems

1. The curb riding height measurement location varies depending on the vehicle. Always measure this height at the manufacturer's recommended location shown in the service manual.
2. The curb riding height is critical because it affects most other front suspension alignment angles.

3. On a torsion bar front suspension, the torsion bars may be adjusted to correct improper curb riding height.
4. Some ball joints have wear indicators, and on these applications the grease fitting position indicates ball joint condition.
5. The spring tension must be unloaded from a ball joint before the joint wear is measured.
6. Steering knuckles should always be checked for wear and out-of-round in the ball joint and tie-rod end openings.
7. Ball joints may be pressed, threaded, bolted, or riveted to the control arm.
8. When installing a cotter pin in a ball joint or tie-rod end nut, never back off the nut to install the cotter pin, and always use a new cotter pin.
9. After replacement of front suspension components such as control arms, control arm bushings, springs, and strut rods, front suspension alignment should be checked.
10. Bent control arms or worn control arm bushings affect curb riding height and front suspension alignment angles.
11. A weak stabilizer bar or worn stabilizer bar bushings causes harsh riding, excessive body sway, and suspension noise.
12. Excessive wear on strikeout bumpers may be caused by improper curb riding height or worn-out shock absorbers.
13. Some control arm bushings require flaring after installation to secure the bushing.
14. Bent strut rods or worn strut rod bushings cause improper front suspension alignment and reduced directional stability.
15. Worn strut rod bushings or loose mounting bolts cause suspension noise.
16. Squeaking front leaf springs may be caused by worn-out silencers between the spring leafs.

## CASE STUDY

A customer complained about steering pull to the left while braking on a 1993 Oldsmobile Toronado. The technician questioned the customer about other symptoms, but the customer stated that the steering did not pull while driving, and the car had no other problems. Further questioning of the customer revealed that extensive brake work had been done in an attempt to correct the problem. The front brake pads had been replaced and the front brake rotors had been turned.

While performing a road test, the technician discovered the car had a definite pull to the left while braking, but the steering was normal while driving. During the road test, no other problems were evident. The technician removed the front wheels and brake rotors. A careful examination of the brake linings and rotor surfaces indicated these components were in excellent condition. The technician checked the rotor surfaces to be sure they were both in the same condition. The pistons in both front calipers moved freely. A check of the brake lines and hoses did not reveal any visible problems, and a pressure check at each front brake caliper indicated equal pressures at both front wheels during a brake application.

Next, the technician made a visual check of the steering and suspension components, and discovered a worn, loose left front strut rod bushing. This bushing was replaced, and the strut rod nut tightened to the specified torque. The strut rod was in satisfactory condition. A road test of the car indicated no steering pull during brake application.

## Terms to Know

Curb riding height
Caster angle
Camber angle
Spring sag
Steering effort
Directional stability
Ball joint wear indicator
Ball joint unloading
Ball joint axial movement
Ball joint radial movement
Stabilizer, or sway, bar
Strut, or radius, rod
Strikeout, or rebound, bumper

## ASE Style Review Questions

1. While discussing curb riding height:
   *Technician A* says worn control arm bushings reduce curb riding height.
   *Technician B* says incorrect curb riding height affects most other front suspension angles.
   Who is correct?
   **A.** A only
   **B.** B only
   **C.** Both A and B
   **D.** Neither A nor B

2. The shoulder on the ball joint grease fitting is inside the ball joint cover:
   *Technician A* says the ball joint should be replaced.
   *Technician B* says the ball joint is satisfactory.
   Who is correct?
   **A.** A only
   **B.** B only
   **C.** Both A and B
   **D.** Neither A nor B

3. While discussing ball joint unloading on a long and short arm suspension system with the coil springs between the lower control arm and the chassis:
   *Technician A* says a floor jack should be placed under the vehicle frame to unload the ball joints.
   *Technician B* says a floor jack should be placed under the lower control arm to unload the ball joints.
   Who is correct?
   **A.** A only
   **B.** B only
   **C.** Both A and B
   **D.** Neither A nor B

4. While discussing ball joint radial measurement:
   *Technician A* says the dial indicator should be positioned against the ball joint housing.
   *Technician B* says the front wheel bearing adjustment does not affect the ball joint radial measurement.
   Who is correct?
   **A.** A only
   **B.** B only
   **C.** Both A and B
   **D.** Neither A nor B

5. While discussing a ball joint that is threaded into the control arm:
   *Technician A* says if 100 foot-pounds of torque can be obtained, the control arm threads are satisfactory.
   *Technician B* says if 125 foot-pounds of torque cannot be obtained, the control arm threads are defective and control arm replacement is necessary.
   Who is correct?
   **A.** A only
   **B.** B only
   **C.** Both A and B
   **D.** Neither A nor B

6. While discussing ball joint installation:
   *Technician A* says the ball joint nut may be backed off to install the cotter key.
   *Technician B* says only the ball joint threads should extend above the opening in the steering knuckle.
   Who is correct?
   **A.** A only
   **B.** B only
   **C.** Both A and B
   **D.** Neither A nor B

7. While discussing strikeout bumpers:
*Technician A* says worn strikeout bumpers may be caused by sagged springs.
*Technician B* says worn strikeout bumpers may be caused by worn-out shock absorbers.
Who is correct?
**A.** A only
**B.** B only
**C.** Both A and B
**D.** Neither A nor B

8. While discussing control arm bushing replacement:
*Technician A* says when a bushing that is contained in a steel sleeve is replaced, press only on the outer steel sleeve.
*Technician B* says a spacer should be installed between the control arm support lugs during bushing removal and installation.
Who is correct?
**A.** A only
**B.** B only
**C.** Both A and B
**D.** Neither A nor B

9. While discussing stabilizer bars:
*Technician A* says a weak stabilizer bar causes excessive body sway when driving on irregular road surfaces.
*Technician B* says a weak stabilizer bar causes wear on the inside edges of the front tire treads.
Who is correct?
**A.** A only
**B.** B only
**C.** Both A and B
**D.** Neither A nor B

10. While discussing strut rods:
*Technician A* says worn strut rod grommets may affect some front suspension alignment angles.
*Technician B* says a bent strut rod on one side of a front suspension may cause the steering to pull to one side.
Who is correct?
**A.** A only
**B.** B only
**C.** Both A and B
**D.** Neither A nor B

# Table 6-1 ASE TASK

Diagnose front suspension system noises, body sway, and ride height problems; determine needed repairs.

| Problem Area | Symptoms | Possible Causes | Classroom Manual | Shop Manual |
|---|---|---|---|---|
| **NOISES** | Rattles on road irregularities | 1. Worn control arm bushings | 113 | 128 |
| | | 2. Worn stabilizer bar bushings | 117 | 129 |
| | | 3. Worn strut rod grommets | 111 | 130 |
| | | 4. Worn leaf spring shackles and bushings | 107 | 131 |
| | | 5. Worn torsion bars, anchors, and bushings | 106 | 132 |
| | | 6. Broken springs or spring insulators | 107 | 132 |
| **BODY SWAY** | Excessive sway on road irregularities | Weak stabilizer bar, worn bushings | 111 | 129 |
| **RIDE HEIGHT PROBLEMS** | Steering pulls to one side | Low ride height, one side of front suspension | 126 | 111 |
| | Reduced directional stability | Low ride height on front suspension | 124 | 111 |
| | Increased steering effort | Low ride height on rear suspension | 125 | 111 |
| | Rapid steering wheel return | Low ride height on rear suspension | 125 | 111 |

## Table 6-2 ASE TASK

Measure vehicle ride height; determine needed repairs.

| Problem Area | Symptoms | Possible Causes | Classroom Manual | Shop Manual |
|---|---|---|---|---|
| RIDE QUALITY | Harsh riding | Low ride height | 124 | 111 |
| DIRECTIONAL STABILITY | Steering pull | Low ride height one side of front suspension | 126 | 111 |
| STEERING QUALITY | Reduced directional stability | Low ride height on front suspension | 124 | 111 |
| | Increased steering effort | Low ride height on rear suspension | 125 | 111 |
| | Rapid steering wheel return | Low ride height on rear suspension | 125 | 111 |

## Table 6-3 ASE TASK

Inspect and replace upper and lower control arms, bushings, shafts, and rebound bumpers.

| Problem Area | Symptoms | Possible Causes | Classroom Manual | Shop Manual |
|---|---|---|---|---|
| NOISES | Rattles on road irregularities | 1. Worn control arm bushings or shafts | 113 | 123 |
| | | 2. Worn strut rod grommets | 111 | 130 |
| | | 3. Worn stabilizer bar bushings | 111 | 129 |
| | Squeaks on road irregularities | 1. Dry or worn control arm bushings | 113 | 123 |
| | | 2. Dry or worn stabilizer bushings | 111 | 129 |
| RIDE QUALITY | Harsh riding | Reduced curb riding height | 124 | 111 |
| STEERING QUALITY | Reduced directional stability | 1. Worn control arm bushings | 114 | 123 |
| | | 2. Reduced curb riding height | 124 | 111 |
| TIRE LIFE | Tire tread wear | Worn control arm bushings | 114 | 123 |

## Table 6-4 ASE TASK

Inspect, adjust, and replace strut rods/radius arm (compression/tension), and bushings.

| Problem Area | Symptoms | Possible Causes | Classroom Manual | Shop Manual |
|---|---|---|---|---|
| NOISE | Rattles on road irregularities | Worn strut rod grommets | 111 | 130 |
| STEERING QUALITY | Steering pull | 1. Worn strut rod bushings | 111 | 130 |
| | | 2. Bent strut rod | 111 | 130 |

## Table 6-5 ASE TASK

Inspect and replace upper and lower ball joints.

| Problem Area | Symptoms | Possible Causes | Classroom Manual | Shop Manual |
|---|---|---|---|---|
| STEERING QUALITY | Reduced directional stability | Worn ball joints | 108 | 121 |
| TIRE LIFE | Tire tread wear | Worn ball joints | 108 | 121 |

## Table 6-6 ASE TASK

Inspect and replace steering knuckle/spindle assemblies.

| Problem Area | Symptoms | Possible Causes | Classroom Manual | Shop Manual |
|---|---|---|---|---|
| STEERING QUALITY | Steering looseness | Worn tie-rod end opening in knuckle | 115 | 117 |
| | Reduced directional stability | 1. Worn ball joint stud opening in knuckle | 115 | 117 |
| | | 2. Bent steering knuckle | 115 | 117 |
| TIRE LIFE | Tire tread wear | Bent steering knuckle | 115 | 117 |

## Table 6-7 ASE TASK

Inspect and replace front suspension system coil springs and spring insulators (silencers).

| Problem Area | Symptoms | Possible Causes | Classroom Manual | Shop Manual |
|---|---|---|---|---|
| NOISE | Rattling on road irregularities | 1. Broken coil springs | 103 | 125 |
| | | 2. Worn or broken spring insulators | 103 | 125 |
| STEERING QUALITY | Reduced directional stability | Weak or broken coil springs | 103 | 125 |
| TIRE LIFE | Tire tread wear | Weak or broken coil springs | 103 | 125 |
| RIDE QUALITY | Harsh riding | Weak or broken spring | 103 | 125 |

## Table 6-8 ASE TASK

Inspect and replace front suspension system leaf spring(s), leaf spring insulators (silencers), shackles, brackets, bushings, and mounts.

| Problem Area | Symptoms | Possible Causes | Classroom Manual | Shop Manual |
|---|---|---|---|---|
| NOISE | Squeaking on road irregularities | Worn leaf spring silencers | 106 | 131 |
| | Rattling on road irregularities | 1. Broken spring leafs | 106 | 131 |
| | | 2. Worn shackles, bushings, mounts | 106 | 131 |
| STEERING QUALITY | Reduced directional stability | 1. Weak or broken spring | 106 | 131 |
| | | 2. Broken spring center bolt | 106 | 131 |
| TIRE LIFE | Tire tread wear | Weak or broken spring | 106 | 131 |
| RIDE QUALITY | Harsh riding | Weak or broken spring | 106 | 131 |

## Table 6-9 ASE TASK

Inspect, replace, and adjust front suspension system torsion bars; inspect mounts.

| Problem Area | Symptoms | Possible Causes | Classroom Manual | Shop Manual |
|---|---|---|---|---|
| RIDE QUALITY | Harsh riding | Reduced curb riding height | 121 | 132 |
| NOISE | Rattling on road irregularities | 1. Worn torsion bar pivot bushings | 121 | 132 |
| | | 2. Worn torsion bar hex and anchor | 121 | 132 |

## Table 6-10 ASE TASK

Inspect and replace stabilizer bar (sway bar) bushings, brackets, and links.

| Problem Area | Symptoms | Possible Causes | Classroom Manual | Shop Manual |
|---|---|---|---|---|
| RIDE QUALITY | Excessive body sway | 1. Weak or broken torsion bar | 117 | 129 |
| | | 2. Worn torsion bar bushings | 117 | 129 |
| NOISE | Rattling on road irregularities | Worn torsion bar bushings, links | 117 | 129 |

CHAPTER 7

# Rear Suspension Service

Upon completion and review of this chapter, you should be able to:

- ❑ Diagnose rear suspension noises.
- ❑ Diagnose rear suspension sway and lateral movement.
- ❑ Measure and correct rear suspension curb riding height.
- ❑ Remove and replace rear coil springs.
- ❑ Inspect rear springs, insulators, and seats.
- ❑ Inspect strut or shock absorber bushings and upper mount, replace strut cartridge.
- ❑ Remove, inspect, and replace lower control arms.
- ❑ Inspect, remove, and replace rear ball joints.
- ❑ Inspect, remove, and replace suspension adjustment links.
- ❑ Diagnose, remove, and replace rear leaf springs.
- ❑ Diagnose, remove, and replace stabilizer bars.
- ❑ Diagnose, remove, and replace track bars.
- ❑ Inspect, remove, and replace tie rods.

## Rear Suspension Diagnosis and Service

**Basic Tools**

Basic technician's tool set

Service manual

Machinist's rule

Floor jack

Jack stands

Transmission jack

Foot-pound torque wrench

Pry bar

### Noise Diagnosis

A squeaking noise in the rear suspension may be caused by a suspension bushing, or a defective strut or shock absorber. If a rattling noise occurs in the rear suspension, check these components:

1. worn or missing suspension bushings, such as control arm bushings, track bar bushings, stabilizer bar bushings, trailing arm bushings, and strut rod bushings
2. worn strut or shock absorber bushings or mounts
3. defective struts or shock absorbers
4. broken springs or worn spring insulators

### Sway and Lateral Movement Diagnosis

Excessive body sway or roll on road irregularities may be caused by a weak stabilizer bar or loose stabilizer bar bushings. If lateral movement is experienced on the rear of the chassis, the track bar or track bar bushings may be defective.

### Curb Riding Height Measurement

**SERVICE TIP:** Proper curb riding height must be maintained to provide normal steering quality and tire wear.

Regular inspection and proper maintenance of suspension systems is extremely important to maintaining vehicle safety. The curb riding height is determined mainly by spring condition. Other suspension components, such as control arm bushings, affect curb riding height if they are worn. *Since incorrect curb riding height affects most of the other suspension angles, this measurement is critical.* Reduced rear suspension height increases steering effort and causes rapid steering wheel return after turning a corner. Harsh riding occurs when the curb riding height is less than specified. The curb riding height must be measured at the vehicle manufacturer's specified location, which varies depending on the type of vehicle and suspension system. On some vehicles, the curb riding height on the rear suspension is measured from the floor to the center of the strut rod mounting bolt when the vehicle is on a level floor or an alignment rack (Figure 7-1). Photo Sequence 5 shows a typical procedure for measuring front and rear curb riding height.

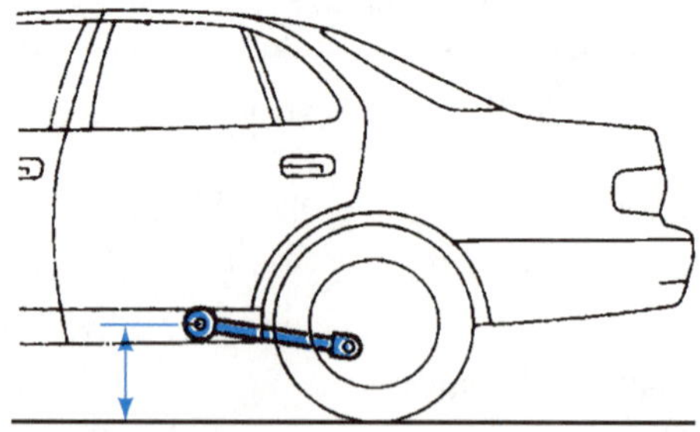


**Figure 7-1** Curb riding height measurement, rear suspension (Courtesy of Toyota Motor Corporation)

## Rear Strut, Coil Spring, and Upper Mount Service and Diagnosis

**WARNING:** When the rear coil springs are mounted on the struts, never loosen the upper strut nut until all the spring tension is removed from the upper support with a spring compressor. If this nut is removed with spring tension on the upper support, the spring tension turns the spring and the upper support into a dangerous projectile, which may cause serious personal injury and/or property damage.

Weak springs cause harsh riding and reduced curb riding height. Broken springs or spring insulators cause a rattling noise while driving on road irregularities. Worn-out struts or shock absorbers cause excessive chassis oscillations and harsh riding. (Refer to Chapter 5 for strut and shock absorber service.) Loose or worn strut or shock absorber bushings cause a rattling noise on road irregularities. The rear coil spring removal and replacement procedure varies depending on the type of rear suspension system. Always follow the vehicle manufacturer's recommended procedure in the service manual. Following is a typical rear strut and spring removal and replacement procedure on a MacPherson strut independent rear suspension system with the coil springs mounted on the struts:

1. Remove the rear seat and the package trim tray (Figure 7-2).
2. Remove the wheel cover and loosen the wheel nuts.
3. Lift the vehicle with a floor jack, and lower the chassis onto jack stands so the rear suspension is allowed to drop downward.
4. Place a wooden block between the floor jack and the rear spindle on the side where the strut and spring removal is taking place. Raise the floor jack to support some of the suspension system weight (Figure 7-3).
5. Remove the rear wheel, disconnect the nut from the small spring in the lower arm (Figure 7-4), and remove the brake hose and antilock brake system (ABS) wire from the strut (Figure 7-5).
6. Remove the stabilizer bar link from the strut (Figure 7-6), and loosen the strut-to-spindle mounting bolts (Figure 7-7).
7. Remove three upper support nuts under the package tray trim (Figure 7-8), and lower the floor jack to remove the strut from the knuckle (Figure 7-9). Remove the strut from the chassis.

**SERVICE TIP:** If the plastic coating on a coil spring is chipped, the spring may break prematurely. The spring may be taped in the compressing tool contact areas to prevent chipping.

## Photo Sequence 5
# Typical Procedure for Measuring Front and Rear Curb Riding Height

**P5-1** Check the trunk for extra weight.

**P5-2** Check the tires for normal inflation pressure.

**P5-3** Park the car on a level shop floor or alignment track.

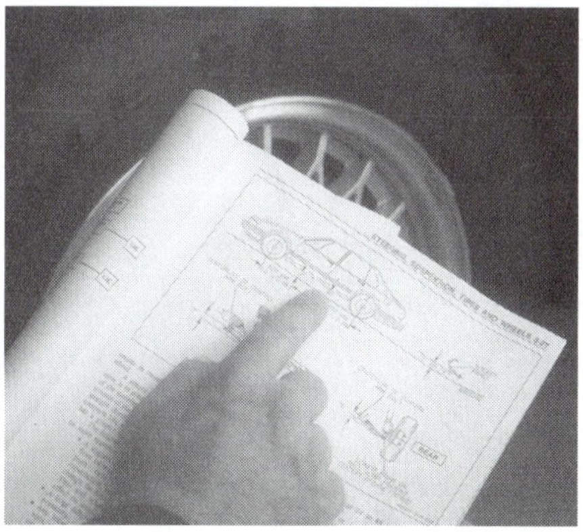

**P5-4** Find the vehicle manufacturer's specified curb riding height measurement locations in the service manual.

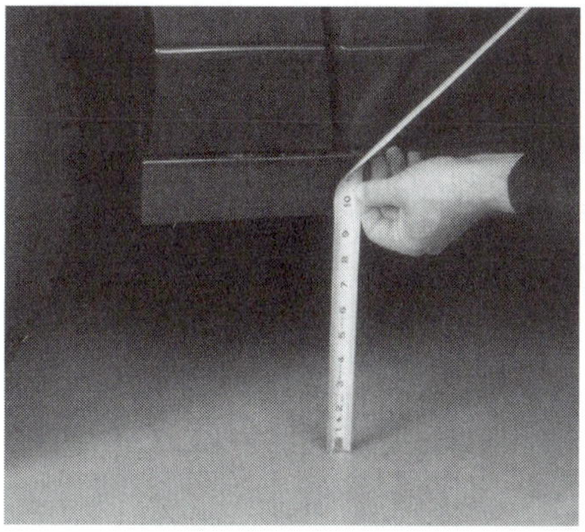

**P5-5** Measure and record the right front curb riding height.

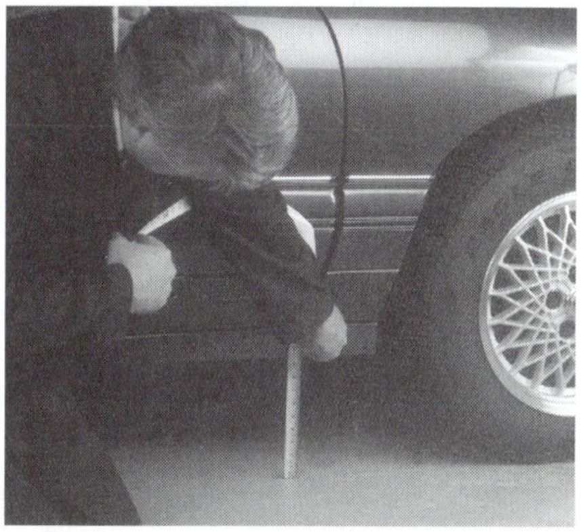

**P5-6** Measure and record the left front curb riding height.

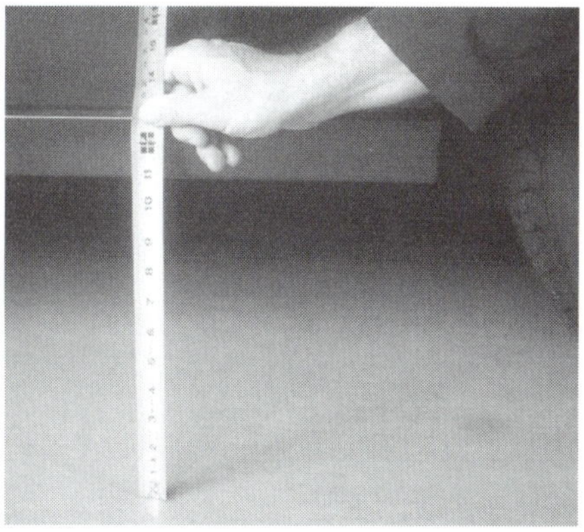

**P5-7** Measure and record the right rear curb riding height.

**P5-8** Measure and record the left rear curb riding height.

**P5-9** Compare the measurement results to the specified curb riding height in the service manual.

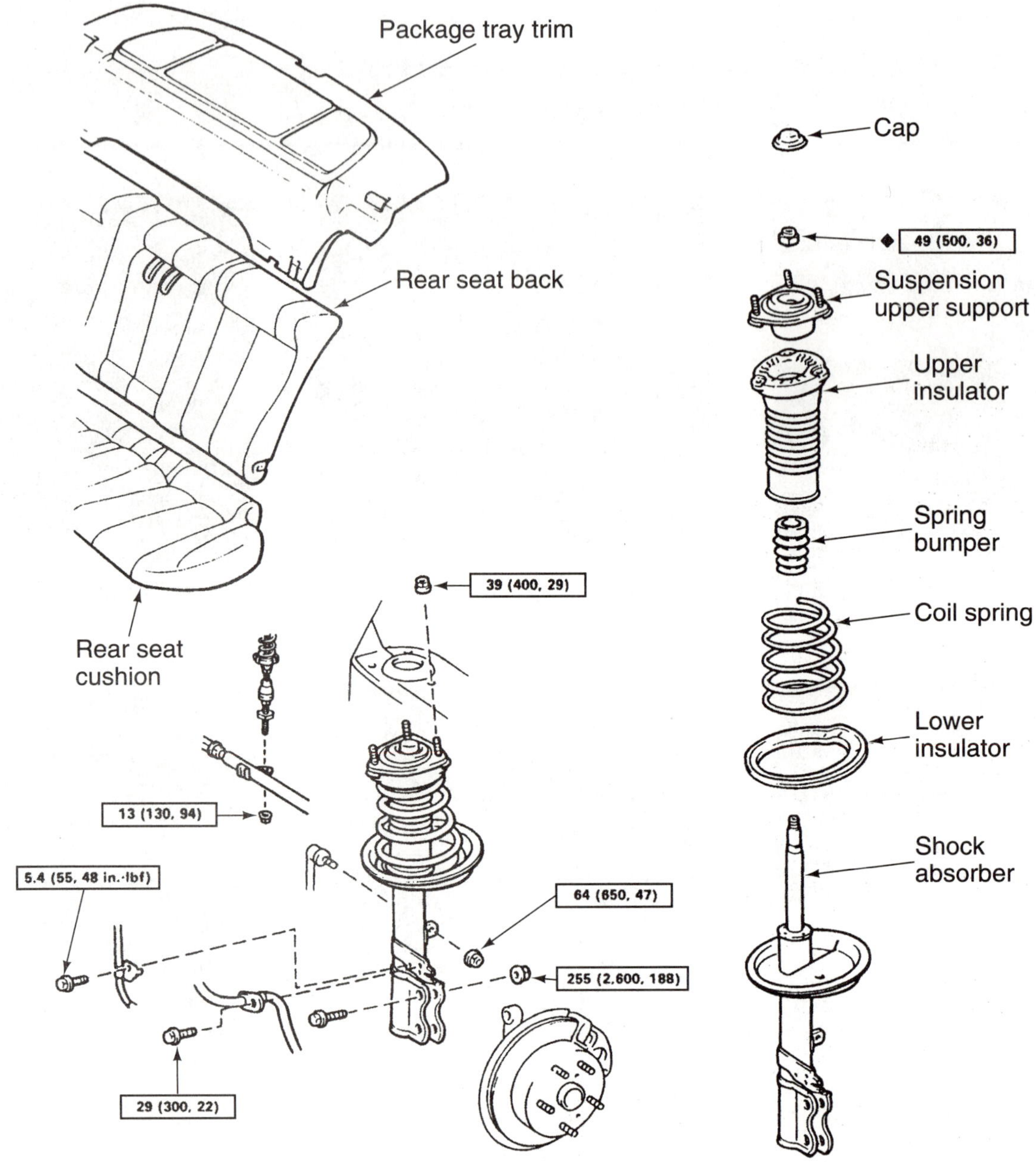


**Figure 7-2** Rear suspension with rear seat and package trim tray (Courtesy of Toyota Motor Corporation)

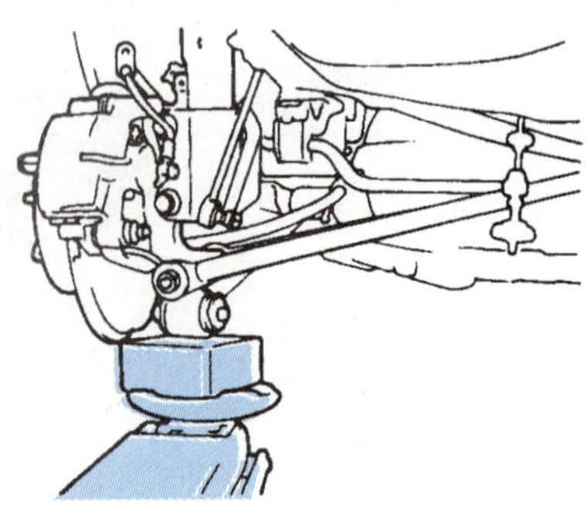

**Figure 7-3** Floor jack supporting some of the rear suspension weight (Courtesy of Toyota Motor Corporation)

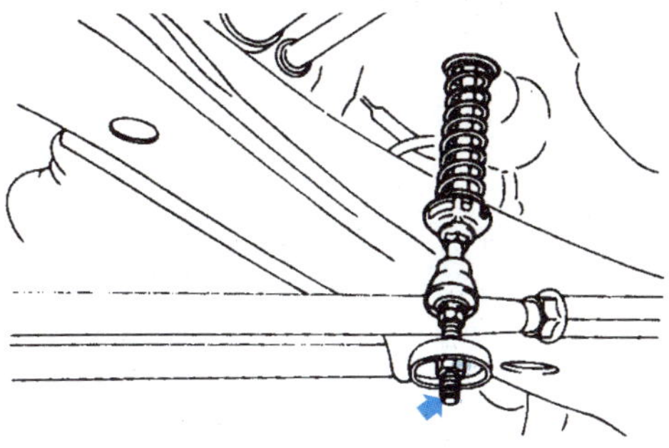

**Figure 7-4** Disconnecting nut from small spring in lower arm (Courtesy of Toyota Motor Corporation)

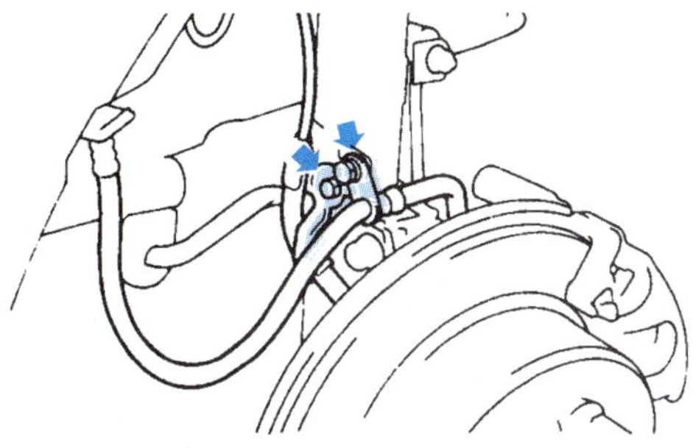

**Figure 7-5** Disconnecting brake hose and ABS wire from rear strut (Courtesy of Toyota Motor Corporation)

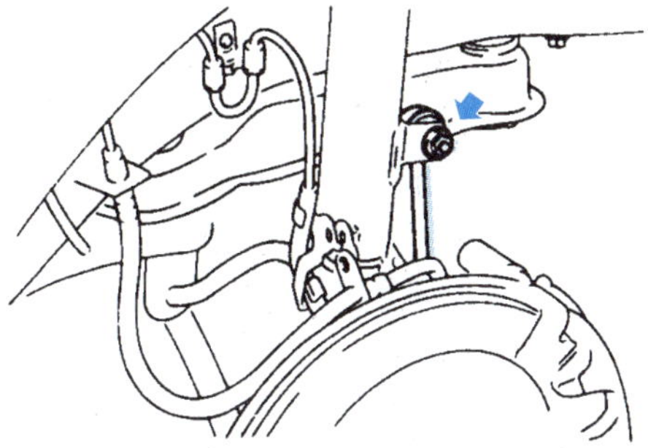

**Figure 7-6** Removing stabilizer bar from rear strut (Courtesy of Toyota Motor Corporation)

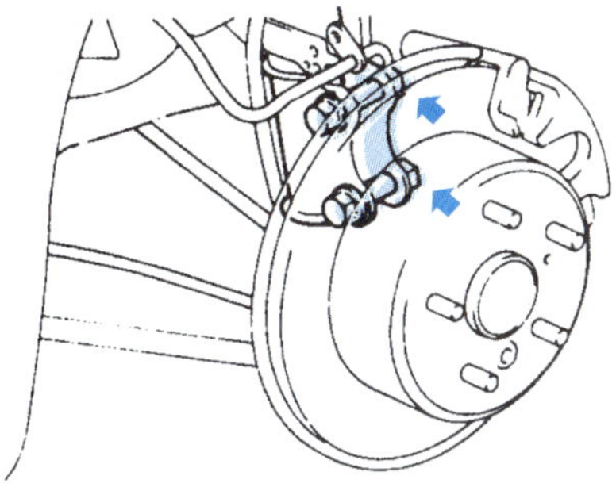

**Figure 7-7** Loosening strut-to-rear spindle bolts (Courtesy of Toyota Motor Corporation)

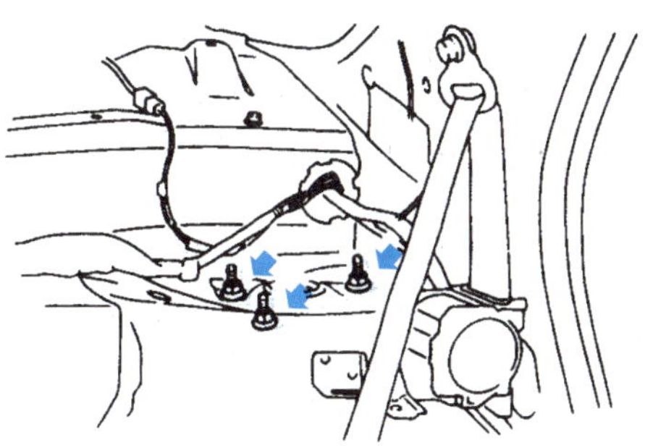

**Figure 7-8** Removing upper mount nuts (Courtesy of Toyota Motor Corporation)

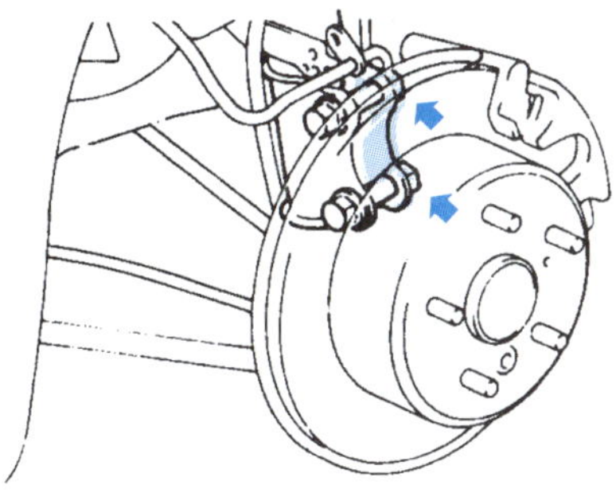

**Figure 7-9** Removing strut from rear spindle (Courtesy of Toyota Motor Corporation)

**8.** Following the car or equipment manufacturer's recommended procedure, install a spring compressor on the coil spring, and tighten the spring compressor until all the spring tension is removed from the upper support (Figure 7-10).

**9.** Install a bolt in the upper strut-to-spindle bolt hole and tighten two nuts on the end of this bolt. One nut must be on each side of the strut bracket (Figure 7-11). Clamp this bolt in a vise to hold the strut, coil spring, and spring compressor.

**10.** Using the special tool to hold the spring and strut from turning, loosen the nut from the strut rod nut (Figure 7-12).

**11.** Remove the strut rod nut, upper support, and upper insulator.

**12.** Remove the strut from the lower end of the spring.

**13.** If the spring is to be replaced, rotate the compressor bolt until all the spring tension is removed from the compressing tool, and then remove the spring from the tool.

**Special Tools**

Coil-spring compressor

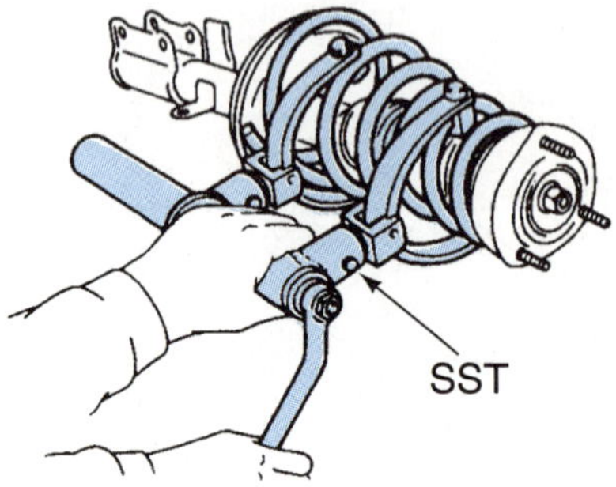


**Figure 7-10** Spring compressor installed on rear spring (Courtesy of Toyota Motor Corporation)

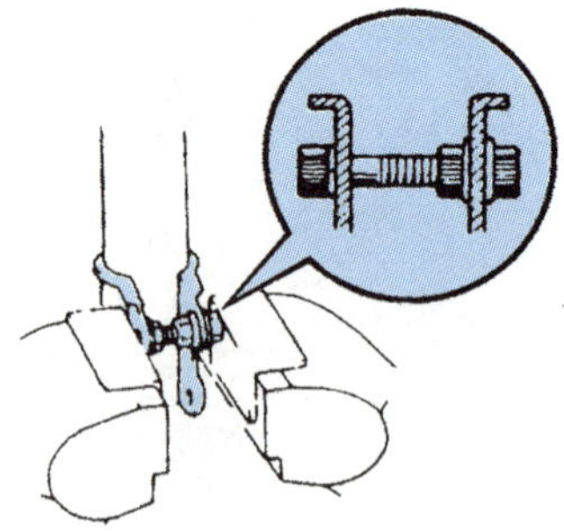

**Figure 7-11** Strut, coil spring, and spring compressor retained in a vise with a bolt and two nuts in upper strut-to-spindle bolt hole (Courtesy of Toyota Motor Corporation)

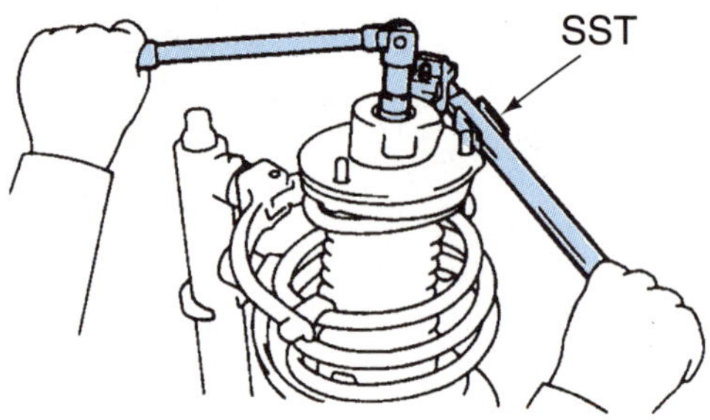


**Figure 7-12** Loosening strut rod nut (Courtesy of Toyota Motor Corporation)

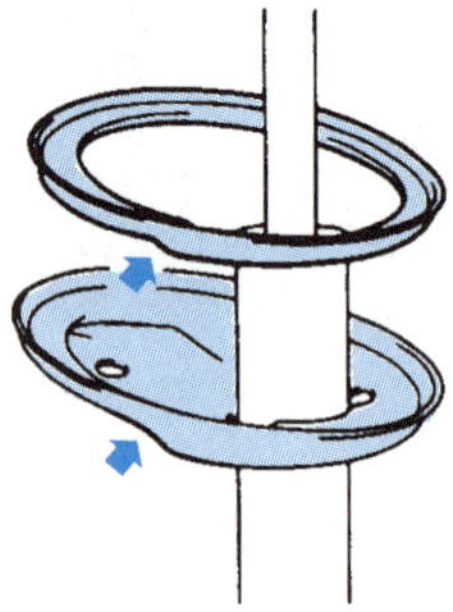

**Figure 7-13** Inspecting lower spring seat and insulator (Courtesy of Toyota Motor Corporation)

**14.** Inspect the lower insulator and spring seat on the strut (Figure 7-13). If the spring seat is warped or damaged, replace the strut. A new cartridge may be installed in some rear struts. (The strut cartridge replacement procedure is explained in Chapter 5.)

**15.** Visually inspect the upper mount, insulator, and spring bumper. If any of these components are damaged, worn, or distorted, replacement is necessary. Worn upper mounts and insulators, or damaged spring seats cause suspension noise while driving on road irregularities. Assemble the spring bumper, spring, upper mount, and insulator (Figure 7-14), and reverse steps 1 through 13 to reinstall the coil spring and strut. Tighten all fasteners to the specified torque.

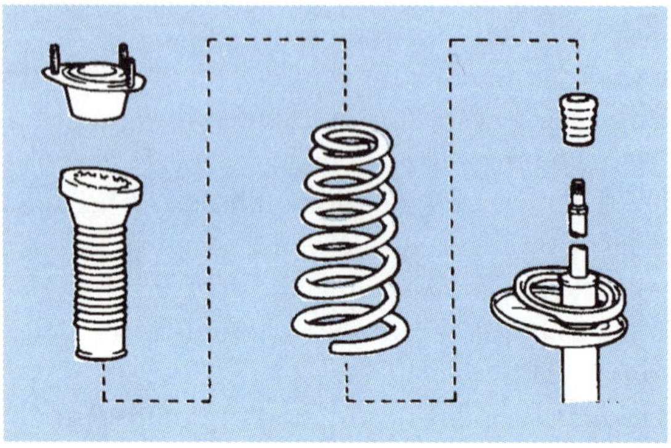

**Figure 7-14** Assembly of rear strut, spring bumper, spring, upper mount, and insulator (Courtesy of Toyota Motor Corporation)

**CUSTOMER CARE:** When talking to customers, always remember the two Ps, pleasant and polite. There may be many days when we do not feel like being pleasant and polite. Perhaps we have several problem vehicles in the shop with symptoms that are difficult to diagnose and correct. Some service work may be behind schedule, and customers may be irate because their vehicles are not ready on time. However, we should always remain pleasant and polite with customers. Our attitude does much to make the customer feel better and to realize their business is appreciated. A customer may not feel very happy about an expensive repair bill, but a pleasant attitude on our part may help to improve the customer's feelings. When the two Ps are remembered by service personnel, customer relations are enhanced, and the customer will return to the shop. Conversely, if service personnel have a grouchy, indifferent attitude, the customer may be turned off, and take his or her business to another shop.

**Classroom Manual**
Chapter 7, page 136

# Lower Control Arm and Ball Joint Diagnosis and Replacement

Worn bushings on the lower control arms may cause incorrect rear wheel camber or toe, which results in rear tire wear and steering pull. Bent lower control arms must be replaced. When ball joints with wear indicators are in normal condition there is 0.050 in (1.27 mm) between the grease nipple shoulder and the cover. If the ball joint is worn, the grease nipple shoulder is flush with or inside the cover (Figure 7-15). A worn ball joint causes improper rear wheel toe and/or camber, which may result in tire tread wear or steering pull.

Directional stability refers to the tendency of the vehicle steering to remain in the straight-ahead position when driven straight ahead on a reasonably smooth, level road surface.

The lower control arm removal and replacement procedure varies depending on the vehicle and the type of suspension. Always follow the vehicle manufacturer's recommended procedure in the service manual. Following is a typical lower control arm removal procedure:

1. Lift the vehicle on a hoist with the chassis supported on the hoist and the control arms dropped downward. The vehicle may be lifted with a floor jack, and the chassis supported on jack stands.
2. Remove the wheel and tire assembly.
3. Remove the stabilizer bar from the knuckle bracket.
4. Remove the parking brake cable retaining clip from the lower control arm.
5. If the car has electronic level control (ELC), disconnect the height sensor link from the control arm.

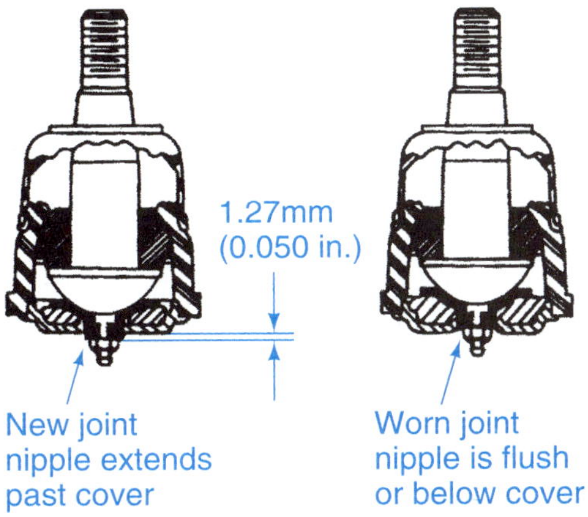


**Figure 7-15** Ball joint wear indicator (Courtesy Oldsmobile Division, General Motors Corporation)

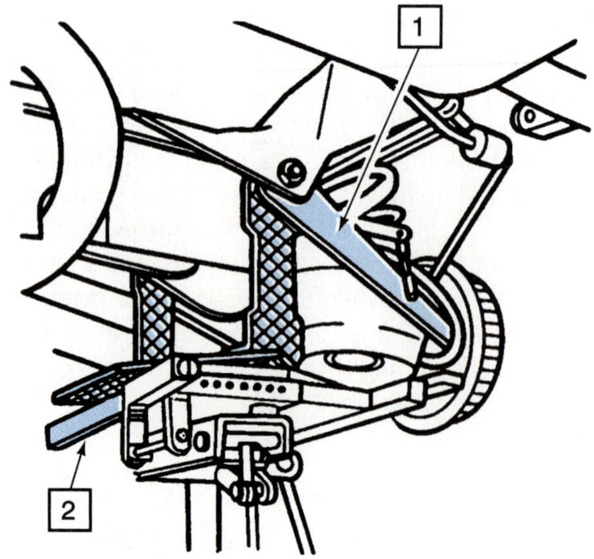


1 Control arm

2 Transmission jack

**Figure 7-16** Special tool installed to support the inner end of the control arm (Courtesy of Oldsmobile Division, General Motors Corporation)

**Special Tools**

Lower control arm support tool

6. Install a special tool to support the lower control arm in the bushing areas (Figure 7-16).
7. Place a transmission jack under the special tool and raise the jack enough to remove the tension from the control arm bushing retaining bolts. If the car was lifted with a floor jack and supported on jack stands, place a floor jack under the special tool.
8. Place a safety chain through the coil spring and around the lower control arm.
9. Remove the bolt from the rear control arm bushing.
10. Be sure the jack is raised enough to relieve the tension on the front bolt in the lower control arm and remove this bolt.
11. Lower the jack slowly and allow the control arm to pivot downward. When all the tension is released from the coil spring, remove the safety chain, coil spring, and insulators.

Check the coil spring for distortion and proper free length. If the spring has a vinyl coating, check this coating for scratches and nicks. Check the spring insulators for cracks and wear. After the coil spring is removed, follow these steps to remove the lower control arm:

1. Remove the nut on the inner end of the suspension adjustment link, and disconnect this link from the lower control arm (Figure 7-17).
2. Remove the cotter key from the ball joint nut, and loosen, but do not remove, the nut from the ball joint stud.
3. Use a special ball joint removal tool to loosen the ball joint in the knuckle.
4. Remove the ball joint nut and the lower control arm (Figure 7-18).

**Special Tools**

Ball joint stud removal tool

Ball joint pressing tool

Check the lower control arm for bends, distortion, and worn bushings. A special ball joint pressing tool is used to press the ball joint from the lower control arm (Figure 7-19), and the same pressing tool with different adapters is used to press the new ball joint into the control arm (Figure 7-20).

**CAUTION:** When disconnecting a linkage joint, such as the ends of the suspension adjustment link, do not use a wedge-type tool between the joint and the attached part. This method of joint removal may damage the joint seal.

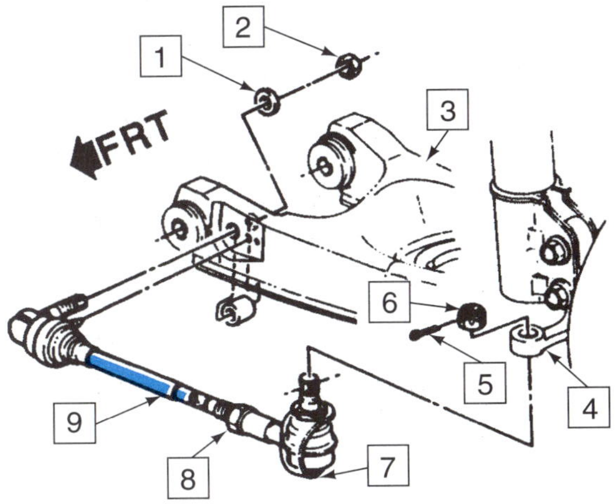


| | | | |
|---|---|---|---|
| 1 | Washer | 6 | Castellated nut 45 N·m (33 ft. lb.) 70 N·m (52 ft. lb.) max to install cotter pin |
| 2 | Retaining nut 85 N·m (63 ft. lb.) | 7 | Boot, left side only |
| 3 | Control arm | 8 | Nut 64 N·m (47 ft. lb.) |
| 4 | Knuckle | 9 | Suspension adjustment link |
| 5 | Cotter pin | 10 | Spacer |

**Figure 7-17** Suspension adjustment link removal from the lower control arm (Courtesy of Oldsmobile Division, General Motors Corporation)

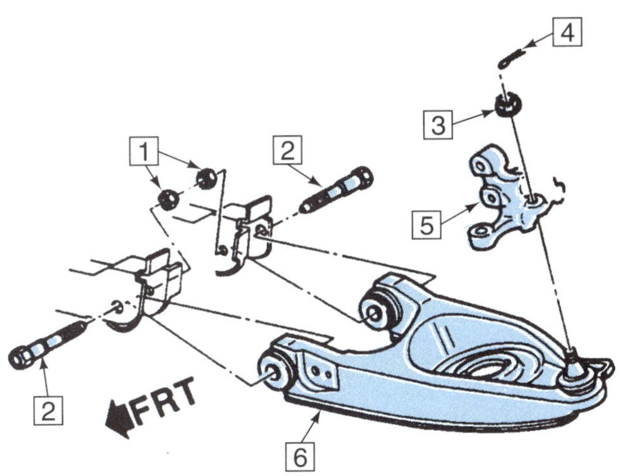


| | | | |
|---|---|---|---|
| 1 | Pivot nuts 115 N·m (85 ft. lb.) | 4 | Cotter pin |
| 2 | Pivot bolts 187 N·m (138 ft. lb.) | 5 | Knuckle |
| 3 | Castellated nut | 6 | Control arm |

**Figure 7-18** Removing lower control arm and ball joint (Courtesy of Oldsmobile Division, General Motors Corporation)

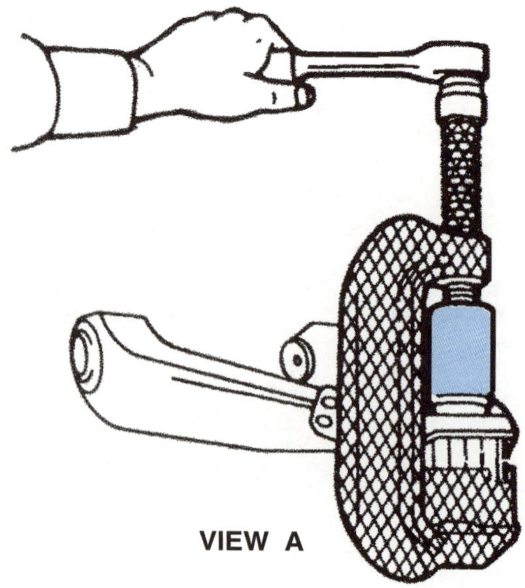


**Figure 7-19** Ball joint removal from lower control arm (Courtesy of Oldsmobile Division, General Motors Corporation)

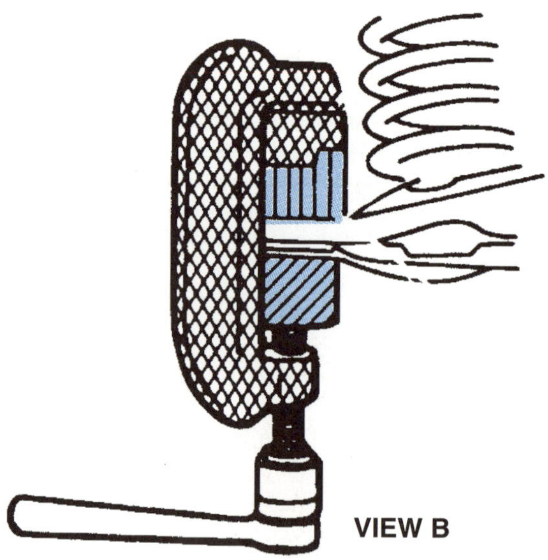


**Figure 7-20** Ball joint installation in lower control arm (Courtesy of Oldsmobile Division, General Motors Corporation)

**Special Tools**

Suspension adjustment link removal tool

Remove the nut and cotter key on the outer end of the suspension adjustment link, and use a special puller to remove this link from the knuckle (Figure 7-21). Remove the suspension adjustment link and inspect the joints. If the joints are loose or the seals are damaged, joint replacement is necessary. The joint studs must fit snugly in the knuckle and lower control arm openings. When the joint studs are worn, joint replacement is necessary, and worn stud openings in the knuckle or lower control arm require component replacement.

Follow this procedure to install the lower control arm and spring:

**CAUTION:** Never loosen a ball joint nut to install a cotter key, because this action causes improper torquing of the nut.

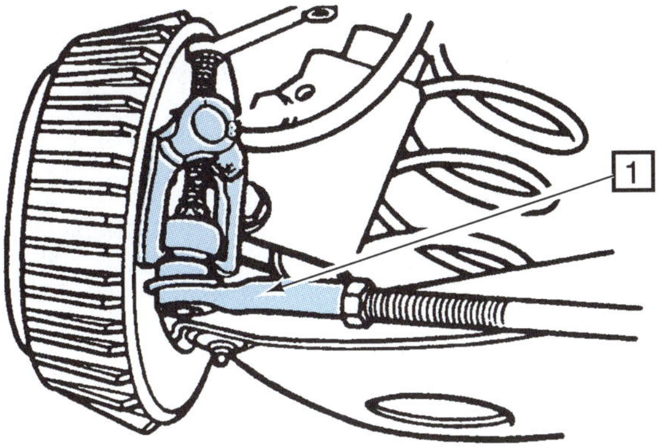


1 Outer suspension adjustment link

Figure 7-21 Suspension adjustment link removal from the knuckle (Courtesy of Oldsmobile Division, General Motors Corporation)

1. Install the ball joint stud in the knuckle and install the nut on the ball joint stud. Tighten the ball joint nut to the specified torque, and then tighten the nut an additional 2/3 turn. If necessary, tighten the nut slightly to align the nut castellations with the cotter key hole in the ball joint stud, and install the cotter key.
2. Snap the upper insulator on the coil spring. Install the lower spring insulator and the spring in the lower control arm (Figure 7-22).

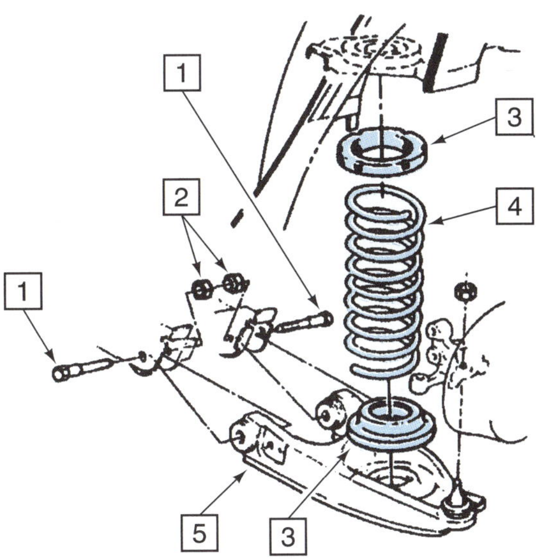


1 Pivot bolt 187 N·m (138 ft. lb.)

2 Pivot nut 115 N·m (85 ft. lb.)

3 Insulator

4 Coil spring

5 Control arm

Figure 7-22 Installation of spring and insulators (Courtesy of Oldsmobile Division, General Motors Corporation)

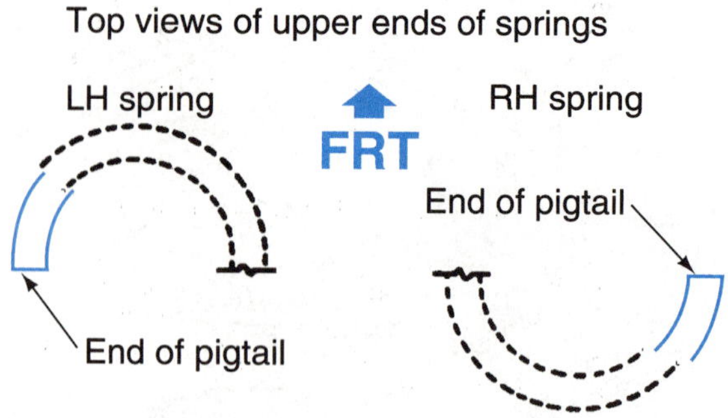


**Figure 7-23** Proper location of upper coil spring ends in relation to the front of the vehicle (Courtesy of Oldsmobile Division, General Motors Corporation)

3. Be sure the top of the coil spring is properly positioned in relation to the front of the vehicle (Figure 7-23).
4. Install the special tool on the inner ends of the control arm, and place the transmission jack under the special tool.
5. Slowly raise the transmission jack until the control arm bushing openings are aligned with the openings in the chassis.

**CAUTION:** The pivot bolts and nuts in the inner ends of the lower control arm must be tightened to the specified torque with the vehicle weight supported on the wheels, and the suspension at normal curb height. Failure to follow this procedure may adversely affect ride quality and steering characteristics.

6. Install the bolts and nuts in the inner ends of the control arm. Do not torque these bolts and nuts at this time.
7. Install the stabilizer-bar-to-knuckle bracket and tighten the fasteners to the specified torque.
8. Install the parking brake retaining clip.
9. If the vehicle has ELC, install the height sensor link, and tighten the fastener to the specified torque.
10. Install the suspension adjustment link and tighten the retaining nuts to the specified torque. Install cotter keys as required.
11. Remove the transmission jack, and install the wheel and tire assembly.
12. Lower the vehicle onto the floor and tighten the wheel hub nuts and lower control arm bolts and nuts to the specified torque.

## Rear Leaf-Spring Diagnosis and Replacement

This leaf-spring discussion applies to multiple-leaf springs on rear suspension systems that have two springs mounted longitudinally in relation to the chassis. Many leaf springs have plastic silencers between the spring leafs. If these silencers are worn out, creaking and squawking noises are heard when the vehicle is driven over road irregularities at low speeds.

When the silencers require checking or replacement, lift the vehicle with a floor jack and support the frame on jack stands so the rear suspension moves downward. With the vehicle weight no longer applied to the springs, the spring leafs may be pried apart with a pry bar to remove and replace the silencers.

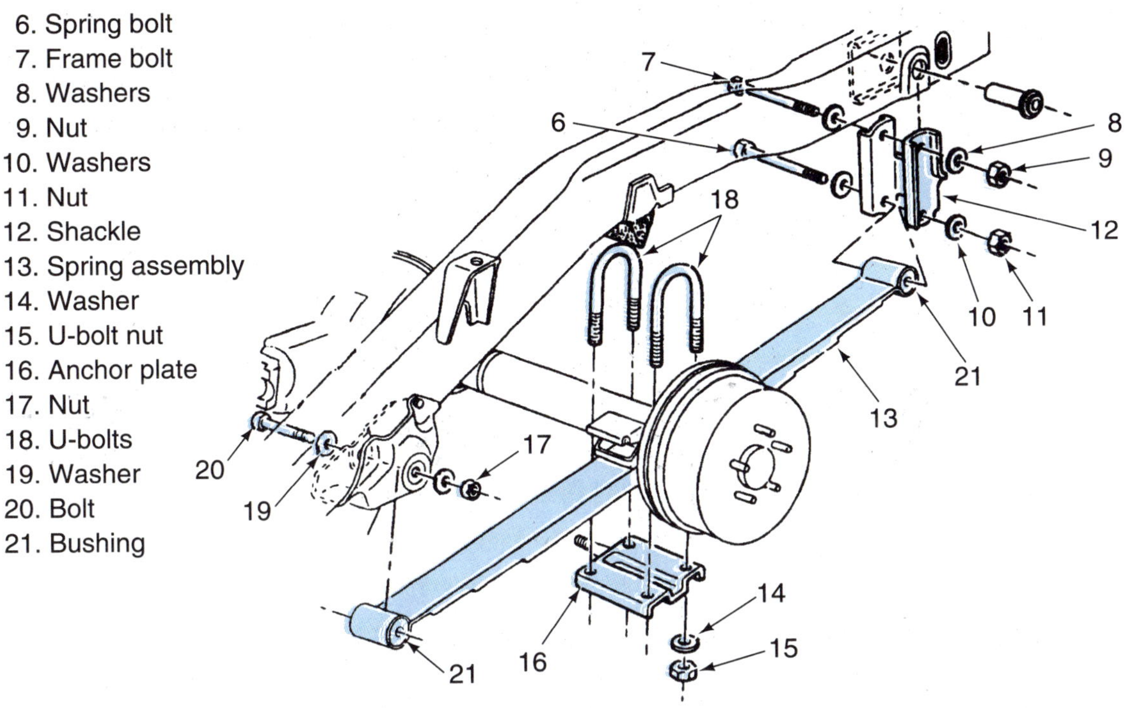


**Figure 7-24** Rear leaf-spring removal (Courtesy of Chevrolet Motor Division, General Motors Corporation)

Worn shackle bushings, brackets, and mounts cause excessive chassis lateral movement and rattling noises. With the normal vehicle weight resting on the springs, insert a pry bar between the rear outer end of the spring and the frame. Apply downward pressure on the bar and observe the rear shackle for movement. Shackle bushings, brackets, or mounts must be replaced if there is movement in the shackle. The same procedure may be followed to check the front bushing in the main leaf. A broken spring center bolt may allow the rear axle assembly to move rearward on one side. This movement changes rear wheel tracking, which results in handling problems, tire wear, and reduced directional stability. Sagged rear springs reduce the curb riding height. Spring replacement is necessary if the springs are sagged. When rear leaf-spring replacement is necessary, proceed as follows:

1. Lift the vehicle with a floor jack and place jack stands under the frame. Lower the vehicle weight onto the jack stands, and leave the floor jack under the differential housing to support the rear suspension weight.
2. Remove the nuts from the spring U-bolts, and remove the U-bolts and lower spring plate (Figure 7-24). The spring plate may be left on the rear shock absorber and moved out of the way.
3. Be sure the floor jack is lowered sufficiently to relieve the vehicle weight from the rear springs.
4. Remove the rear shackle nuts, plate, shackle, and bushings.
5. Remove the front spring mounting bolt and remove the spring from the chassis. Check the spring center bolt to be sure it is not broken.
6. Check the front hanger, bushing, and bolt, and replace as necessary.
7. Check the rear shackle, bushings, plate, and mount, and replace the worn components.
8. Reverse steps 1 through 5 to install the spring. Tighten all bolts and nuts to the specified torque.

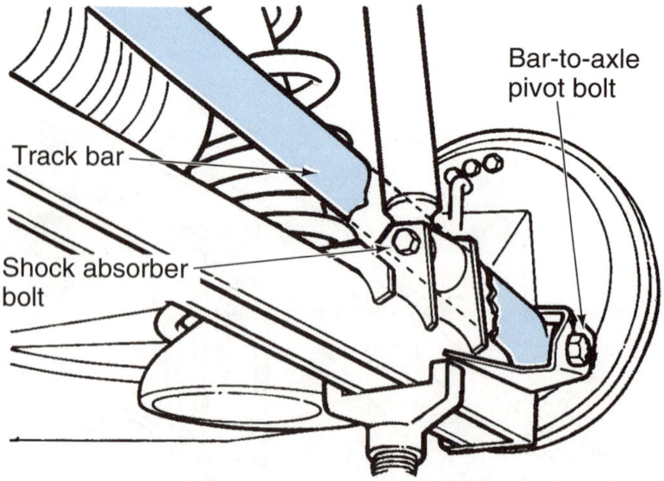


**Figure 7-25** Checking track bar bushings (Courtesy of Chrysler Corporation)

## Track Bar Diagnosis and Replacement

Some rear suspension systems have a track bar to control lateral chassis movement. Rubber mounting bushings insulate the track bar from the chassis components. Worn track bar mounts and bushings may cause rattling and excessive lateral chassis movement.

When the track bar is checked, lift the vehicle on a hoist or floor jack, with the rear suspension in the normal riding height position. If the vehicle is lifted with a floor jack, place safety stands under the rear axle to support the vehicle weight. Grasp the track bar firmly and apply vertical and horizontal force. If there is movement in the track bar mountings, track bar, or bushing, replacement is essential (Figure 7-25).

Another track bar checking method is to leave the vehicle on the shop floor and observe the track bar mounts as an assistant applies side force to the chassis or rear bumper. If there is lateral movement in the track bar bushings or brackets, replace the bushings and check the bracket bolts. Bent track bars must be replaced.

When the track bar is replaced, remove the mounting bolts, bushings, grommets, and track bar. Inspect the mounting bolt holes in the chassis for wear. After the track bar is installed with the proper grommets and bushings, tighten the mounting bolts to the specified torque.

## Stabilizer Bar Diagnosis and Service

A stabilizer bar may be referred to as a sway bar.

Worn stabilizer bar mounting bushings, grommets, or mounting bolts cause a rattling noise as the vehicle is driven on irregular road surfaces. A weak stabilizer bar, or worn bushings and grommets cause harsh riding and excessive body sway while driving on irregular road surfaces. Worn or very dry stabilizer bar bushings may cause a squeaking noise on irregular road surfaces. All stabilizer bar components should be visually inspected for wear. Stabilizer bar removal and replacement procedures vary depending on the vehicle. Always follow the vehicle manufacturer's recommended procedure in the service manual. Following is a typical rear stabilizer bar removal and replacement procedure:

1. Lift the vehicle on a hoist and allow both sides of the rear suspension to drop downward as the vehicle chassis is supported on the hoist.
2. Remove the mounting bolts at the outer ends of the stabilizer bar and remove the bushings, grommets, brackets, or spacers (Figure 7-26).

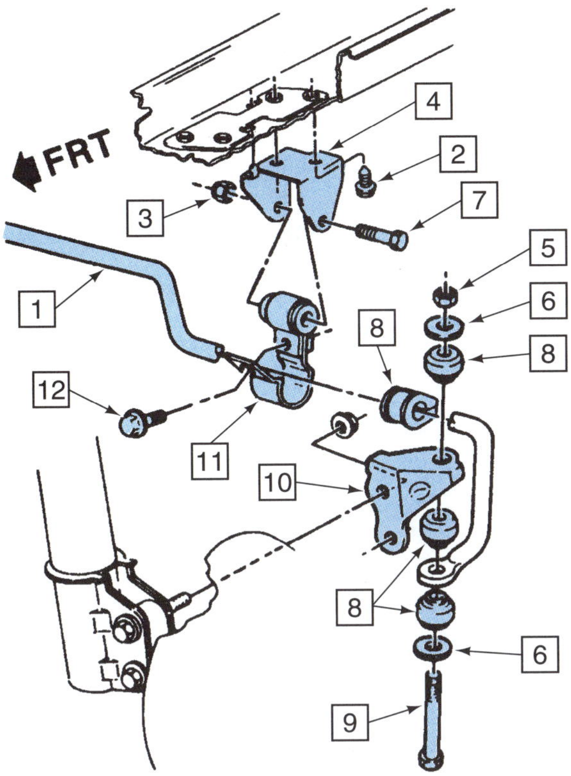


| | | | |
|---|---|---|---|
| 1 | Stabilizer shaft | 7 | Bolt 50 N·m (37 ft. lb.) |
| 2 | Bolt 18 N·m (14 ft. lb.) | 8 | Insulator |
| 3 | Nut 47 N·m (35 ft. lb.) | 9 | Bolt |
| 4 | Bracket | 10 | Stabilizer shaft bracket |
| 5 | Nut 18 N·m (14 ft. lb.) | 11 | Link |
| 6 | Retainer | 12 | Bolt 50 N·m (37 ft. lb.) |

**Figure 7-26** Stabilizer bar, bushings, grommets, spacers, and brackets (Courtesy of Oldsmobile Division, General Motors Corporation)

3. Remove the mounting bolts in the center area of the stabilizer bar.
4. Remove the stabilizer bar from the chassis.
5. Visually inspect all stabilizer bar components such as bushings, bolts, and spacer sleeves. Replace the stabilizer bar, grommets, bushings, brackets, or spacers as required. Split bushings may be removed over the stabilizer bar. Bushings that are not split must be pulled from the bar.
6. Reverse steps 2 through 4 to install the stabilizer bar. Make sure all stabilizer bar components are installed in the original position, and tighten all fasteners to the specified torque.

# Rear Suspension Tie Rod Inspection and Replacement

Rear tie rods should be inspected for worn grommets, loose mountings, and bent conditions. Loose tie rod bushings or a bent tie rod will change the rear wheel tracking and result in reduced directional stability. Worn tie rod bushings also cause a rattling noise on road irregularities. When

**Classroom Manual**
Chapter 7, page 138

the rear tie rod is replaced, remove the front and rear rod mounting nuts. The lower control arm or rear axle may have to be pried rearward to remove the tie rod. Check the tie rod grommets and mountings for wear, and replace parts as required. When the tie rod is reinstalled, tighten the mounting bolts to specifications, and check the rear wheel toe.

## Guidelines for Servicing Rear Suspension Systems

1. A squeaking noise may be caused by a suspension bushing, or a defective strut or shock absorber.
2. A rattling noise may be caused by worn suspension bushings, loose shock absorber or strut bushings or mounts, broken springs or spring insulators, or defective struts or shock absorbers.
3. Excessive body sway, or roll, when one wheel strikes a road irregularity may be caused by a defective stabilizer bar or bushings.
4. Excessive lateral movement of the rear chassis may be caused by a defective track bar or bushings.
5. Reduced curb riding height causes harsh riding.
6. Reduced curb riding height may be caused by sagged springs, worn control arm bushings, or bent control arms.
7. Sagged rear springs cause excessive steering effort and rapid steering wheel return after a turn.
8. When the coil spring is mounted on the rear strut, never loosen the upper strut nut until the spring is compressed with a spring compressing tool.
9. If a coil spring has a vinyl coating, the spring should be taped in the compressing tool contact areas to prevent chipping the coating.
10. On some independent rear suspension systems, a ball joint connects the lower control arm to the knuckle. Many of these ball joints have a conventional wear indicator and a press fit in the control arm.
11. When some lower control arms are installed, the control arm bolts should be tightened with the normal vehicle weight supported on the tires.
12. Suspension adjustment link studs should not be loosened with a wedge-type tool.
13. A nut should never be loosened to install a cotter key.
14. A broken leaf-spring center bolt may allow the rear axle to move, which results in improper tracking and reduced directional stability.

### CASE STUDY

A customer complained about steering pull to the left on a Chevrolet Celebrity. The customer also said the problem had just occurred in the last few days. The technician road tested the car and found the customer's description of the complaint to be accurate. A careful inspection of the tires indicated there was no abnormal tire wear on the front or rear tires. The vehicle was inspected for recent collision damage, but there was no evidence of this type of damage. After lifting the vehicle on a hoist, the technician checked all the front suspension components, including ball joints, control arms, control arm bushings, and wheel bearings. However, none of these front suspension components indicated any sign of wear or looseness.

Realizing that improper rear wheel tracking causes steering pull, the technician inspected the rear suspension. A pry bar was used to apply downward and rearward force to the trailing arms on the rear suspension. When this action was taken on the right rear trailing arm, the technician discovered the trailing arm bushing was very loose. This defect had allowed the right side of the rear axle to move rearward a considerable amount, which explained why the steering pulled to the left.

After installing a new trailing arm bushing, the alignment was checked on all four wheels, and the wheel alignment was within specifications. A road test indicated no evidence of steering pull or other steering problems.

## Terms to Know

Body sway
Lateral movement
Directional stability
Curb riding height
Wear indicator
Suspension adjustment link
Spring silencers

## ASE Style Review Questions

1. While discussing lateral movement of the rear chassis:
*Technician A* says this problem may be caused by loose stabilizer bar bushings.
*Technician B* says this problem may be caused by loose track bar bushings.
Who is correct?
**A.** A only
**B.** B only
**C.** Both A and B
**D.** Neither A nor B

2. While discussing excessive rear body sway, or roll, while driving on road irregularities:
*Technician A* says this problem may be caused by worn spring insulators.
*Technician B* says this problem may be the result of sagged rear springs.
Who is correct?
**A.** A only
**B.** B only
**C.** Both A and B
**D.** Neither A nor B

3. While discussing curb riding height:
*Technician A* says reduced rear curb riding height causes harsh riding.
*Technician B* says reduced rear curb riding height increases steering effort.
Who is correct?
**A.** A only
**B.** B only
**C.** Both A and B
**D.** Neither A nor B

4. While discussing curb riding height:
*Technician A* says the curb riding height is measured at the same location on each vehicle.
*Technician B* says the vehicle should be placed on a level floor before checking curb riding height.
Who is correct?
**A.** A only
**B.** B only
**C.** Both A and B
**D.** Neither A nor B

5. While discussing spring removal on a MacPherson strut rear suspension:
*Technician A* says all the spring tension must be removed from the upper mount before the upper strut nut is loosened.
*Technician B* says a vinyl-coated spring should be taped in the compressing tool contact area.
Who is correct?
**A.** A only
**B.** B only
**C.** Both A and B
**D.** Neither A nor B

6. While discussing rear suspension lower control arms and ball joints:
*Technician A* says some rear suspension ball joints have wear indicators.
*Technician B* says worn lower control arm bushings may cause rear tire wear.
Who is correct?
**A.** A only
**B.** B only
**C.** Both A and B
**D.** Neither A nor B

7. While discussing rear suspension lower control arm service:
*Technician A* says on some rear suspensions the vehicle weight must be resting on the tires when the inner bolts are tightened to the specified torque.
*Technician B* says on some rear suspensions the ball joint is pressed into the lower control arm.
Who is correct?
**A.** A only
**B.** B only
**C.** Both A and B
**D.** Neither A nor B

8. While discussing rear suspension adjustment links:
*Technician A* says the suspension adjustment link is used to adjust rear wheel camber.
*Technician B* says ends of this link may be loosened with a wedge-type tool.
Who is correct?
**A.** A only
**B.** B only
**C.** Both A and B
**D.** Neither A nor B

9. While discussing longitudinally mounted rear leaf springs:
*Technician A* says the improper vehicle tracking may be caused by a broken spring center bolt.
*Technician B* says worn shackle bushings may cause excessive lateral movement of the rear chassis.
Who is correct?
**A.** A only
**B.** B only
**C.** Both A and B
**D.** Neither A nor B

10. While discussing rear wheel tracking:
*Technician A* says a worn bent tie rod could result in improper rear wheel tracking.
*Technician B* says improper tracking occurs when both rear springs are sagged the same amount.
Who is correct?
**A.** A only
**B.** B only
**C.** Both A and B
**D.** Neither A nor B

## Table 7-1 ASE TASK

Diagnose suspension system noises, body sway, and ride height problems; determine needed repairs.

| Problem Area | Symptoms | Possible Causes | Classroom Manual | Shop Manual |
|---|---|---|---|---|
| **NOISE** | Rattling on road irregularities | 1. Worn suspension bushings, grommets | 139 | 143 |
| | | 2. Defective struts, shock absorbers | 138 | 143 |
| | Squeaking on road irregularities | Worn, dry suspension bushings, grommets | 139 | 143 |
| **RIDE QUALITY** | Excessive body sway | Defective stabilizer or bushings | 136 | 143 |
| | Harsh riding | Insufficient ride height | 147 | 143 |
| **STEERING QUALITY** | Excessive steering effort | Insufficient rear ride height | 147 | 143 |

## Table 7-2 ASE TASK

Inspect and replace rear suspension system coil springs and spring insulators (silencers).

| Problem Area | Symptoms | Possible Causes | Classroom Manual | Shop Manual |
|---|---|---|---|---|
| NOISE | Rattling on road irregularities | Broken springs or spring insulators | 135 | 144 |
| STEERING QUALITY | Excessive steering effort | Sagged rear springs | 147 | 144 |

## Table 7-3 ASE TASK

Inspect and replace rear suspension system transverse links (track bars), control arms, stabilizer bars (sway bars), bushings, and mounts.

| Problem Area | Symptoms | Possible Causes | Classroom Manual | Shop Manual |
|---|---|---|---|---|
| NOISE | Rattling on road irregularities | Worn bushings in track bar, stabilizer bar, or control arms | 136 | 156 |
| RIDE QUALITY | Excessive body sway | Worn stabilizer bushings, loose brackets | 136 | 156 |
| | Excessive lateral movement, rear chassis | Worn track bar bushings, loose brackets | 136 | 156 |
| STEERING QUALITY | Steering pulls to one side | Bent lower control arm, worn bushings | 140 | 149 |
| TIRE WEAR | Excessive tread wear | Bent lower control arm, worn bushings | 140 | 149 |

## Table 7-4 ASE TASK

Inspect and replace rear suspension system leaf spring(s), leaf-spring insulators (silencers), shackles, brackets, bushings, and mounts.

| Problem Area | Symptoms | Possible Causes | Classroom Manual | Shop Manual |
|---|---|---|---|---|
| NOISE | Squeaking on road irregularities | Worn silencers | 132 | 154 |
| | Rattling on road irregularities | Worn bushings, shackles, brackets | 132 | 155 |
| RIDE QUALITY | Excessive body sway | Worn bushings, shackles, brackets | 132 | 155 |
| | Harsh riding | Reduced ride height | 132 | 155 |

## Table 7-5 ASE TASK

Inspect and replace rear MacPherson strut cartridge or assembly, and upper mount assembly.

| Problem Area | Symptoms | Possible Causes | Classroom Manual | Shop Manual |
|---|---|---|---|---|
| NOISE | Rattling on road irregularities | 1. Loose strut bushings, upper mount | 136 | 148 |
| | | 2. Defective struts | 136 | 148 |
| RIDE QUALITY | Harsh riding, excessive chassis oscillations | Defective struts | 136 | 144 |

## Table 7-6 ASE TASK

Inspect and replace rear ball joints and tie rod assemblies.

| Problem Area | Symptoms | Possible Causes | Classroom Manual | Shop Manual |
|---|---|---|---|---|
| NOISE | Rattling on road irregularities | Worn tie rod bushings | 140 | 152 |
| STEERING QUALITY | Steering pull | 1. Worn tie rod bushings | 140 | 152 |
| | | 2. Bent tie rod | 140 | 152 |
| TIRE LIFE | Tread wear | Worn ball joint | 140 | 152 |

CHAPTER 8

# Computer-Controlled Suspension System Service

Upon completion and review of this chapter, you should be able to:

- ❑ Diagnose electronic air suspension systems.
- ❑ Remove, replace, and inflate air springs.
- ❑ Adjust front and rear trim height on electronic air suspension systems.
- ❑ Service and repair nylon air lines.
- ❑ Adjust trim height on a rear load-leveling air suspension system.
- ❑ Diagnose rear load-leveling air suspension systems.
- ❑ Diagnose programmed ride control (PRC) systems.
- ❑ Diagnose computer command ride (CCR) systems.
- ❑ Diagnose automatic air suspension systems.
- ❑ Diagnose air suspension systems with rear shock damping control systems.
- ❑ Diagnose road-sensing suspension systems.

## Electronic Air Suspension Service and Diagnosis

**Basic Tools**

Basic technician's tool set

Service manual

Floor jack

Jack stands

Utility knife

Machinist's ruler, or tape measure

### Diagnostic Procedure

If the air suspension warning lamp is illuminated with the engine running, the control module has detected a defect in the electronic air suspension system. The electronic air suspension diagnostic and service procedures vary depending on the vehicle. Always follow the vehicle manufacturer's recommended procedures in the service manual. The following diagnostic and service procedures apply to electronic air suspension systems on some Lincoln Continentals and Mark VIIs. When the air suspension warning lamp indicates a system defect, the diagnostic procedure may be entered as follows:

1. Be sure the air suspension system switch is turned on.
2. Turn the ignition switch on for 5 seconds and then turn it off. Leave the driver's door open and the other doors closed.
3. Ground the diagnostic lead located near the control module, and close the driver's door with the window down.
4. Turn the ignition switch on. The warning lamp should blink continuously at 1.8 times per second to indicate that the system is in the diagnostic mode.

There are 10 tests in the diagnostic procedure. The control module switches from one test to the next when the driver's door is opened and closed. The first three tests in the diagnostic procedure are the following:

1. rear suspension
2. right front suspension
3. left front suspension

During these three tests, each suspension location should be raised for 30 seconds, lowered for 30 seconds, and raised for 30 seconds. For example, in test 2 this procedure should be followed on the right front suspension. If the expected signal or an illegal signal is received during the test procedure, the test stops and the air suspension warning lamp is illuminated. If all the signals and commands are normal during the first three tests, the warning lamp continues to flash at 1.8 times per second.

While tests 4 through 10 are being performed, the air suspension warning lamp flashes the test number. For example, during test 4 the warning lamp flashes 4 times followed by a pause and

4 more flashes. This flash sequence continues while test 4 is completed. The driver's door must be opened and closed to move to the next test. During tests 4 through 10, the technician must listen to or observe various components to detect abnormal operation. The warning lamp only indicates the test number during these tests. Actions performed by the control module during tests 4 through 10 are as follows:

4. The compressor is cycled on and off at 0.25 cycle per second. This action is limited to 50 cycles.
5. The vent solenoid is opened and closed at 1 cycle per second.
6. The left front air valve is opened and closed at 1 cycle per second, and the vent solenoid is opened. When this occurs, the left front corner of the vehicle should drop slowly.
7. The right front air valve is opened and closed at 1 cycle per second, and the vent solenoid is opened. This action causes the right front corner of the vehicle to drop slowly.
8. During this test, the right rear air valve is opened and closed at 1 cycle per second and the vent valve is opened. This action should cause the right rear corner of the vehicle to drop slowly.
9. The left rear solenoid is opened and closed at 1 cycle per second and the vent valve is opened, which should cause the left rear corner of the vehicle to drop slowly.
10. Return the module from the diagnostic mode to normal operation by disconnecting the diagnostic lead from ground. This mode change also occurs if the ignition switch is turned off, or when the brake pedal is depressed.

If defects are found during the test sequence, specific electrical tests may be performed on air valve or vent valve windings and connecting wires to locate the problem.

## Air Spring Removal and Installation

**CAUTION:** The system control switch must be in the off position when system components are serviced.

**WARNING:** The system control switch must be turned off prior to hoisting, jacking, or towing the vehicle. If the front of the chassis is lifted with a bumper jack, the rear suspension moves downward. The electronic air suspension system will attempt to restore the rear trim height to normal, and this action may cause the front of the chassis to fall off the bumper jack, resulting in personal injury or vehicle damage.

**WARNING:** When air spring valves are being removed, always rotate the valve to first stage until all the air escapes from the air spring. Never turn the valve to the second, release, stage until all the air is released from the spring.

Many components in an electronic air suspension system, such as control arms, shock absorbers, and stabilizer bars, are diagnosed and serviced in the same way as the components in a conventional suspension system. However, the air spring service procedures are different compared to coil-spring service procedures on a conventional suspension system.

Follow these steps for air spring removal:

1. Turn off the electronic air suspension switch in the trunk.
2. Hoist the vehicle and allow the suspension to drop downward, or lift the vehicle with a floor jack and place jack stands under the chassis. Lower the vehicle onto the jack stands and allow the suspension to drop downward.
3. Disconnect the nylon air line from the spring solenoid valve, and rotate the valve to the first stage to allow the air to escape from the spring. Never turn the valve to the second stage until all the air is exhausted from the spring.
4. Disconnect the lower spring retainer, and remove the spring from the chassis.

5. Before an air spring is installed, it must be properly folded over the piston at the bottom of the membrane (Figure 8-1).
6. Install the spring in the chassis and connect the lower spring retainer. Be sure the top of the spring is properly seated in the spring seat. When an air spring is installed in the front or rear suspension, the spring must be properly positioned to eliminate folds and creases in the membrane (Figure 8-2).

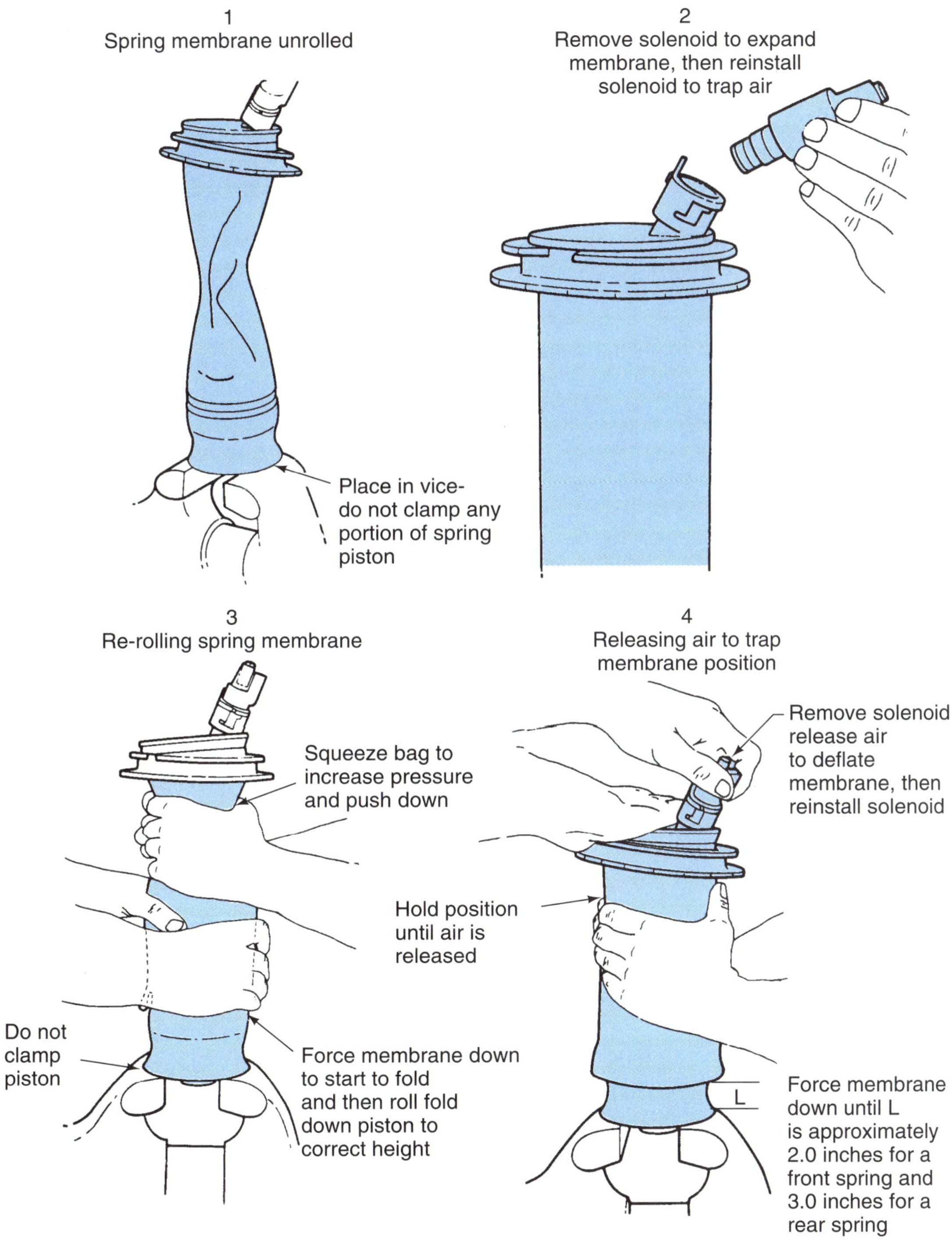


**Figure 8-1** Air spring folding (Courtesy of Ford Motor Company)

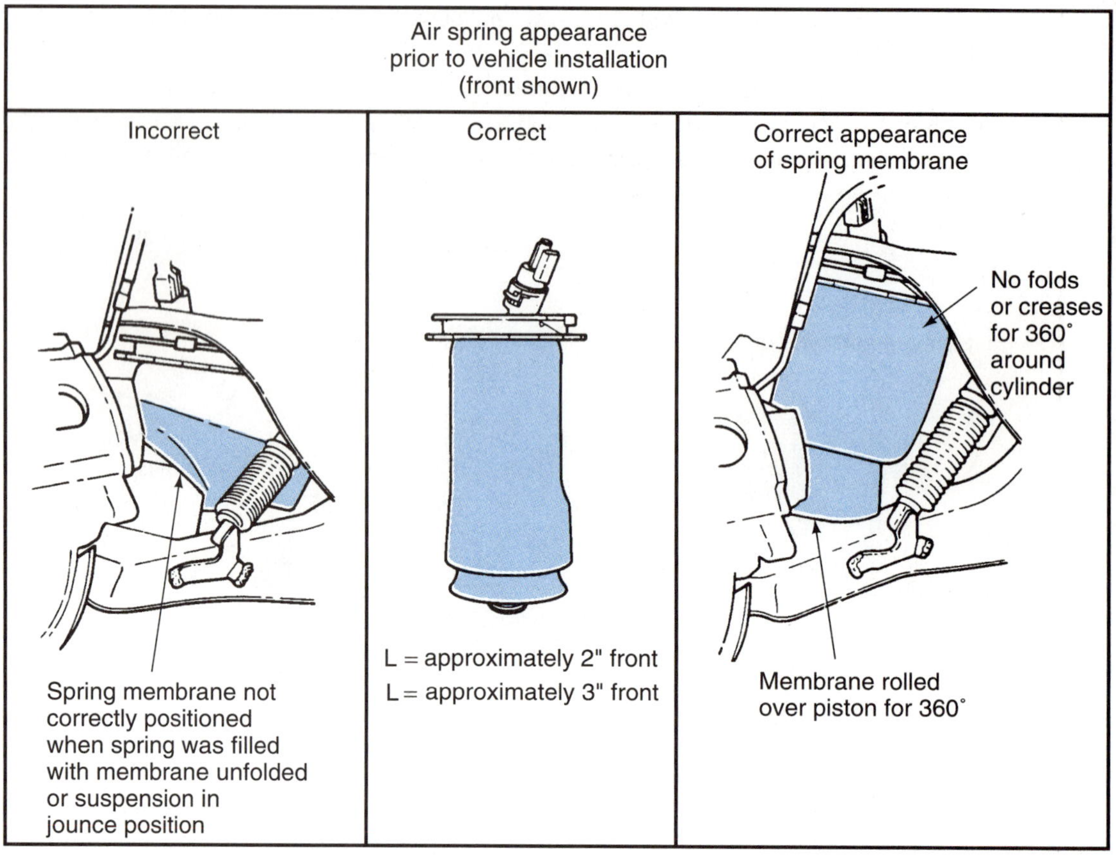


**Figure 8-2** When an air spring is installed in the front or rear suspension, the spring must be properly positioned to eliminate folds and creases in the membrane. (Courtesy of Ford Motor Company)

## Air Spring Inflation

**WARNING:** Do not allow the suspension to compress an air spring until the air spring is inflated. This action may damage the air spring.

The weight of the vehicle must not be allowed to compress an uninflated air spring. When an air spring is being inflated, use this procedure:

1. With the vehicle chassis supported on a hoist, lower the hoist until a slight load is placed on the suspension. Do not lower the hoist until the suspension is heavily loaded.
2. Turn on the air suspension system switch.
3. Turn the ignition switch from off to run for 5 seconds with the driver's door open and the other doors shut. Turn the ignition switch off.
4. Ground the diagnostic lead.
5. Apply the brake pedal and turn the ignition to the run position. The warning lamp will flash every 2 seconds to indicate the fill mode.
6. To fill a rear spring, or springs, close and open the driver's door once. After a 6-second delay, the rear spring will be filled for 60 seconds.
7. To fill a front spring, or springs, close and open the driver's door twice. After a 6-second delay the front spring will be filled for 60 seconds.
8. When front and rear springs require filling, fill the rear springs first. Once the rear springs are filled, close and open the driver's door once to begin filling the front springs.

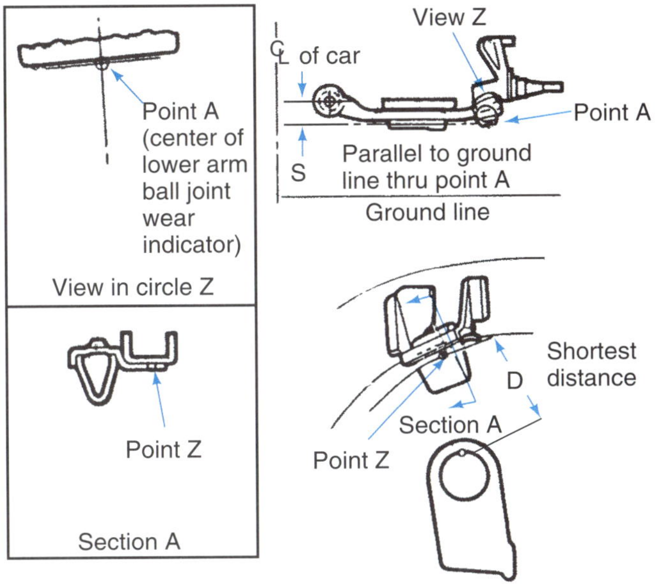


Suspension ride height

| VEHICLE | | S | D |
|---|---|---|---|
| MARK VII/ CONTINENTAL | INCHES | 0.24 | 5.06 |
| | MM | 6.0 | 128.6 |

**Figure 8-3** Trim height measurement locations (Courtesy of Ford Motor Company)

9. The spring fill mode is terminated if the diagnostic lead is disconnected from ground. This action is also obtained if the ignition switch is turned off, or the brake pedal is applied.

## Trim Height Adjustment

The trim height should be measured on the front and rear suspension at the locations specified by the vehicle manufacturer (Figure 8-3).

If the rear suspension trim height is not within specifications, it may be adjusted by loosening the attaching bolt on the top height sensor bracket (Figure 8-4). When the bracket is moved one index mark up or down, the ride height is lowered or raised 0.25 in (6.35 mm).

The front suspension trim height may be adjusted by loosening the lower height sensor attaching bolt. Three adjustment positions are located in the lower front height sensor bracket (Figure 8-5). If the height sensor attaching bolt is moved one position up or down, the front suspension height is lowered or raised 0.5 in (12.7 mm).

Front and rear trim height is the distance between the vehicle chassis and the road surface measured at locations specified by the vehicle manufacturer.

## Line Service

Nylon lines on the electronic air suspension system have quick disconnect fittings. These fittings should be released by pushing downward and holding the plastic release ring, and then pulling outward on the nylon line (Figure 8-6). Simply push the nylon line into the fitting until it seats to reconnect an air line.

If a line fitting is damaged, it may be removed by looping the line around your fingers and pulling on the line without pushing on the release ring (Figure 8-7). When a new collet and release ring is installed, the O-ring under the collet must be replaced. If a leak occurs in a nylon line, a sharp knife may be used to cut the defective area out of the line. A service fitting containing a collet fitting in each end is available for line repairs. After the defective area is cut out of the line, push the two ends of the line into the service fitting.

**Classroom Manual**
Chapter 8, page 154

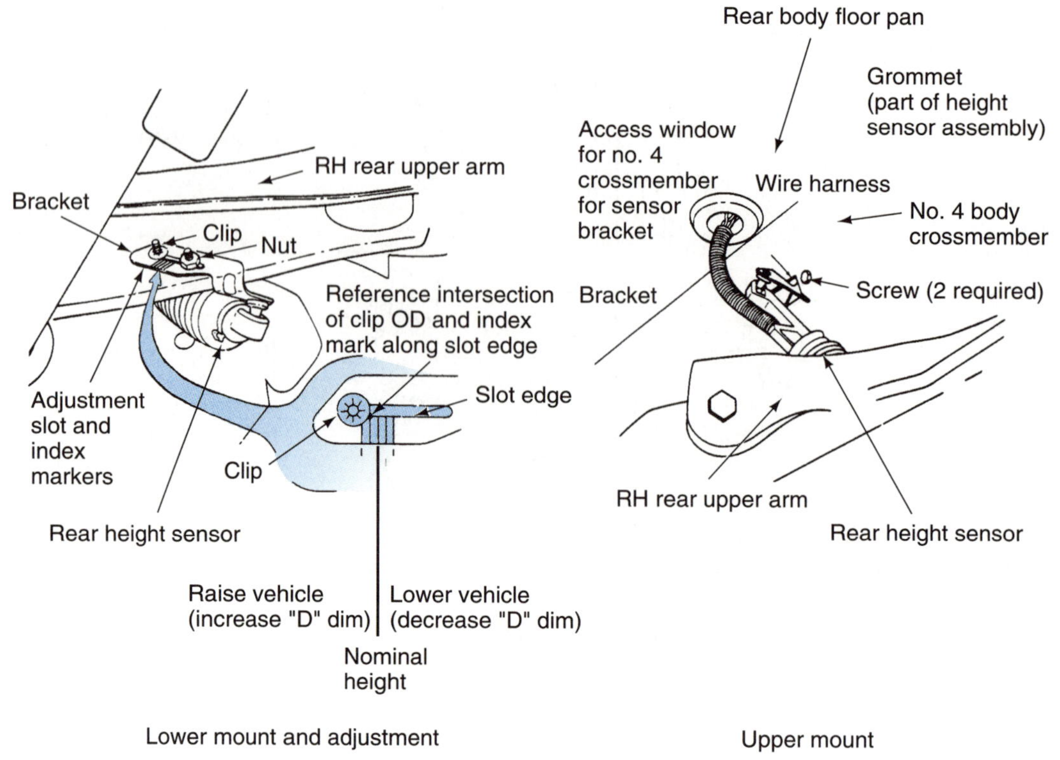


**Figure 8-4** Rear trim height adjustment (Courtesy of Ford Motor Company)

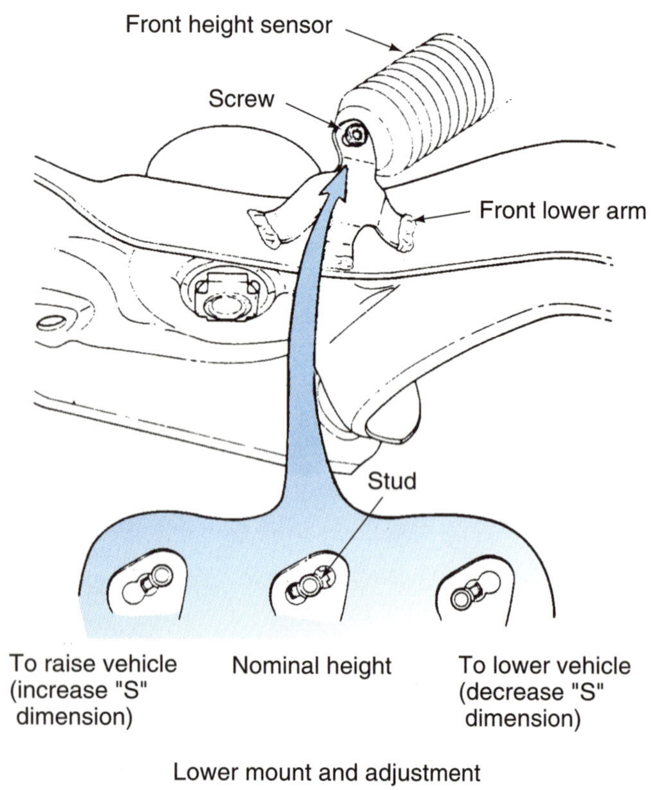


**Figure 8-5** Front trim height adjustment (Courtesy of Ford Motor Company)

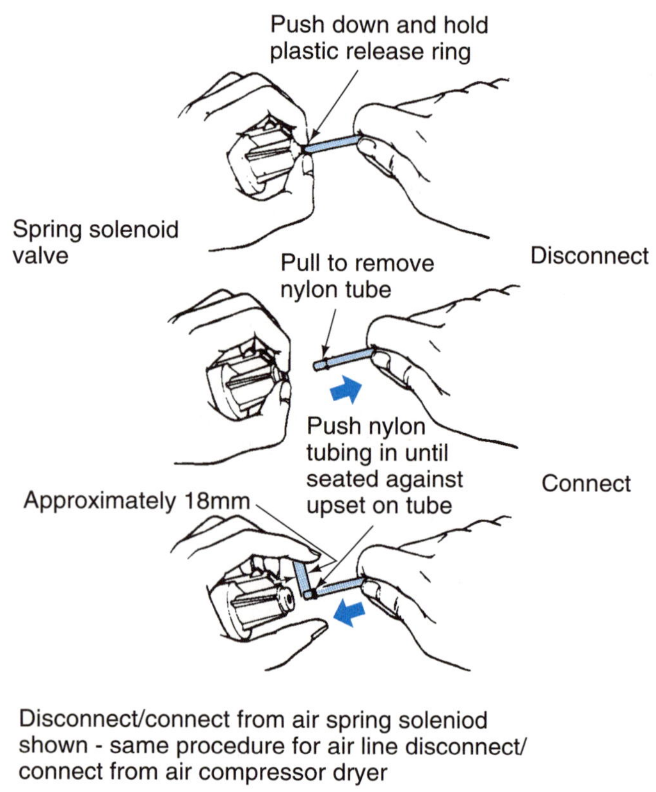


**Figure 8-6** Air line removal from air spring valves or compressor outlets (Courtesy of Ford Motor Company)

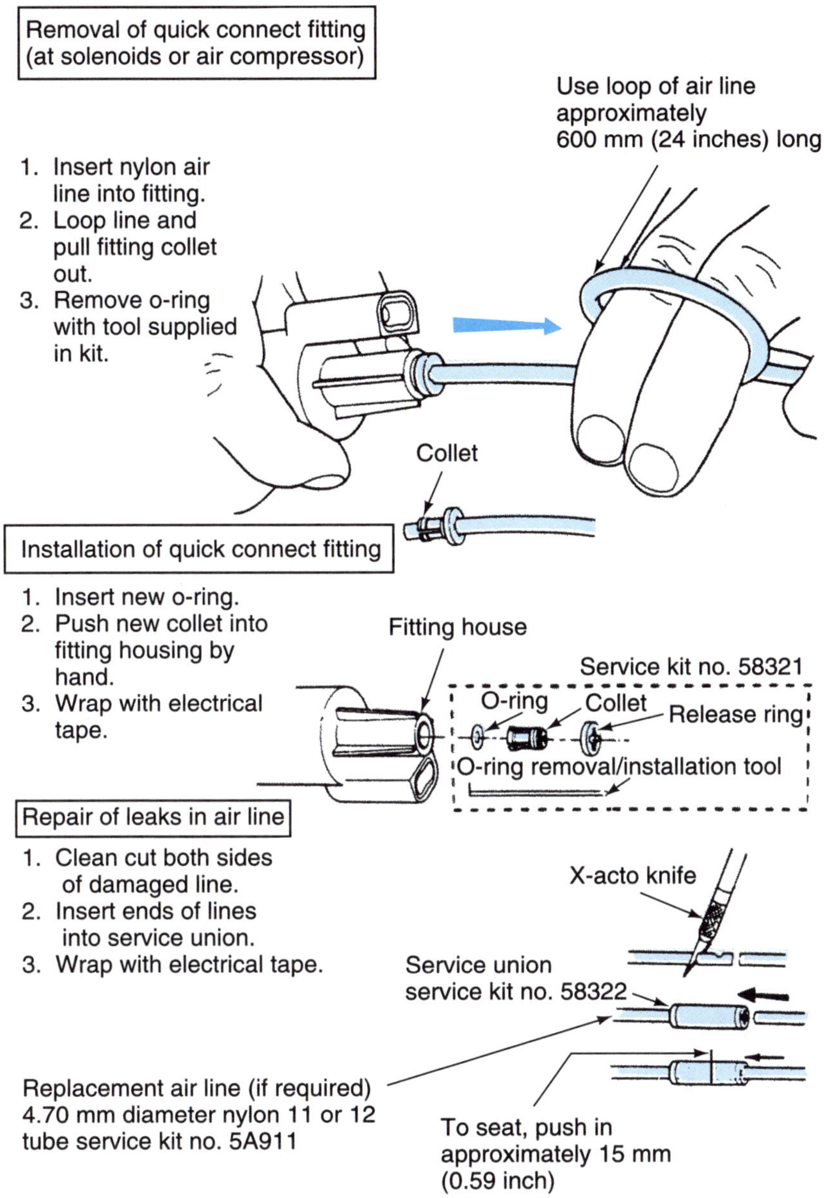


Figure 8-7 Removal of quick disconnect fittings, and air line repair procedure (Courtesy of Ford Motor Company)

# Rear Load-Leveling Air Suspension System Service and Diagnosis

## Trim Height Adjustment

The trim height is measured between the top of the rear axle housing and the frame (Figure 8-8). The upper height sensor bracket mounting holes have long slots for adjustment purposes. If a trim height adjustment is required, the upper height sensor mounting bolts may be loosened and the bracket moved downward to increase trim height, or upward to decrease trim height (Figure 8-9).

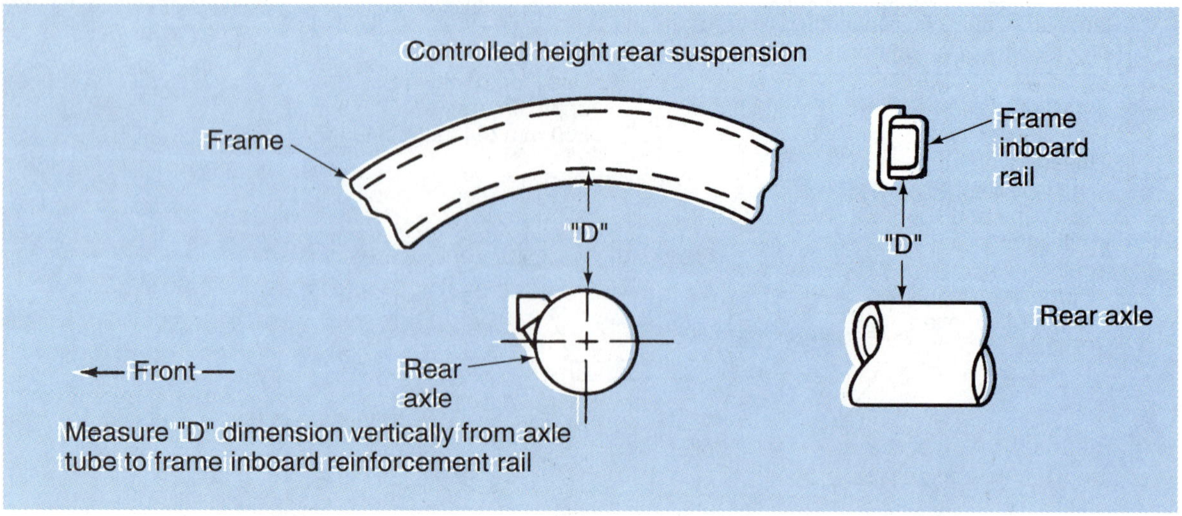


**Figure 8-8** Trim height measurement location (Courtesy of Ford Motor Company)

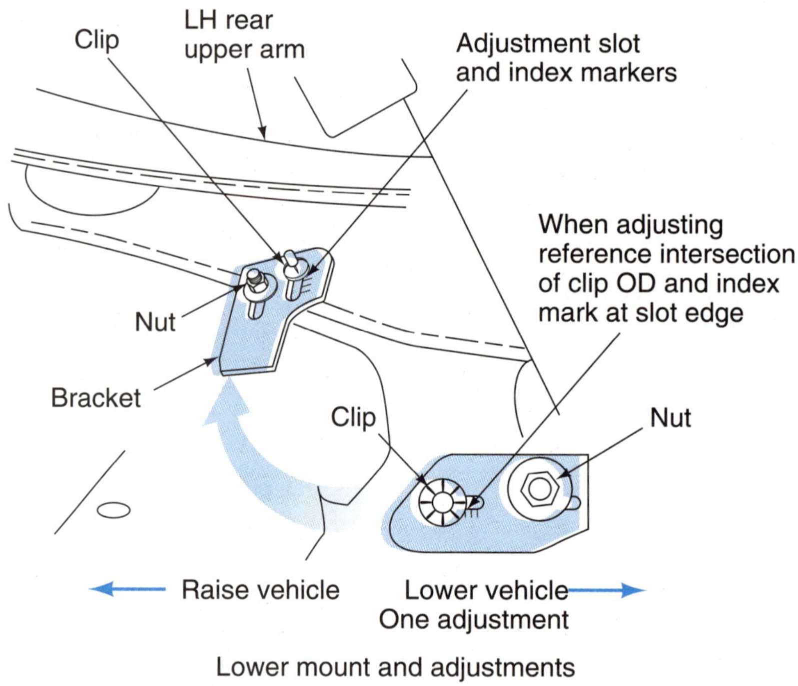


**Figure 8-9** Trim height adjustment procedure (Courtesy of Ford Motor Company)

**Special Tools**

Scan tester

A scan tester is a digital tester that may be used to diagnose various on-board computer systems.

## Diagnosis

**SERVICE TIP:** It is difficult to diagnose a rear load-leveling suspension system without a scan tester.

A scan tester with air suspension test capabilities may be connected to the test connector in the trunk to obtain fault codes from the control module in the combined EVO and rear load-leveling air suspension system (Figure 8-10).

When the scan tester is connected to the diagnostic connector, these tests may be completed:

1. auto/manual diagnostic checks
2. fault code display

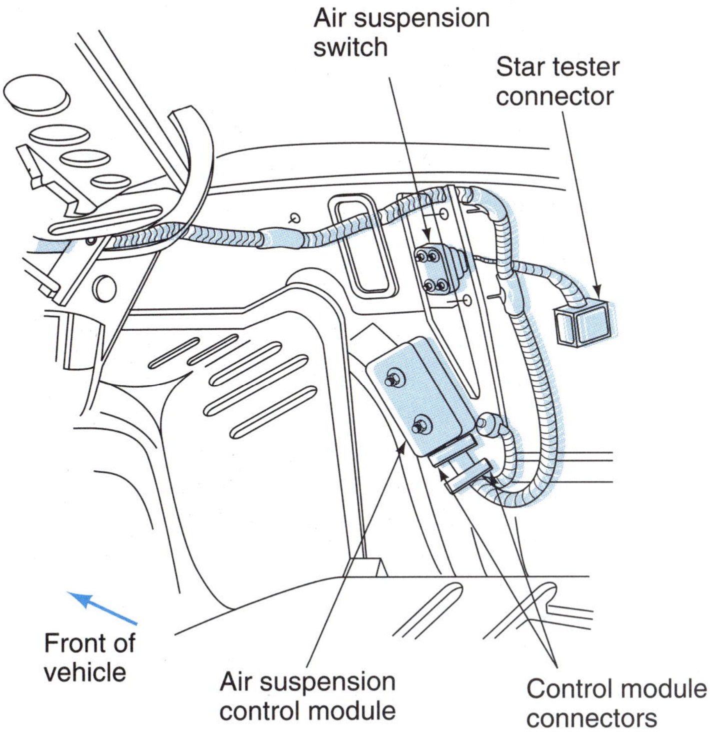


**Figure 8-10** Diagnostic connector for rear load-leveling air suspension system (Courtesy of Ford Motor Company)

The same control module operates the rear load-leveling air suspension system and the electronic variable orifice (EVO) steering system. In the EVO steering system, the control module operates a solenoid in the power steering pump so that less effort is required to turn the steering wheel at low speeds, and more steering effort is required at higher speeds.

3. pinpoint tests
4. functional test

During the auto/manual tests, the control module checks the system components for electrical defects. After this electrical check, the control module raises and lowers the vehicle to test all three height sensor states. This height sensor check is followed by a manual test in which the control module monitors the door switches and steering sensor as each door is opened, and the steering wheel is turned about one-half turn in each direction.

The fault codes are obtained after the auto/manual test is completed. If there are no codes in the control module, "code 11 system pass" is displayed (Figure 8-11).

In the manufacturer's service information, a pinpoint test is provided for each fault code. These pinpoint tests are voltmeter and ohmmeter tests that may be followed to locate the exact cause of a fault code.

The functional test may be used at the end of the auto/manual tests to cycle the component represented by a fault code. Never enter the functional test unless a pinpoint test directs this action. If a functional test is entered without this specified direction, control module damage may result. Detailed test procedures are provided in the manufacturer's service manuals.

**Classroom Manual**
Chapter 8, page 164

# Programmed Ride Control (PRC) System Diagnosis

## General Diagnosis

The PRC control module monitors the complete system while the vehicle is driven. If a defect occurs in the PRC system, the mode indicator light starts flashing. When the mode selector switch is moved to the opposite position and returned to the same position, erroneous codes are cleared from the control module memory. If the mode indicator light continues flashing, the self-test should be performed.

**AIR SUSPENSION DIAGNOSTICS**
**SERVICE CODES**

| SERVICE CODE | PINPOINT TEST | DESCRIPTION | SERVICE PRIORITY |
|---|---|---|---|
| 10 | | DIAGNOSTICS ENTERED, AUTO TEST IN PROGRESS | |
| 11 | | VEHICLE PASSES. IF VEHICLE IS STILL LOW OR HIGH IN REAR, CHECK REAR RIDE HEIGHTS. | |
| 12 | | AUTO TEST PASSED, PERFORM MANUAL INPUTS | |
| 13 | | AUTO TEST FAILED, PERFORM MANUAL INPUTS | |
| 16 | * | EVO SHORT CIRCUIT | |
| 17 | * | EVO OPEN CIRCUIT | |
| 18 | * | EVO ACTUATOR VALVE | |
| 39 | A | COMPRESSOR RELAY CIRCUIT SHORTED TO BATTERY | 2ND |
| 42 | B | AIR SPRING SOLENOID CIRCUIT SHORTED TO GROUND | 2ND |
| 43 | C | AIR SPRING SOLENOID CIRCUIT SHORTED TO BATTERY | 2ND |
| 44 | D | VENT SOLENOID CIRCUIT SHORTED TO BATTERY | 2ND |
| 45 | E | AIR COMPRESSOR RELAY CIRCUIT SHORTED TO GROUND OR VENT SOLENOID CIRCUIT SHORTED TO GROUND | 2ND |
| 46 | F | HEIGHT SENSOR POWER SUPPLY CIRCUIT SHORTED TO GROUND OR BATTERY | 2ND |
| 51 | G | UNABLE TO DETECT LOWERING OF REAR | 3RD |
| 54 | H | UNABLE TO DETECT RAISING OF REAR | 3RD |
| 68 | J | HEIGHT SENSOR OUTPUT CIRCUIT SHORTED TO GROUND | 2ND |
| 70 | K | REPLACE AIR SUSPENSION/EVO MODULE | 1ST |
| 71 | L | OPEN HEIGHT SENSOR CIRCUIT | 3RD |
| 72 | M | FOUR OPEN AND CLOSED DOOR SIGNALS NOT DETECTED | 4TH |
| 74 | * | STEERING WHEEL ROTATION NOT DETECTED | |
| 80 | N | INSUFFICIENT BATTERY VOLTAGE TO RUN DIAGNOSTICS | 1ST |

| CODE | FUNCTIONAL TEST | NOTE |
|---|---|---|
| 23 | FUNCTIONAL TEST, VENT REAR | AIR SUSPENSION FUNCTIONAL TESTS, NEVER PERFORM A FUNCTIONAL TEST UNLESS DIRECTED TO DO SO IN A PINPOINT TEST. |
| 26 | FUNCTIONAL TEST, COMPRESS REAR | |
| 31 | FUNCTIONAL TEST, TOGGLE COMPRESSOR | |
| 32 | FUNCTIONAL TEST, VENT SOLENOID TOGGLE | |
| 33 | FUNCTIONAL TEST, SPRING SOLENOID TOGGLE | |

*FOR CODES 16, 17, 18 AND 74, REFER TO PINPOINT TEST STEP A1 IN SECTION 13-54. FOR ALL OTHER CODES, REFER TO SECTION 14-40 FOR THE TEST INDICATED ON THIS CHART.

**Figure 8-11** Fault codes for rear load-leveling air suspension system and electronic variable orifice (EVO) steering system (Courtesy of Ford Motor Company)

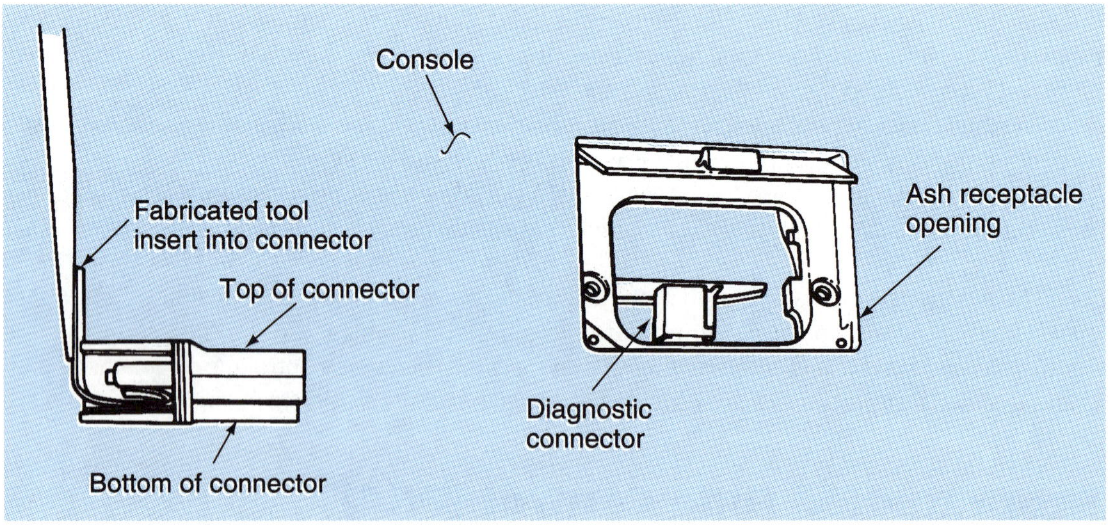


**Figure 8-12** Self-test terminals, programmed ride control system (Courtesy of Ford Motor Company)

## Self-test

The self-test connector is located under the ash tray (Figure 8-12). This connector has two terminals, which must be connected together during the self-test. A jumper tool should be fabricated to

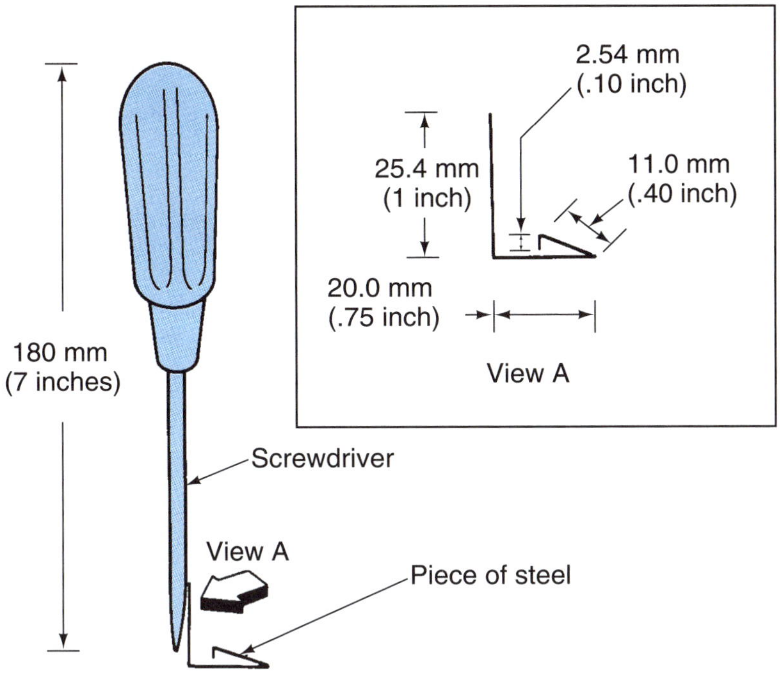


**Figure 8-13** Fabricated tool to connect across self-test terminals (Courtesy of Ford Motor Company)

connect across these terminals, and this tool should be soldered to the end of a 7-inch slotted screwdriver blade to access the self-test terminals (Figure 8-13).

When the self-test is performed, follow this procedure:

1. Turn the ignition switch off and be sure the headlights and parking lights are off. These lights must remain off during the self-test.
2. Position the mode select switch in the AUTO position.
3. Remove the ash tray and insert the fabricated tool in the self-test connector terminals.
4. Start the engine and leave the mode select switch in the AUTO position.
5. When the engine has been running for 20 seconds or more, remove the tool from the self-test connector.
6. Count the mode indicator light flashes to obtain the trouble codes. For example, if the light flashes six times, code 6 has been provided. The light flashes each code four times at nine-second intervals. The mode indicator light is positioned in the tachometer (Figure 8-14).

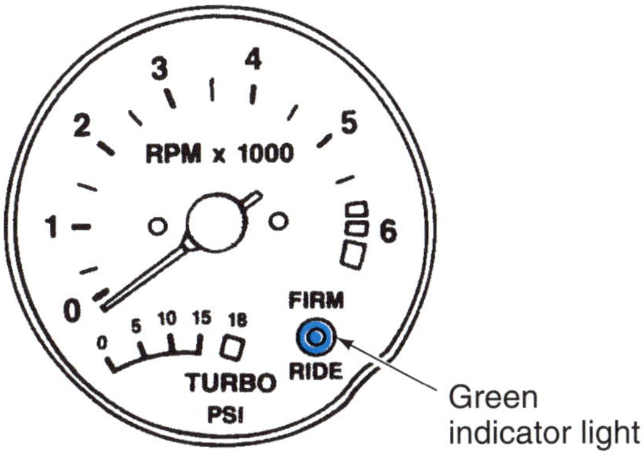


**Figure 8-14** Mode indicator light in tachometer flashes trouble codes (Courtesy of Ford Motor Company)

## Trouble Codes

**SERVICE TIP:** Trouble codes together with the vehicle manufacturer's service manual diagnostic procedures will usually locate the problem in a PRC system.

These codes are available on the PRC system (Table 8-1):

**Table 8-1 PROGRAMMED RIDE CONTROL (PRC) SYSTEM CODES**

| Code | Defect |
|---|---|
| 6 | No problem |
| 1 | Left rear actuator circuit |
| 2 | Right rear actuator circuit |
| 3 | Right front actuator circuit |
| 4 | Left front actuator circuit |
| 5 | Soft relay control circuit shorted |
| 7 | PRC control module |
| 13 | Firm relay control circuit shorted |
| 14 | Relay control circuit |

(Courtesy of Ford Motor Company)

**Classroom Manual**
Chapter 8, page 166

The fault codes vary depending on the vehicle make and year. Always use the fault code list in the manufacturer's service manual. A trouble code indicates a defect in a specific area. For example, if code 2 is received, the defect may be in the right front actuator or the connecting wires.

After a PRC system defect has been corrected, the codes should be cleared, and the system should be checked for codes again to make sure there are no other faults in the system.

A light emitting diode (LED) is a special type of diode from which light is emitted when the diode conducts current.

A nonvolatile electronically erasable programmable read only memory (EEPROM) is an electronic chip that retains information in memory when ignition voltage is removed from the chip.

**Special Tools**

Jumper wires

# Diagnosis of Computer Command Ride (CCR) System

## Fault Code Diagnosis

If a defect occurs in the CCR system, both driver-select switch LEDs are illuminated. Should the defect correct itself, the LEDs will go out, but a fault code should be retained in the control module nonvolatile electronically erasable programmable read only memory (EEPROM). When a code is set in the control module memory, the control module performs a 1-second retest at 3-minute intervals, or when the driver-select switch position is changed. Both switch LEDs go out if the defect is corrected. If the defect is still present, the LEDs remain illuminated.

A jumper wire may be connected between terminals A and C in the data link connector (DLC) to obtain the fault codes from the control module in the computer command ride system (Figure 8-15).

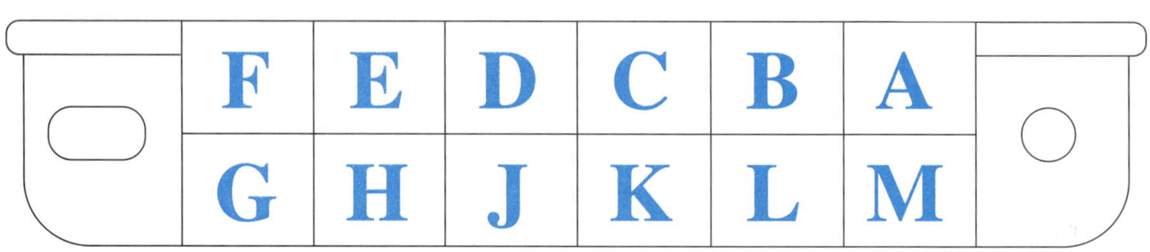


**Figure 8-15** Data link connector (DLC)

After the jumper wire is installed across terminals A and C in the DLC, the ignition switch is turned on, and both the driver-select switch LEDs blink at the same time to provide the fault codes. If these LEDs flash twice followed by a pause and five more flashes, code 25 is indicated. Code 12 is flashed first. This code indicates the beginning of the code sequence. Each code is flashed three times, and the codes are displayed in numerical order. The code sequence keeps repeating until the ignition switch is turned off or the jumper wire is removed. A fault code indicates a problem in a specific area, but voltmeter and ohmmeter tests may be necessary to locate the exact cause of the problem. These tests are explained in detail in the manufacturer's service manuals.

After a defect is corrected, the code may be cleared by alternately connecting and disconnecting the jumper wire between DLC terminals A and C three times with the ignition switch on. Each jumper wire connection must be held for 1 second, and there should be a 1 second pause between each connection. When the code erasing procedure is completed, turn the ignition switch off and connect a jumper wire between terminals A and C in the DLC. Turn the ignition switch on and check the driver-select switch LED flashes. If the defect is corrected and the codes are erased, only code 12 is flashed. A CCR system fault code list is provided in Table 8-2.

A test terminal called the data link connector (DLC) is located under the center or left side of the instrument panel on many General Motors products. Test connections are completed at this terminal to diagnose many of the electronic systems on the vehicle.

**Table 8-2 COMPUTER COMMAND RIDE (CCR) SYSTEM FAULT CODES**

| Code | Interpretation |
|---|---|
| 12 | Initiation |
| 13 | Left front actuator over current |
| 14 | Right front actuator over current |
| 15 | Left rear actuator over current |
| 16 | Right rear actuator over current |
| 23 | Left front actuator position error |
| 24 | Right front actuator position error |
| 25 | Left rear actuator position error |
| 26 | Right rear actuator position error |
| 32 | Mercury switch accelerometer error |
| 33 | Driver-select switch input error |
| 34 | Vehicle speed signal error |

(Courtesy of Oldsmobile Division, General Motors Corporation)

The Society of Automotive Engineers (SAE) J1930 terminology is an attempt to standardize electronics terminology in the automotive industry.

In the SAE J1930 terminology, the term data link connector (DLC) replaces previous terms such as ALDL.

## Continual Strut Cycling

When diagnosing a CCR system, always follow the service procedures in the vehicle manufacturer's service manual. See Photo Sequence 6 for a typical procedure for diagnosing a CCR system. If a 6.6-Ω resistor is connected between terminals A and C in the DLC, the CCR control module enters a continual strut cycling mode. In this mode, the control module cycles all the struts from the normal position to the firm position every 2 seconds. When a defect is present in a strut, that strut does not cycle until a retest is performed. If the defect has corrected itself before the 3-minute retest, the strut cycles when the retest is completed.

## Strut Connector Precautions

When a strut connector is removed, the connector cavity must be clean and dry before the connector is reconnected. The connector should be inspected for corroded terminals and damaged wires. The strut connector alignment groove must be properly aligned with the strut alignment pin as the connector is installed (Figure 8-16). Marking the alignment pin position on the outside of the strut with a wax pencil provides easier groove-to-pin alignment.

**Classroom Manual**
Chapter 8, page 172

# Photo Sequence 6
## Typical Procedure for Diagnosing Computer Command Ride (CCR) System

**P6-1** Road test the car to check CCR operation and warm up the CCR components.

**P6-2** Park vehicle in shop service bay.

**P6-3** Visually inspect wires and terminals on the strut actuators.

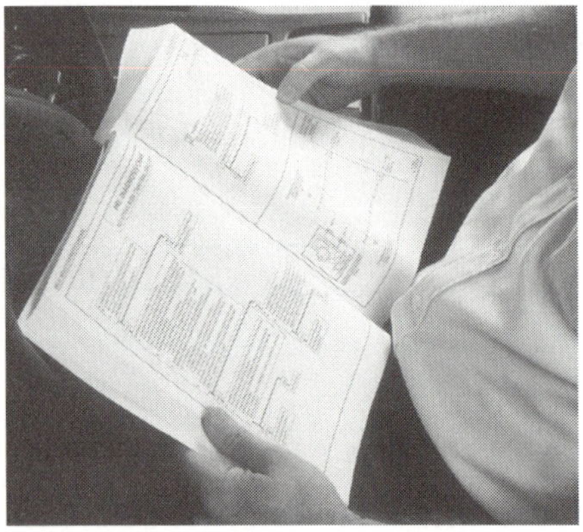

**P6-4** Look up diagnostic procedure and trouble codes in the service manual.

**P6-5** Locate the data link connector (DLC) under the instrument panel.

**P6-6** Connect a jumper tool or wire between terminals A and C in the DLC.

**P6-7** Turn the ignition switch on.

**P6-8** Observe the CCR indicator light flashes to read the trouble codes.

**P6-9** Turn off the ignition switch, and remove the jumper tool in the DLC.

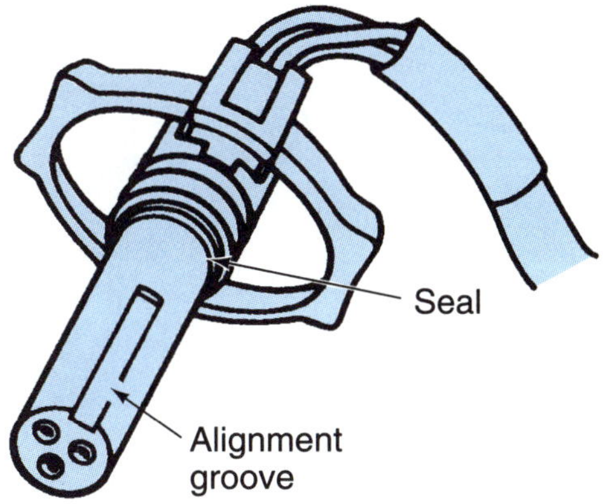


**Figure 8-16** Strut connector (Courtesy of Oldsmobile Division, General Motors Corporation)

# Automatic Air Suspension Service and Diagnosis

## Drive Cycle Diagnostics

The drive cycle diagnostic procedure displays any fault codes that have occurred during the last drive cycle. This test procedure detects intermittent faults that may not appear during the other test procedures.

The car should be driven for a minimum of 5 minutes at different speeds and road conditions, or until a ride control message is illuminated in the instrument panel. After this driving period, the car should be driven directly to the service area and parked with the ignition switch off. The fault codes are now stored in the air suspension control module for 1 hour of ignition off time, or until the ignition switch is turned on. If the ignition switch is turned on during this 1-hour period, the fault codes are erased in the air suspension module memory. The automatic air suspension on/off switch in the trunk must be on during the test procedure.

A scan tester with air suspension test capabilities should be connected to the diagnostic connector near the air suspension module in the trunk (Figure 8-17). Always follow the instructions in the vehicle manufacturer's service manual and in the scan tester manufacturer's manual. The scan tester displays a pass code or one or more of 32 possible fault codes. Codes are displayed continuously for 1 hour or until the ignition switch is turned on. The fault codes should be written down for later reference.

## Service Bay Diagnostics

**Auto/Manual Diagnostic Check.** During the auto/manual test, the air suspension control module verifies itself and checks other system components. After this check, the scan tester displays "12/okay do manual checks," or "13/faults detected do manual checks." The manual tests allow the technician to test the system inputs to the air suspension module.

A hard fault code indicates a fault that is present all the time, including the time of testing.

**Fault Code Display.** After the auto/manual test procedure, the fault codes may be displayed by pressing the appropriate scan tester button (Figure 8-18). Each fault code is displayed for approximately 15 seconds, and the fault code display continues until the next test mode is entered. The fault codes obtained during the drive cycle diagnosis should be compared to fault codes that

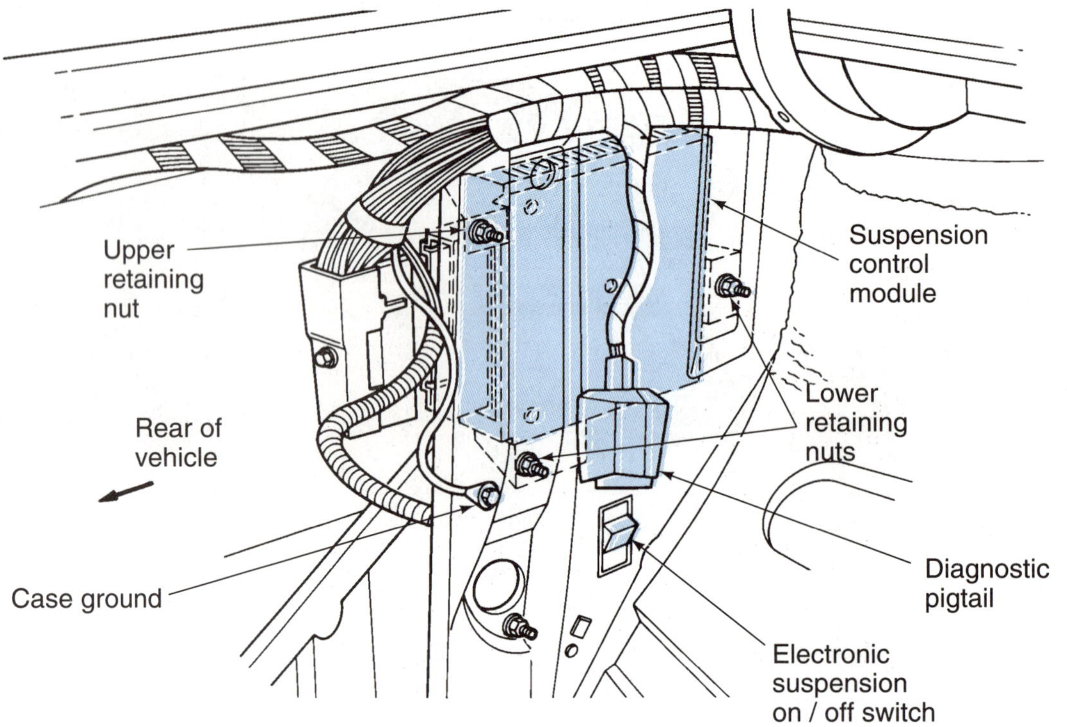


**Figure 8-17** Automatic air suspension control module and diagnostic connector (Courtesy of Ford Motor Company)

An intermittent fault code indicates a fault that occurs and then disappears.

appear during the service bay diagnostics. When a code appears during both test modes, the code represents a hard fault. If the code is only present during the drive cycle diagnosis, the code indicates an intermittent fault.

**Pinpoint Tests.** Many of the fault codes have a pinpoint procedure represented by two letters. Detailed pinpoint test procedures must be followed in the car manufacturer's service manual. The pinpoint tests are designed to locate the exact cause of a fault code. For example, fault code 40 may be obtained, indicating a problem in the left front air spring solenoid valve circuit. Pinpoint test EA will inform the technician to perform specific tests to prove if the trouble is in the solenoid valve, connecting wires, or module (Figure 8-19).

The pinpoint tests are prioritized with 1 being the highest priority, and 7 being the lowest priority. One fault may result in the display of a second fault. The pinpoint tests must be performed in the order of priority, starting with the highest priority.

**SERVICE TIP:** When fault codes are provided on a priority basis, always recheck the system for codes after a fault is corrected. Some computers are only capable of storing a specific number of codes in order of priority. If only one code is displayed, there could be another fault of lesser priority in the system.

## Spring Fill Diagnostics

**Classroom Manual**
Chapter 8, page 175

The spring fill diagnostics allow the technician to force the automatic air suspension control module to fill and vent each air spring. This procedure indicates inoperative components such as air spring solenoids or the module. During the spring fill procedure, the technician can check for air line leaks.

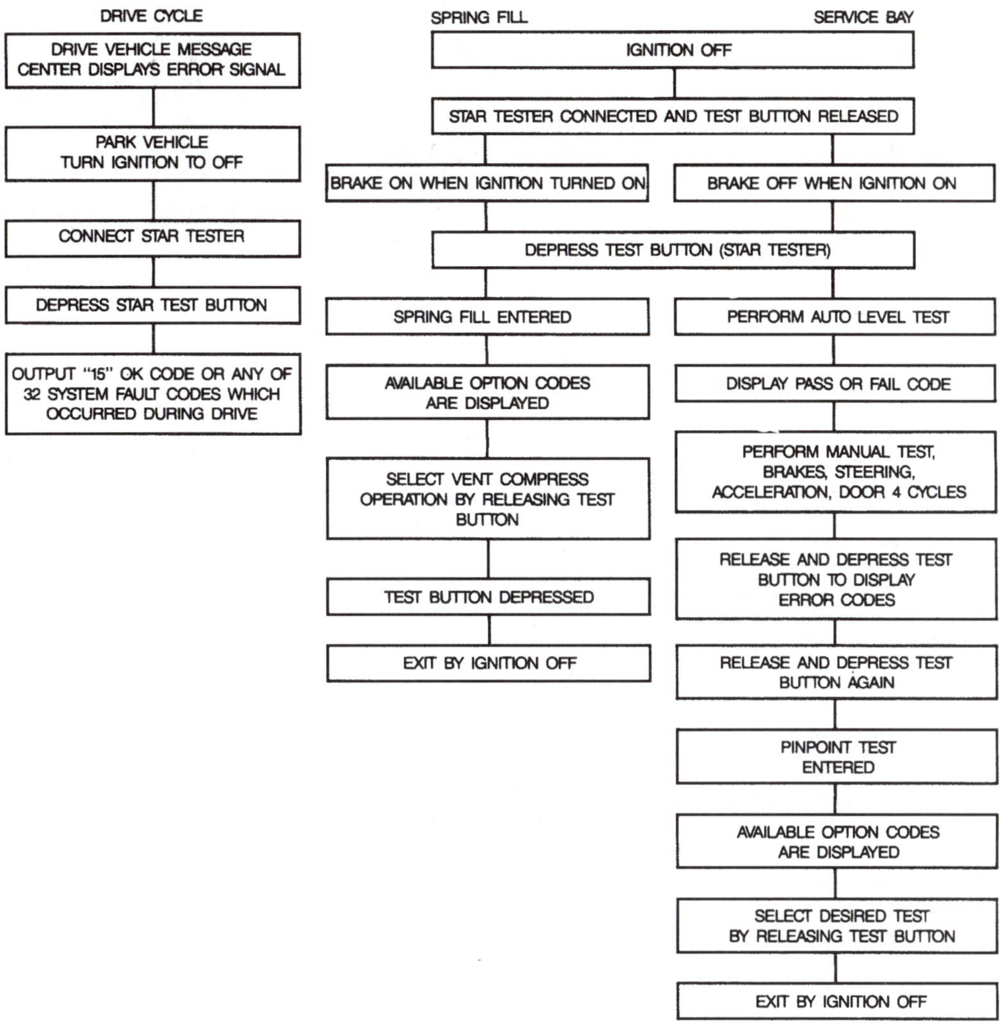


**Figure 8-18** Drive cycle, service bay, and spring fill diagnostic procedures (Courtesy of Ford Motor Company)

# Diagnosis of Air Suspension with Rear Shock Damping Control System

## Functional Test 211

The functional test is always performed first in the diagnostic procedure. Since the test procedures vary depending on the vehicle model and year, always follow the procedure in the vehicle manufacturer's service manual. The following procedures are typical. A scan tester must be connected to the suspension diagnostic connector located on the right front shock absorber tower. During the functional test, codes stored in the suspension control module memory are displayed. A code 15 indicates there are no codes stored in the module memory. The auto test is performed next to retrieve fault codes representing existing conditions.

## Auto Test

The electronic variable orifice (EVO) steering actuator is a solenoid that is controlled by the suspension control module, which varies power steering assist in relation to vehicle speed.

When the auto test procedure is entered on the scan tester, the suspension control module checks the inputs and outputs for errors. If no errors are found, the module cycles the compressor

| STAR CODE | PINPOINT PROCEDURE | DESCRIPTION | SERVICE PRIORITY |
|---|---|---|---|
| 10 | | Service Bay Diagnostics Entered | |
| 11 | | System Checked Out Okay | |
| 12 | | Automatic Test Completed — No Faults Detected — Perform Manual Inputs | |
| 13 | | Automatic Test Completed — Faults Detected — Perform Manual Inputs | |
| 15 | | No Faults Detected | |
| 21 | | Vent Right Front Air Spring | |
| 22 | | Vent Left Front Air Spring | |
| 23 | | Vent Right Rear Air Spring | |
| 24 | | Inflate Right Front Air Spring | |
| 25 | | Inflate Left Front Air Spring | |
| 26 | | Inflate Right Rear Air Spring | |
| 27 | | Vent Left Rear Air Spring | |
| 28 | | Inflate Left Rear Air Spring | |
| 31 | | Air Compressor Toggle | |
| 32 | | Vent Solenoid Valve Toggle | |
| 33 | | Air Spring Solenoid Valve Toggle | |
| 34 | | Shock Actuator Toggle (Firm/Soft) | |
| 35 | | Door Open & Door Closed Detection | |
| 40 | EA | Short — Left Front Air Spring Solenoid Valve Circuit | 2nd |
| 41 | EB | Short — Right Front Air Spring Solenoid Valve Circuit | 2nd |
| 42 | EC | Short — Left Rear Air Spring Solenoid Valve Circuit | 2nd |
| 43 | ED | Short — Right Rear Air Spring Solenoid Valve Circuit | 2nd |
| 44 | EE | Short — Vent Solenoid Valve Circuit | 2nd |
| 45 | EF | Short — Air Compressor Relay Circuit | 2nd |
| 46 | EG | Short — Height Sensor Power Supply Circuit | 2nd |
| 47 | EH | Short — Soft Shock Actuator Relay Circuit | 2nd |
| 48 | EI | Short — Firm Shock Actuator Relay Circuit | 2nd |
| 49 | HA | Unable to Detect Lowering of Right Front Corner | 5th |
| 50 | HB | Unable to Detect Lowering of Left Front Corner | 5th |
| 51* | HC | Unable to Detect Lowering of Right Rear Corner | 5th |
| 51* | | Unable to Detect Lowering of Rear of Vehicle | 5th |
| 52 | IA | Unable to Detect Raising of Right Front Corner | 6th |
| 53 | IB | Unable to Detect Raising of Left Front Corner | 6th |
| 54* | IC | Unable to Detect Raising of Right Rear Corner | 6th |
| 54* | | Unable to Detect Raising of Rear of Vehicle | 6th |
| 55 | JA | Speed Greater Than 15 mph Not Detected | 7th |
| 56 | GA | Soft Not Detected — Left Rear Shock Actuator Circuit | 4th |
| 57 | GB | Soft Not Detected — Right Front Shock Actuator Circuit | 4th |
| 58 | GC | Soft Not Detected — Left Front Shock Actuator Circuit | 4th |
| 59 | GD | Soft Not Detected — Right Rear Shock Actuator Circuit | 4th |
| 60 | GA | Firm Not Detected — Left Rear Shock Actuator Circuit | 4th |
| 61 | GB | Firm Not Detected — Right Front Shock Actuator Circuit | 4th |
| 62 | GC | Firm Not Detected — Left Frt Shock Actuator Circuit | 4th |
| 63 | GD | Firm Not Detected — Right Rear Shock Actuator Circuit | 4th |
| 64 | GE | Soft Not Detected — All Shock Actuator Circuits | |
| 65 | GE | Firm Not Detected — All Shock Actuator Circuits | 4th |
| 66 | EJ | Short — Right Front Height Sensor Circuit | 2nd |
| 67 | EK | Short — Left Front Height Sensor Circuit | 2nd |
| 68 | EL | Short — Rear Height Sensor Circuit | 2nd |
| 69 | FA | Open — Rilght Front Height Sensor Circuit | 3rd |
| 70 | FB | Open — Left Front Height Sensor Circuit | 3rd |
| 71 | FC | Open — Rear Height Sensor Circuit | 3rd |
| 72 | JB | At Least Four Open & Closed Door Signals Not Detected | 7th |
| 73 | JC | Brake Pressure Switch Activation Not Detected | 7th |
| 74 | JD | Steering Wheel Rotations Not Detected | 7th |
| 75 | JE | Acceleration Signal Not Detected | 7th |
| 78 | HD | Unable to Detect Lowering of Left Rear Corner | 5th |
| 79 | ID | Unable to Detect Raising of Left Rear Corner | 6th |
| 80 | DA | Insufficient Battery Voltage to Run Diagnostics | 1st |

**NOTE:** System faults have been prioritized for repair. Start with those codes identified with a service priority of: 1st, then 2nd, then 3rd, . . . and finally 7th.

**Figure 8-19** Fault codes and pinpoint tests (Courtesy of Ford Motor Company)

The electronic variable orifice (EVO) steering system is similar to the speed sensitive steering (SSS) explained in Chapter 8 of the Classroom Manual.

and system solenoids, including the electronic variable orifice (EVO) steering actuator. The EVO steering system is explained later in the Classroom Manual chapter on power rack and pinion steering gears.

The suspension control module then pumps and vents each air spring to verify component operation. After this procedure, the module performs a vehicle trim height adjustment.

The module then requests manual inputs from the steering wheel and the doors. To provide these inputs, each door must be opened and closed, and the steering wheel must be rotated at least one-quarter turn in each direction. This action checks the door switches and the steering sensor. After all the tests are completed, code 11 is displayed if there are no faults in the system. If the tests are unsuccessful, codes are displayed indicating the system faults. If the vehicle suspension

**FUNCTION TEST INDEX**

| Code | Description |
|---|---|
| 211 | Display All Diagnostic Trouble Codes In Memory |
| 212 | LF Pump with Audible Sensor Check |
| 213 | LF Vent with Audible Sensor Check |
| 214 | RF Pump with Audible Sensor Check |
| 215 | RF Vent with Audible Sensor Check |
| 216 | LR Pump with Audible Sensor Check |
| 217 | LR Vent with Audible Sensor Check |
| 218 | RR Pump with Audible Sensor Check |
| 219 | RR Vent with Audible Sensor Check |
| 221 | Compressor Run |
| 222 | Actuator Output Test (Cycles All Solenoids and EVO Actuator) |
| 223 | LF Height Sensor Trim Detection, Audible Output |
| 224 | RF Height Sensor Trim Detection, Audible Output |
| 225 | LR Height Sensor Trim Detection, Audible Output |
| 226 | Speed Sensor Detection |
| 227 | Pulse EVO Actuator through Duty Cycle |
| 228 | Erase All Codes Stored in Module Memory |

**Figure 8-20** Functional tests with audible tone in the pinpoint tests (Courtesy of Ford Motor Company)

was below trim height when the auto test was started, a pass code 11 may be received even when the correct trim height was not achieved. When this situation occurs, cycle the ignition switch to reset the trim height, and repeat the auto test.

After any air suspension system component is serviced or replaced, the auto test should be performed to be sure the service has corrected the problem, and any stored fault codes should be erased.

## Pinpoint Tests

Pinpoint tests are grouped by the components within the system. The scan tester is used during the pinpoint tests, and the technician is requested to perform additional functional tests during these tests. These functional tests include many system components, including the height sensors. During the pinpoint functional tests, the air suspension switch must be in the off position. Always follow the exact test procedure in the vehicle manufacturer's service manual. Many of these functional tests are audible tests in which the suspension module produces an audible tone through the scan tester speaker. The tone during the audible tests is a steady tone when the height sensor is at trim height. The tone pulse becomes slower as the sensor moves away from trim height. The last functional test in the pinpoint tests is an erase code function (Figure 8-20).

**Classroom Manual**
Chapter 8, page 181

# Diagnosis of Road Sensing Suspension System

## Trouble Code Display

**WARNING:** Never remove or install the wiring connector on a computer or computer system component with the ignition switch on. This action may result in computer damage.

**WARNING:** Do not supply voltage to, or ground, any circuit or component in a computer system unless instructed to do so in the vehicle manufacturer's service manual. This action may damage computer system components.

**WARNING:** During computer system diagnosis, use only the test equipment recommended in the vehicle manufacturer's service manual to prevent damage to computer system components.

**SERVICE TIP:** When removing, replacing, or servicing an electronic component on a vehicle, always disconnect the negative battery cable before starting the service procedure. If the vehicle is equipped with an air bag, or bags, wait 1 minute after the battery negative cable is removed to prevent accidental air bag deployment. Many air bag computers have a backup power supply that is capable of deploying the air bag for a specific length of time after the battery is disconnected.

The powertrain control module (PCM) controls such output functions as electronic fuel injection, spark advance, and emission devices.

The instrument panel cluster (IPC) module controls most instrument panel displays.

The air conditioning programmer (ACP) controls the air conditioning output functions.

The supplemental inflatable restraint (SIR) module controls air bag inflation.

The traction control system (TCS) module is combined with the antilock brake system (ABS) module. This module controls wheel lockup during hard braking and wheel spinning during acceleration.

The real time damping (RTD) module controls the road sensitive suspension (RSS) system. As mentioned in the Classroom Manual, the term RTD is used in place of RSS in the system diagnosis.

To enter the diagnostic mode, turn the ignition switch on, and press the off and warmer buttons simultaneously on the climate control center (CCC). Hold these buttons until the segment display occurs in the instrument panel cluster (IPC). The segment display verifies all the segments are operational in the IPC. The turn signal indicators are not illuminated during this check. Do not proceed with the diagnostics unless all IPC segments are illuminated during the segment display. If any of the IPC segments are not illuminated, erroneous diagnosis may occur. The IPC must be replaced if any segments do not illuminate.

Once the diagnostics are entered, the HI and LO buttons in the CCC may be used to select or reject test displays. Pressing the HI button may be compared to a yes input, while pressing the LO button may be considered a no input. After the diagnostics are entered, the technician may select diagnostic code displays from these computers:

1. powertrain control module (PCM)
2. instrument panel cluster (IPC)
3. air conditioning programmer (ACP)
4. supplemental inflatable restraint (SIR)
5. traction control system (TCS)
6. real time damping (RTD)

The abbreviation for each computer appears in the instrument panel display, and the technician presses the HI button to select fault codes from the computer displayed. If the technician does not want fault codes from a computer, the LO button is pressed, and the display moves to the next computer abbreviation (Figure 8-21).

The fault codes from any computer are a three-digit number with a one-letter prefix and a one-letter suffix. The fault codes have these prefixes:

1. PCM - P
2. IPC - I
3. APC - A
4. SIR - R
5. TCS - T
6. RTD - S

Therefore, any fault codes in the RTD module are prefixed by the letter S. The suffix is a C or an H. When a code has a C suffix, it is a current code that is present at the time of diagnosis. An H suffix indicates the fault code is a history, or intermittent, code. The fault codes are displayed in numerical order.

If there are no fault codes in the RTD module, NO S CODES appears in the IPC display. If NO S DATA is displayed, the RTD module cannot communicate with the IPC display. When the AUTO button is pressed any time during the fault code display, the system changes from the diagnostic mode to normal operation.

## Optional Diagnostic Modes

After the fault code display, the technician may select these options:

1. RTD data
2. RTD inputs

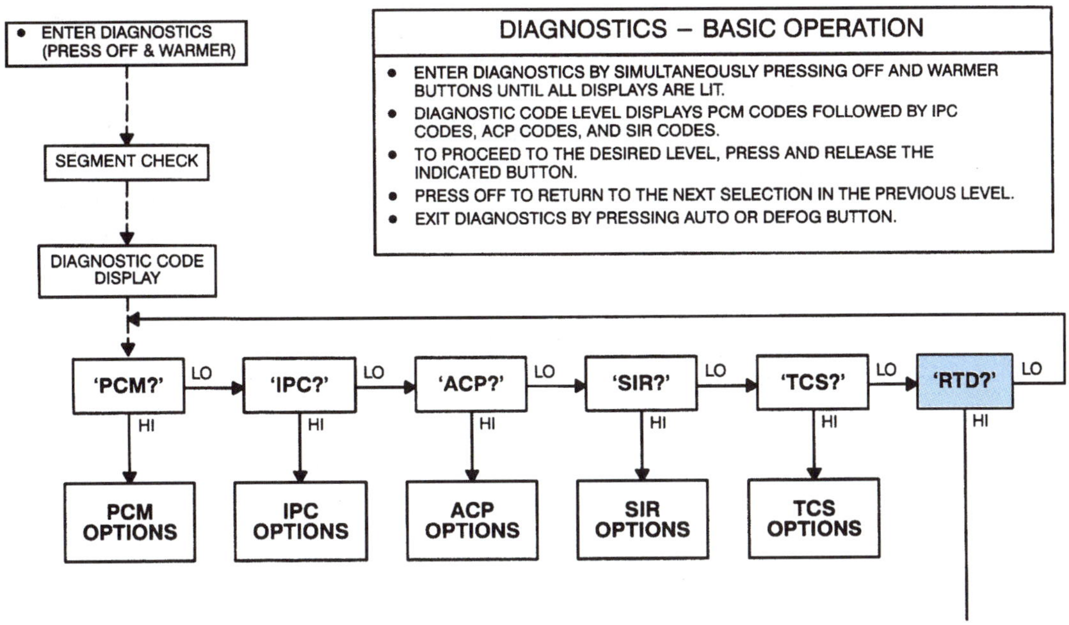


**Figure 8-21** Selecting fault code displays from various computers during the diagnostic procedure (Courtesy of Cadillac Motor Car Division, General Motors Corporation)

3. RTD outputs
4. RTD clear codes

When one of these options is displayed in the IPC, the technician selects the option by pressing the HI button. The technician rejects the displayed option by pressing the LO button, and the display moves to the next option.

If RTD data is selected, the data from a number of inputs may be displayed in numerical order, beginning with number SD01, R.F. accelerometer (Figure 8-22). This means the car must be driven to obtain a variable input from this sensor. This sensor input appears as a numerical range from -100 to 100%.

As the vehicle is accelerated and decelerated, this sensor input should change within the specified range. Always consult the car manufacturer's service manual for more detailed diagnosis and specifications. The next test in the data mode is selected by pressing the HI button. The display goes back to the previous test parameter when the LO button is pressed.

If the technician selects the RTD inputs, the lift/dive discrete input may be checked. When the technician selects RTD outputs, the RTD module cycles the displayed output. The next output in the test sequence is selected by pressing the HI button.

When CLEAR CODES is selected, the codes are erased, and RTD CODES CLEAR is displayed. The AUTO, or DEFOG, button may be pressed to exit the diagnostics.

**Classroom Manual**
Chapter 8, page 183

**CUSTOMER CARE:** While discussing computer-controlled suspension systems with customers, remember that the average customer is not familiar with automotive electronics terminology. Always use basic terms that customers can understand when explaining electronic suspension problems. Most customers appreciate a few minutes spent by service personnel to explain their automotive electronic problems. It is not necessary to provide customers with a lesson in electronics, but it is important that customers understand the basic cause of the problem with their car so they feel satisfied that the repair expenditures are necessary. A satisfied customer is usually a repeat customer!

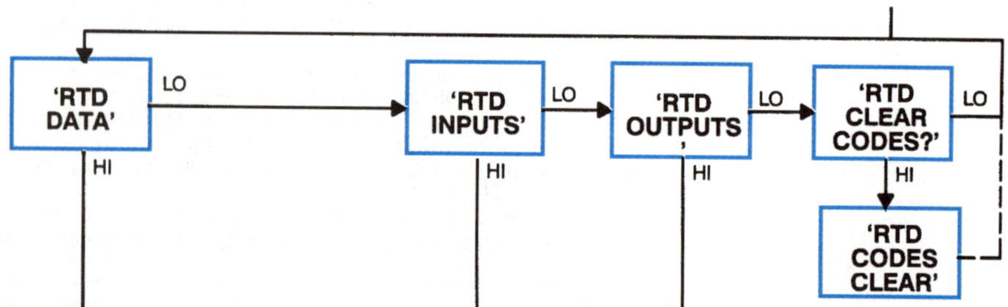


**RTD DATA**

| PARAMETER NUMBER | PARAMETER | DISPLAY RANGE | UNITS |
|---|---|---|---|
| SD01 | R.F. Accelerometer | −100 to 100% | Travel |
| SD02 | L.F. Accelerometer | −100 to 100% | Travel |
| SD03 | R.R. Accelerometer | −100 to 100% | Travel |
| SD04 | L.R. Accelerometer | −100 to 100% | Travel |
| SD05 | SSS Solenoid PWM | 0 to 100 | % |
| SD06 | SSS Solenoid Feedback | 0 to 100 | % |
| SD11 | R.F. Position Sensor | −100 to 100% | Travel |
| SD12 | L.F. Position Sensor | −100 to 100% | Travel |
| SD13 | R.R. Position Sensor | −100 to 100% | Travel |
| SD14 | L.R. Position Sensor | −100 to 100% | Travel |
| SD20 | Vehicle Speed | 0 to 159 | MPH |
| SD50 | Battery Volts | 0 to 16.3 | Volts |
| SD96 | HPC PROM ID | | Code |
| SD97 | COP PROM ID | | Code |
| SD98 | TMS320 PROM ID | | Code |
| SD99 | EEPROM ID | | Code |

**RTD INPUT**

| INPUT NUMBER | INPUT |
|---|---|
| SI01 | Lift/ Dive Discrete |

**RTD OUTPUTS**

| OUTPUT NUMBER | OUTPUT |
|---|---|
| SO00 | No Outputs |
| SO01 | R.F. Damper |
| SO02 | L.F. Damper |
| SO03 | R.R. Damper |
| SO04 | L.R. Damper |
| SO05 | SSS Solenoid |
| SO06 | ELC Compressor |
| SO07 | ELC Exhaust Valve |
| SO08 | Cycle All |

**KEY**

To Select Another Test Press:
Hi – To Increment
Lo – To Decrement

- *PROM ID Code Number Identifies An Individual Calibration And Is Periodically Updated; Refer To Latest Service Publication For Correct ID Number.*

**Figure 8-22** Optional test modes including RTD data, inputs, outputs, and clear codes (Courtesy of Cadillac Motor Car Division, General Motors Corporation)

## Guidelines for Servicing Computer-Controlled Suspension Systems

1. When an electronic air suspension system or rear load-leveling air suspension system is serviced, always turn off the air suspension switch in the trunk.
2. When hoisting, jacking, towing, or lifting a vehicle equipped with an electronic air suspension system or a rear load-leveling air suspension system, always turn off the air suspension switch in the trunk.
3. If a defect occurs in an electronic air suspension system, the control module illuminates the suspension warning lamp with the engine running.
4. There are 10 steps in the electronic air suspension diagnostic procedure. During the first 3 tests the suspension warning lamp flashes at 1.8 times per second if the system is normal. While tests 4 through 10 are performed, the suspension warning lamp flashes the test number. The driver's door is opened and closed to move to the next test in the sequence.
5. To release air from an air spring, the air spring valve should be rotated to the first stage. Never rotate an air spring valve to the second stage until all the air is exhausted from the spring.
6. Prior to installation, an air spring membrane must be folded properly onto the lower piston.
7. After an air spring is installed, there must be no creases in the spring membrane.
8. Never lower the full chassis weight onto an air spring with no air pressure in the spring.
9. In an electronic air suspension system, the front and rear height sensors may be adjusted to obtain the proper trim height.

10. A service fitting is available to splice damaged nylon air lines.
11. The rear height sensor may be adjusted to correct the trim height on a rear load-leveling air suspension system.
12. A scan tester may be connected to a test connector in the trunk to obtain fault codes in a rear load-leveling suspension system.
13. Fault codes may be obtained on a programmed ride control (PRC) system by connecting the two wires in a diagnostic connector behind the ash tray.
14. In a PRC system, the fault codes are flashed by the mode indicator light.
15. In a computer command ride (CCR) system, fault codes are obtained by connecting a jumper wire across terminals A and C in the DLC.
16. In a CCR system, both mode indicator lights flash at the same time to indicate the fault codes.
17. The test modes in the automatic air suspension diagnosis are drive cycle diagnosis, service bay diagnosis, and spring fill diagnosis.
18. When diagnosing the air suspension system with speed leveling capabilities, the test modes are functional test 211, auto test, and pinpoint tests.
19. To enter the diagnostics on the road sensing suspension system, the off and warmer buttons in the climate control center are pressed simultaneously with the ignition switch on.
20. When diagnosing the road sensing suspension system, the HI and LO buttons in the climate control center are used as yes and no inputs to select fault code displays from various on-board computers.
21. When diagnosing the road sensing suspension system, the HI and LO buttons are used to select various test options and move ahead or back up within the parameters in a specific test option.

## CASE STUDY

A customer complained about excessive body lean while cornering on the left front suspension of a 1993 Thunderbird turbo coupe equipped with a programmed ride control (PRC) system. The customer also said that the programmed ride control lamp in the tachometer started flashing when the problem occurred, but this light would stop flashing after a while. Further questioning of the customer indicated that the problem was intermittent. The problem did not always occur during hard cornering. The service writer asked the customer if there was any evidence of noise from the suspension, and the customer indicated that the flashing mode indicator lamp and the body lean on the left front suspension were the only symptoms.

Since the PRC mode indicator lamp was not flashing when the customer brought the car into the shop, the technician took the car for a road test. The technician brought along a digital volt-ohmmeter and a fabricated tool to connect the terminals in the PRC self-test connector. After several attempts at hard cornering, the car displayed the symptoms explained by the customer. While the PRC mode indicator light was still flashing, the technician drove the car into a parking lot and performed a self-test on the PRC system. A code 4 was present during the self-test, indicating a problem in the left front strut actuator circuit. The technician disconnected the wires from the left front actuator and connected an ohmmeter across the strut motor terminals connected to the firm and soft relays. A normal reading was obtained at these terminals. Careful examination of the wiring harness near the strut tower indicated the harness had been punctured with something. The car had a new paint job and closer examination indicated extensive repairs on the front body structure. It was logical to assume that the wiring harness had been punctured during collision damage.

The technician removed some of the tape from the wiring harness and found that the insulation was damaged on the wires from the firm and soft relays to the strut actuator, and these wires were touching together at times.

The technician drove the car back to the shop and repaired the wires. Another road test indicated that the PRC system worked perfectly.

## Terms to Know

Trim height
Scan tester
Pinpoint tests
Fault codes
Programmed ride control (PRC) system
Computer command ride (CCR) system
Light emitting diode (LED)
Electronically erasable programmable read only memory (EEPROM)
Data link connector (DLC)
Continual strut cycling
Drive cycle diagnosis
Service bay diagnostics
Auto/manual diagnostic check
Hard fault code
Intermittent fault code
Pinpoint tests
Spring fill diagnostics
Functional test 211
Auto test
Powertrain control module (PCM)
Instrument panel cluster (IPC)
Air conditioning programmer (ACP)
Supplemental inflatable restraint (SIR)
Traction control system (TCS)
Real time damping (RTD)
Road sensing suspension (RSS)
Current code
History code

## ASE Style Review Questions

1. When a deflated air spring is installed:
*Technician A* says the complete vehicle weight should be applied to the spring prior to spring inflation.
*Technician B* says the vehicle weight should be supported on a hoist so that a slight load is placed on the air spring prior to spring inflation.
Who is correct?
**A.** A only
**B.** B only
**C.** Both A and B
**D.** Neither A nor B

2. On a vehicle with an electronic air suspension system, a bumper jack is used to lift one corner of the vehicle to change a tire:
*Technician A* says the electronic air suspension switch should be in the on position.
*Technician B* says the air spring should be deflated on the corner of the vehicle being lifted.
Who is correct?
**A.** A only
**B.** B only
**C.** Both A and B
**D.** Neither A nor B

3. While discussing electronic air suspension system service:
*Technician A* says the suspension warning lamp flashes 1.8 times per second during the first 3 tests in the test sequence.
*Technician B* says the suspension warning lamp flashes the test number during tests 4 through 10 in the test sequence.
Who is correct?
**A.** A only
**B.** B only
**C.** Both A and B
**D.** Neither A nor B

4. While discussing rear load-leveling air suspension system service:
*Technician A* says a scan tester may be connected to a diagnostic connector in the trunk to obtain fault codes.
*Technician B* says if there are no faults in the control module memory, a code 14 is provided.
Who is correct?
**A.** A only
**B.** B only
**C.** Both A and B
**D.** Neither A nor B

5. While discussing PRC system service:
*Technician A* says a glowing mode indicator light indicates a system defect.
*Technician B* says a flashing mode indicator light indicates a system defect.
Who is correct?
**A.** A only
**B.** B only
**C.** Both A and B
**D.** Neither A nor B

6. While discussing PRC system service and diagnosis:
*Technician A* says the self-test connector is located behind the ash tray.
*Technician B* says the self-test connector is mounted beside the fuse panel.
Who is correct?
**A.** A only
**B.** B only
**C.** Both A and B
**D.** Neither A nor B

7. While discussing computer command ride (CCR) system service and diagnosis:
*Technician A* says that terminals A and B should be connected in the data link connector (DLC) to obtain fault codes.
*Technician B* says that terminals A and H should be connected in the ALDL to obtain fault codes.
Who is correct?
**A.** A only
**B.** B only
**C.** Both A and B
**D.** Neither A nor B

8. While discussing CCR system service and diagnosis:
*Technician A* says if a defect occurs in the system, the control module illuminates both driver-select switch light emitting diodes (LEDs).
*Technician B* says when this system is in the diagnostic mode, both driver-select switch LEDs flash the fault codes.
Who is correct?
**A.** A only
**B.** B only
**C.** Both A and B
**D.** Neither A nor B

9. While discussing road sensing suspension system service and diagnosis:
*Technician A* says the diagnostics are entered by pressing the AUTO and DEFOG buttons simultaneously with the ignition switch on.
*Technician B* says the HI button in the climate control center may be pressed to move to the next parameter in a test option.
Who is correct?
**A.** A only
**B.** B only
**C.** Both A and B
**D.** Neither A nor B

10. While discussing road sensing suspension system service and diagnosis:
*Technician A* says in the output test option the road sensing suspension module cycles various outputs.
*Technician B* says a history code represents an intermittent fault.
Who is correct?
**A.** A only
**B.** B only
**C.** Both A and B
**D.** Neither A nor B

## Table 8-3 ASE TASK

Diagnose, inspect, adjust, repair, or replace components of electronically controlled suspension systems.

| Problem Area | Symptoms | Possible Causes | Classroom Manual | Shop Manual |
|---|---|---|---|---|
| **RIDE CONTROL** | Body sway, dive, or lift | Defective electronic suspension system | 153 | 163 |
| **RIDE HEIGHT** | Low or high ride height | 1. Defective electronic air suspension system | 153 | 163 |
| | | 2. Trim height adjustment | 153 | 167 |

CHAPTER 9

# Steering Column and Linkage Diagnosis and Service

Upon completion and review of this chapter, you should be able to:

- ❑ Diagnose steering columns.
- ❑ Remove and replace steering wheels on air-bag-equipped vehicles and non-air-bag-equipped vehicles.
- ❑ Remove and replace air bag deployment modules and clock spring electrical connectors.
- ❑ Remove and replace steering columns.
- ❑ Inspect collapsible steering columns for damage.
- ❑ Disassemble steering columns.
- ❑ Inspect steering column components and replace necessary parts.
- ❑ Assemble steering columns.
- ❑ Diagnose and service flexible couplings and universal joints.
- ❑ Diagnose steering linkage mechanisms.
- ❑ Diagnose, remove, and replace tie rod ends.
- ❑ Diagnose, remove, and replace pitman arms.
- ❑ Diagnose, remove, and replace center links.
- ❑ Diagnose, remove, and replace idler arms.
- ❑ Diagnose, remove, and replace steering dampers.
- ❑ Diagnose steering arms.

## Air Bag Deployment Module, Steering Wheel, and Clock Spring Electrical Connector Removal and Replacement

**Basic Tools**

Basic technician's tool set

Service manual

Steering wheel puller

Foot-pound torque wrench

Machinist's rule

Electric drill, 1/4 inch

Set of drill bits

Screw extractors

**WARNING:** Always disconnect the negative battery terminal and wait one minute before diagnosing or servicing any air bag system component. Failure to observe this precaution may result in accidental air bag deployment. Air bag service precautions vary depending on the vehicle. Always follow the vehicle manufacturer's air bag service precautions in the service manual.

**WARNING:** On an air-bag-equipped vehicle, the wait time prior to servicing electrical components after the negative battery terminal is disconnected varies depending on the vehicle make and model year. Always follow the wait time and all other service precautions recommended in the vehicle manufacturer's service manual.

**WARNING:** Always use the vehicle manufacturer's recommended tools when diagnosing or servicing an air bag system or the electrical system on an air-bag-equipped vehicle. Do not use such equipment as a 12-V test light or a self-powered test light to diagnose air bag systems or electrical systems on air-bag-equipped vehicles. Accidental air bag deployment may result from improper use of tools.

**SERVICE TIP:** Before disconnecting the battery negative cable, note how the customer has the radio stations programmed, and reset the radio and clock after the battery negative cable is reconnected.

Prior to working on an air bag system, always disconnect the negative battery cable and wait one minute before proceeding with the diagnostic or service work. Many air bag systems have a backup power supply circuit designed into the air bag computer or located in a separate module.

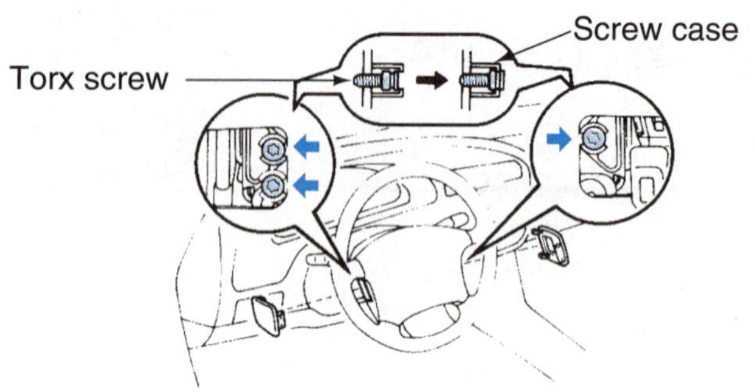


**Figure 9-1** Three air-bag-retaining Torx screws (Courtesy of Toyota Motor Corporation)

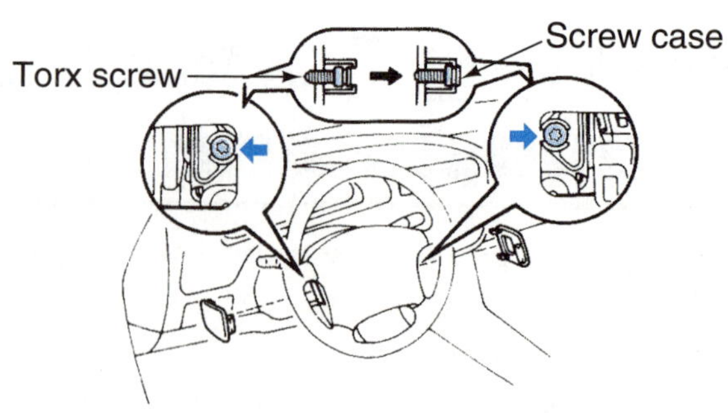


**Figure 9-2** Two air-bag-retaining Torx screws (Courtesy of Toyota Motor Corporation)

This backup power supply provides power to deploy the air bag for a specific length of time after the battery power is disconnected in a collision. One minute after the negative battery terminal is disconnected, this backup power supply is powered down, and no power is available to deploy the air bag while the battery remains disconnected.

An air bag warning light in the instrument panel indicates the status of the air bag system. On some vehicles, this warning light should be illuminated for a few seconds when the ignition switch is turned on. The warning light should remain off while cranking the engine, and it should be on for a few seconds after the engine starts. After the engine has been running for a few seconds, the air bag warning light should remain off. On other vehicles, the air bag warning light flashes 7 to 9 times when the ignition switch is turned on and after the engine is started. The air bag warning light operation varies depending on the make and model year of the vehicle. Always check the vehicle manufacturer's service manual for the exact air bag warning light operation. When the air bag warning light does not operate as specified by the vehicle manufacturer, the air bag system is defective, and the air bag, or bags, will probably not deploy if the vehicle is involved in a collision.

Air bag module and steering wheel removal and replacement procedures vary depending on the vehicle. Always follow the vehicle manufacturer's recommended procedure in the service manual. Following is a typical air bag module and steering wheel removal and replacement procedure:

1. Turn the ignition switch to the lock position and place the front wheels facing straight ahead.
2. Remove the negative battery terminal and wait one minute.
3. Loosen the three air-bag-retaining Torx screws under the steering wheel (Figure 9-1).
4. Loosen the other two air-bag-retaining Torx screws under the steering wheel (Figure 9-2). Loosen all five Torx screws until the groove along the screw circumference catches on the screw case.

An air bag deployment module may be referred to as a steering wheel pad.

**CAUTION:** When an air bag deployment module is temporarily stored on the work bench, always place this module face upward. If the air bag deployment module accidentally deployed when facing downward, the module would become a projectile, and personal injury might result.

5. Pull the air bag deployment module from the steering wheel, and disconnect the air bag module electrical connector (Figure 9-3). Do not pull on the air bag wires in the steering column. Place the air bag deployment module face upward on the work bench.
6. Disconnect the air bag wiring retainer in the steering wheel (Figure 9-4).
7. Use the proper size socket and a ratchet to remove the steering wheel retaining nut.

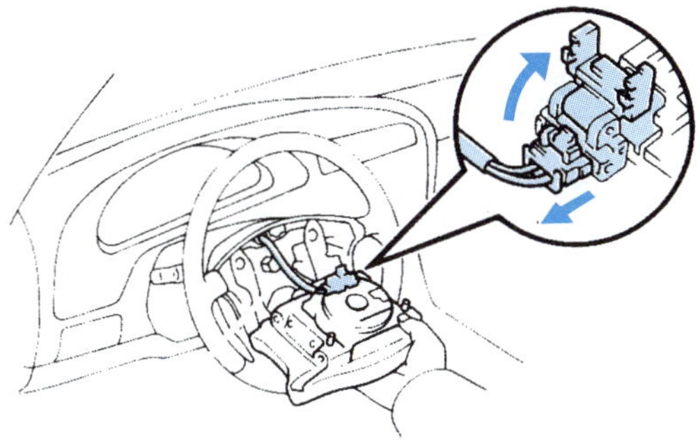

Figure 9-3 Disconnecting air bag module electrical connector (Courtesy of Toyota Motor Corporation)

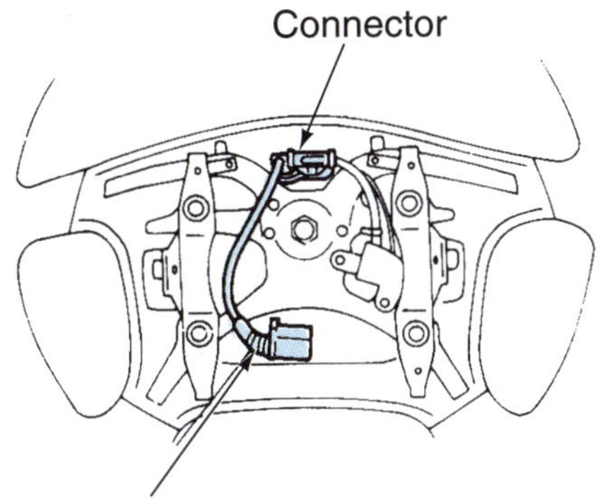


Figure 9-4 Disconnecting the air bag wiring retainer (Courtesy of Toyota Motor Corporation)

8. Observe the matching alignment marks on the steering wheel and the steering shaft. If these alignment marks are not present, place alignment marks on the steering wheel and steering shaft with a center punch and a hammer.

**WARNING:** Do not hammer on the top of the steering shaft to remove the steering wheel. This action may damage the shaft.

**CAUTION:** Do not pull on the steering wheel in an attempt to remove it from the steering shaft. Photo Sequence 7 shows a typical procedure for removing a steering wheel. The steering wheel may suddenly come off, resulting in personal injury, or the steering wheel may be damaged by the pulling force.

9. Install a steering wheel puller with the puller bolts threaded into the bolt holes in the steering wheel. Tighten the puller nut to remove the steering wheel (Figure 9-5). Visually check the steering wheel condition. If the steering wheel is bent or cracked, replace the wheel.
10. Disconnect the four retaining screws and remove the clock spring electrical connector (Figure 9-6).
11. Be sure the front wheels are facing straight ahead. Turn the clock spring electrical connector counterclockwise by hand until becomes harder to turn as it becomes fully wound in that direction.

A clock spring electrical connector may be called a coil, spiral cable, or cable reel.

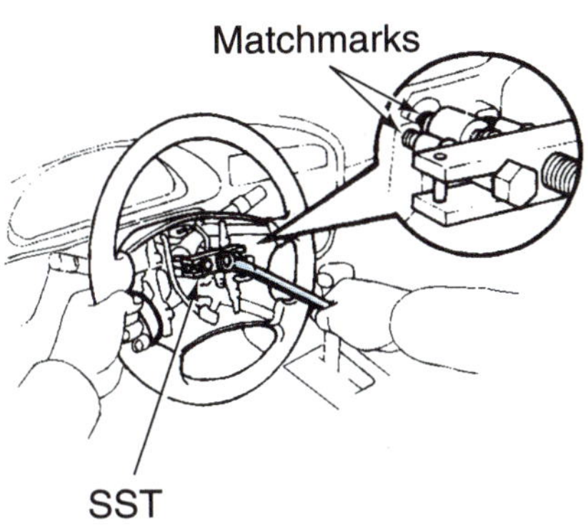


Figure 9-5 Removing steering wheel with the proper steering wheel puller (Courtesy of Toyota Motor Corporation)

## Photo Sequence 7

# Typical Procedure for Removing a Steering Wheel

**P7-1** Check and record the radio stations programmed in the stereo system.

**P7-2** Look up the car manfacturer's steering wheel removal procedure in the service manual.

**P7-3** Disconnect the negative battery cable, and wait for the car manufacturer's specified length of time.

**P7-4** Remove the air bag deployment module retaining screws under the steering wheel.

**P7-5** Lift the air bag deployment module upward from the steering wheel, and disconnect the module wiring connector.

**P7-6** Set the air bag deployment module face upward on the work bench.

**P7-7** Loosen and remove the steering wheel retaining nut.

**P7-8** Observe the alignment marks on the steering wheel and shaft.

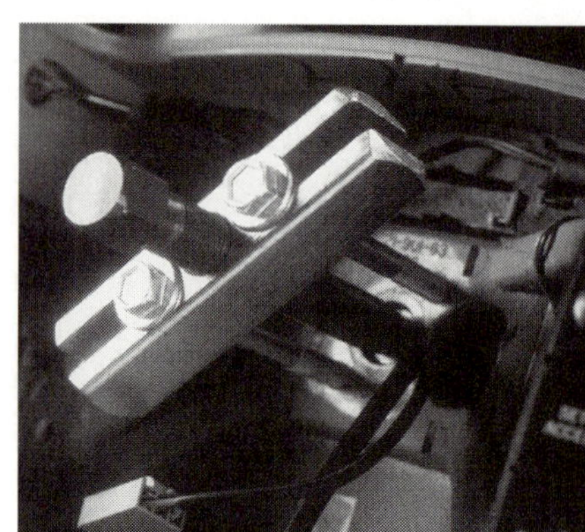

**P7-9** Connect the proper steering wheel puller to the steering wheel.

# Typical Procedure for Removing a Steering Wheel (continued)

**P7-10** Turn the puller screw to loosen the steering wheel on the shaft.

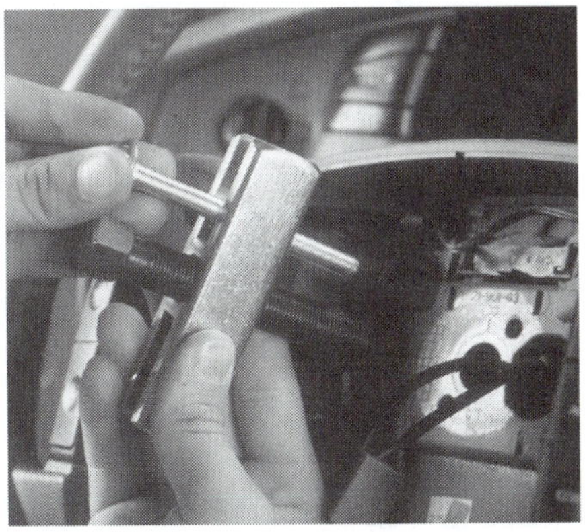

**P7-11** Remove the puller from the steering wheel.

**P7-12** Lift the steering wheel off the shaft.

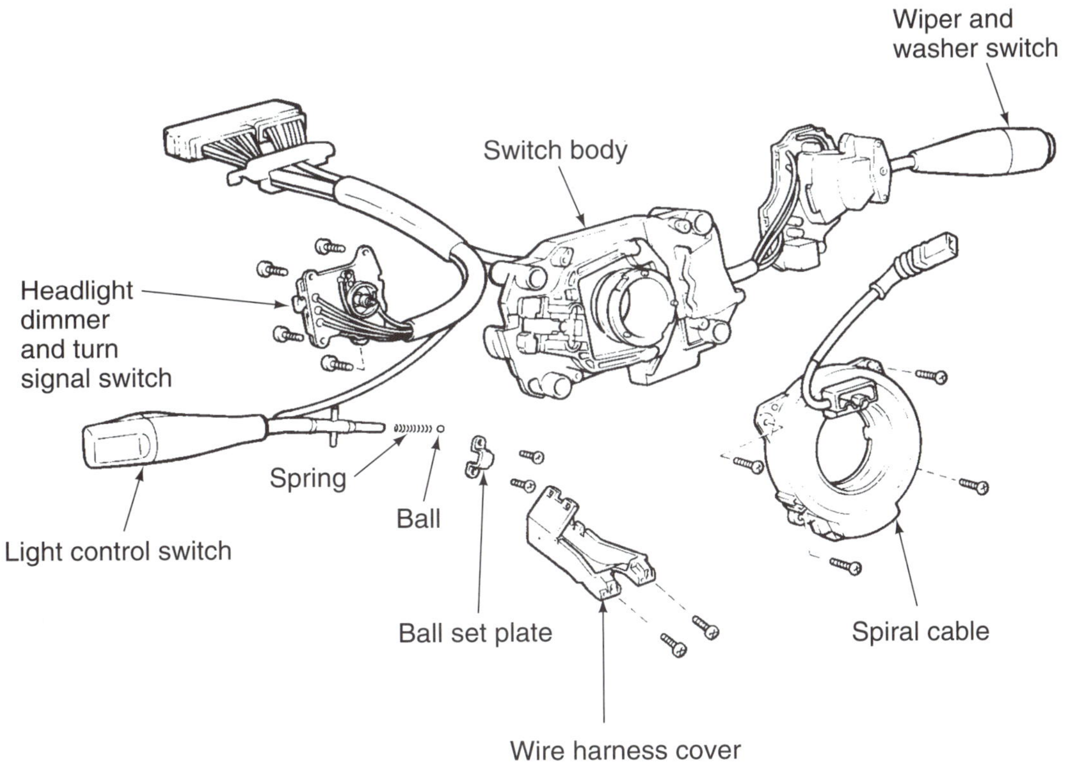


**Figure 9-6** Removing clock spring (Courtesy of Toyota Motor Corporation)

**WARNING:** Failure to center a clock spring prior to installation may cause a broken conductive tape in the clock spring.

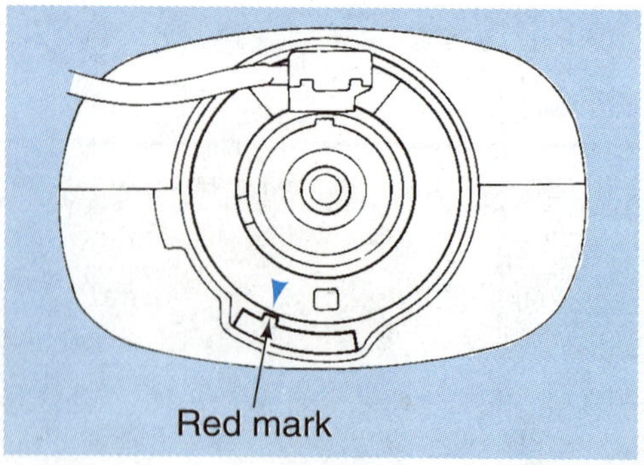


**Figure 9-7** Centering clock spring (Courtesy of Toyota Motor Corporation)

12. Turn the clock spring electrical connector clockwise three turns, and align the red mark on the center part of the spring face with the notch in the cable circumference (Figure 9-7). This action centers the clock spring electrical connector.
13. Install the clock spring electrical connector and tighten the four retaining screws to the specified torque.
14. Align the marks on the steering wheel and the steering shaft, and install the steering wheel on the shaft.
15. Install the steering wheel retaining nut and tighten this nut to the specified torque.
16. Install the air bag wiring retainer in the steering wheel.
17. Hold the air bag deployment module near the top of the steering wheel, and connect the air bag module connector.
18. Install the air bag deployment module in the top of the steering wheel, and tighten the five retaining Torx screws.
19. Reconnect the negative battery cable.
20. Reset the clock and radio.

If the vehicle is not equipped with an air bag, the steering wheel removal and replacement procedure is basically the same, but all steps pertaining to the air bag module and clock spring are not required. On a non-air bag vehicle, the center steering wheel cover must be removed to access the steering wheel retaining nut.

**CUSTOMER CARE:** While servicing a vehicle, always check the operation of the indicator lights or gauges in the instrument panel. These lights or gauges may indicate a problem that the customer has been ignoring. For example, if the air bag warning light is not operating properly, the air bag, or bags, may not be deployed in a collision, resulting in serious injury to the driver and/or passenger. If the air bag warning light is not working properly, always advise the customer that he or she will not be protected by the air bag in a collision, and the vehicle should not be driven under this condition.

## Steering Column Removal and Replacement

Steering column removal and replacement procedures vary depending on the vehicle make, type of steering column, and gearshift lever position. Always follow the vehicle manufacturer's recommended procedure in the service manual. Following is a typical steering column removal and replacement procedure:

1. Disconnect the battery negative cable. If the vehicle is equipped with an air bag, wait one minute.
2. Install a seat cover on the front seat.

3. Place the front wheels in the straight-ahead position, and remove the ignition key from the switch to lock the steering column.
4. Remove the cover under the steering column, and remove the lower finish panel if necessary.
5. Disconnect all wiring connectors from the steering column.
6. If the vehicle has a column-mounted gearshift lever, disconnect the gearshift linkage at the lower end of the steering column.
7. Remove the retaining bolt, or bolts, in the lower universal joint, or flexible coupling.
8. Remove the steering-column-to-instrument-panel mounting bolts.
9. Carefully remove the steering column from the vehicle. Be careful not to damage upholstery or paint.
10. Install the steering column under the instrument panel, and insert the steering shaft into the lower universal joint.
11. Install the steering-column-to-instrument-panel mounting bolts. Be sure the steering column is properly positioned, and tighten these bolts to the specified torque.
12. Install the retaining bolt, or bolts, in the lower universal joint, or flexible coupling, and tighten this bolt to the specified torque.
13. Connect the gearshift linkage, if the vehicle has a column-mounted gearshift.
14. Connect all the wiring harness connectors to the steering column connectors.
15. Install the steering column cover and the lower finish panel.
16. Reconnect the negative battery cable.
17. Road test the vehicle and check for proper steering column operation.

## Collapsible Steering Column Inspection

Since steering column design varies depending on the vehicle, the collapsible steering column inspection procedure should be followed in the vehicle manufacturer's service manual. Following is a typical collapsible steering column inspection procedure:

1. Measure the clearance between the capsules and the slots in the steering column bracket (Figure 9-8). If this measurement is not within specifications, replace the bracket.
2. Check the contact between the bolt head and the bracket (Figure 9-9). If the bolt head contacts the bracket, the shear load is too high, and the bracket must be replaced.

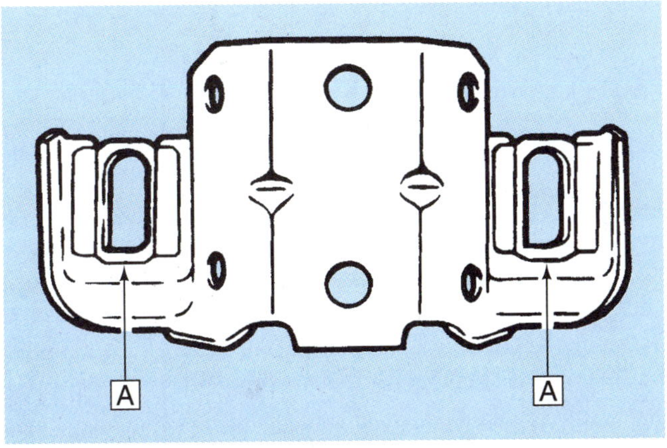


**Figure 9-8** Capsules in steering column bracket (Courtesy of Oldsmobile Division, General Motors Corporation)

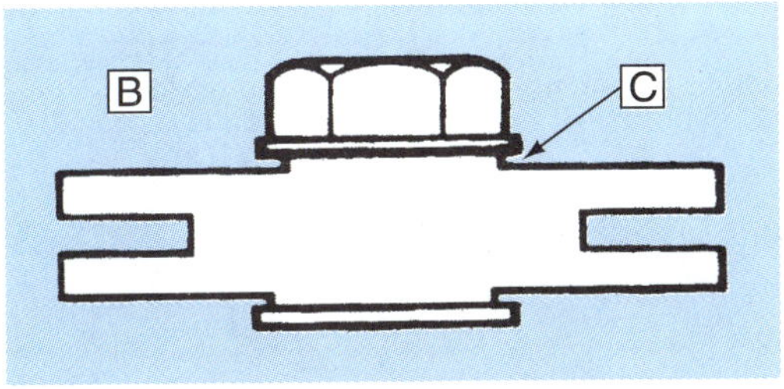


**Figure 9-9** Bolt head to bracket clearance (Courtesy of Oldsmobile Division, General Motors Corporation)

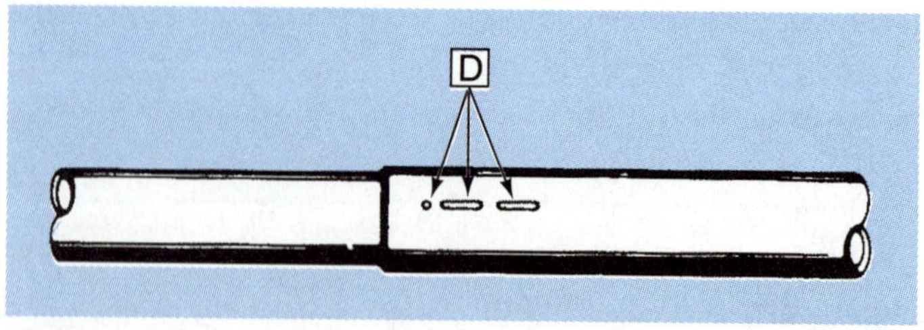


**Figure 9-10** Checking for sheared injected plastic in jacket openings (Courtesy of Oldsmobile Division, General Motors Corporation)

3. Check the steering column jacket for sheared injected plastic in the openings on the side of the jacket (Figure 9-10). If sheared plastic is present, the column is collapsed. Measure the distance from the end of the bearing assembly to the lower edge of the upper steering column jacket (Figure 9-11). If this distance is not within the vehicle manufacturer's specification, a new jacket must be installed.
4. Visually check the gearshift tube for sheared injected plastic. If the gearshift tube indicates sheared plastic, replace the tube.
5. Remove the intermediate steering shaft between the column and the steering gear. Position a dial indicator stem against the lower end of the steering shaft and rotate the steering wheel. If the runout on the dial indicator exceeds the vehicle manufacturer's specification, the steering shaft is bent and must be replaced.
6. If the steering shaft is not bent, but shows sheared injected plastic (Figure 9-12), this shaft may be repaired with a service repair package.

## Tilt Steering Column Disassembly

Since there are many variations in steering column design, always follow the vehicle manufacturer's recommended steering column disassembly procedure in the service manual. Following is a typical tilt steering column disassembly procedure:

1. Remove the air bag deployment module, steering wheel, and spiral cable as mentioned previously in this chapter.
2. Remove the ignition key cylinder illumination mounted on top of the ignition switch cylinder (Figure 9-13).

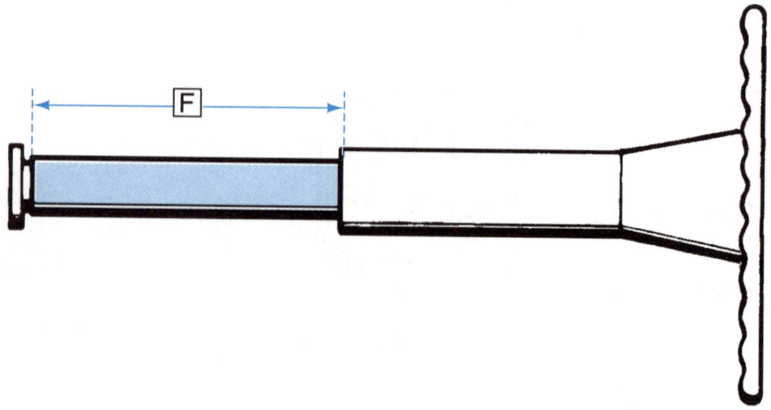


**Figure 9-11** Measuring distance from the end of the bearing assembly to the upper steering column jacket (Courtesy of Oldsmobile Division, General Motors Corporation)

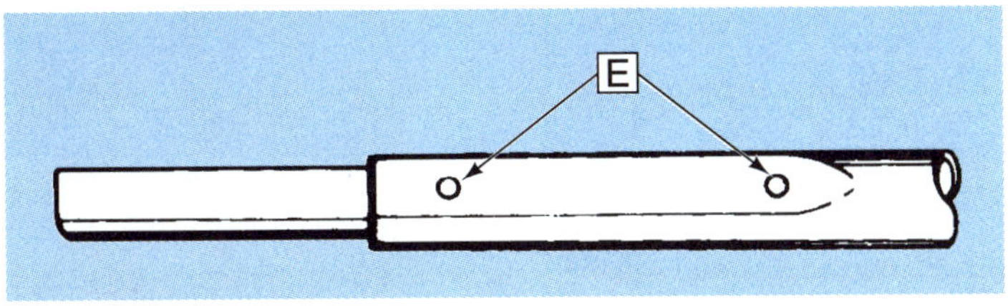


**Figure 9-12** Checking for sheared injected plastic in steering shaft (Courtesy of Oldsmobile Division, General Motors Corporation)

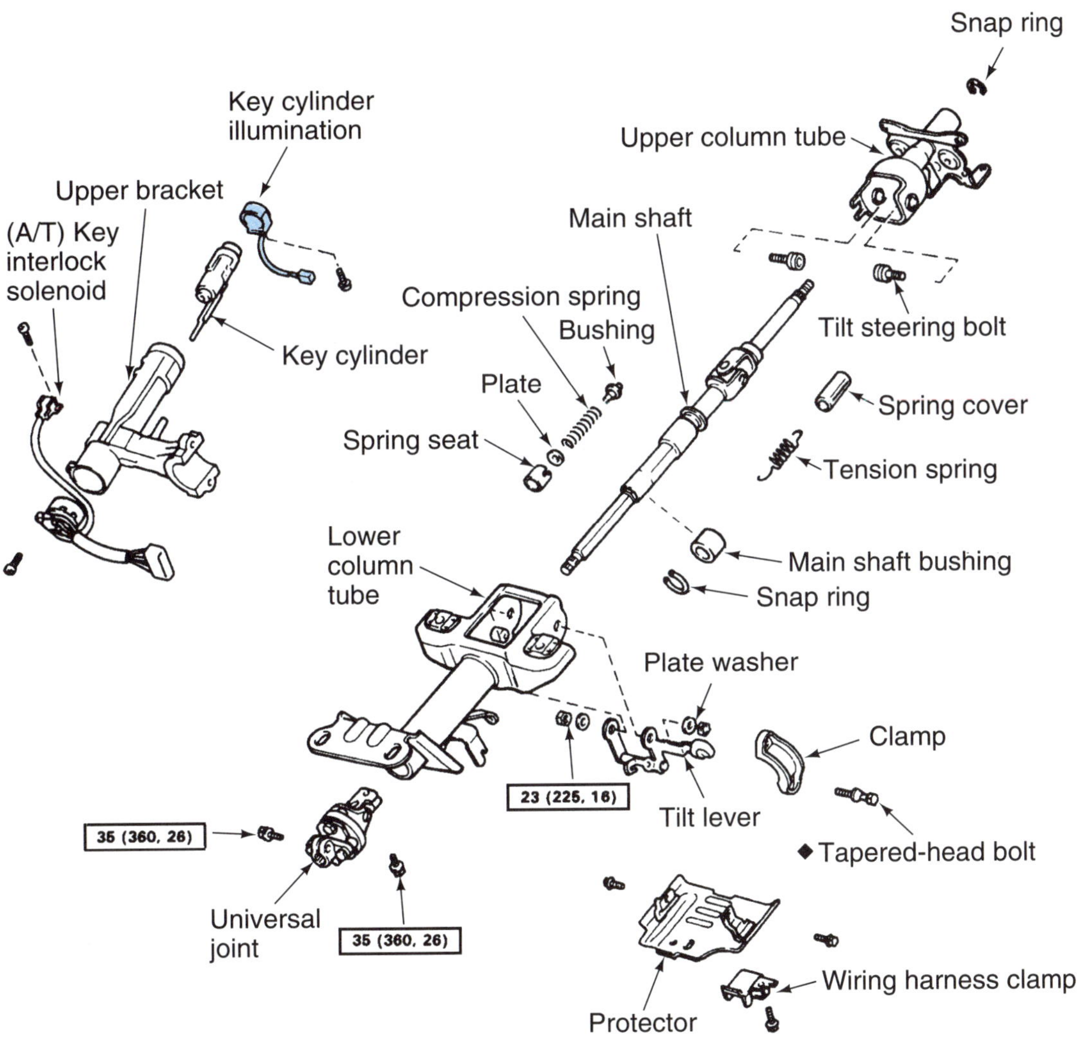


**Figure 9-13** Ignition key cylinder illumination mounted above ignition switch cylinder (Courtesy of Toyota Motor Corporation)

3. Remove the universal joint from the lower end of the steering shaft.
4. Remove the steering column protector and the wiring harness clamp mounted under the steering column.
5. Remove the steering damper mounted near the top of the column.

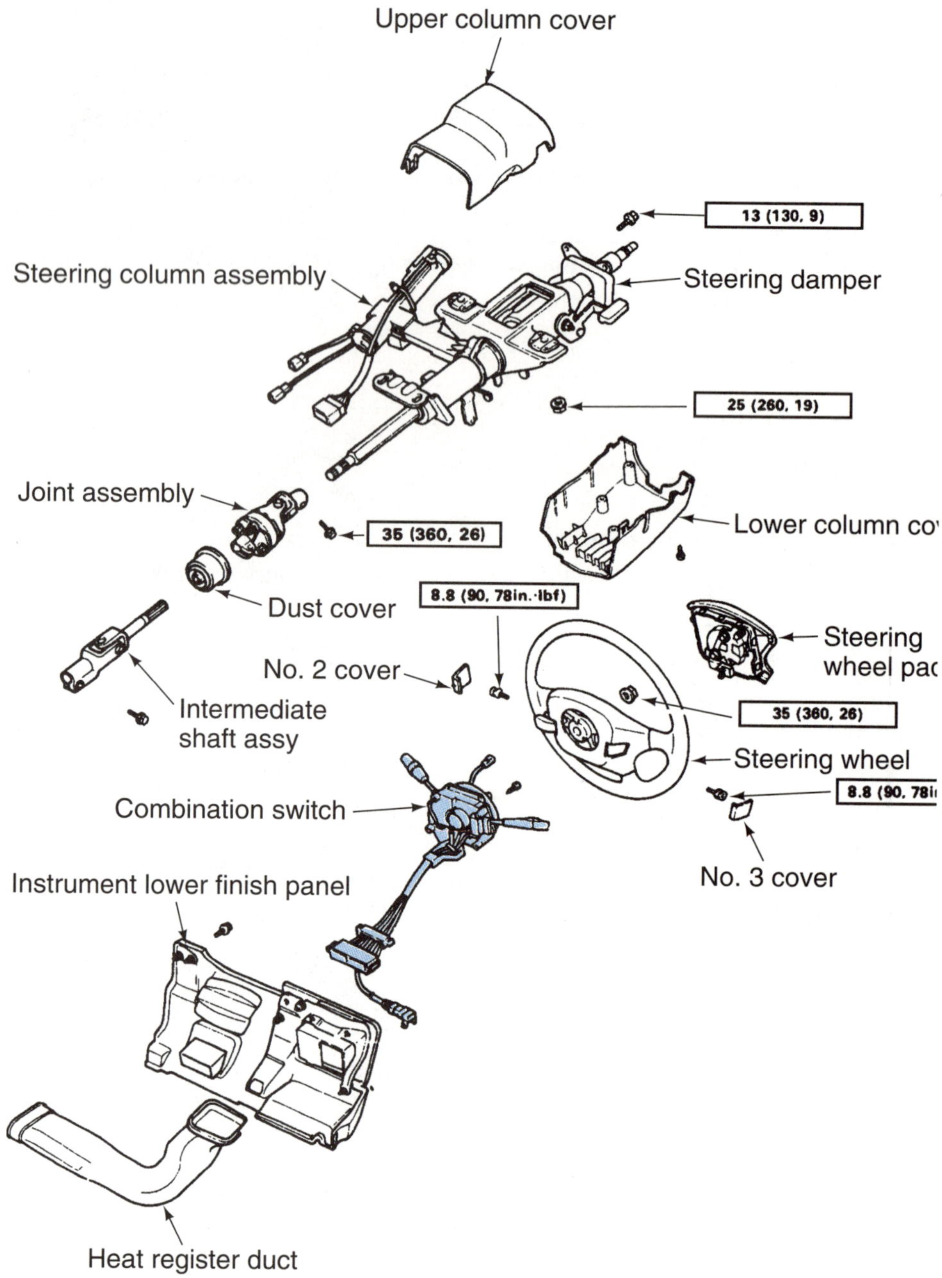


**Figure 9-14** Removing combination switch and wiring harness (Courtesy of Toyota Motor Corporation)

6. Remove the retaining screw and the combination switch with the wiring harness (Figure 9-14).
7. Mark the center of the tapered-head bolts on the steering column with a center punch. Use a 0.120 to 0.160 in (3 to 4 mm) drill bit to drill into the tapered-head bolts (Figure 9-15).
8. Remove the tapered-head bolts with a screw extractor and separate the upper bracket and column tube (Figure 9-16).
9. Grasp the compression spring seat with a pair of pliers and turn the seat to release the seat, compression spring, and bushing (Figure 9-17).

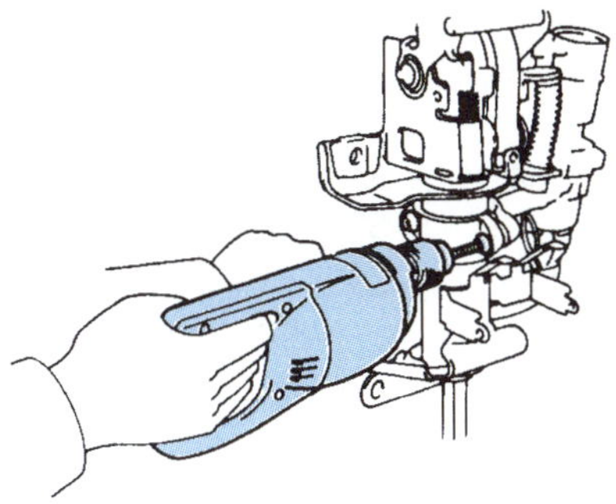

**Figure 9-15** Drilling tapered-head bolts (Courtesy of Toyota Motor Corporation)

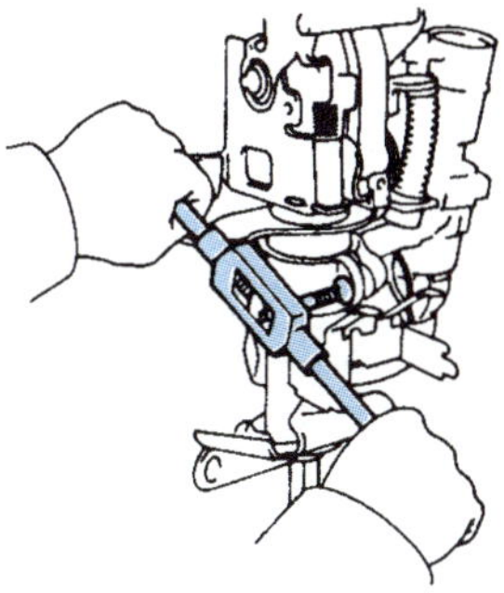

**Figure 9-16** Removing tapered-head bolts with a screw extractor (Courtesy of Toyota Motor Corporation)

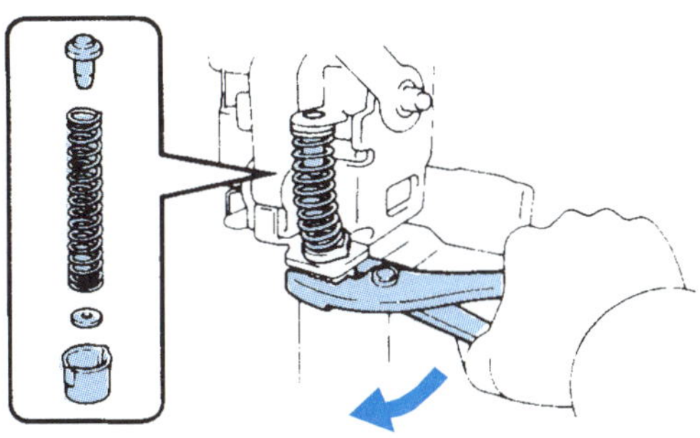

**Figure 9-17** Removing compression spring, seat, and bushing (Courtesy of Toyota Motor Corporation)

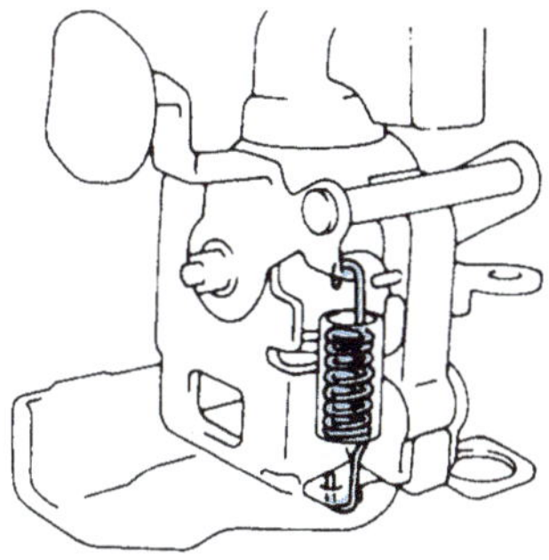

**Figure 9-18** Removing tension spring (Courtesy of Toyota Motor Corporation)

**10.** Grasp the tension spring with a pair of pliers and extend this spring to remove it from the column (Figure 9-18).

**11.** Remove the snap ring from the upper column tube with snap ring pliers (Figure 9-19).

**12.** Use a soft hammer and a screwdriver to loosen the staked parts of the upper column tube (Figure 9-20).

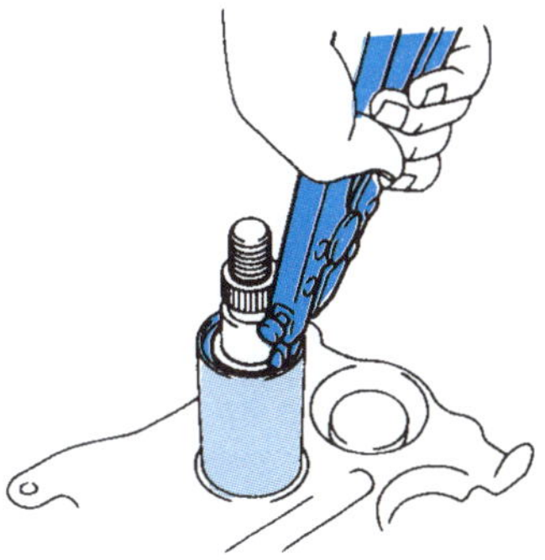

**Figure 9-19** Removing snap ring from upper column tube (Courtesy of Toyota Motor Corporation)

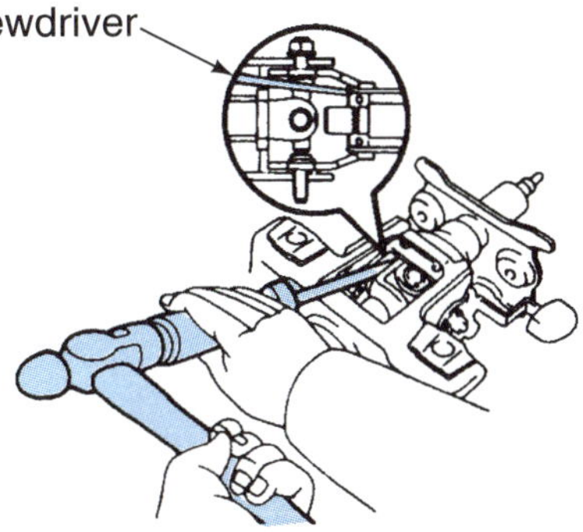


**Figure 9-20** Loosening staked parts in upper column tube (Courtesy of Toyota Motor Corporation)

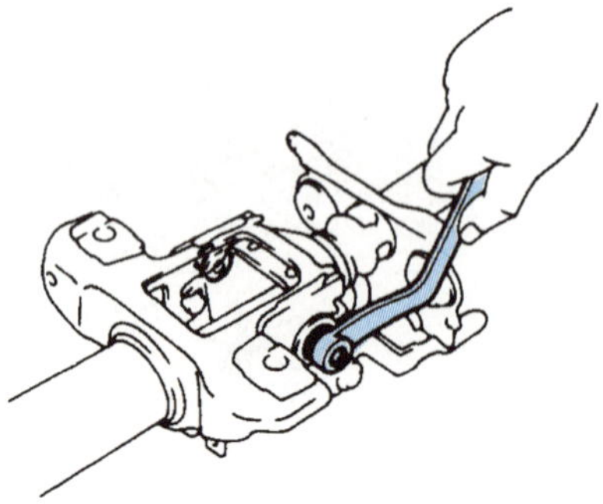

**Figure 9-21** Removing nuts on upper column pivot bolts (Courtesy of Toyota Motor Corporation)

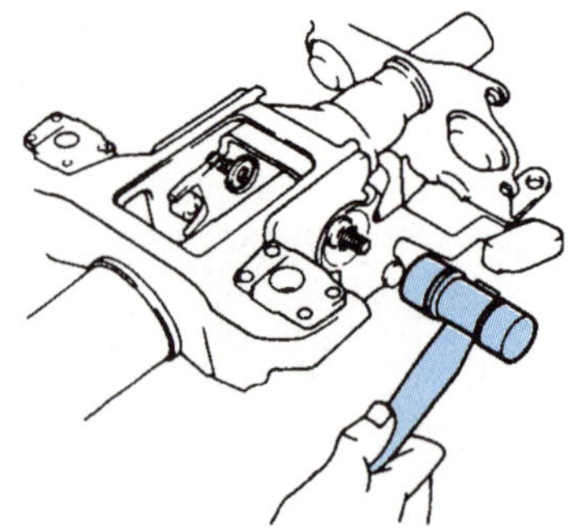

**Figure 9-22** Removing pivot bolts from column (Courtesy of Toyota Motor Corporation)

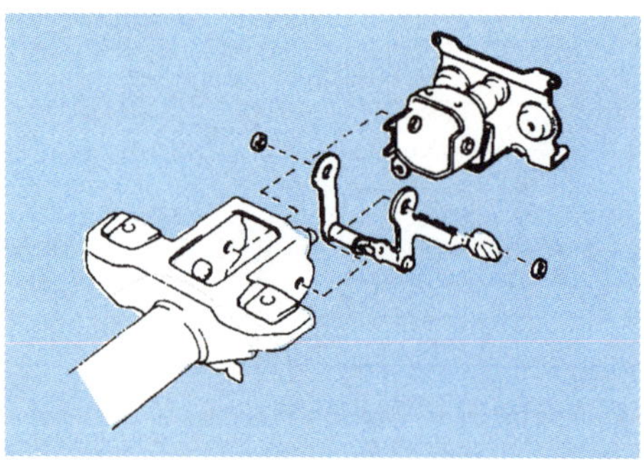

**Figure 9-23** Removing upper column tube, tilt lever assembly, and plate washers (Courtesy of Toyota Motor Corporation)

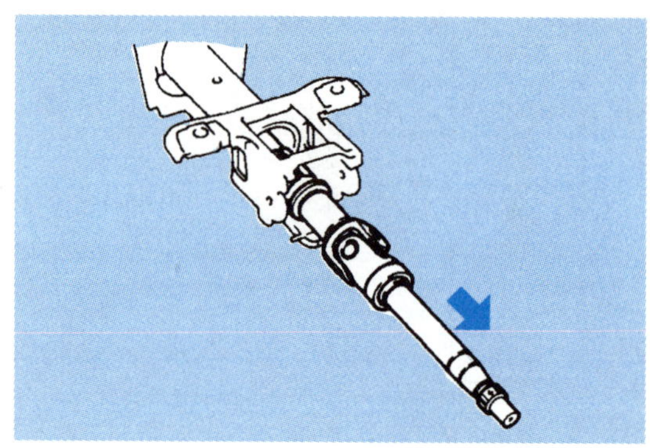

**Figure 9-24** Removing steering shaft from the steering column (Courtesy of Toyota Motor Corporation)

**13.** Remove the two nuts on the upper column pivot bolts (Figure 9-21).
**14.** Use a plastic hammer to tap the pivot bolts out of the column (Figure 9-22).
**15.** Remove the upper column tube, tilt lever assembly, and plate washers (Figure 9-23).
**16.** Remove the steering shaft from the column (Figure 9-24).
**17.** Use snap ring pliers to remove the snap ring above the steering shaft bushing (Figure 9-25). Remove the steering shaft bushing.

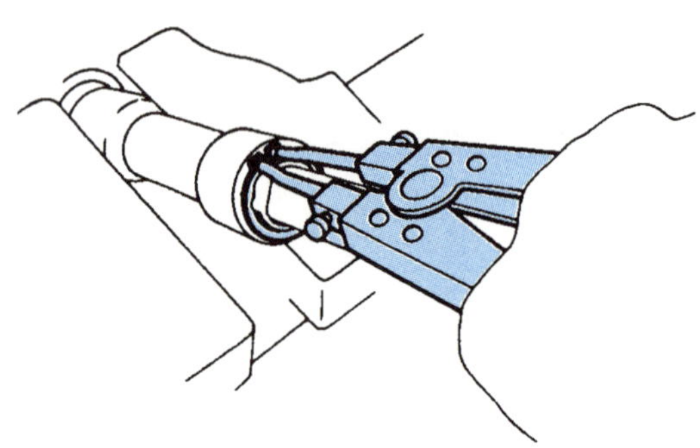

**Figure 9-25** Removing snap ring above steering shaft bushing (Courtesy of Toyota Motor Corporation)

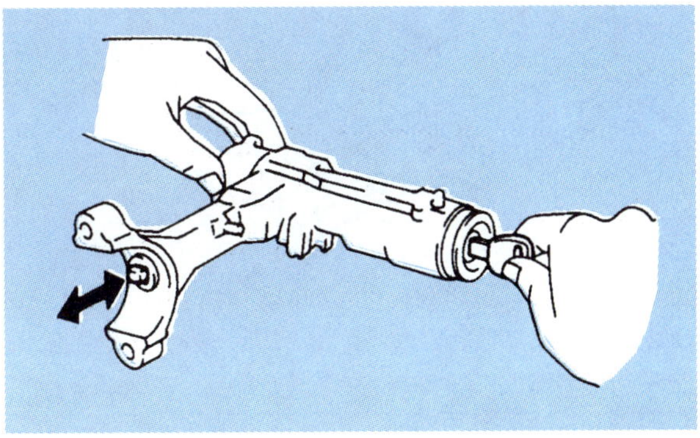

Figure 9-26 Checking ignition switch operation (Courtesy of Toyota Motor Corporation)

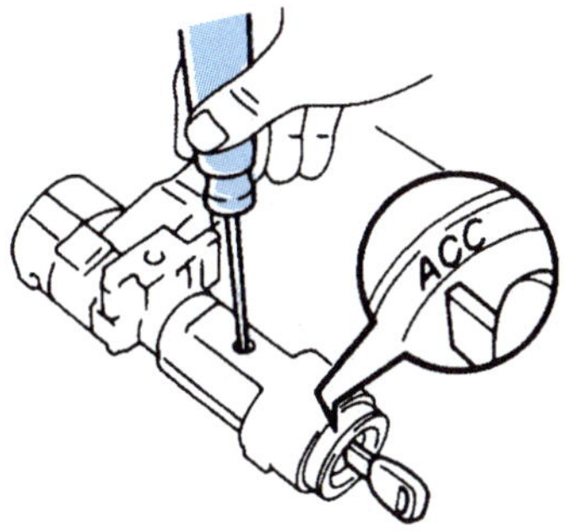


Figure 9-27 Depressing stop pin to release ignition key cylinder (Courtesy of Toyota Motor Corporation)

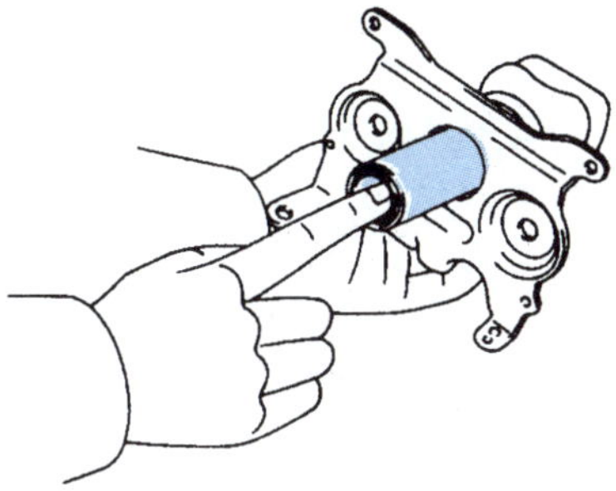

Figure 9-28 Checking upper steering shaft bearing (Courtesy of Toyota Motor Corporation)

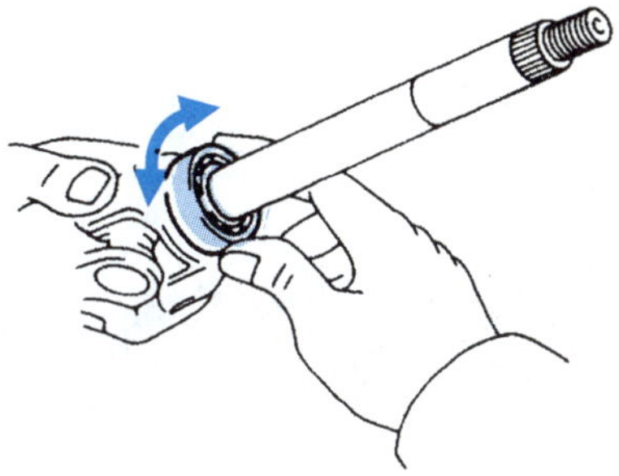

Figure 9-29 Checking lower steering shaft bearing (Courtesy of Toyota Motor Corporation)

## Tilt Steering Column Inspection and Parts Replacement

1. Place the ignition key in the switch and move the key through all the switch positions. Be sure the operating pin moves properly in the switch mechanism (Figure 9-26). If the ignition key cylinder must be replaced, turn the key to the accessory (ACC) position. Use a small steel rod to push down on the stop pin near the bottom of the key cylinder (Figure 9-27), and pull the cylinder out of the housing. Install the new key cylinder with the key in the ACC position.
2. Rotate the upper steering shaft bearing and check for noise, looseness, and wear (Figure 9-28). If any of these conditions are present, replace the upper tube.
3. Check the lower steering shaft bearing for noise, looseness, and wear (Figure 9-29). Replace this bearing if necessary.
4. Inspect the ignition key interlock solenoid for damaged wires and loose mounting screws (Figure 9-30). Repair or replace this solenoid as required.

## Tilt Steering Column Assembly

1. Coat all rubbing parts with molybdenum disulphide lithium-based grease. Install the steering shaft bushing and the bushing snap ring (Figure 9-31).

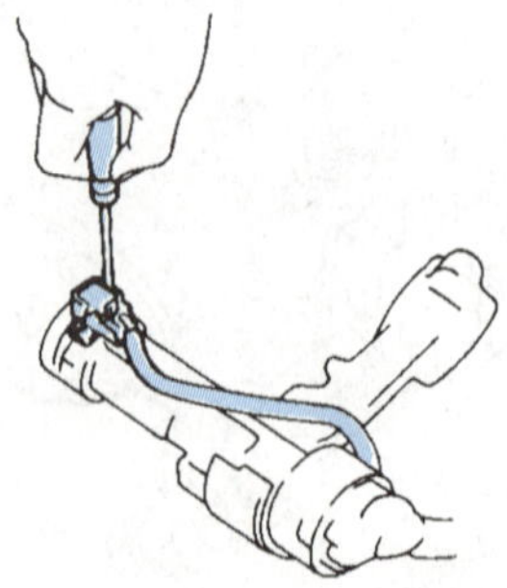

**Figure 9-30** Checking ignition key interlock solenoid (Courtesy of Toyota Motor Corporation)

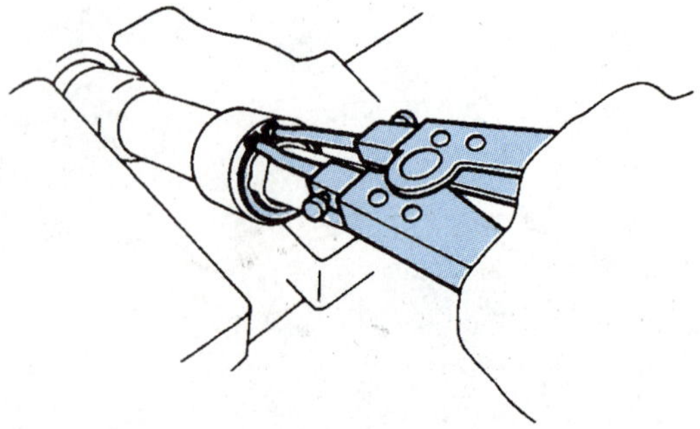

**Figure 9-31** Installing the steering shaft and snap ring (Courtesy of Toyota Motor Corporation)

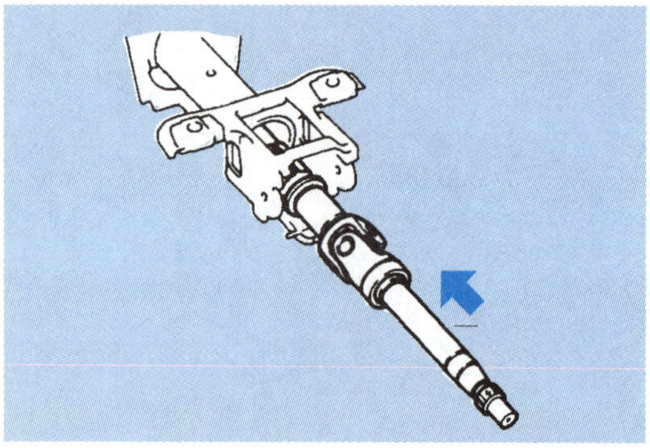

**Figure 9-32** Installing the steering shaft in the lower tube (Courtesy of Toyota Motor Corporation)

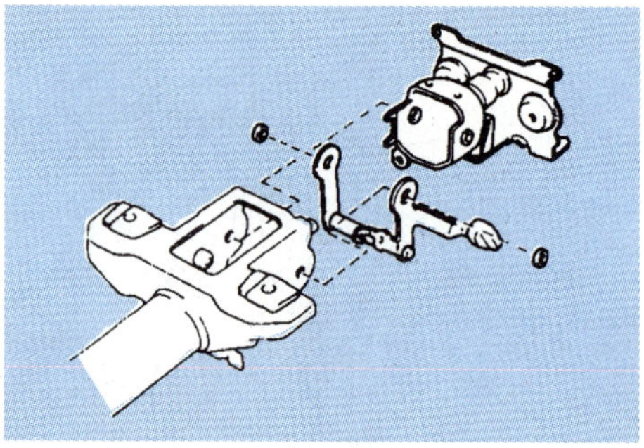

**Figure 9-33** Installing upper column tube, tilt lever mechanism, and two plate washers (Courtesy of Toyota Motor Corporation)

2. Install the steering shaft in the lower tube (Figure 9-32).
3. Install the upper column tube, tilt lever mechanism, and two plate washers (Figure 9-33). Install the two pivot bolts.
4. Install the pivot bolt nuts, and tighten these nuts to the specified torque (Figure 9-34).
5. Use a brass bar and a hammer to tap the steering shaft into the upper column tube (Figure 9-35).

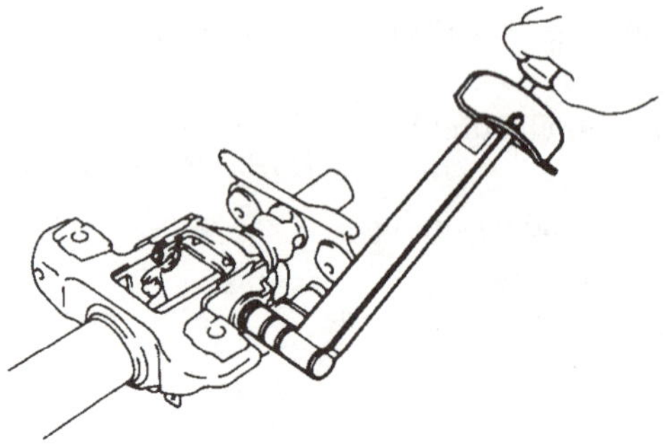

**Figure 9-34** Tightening pivot bolt nuts (Courtesy of Toyota Motor Corporation)

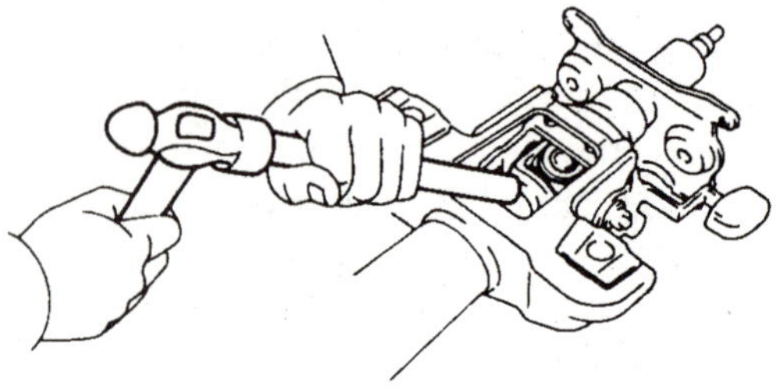

**Figure 9-35** Tapping steering shaft into upper column tube (Courtesy of Toyota Motor Corporation)

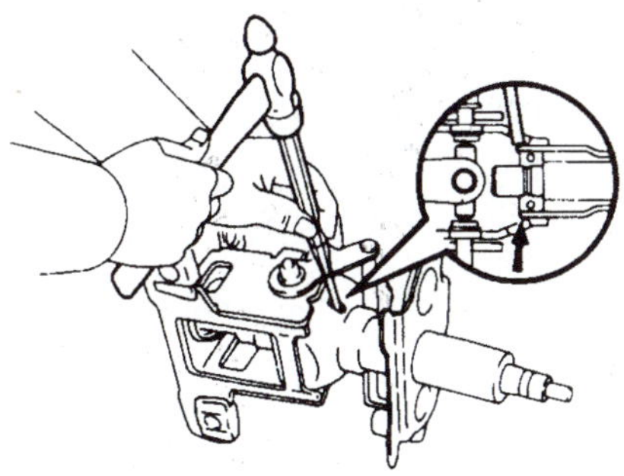

**Figure 9-36** Staking upper column tube (Courtesy of Toyota Motor Corporation)

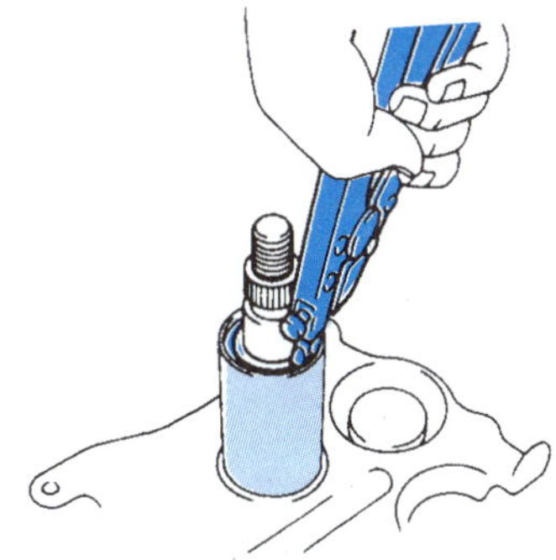

**Figure 9-37** Installing snap ring in upper column tube (Courtesy of Toyota Motor Corporation)

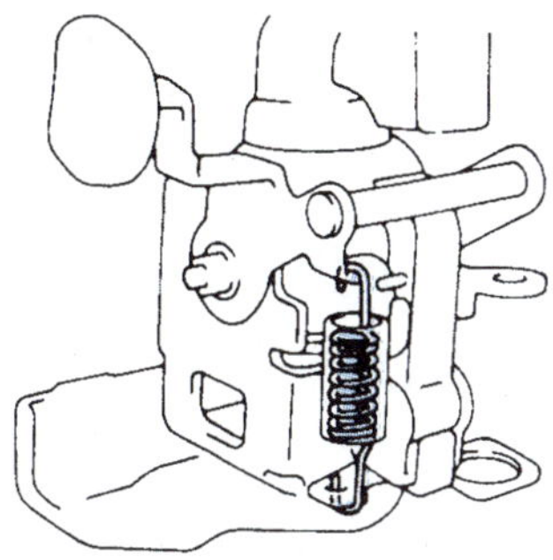

**Figure 9-38** Installing tension spring (Courtesy of Toyota Motor Corporation)

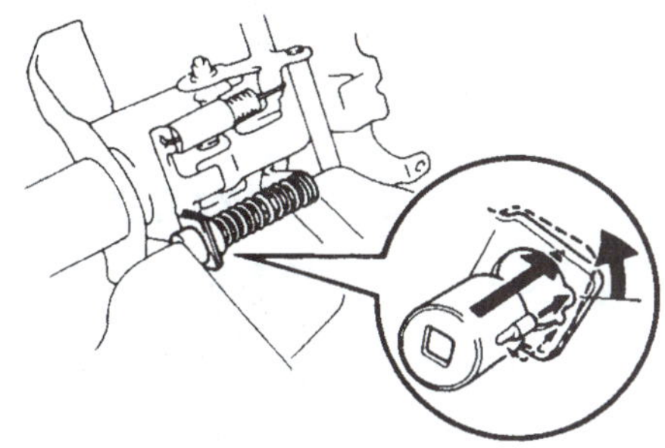

**Figure 9-39** Installing compression spring (Courtesy of Toyota Motor Corporation)

6. Stake the upper column tube with a pin punch and a hammer (Figure 9-36).
7. Install the snap ring in the top of the upper column tube (Figure 9-37).
8. Install the tension spring (Figure 9-38). Assemble the compression spring, bushing, plate, and seat. Use a vise to install the compression spring and related components (Figure 9-39).
9. Install the upper bracket with two new tapered-head bolts. Tighten the tapered-head bolts until the tapered head breaks off (Figure 9-40).
10. Install the protector and wiring harness clamp.
11. Install the steering damper and ignition key illumination.
12. Install the universal joint on the bottom of the steering shaft, and tighten the retaining bolt to the specified torque.

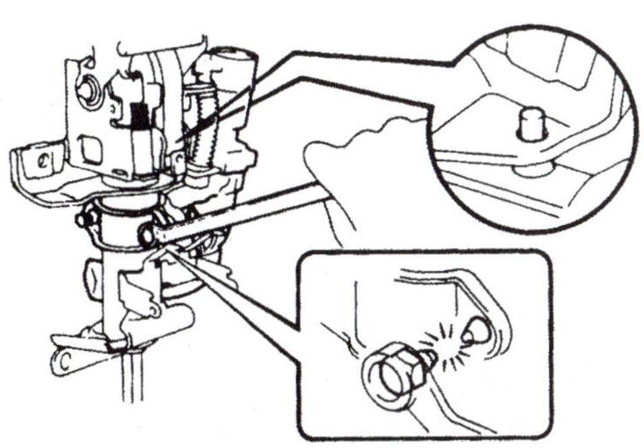

**Figure 9-40** Tightening tapered-head bolts in upper bracket (Courtesy of Toyota Motor Corporation)

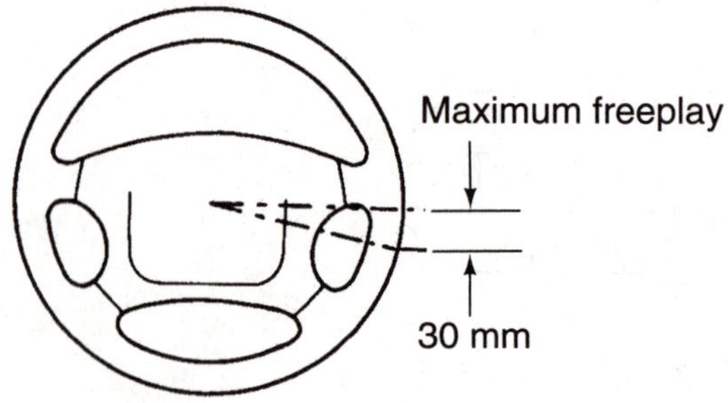


**Figure 9-41** Measuring steering freeplay (Courtesy of Toyota Motor Corporation)

13. Install the combination switch and wiring harness, and tighten the retaining screw to the specified torque.
14. Install the spiral cable, steering wheel, and air bag deployment module as mentioned previously in this chapter.

# Steering Column Flexible Coupling and Universal Joint Service and Diagnosis

## Checking Steering Wheel Freeplay

Steering wheel freeplay is the amount of steering wheel movement before the front wheels start to turn.

With the engine stopped and the front wheels in the straight-ahead position, move the steering wheel in each direction with light finger pressure. Measure the amount of steering wheel movement before the front wheels begin to turn (Figure 9-41). This movement is referred to as steering freeplay. On some vehicles, this measurement should not exceed 1.18 in (30 mm). Always refer to the vehicle manufacturer's specifications. Excessive steering freeplay is caused by worn steering shaft universal joints, or flexible coupling. Other causes of excessive steering wheel freeplay include worn steering linkage mechanisms, and a worn, or out of adjustment, steering gear.

A worn universal joint, or flexible coupling, in the steering column may also cause rattling noises. The rattling noises may occur while driving the vehicle straight ahead on irregular road surfaces. With the normal vehicle weight resting on the front suspension, observe the flexible coupling, or universal joint, as an assistant turns the steering wheel 1/2 turn in each direction. If the vehicle has power steering, the engine should be running with the gear selector in park. The flexible coupling, or universal joint, must be replaced if there is freeplay in this component.

## Flexible Coupling Replacement

If the flexible coupling must be replaced, loosen the coupling-to-steering-gear-stub-shaft bolt. Disconnect the steering column from the instrument panel, and move the column rearward until the flexible coupling can be removed from the steering column shaft. Remove the coupling-to-steering-shaft bolts and disconnect the coupling from the shaft. When the new coupling and the steering column are installed on some vehicles, the clearance between the coupling clamp and the steering gear adjusting plug should be 1/16 in (1.5 mm) (Figure 9-42). This specification may vary depending on the vehicle. Always use the vehicle manufacturer's specifications in the service manual.

# Steering Column Diagnosis

**Classroom Manual**
Chapter 9, page 197

There are variations in steering columns depending on the vehicle, type of transmission, and the transmission gear selector position. Thus different column diagnostic procedures may be required. See Tables 9-1, 9-2, and 9-3 for a typical steering column diagnosis.

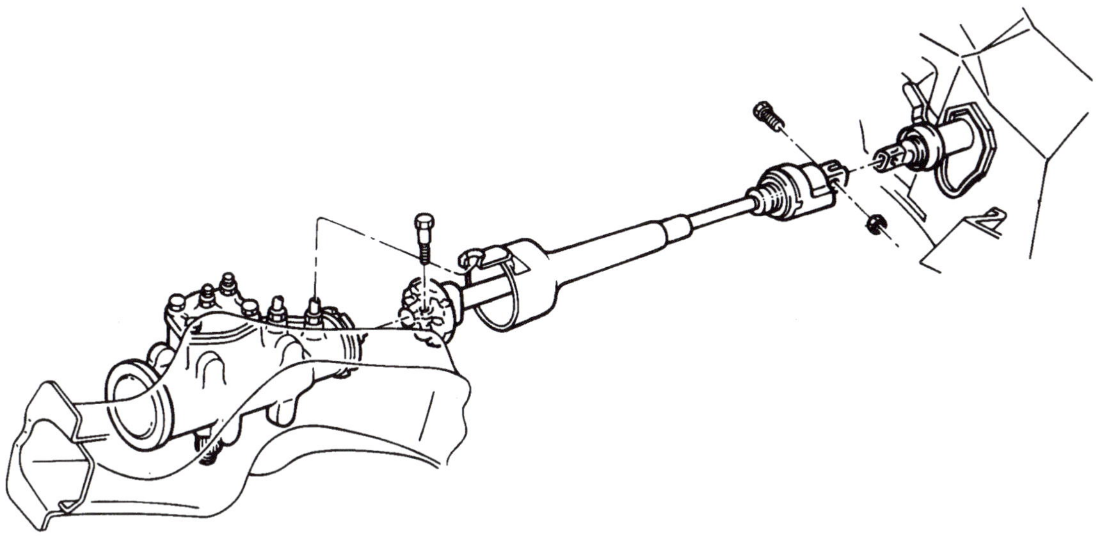

Figure 9-42 Flexible coupling installation (Courtesy of Chevrolet Motor Division, General Motors Corporation)

**Table 9-1 AUTOMATIC TRANSMISSION—STEERING COLUMN DIAGNOSIS**

| Condition | Possible Cause | Correction |
|---|---|---|
| Lock system—will not unlock | 1. Lock bolt damaged | 1. Replace lock bolt. |
| | 2. Defective lock cylinder | 2. Replace or repair lock cylinder. |
| | 3. Damaged housing | 3. Replace housing. |
| | 4. Damaged or collapsed sector | 4. Replace sector. |
| | 5. Damaged rack | 5. Replace rack. |
| | 6. Shear flange on sector shaft collapsed | 6. Replace. |
| Lock system—will not lock | 1. Lock bolt spring broken or defective | 1. Replace spring. |
| | 2. Damaged sector tooth, or sector installed incorrectly | 2. Replace, or install correctly. |
| | 3. Defective lock cylinder | 3. Replace lock cylinder. |
| | 4. Burr or lock bolt or housing | 4. Remove burr. |
| | 5. Damaged housing | 5. Replace housing. |
| | 6. Transmission linkage adjustment incorrect | 6. Readjust. |
| | 7. Damaged rack | 7. Replace rack. |
| | 8. Interference between bowl and coupling (tilt) | 8. Adjust or replace as necessary. |
| | 9. Ignition switch stuck | 9. Readjust or replace. |
| | 10. Actuator rod restricted or bent | 10. Readjust or replace. |

**Table 9-1 AUTOMATIC TRANSMISSION—STEERING COLUMN DIAGNOSIS (continued)**

| Condition | Possible Cause | Correction |
|---|---|---|
| Lock system—high effort | 1. Lock cylinder defective<br>2. Ignition switch defective<br>3. Rack preload spring broken or deformed<br>4. Burrs on sector, rack, housing, support, tang of shift gate, or actuator rod coupling<br>5. Bent sector shaft<br>6. Distorted rack<br>7. Misalignment of housing to cover (tilt only)<br>8. Distorted coupling slot in rack (tilt)<br>9. Bent or restricted actuator rod<br>10. Ignition switch mounting bracket bent | 1. Replace lock cylinder.<br>2. Replace switch.<br>3. Replace spring.<br>4. Remove burr.<br>5. Replace shaft.<br>6. Replace rack.<br>7. Replace either or both.<br>8. Replace rack.<br>9. Straighten, remove restriction, or replace.<br>10. Straighten or replace. |
| High effort lock cylinder—between "off" and "off-lock" positions | 1. Burr on tang of shift gate<br>2. Distorted rack | 1. Remove burr.<br>2. Replace rack. |
| Sticks in "start" position | 1. Actuator rod deformed<br>2. Any high effort condition | 1. Straighten or replace.<br>2. Check items under high effort section. |
| Key cannot be removed in "off-lock" position | 1. Ignition switch not set correctly<br>2. Defective lock cylinder | 1. Readjust ignition switch.<br>2. Replace lock cylinder. |
| Lock cylinder can be removed without depressing retainer | 1. Lock cylinder with defective retainer<br>2. Lock cylinder without retainer<br>3. Burr over retainer slot in housing cover | 1. Replace lock cylinder.<br>2. Replace lock cylinder.<br>3. Remove burr. |
| Lock bolt hits shaft lock in "off" and "PARK" positions | Ignition switch not set correctly | Readjust ignition switch. |
| Ignition system—electrical system will not function | 1. Defective fuse in "accessory" circuit<br>2. Connector body loose or defective | 1. Replace fuse.<br>2. Tighten or replace. |

### Table 9-1 AUTOMATIC TRANSMISSION—STEERING COLUMN DIAGNOSIS (continued)

| Condition | Possible Cause | Correction |
|---|---|---|
| | 3. Defective wiring<br>4. Defective ignition switch<br>5. Ignition switch not adjusted properly | 3. Repair or replace.<br>4. Replace ignition switch.<br>5. Readjust ignition switch. |
| Switch will not actuate mechanically | Defective ignition switch | Replace ignition switch. |
| Switch cannot be set correctly | 1. Switch actuator rod deformed<br>2. Sector to rack engaged in wrong tooth (tilt) | 1. Repair or replace switch actuator rod.<br>2. Engage sector to rack correctly. |
| Noise in column | 1. Coupling bolts loose<br>2. Column not correctly aligned<br>3. Coupling pulled apart<br>4. Sheared intermediate shaft plastic joint<br>5. Horn contact ring not lubricated<br>6. Lack of grease on bearings or bearing surfaces<br>7. Lower shaft bearing tight or frozen<br>8. Upper shaft tight or frozen<br>9. Shaft lock plate cover loose<br>10. Lock plate snap ring not seated<br>11. Defective buzzer dog cam on lock cylinder<br>12. One click when in "off-lock" position and the steering wheel is moved | 1. Tighten pinch bolts to specified torgue.<br>2. Realign column.<br>3. Replace coupling and realign column.<br>4. Replace or repair steering shaft and realign column.<br>5. Lubricate with lubriplate.<br>6. Lubricate bearings.<br>7. Replace bearing. Check shaft and replace if scored.<br>8. Replace housing assembly.<br>9. Tighten three screws or, if missing, replace. CAUTION: Use specified screws (15 inch-pounds).<br>10. Replace snap ring. Check for proper seating in groove.<br>11. Replace lock cylinder.<br>12. Normal condition: lock bolt is seating. |
| High steering shaft effort | 1. Column assembly misaligned in vehicle<br>2. Improperly installed or deformed dust seal | 1. Realign.<br>2. Remove and replace. |

**Table 9-1 AUTOMATIC TRANSMISSION—STEERING COLUMN DIAGNOSIS (continued)**

| Condition | Possible Cause | Correction |
|---|---|---|
| | 3. Tight or frozen upper or lower bearing | 3. Replace affected bearing or bearings. |
| | 4. Flash on ID of shift tube from plastic joint | 4. Replace shift tube. |
| High shift effort | 1. Column not aligned correctly in car | 1. Realign. |
| | 2. Improperly installed dust seal | 2. Remove and replace. |
| | 3. Lack of grease on seal or bearing areas | 3. Lubricate bearings and seals. |
| | 4. Burr on upper or lower end of shift tube | 4. Remove burr. |
| | 5. Lower bowl bearing not assembled properly (tilt) | 5. Reassemble properly. |
| | 6. Wave washer with burrs (tilt only) | 6. Replace wave washer. |
| Improper transmission shifting | 1. Sheared shift tube joint | 1. Replace shift tube assembly. |
| | 2. Improper transmission linkage adjustment | 2. Readjust linkage. |
| | 3. Loose lower shift lever | 3. Replace shift tube assembly. |
| | 4. Improper gate plate | 4. Replace with correct part. |
| | 5. Sheared lower shift lever weld | 5. Replace tube assembly. |
| Lash in mounted column assembly | 1. Instrument panel mounting bolts loose | 1. Tighten to specifications (20 foot-pounds). |
| | 2. Broken weld nuts on jacket | 2. Replace jacket assembly. |
| | 3. Instrument panel bracket capsure sheared | 3. Replace bracket assembly. |
| | 4. Instrument panel to jacket mounting bolts loose | 4. Tighten to specifications (15 foot-pounds). |
| | 5. Loose shoes in housing (tilt only) | 5. Replace. |
| | 6. Loose tilt head pivot pins (tilt only) | 6. Replace. |
| | 7. Loose shoe lock pin in support (tilt only) | 7. Replace. |

### Table 9-1 AUTOMATIC TRANSMISSION—STEERING COLUMN DIAGNOSIS (continued)

| Condition | Possible Cause | Correction |
|---|---|---|
| Miscellaneous | 1. Housing loose on jacket noticed with ignition in "off-lock" and a torque applied to the steering wheel. | 1. Tighten four mounting screws (60 inch-pounds). |
| | 2. Shroud loose on shift bowl | 2. Bend tabs on shroud over lugs on bowl. |

### Table 9-2 MANUAL TRANSMISSION—STEERING COLUMN DIAGNOSIS

| Condition | Possible Cause | Correction |
|---|---|---|
| Driver can lock steering in second gear | 1. Defective upper shift lever | 1. Replace shift lever. |
| | 2. Defective shift lever gate | 2. Replace shift lever gate. |
| | 3. Loose relay lever on shift tube | 3. Replace shift tube assembly. |
| | 4. Use of upper shift lever prior to 1969 model year | 4. Replace with current lever. |
| High shift effort | 1. Column not aligned correctly in car | 1. Realign column. |
| | 2. Lower bowl bearing not assembled correctly | 2. Reassemble correctly. |
| | 3. Improperly installed seal | 3. Remove and replace. |
| | 4. Wave washer in lower bowl bearing defective | 4. Replace wave washer. |
| | 5. Improper adjustment of lower shift levers | 5. Readjust. |
| | 6. Lack of grease on seal, bearing areas, or levers | 6. Lubricate seal, levers, and bearings. |
| | 7. Damaged shift tube in bearing areas | 7. Replace shift tube assembly. |
| Improper transmission shifting | Loose relay lever on shift tube | Replace shift tube assembly. |

### Table 9-3 MANUAL TRANSMISSION—TILT COLUMN DIAGNOSIS

| Condition | Possible Cause | Correction |
|---|---|---|
| Housing scraping on bowl | Bowl bent or not concentric with hub | Replace bowl. |
| Steering wheel loose | 1. Excessive clearance between holes in support or housing and pivot pin diameters | 1. Replace either or both. |

**Table 9-3 MANUAL TRANSMISSION—TILT COLUMN DIAGNOSIS (continued)**

| Condition | Possible Cause | Correction |
|---|---|---|
| | 2. Defective or missing antilash spring in spheres | 2. Add spring or replace both. |
| | 3. Upper bearing seat not seating in bearing | 3. Replace both. |
| | 4. Upper bearing inner race seat missing | 4. Install seat. |
| | 5. Loose support screws | 5. Tighten to 60 inch-pounds. |
| | 6. Bearing preload spring missing or broken | 6. Replace preload spring. |
| Steering wheel loose every other tilt position | Loose fit between shoe and shoe pivot pin | Replace both. |
| Noise when tilting column | 1. Upper tilt bumper worn | 1. Replace tilt bumper. |
| | 2. Tilt spring rubbing in housing | 2. Lubricate. |
| Steering column not locking in any tilt position | 1. Shoe seized on its pivot pin | 1. Replace shoe and pivot pin. |
| | 2. Shoe grooves might have burrs or dirt | 2. Replace shoe. |
| | 3. Shoe lock spring weak or broken | 3. Replace lock spring. |
| Steering wheel fails to return to top tilt position | 1. Pivot pins bound up | 1. Replace pivot pins. |
| | 2. Wheel tilt spring defective | 2. Replace tilt spring. |
| | 3. Turn signal switch wires too tight | 3. Reposition wires. |

## Steering Linkage Diagnosis and Service

### Diagnosis of Center Link, Pitman Arm, and Tie Rod Ends

The vehicle should be raised, and safety stands must be positioned under the lower control arms to support the vehicle weight. Use vertical hand force to check for looseness in all the pivots on the tie rod ends and the center link. Check the seals on each tie rod end, and pivot on the center link or pitman arm for damage and cracks. Cracked seals allow dirt to enter the pivoted joints, which results in rapid wear. If looseness or damaged seals is found on any pivoted joint on the tie rods and center link, these components must be replaced.

The second part of this diagnosis is done with the front wheels resting on the shop floor. If the vehicle is equipped with power steering, start the engine and allow the engine to idle with the transmission in park and the parking brake applied. While someone turns the steering wheel one-quarter turn in each direction from the straight-ahead position, observe all the pivoted joints on the tie rod ends and center link. This test allows the technician to check the steering linkage pivots under load. If any of the pivoted joints show a slight amount of play, they must be replaced.

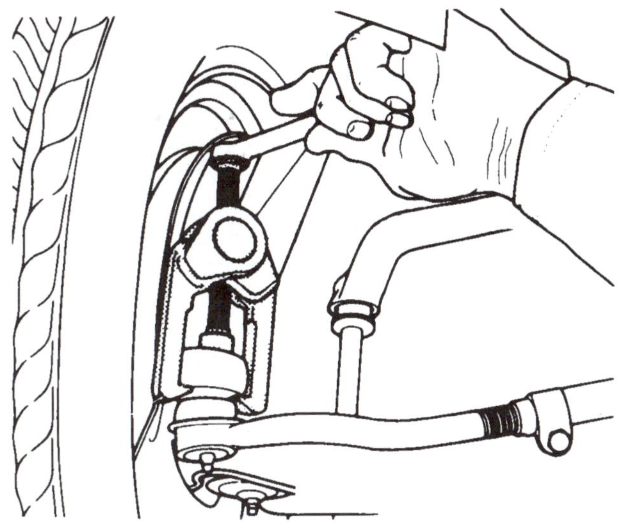

**Figure 9-43** Tie rod end removal (Courtesy of Chevrolet Motor Division, General Motors Corporation)

## Tie Rod End Replacement

Worn tie rod ends result in these problems:

1. excessive steering wheel freeplay
2. incorrect front wheel toe setting
3. tire squeal on turns
4. tread wear on front tires
5. front wheel shimmy
6. rattling noise on road irregularities

The cotter key and nut must be removed prior to tie rod end replacement. A puller is used to remove the tie rod end from the steering arm (Figure 9-43).

The tie rod clamp must be loosened before the tie rod is removed from the sleeve. Count the number of turns required to remove the tie rod end from the sleeve, and install the new tie rod with the same number of turns. Even when this procedure is followed, the toe must be checked after the steering linkage components are replaced. Before the new tie rod end is installed, center the stud in the tie rod end. When the tie rod stud is installed in the steering arm opening, only the threads should be visible above the steering arm surface. If the machined surface of the tie rod stud is visible above the steering arm surface or the stud fits loosely in the steering arm opening, this opening is worn or the tie rod end is not correct for that application. The tie rod nut must be torqued to manufacturer's specifications, and the cotter key must be installed through the tie rod end and nut openings (Figure 9-44).

The tie rod nut must never be loosened from the specified torque to install the cotter key. Another method of positioning replacement tie rod ends is to measure the distance from the center of the tie rod stud to the end of the sleeve prior to removal. When the new tie rod is installed, be sure this measurement is the same. The slots in the tie rod sleeve must be positioned away from the opening in the sleeve clamps (Figure 9-45). Leave the sleeve clamps loose until the front wheel toe is checked, and then tighten the sleeve clamp bolts to the specified torque. A special tool is available to rotate the tie rod sleeves and set the front wheel toe (Figure 9-46).

Front wheel shimmy may be defined as a consistent, fast, side-to-side movement of the front wheels and steering wheel. This movement is usually experienced at speeds above 40 mph (25 kph), and it may occur more frequently while driving on irregular road surfaces.

**Special Tools**

Tie rod end puller

## Pitman Arm Diagnosis and Replacement

Some pitman arms contain a ball socket joint on the outer end. The threaded extension on this ball socket fits into the center link. On other steering linkages, the ball socket joint is in the center

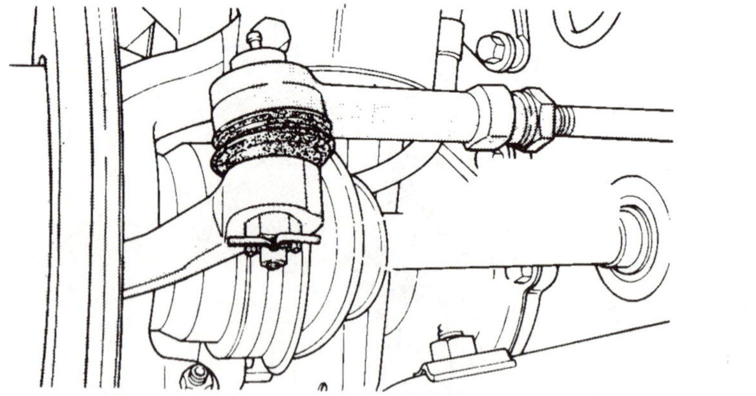

**Figure 9-44** Tie rod nut and cotter key installation (Courtesy of Chrysler Corporation)

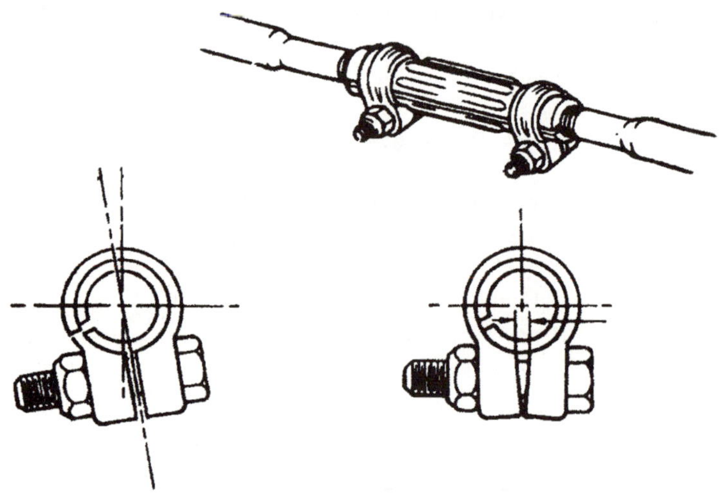

**Figure 9-45** Proper slot and clamp position on tie rod sleeves (Courtesy of Chevrolet Motor Division, General Motors Corporation)

**Figure 9-46** Tie rod sleeve adjusting tool (Courtesy of Mac Tools, Inc.)

link, and the threaded extension fits into the pitman arm opening. If the pitman arm is bent, it must be replaced. If the pitman arm is bent, the tie rod is not parallel to the lower control arm. Under this condition, excessive front wheel toe change occurs on road irregularities, and front tire wear may be excessive. Following is a typical pitman arm replacement procedure:

1. Position the front wheels straight ahead, and remove the cotter key and nut from the ball socket joint on the outer end of the pitman arm.
2. Remove the ball socket extension from the pitman arm, or center link, with a tie rod end puller.

3. Loosen the pitman arm-to-pitman shaft nut.
4. Use a puller to pull the pitman arm loose on the shaft.
5. Remove the nut, lock washer, and pitman arm.
6. Check the pitman shaft splines. If the splines are damaged or twisted, the shaft must be replaced.
7. Reverse steps 1 through 5 to install the pitman arm. The pitman arm-to-shaft nut and the ball socket extension nut must be tightened to the manufacturer's specified torque. Be sure the pitman arm is installed in the correct position on the shaft splines. Install the cotter key in the ball socket extension.

**Special Tools**

Pitman arm puller

## Center Link Diagnosis and Replacement

**WARNING:** Never attempt to straighten steering linkage components. This action may weaken the metal and cause sudden component failure, vehicle damage, and personal injury.

A bent center link must be replaced. Do not attempt to straighten this rod. If the ball socket joints are loose on either end of the rod, center link assembly replacement is necessary. If the ball stud openings in the center link are worn, replace the center link. Follow these steps for a typical center link replacement:

1. Remove the cotter keys from the tie rod-to-center link nuts, and the idler arm and pitman arm to center link nuts.
2. Remove the nuts on the tie rod inner ends, idler arm-to-center link ball socket extension, and the pitman arm-to-center link ball socket extension.
3. Use a tie rod end puller to pull the inner tie rods from the center link. Follow the same procedure to remove the center link-to-pitman arm ball socket extension.
4. Remove the idler arm from the center link, and remove the center link.
5. Reverse steps 1 through 4 to install the center link. Tighten all the ball socket nuts to the manufacturer's specified torque, and install cotter keys in all the nuts. If the ball sockets have grease fittings, lubricate the ball sockets with a grease gun and chassis lubricant.

## Idler Arm Diagnosis

Grasp the center link firmly near the idler arm, and apply a 25-pound vertical load to the idler arm. If the total idler arm vertical movement measured at the outer end of the arm exceeds 1/4 inch, idler arm replacement is necessary. If idler arm vertical movement is excessive, the tie rod is not parallel to the lower control arm. Excessive idler arm vertical movement causes these steering problems:

1. excessive toe change and front tire tread wear
2. excessive steering wheel free play and reduced steering control
3. front end shimmy

**SERVICE TIP:** A binding idler arm may suddenly break off and cause complete loss of steering control, vehicle damage, and personal injury.

Binding idler arm bushings result in these complaints:

1. hard steering
2. squawking noise when the front wheels are turned
3. poor steering wheel returnability

Some idler arms contain steel bushings, whereas others are equipped with rubber bushings. Following is a typical idler arm removal and replacement procedure:

1. Remove the idler arm-to-center link cotter key and nut.
2. Remove the center link from the idler arm.

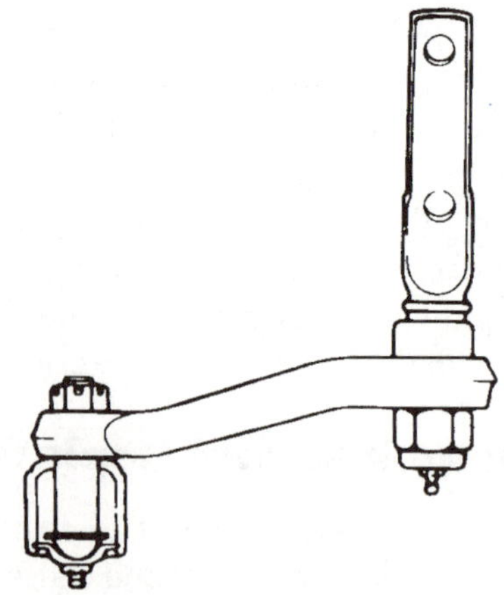

Figure 9-47 Specified clearance between center of lower bracket bolt hole and upper idler arm surface (Courtesy of Chevrolet Motor Division, General Motors Corporation)

3. Remove the idler arm bracket mounting bolts, and remove the idler arm.
4. If the idler arm has a steel bushing, thread the bracket into the idler arm bushing until the specified clearance is obtained between the center of the lower bracket bolt hole and the upper idler arm surface (Figure 9-47).
5. Install the idler arm bracket to frame bolts and tighten the bolts to the specified torque. Be sure that lock washers are installed on the bolts.
6. Install the center link into the idler arm, and tighten the mounting nut to the specified torque. Install the cotter key in the nut.
7. If the idler arm steel bushing, or bushings, contains a grease fitting, lubricate as required.

The idler arm adjustment is very important. If this adjustment is incorrect, front wheel toe is affected. After idler arm replacement, the front wheel toe should be checked.

## Steering Damper Diagnosis and Replacement

Some steering systems have a damper connected between the center link and the chassis. A damper is similar to a small shock absorber. The purpose of the damper is to prevent the transmission of steering shock and vibrations to the steering wheel. A worn-out steering damper may cause excessive steering shock and vibration on the steering wheel, especially on irregular road surfaces. A rattling noise occurs if the damper mounting bolts or brackets are loose. Following is a typical steering damper checking and replacement procedure:

1. Lift the vehicle on a hoist and grasp the damper firmly. Apply vertical and horizontal pressure to the damper and check for movement in the damper mounts. If movement exists, tighten or replace the damper mounting bushings or brackets.
2. Visually inspect the damper for oil leaks. A slight film of oil on the damper body near the shaft seal is acceptable. If there is any indication of oil dripping from the damper, the unit must be replaced.
3. Disconnect one end of the damper and pull the damper back and forth horizontally. The damper should offer a slight equal resistance to movement in either direction. When this resistance is not felt in one or both directions, replace the damper.
4. To replace the damper, remove the mounting bolts from the chassis and the center link.
5. When the new damper is installed, tighten the mounting bolts to the specified torque. Turn the steering wheel fully in each direction and be sure the damper does not restrict linkage movement.

## Steering Arm Diagnosis

If the front rims have been damaged, the steering arms should be checked for a bent condition. Measure the distance from the center of the tie rod stud to the edge of the rim on each side. Unequal readings may indicate a bent steering arm. Bent steering arms must be replaced. Steering linkage diagnosis is summarized in Table 9-4.

**Classroom Manual**
Chapter 9, page 205

**Table 9-4 STEERING LINKAGE DIAGNOSIS**

| Condition | Possible Cause | Correction |
|---|---|---|
| Excessive play or looseness in steering system | 1. Front wheel bearings loosely adjusted | 1. Adjust bearings to obtain proper end play. |
| | 2. Worn steering shaft couplings | 2. Replace part. |
| | 3. Worn upper ball joints | 3. Check and replace, if necessary. |
| | 4. Steering wheel loose on shaft, or loose pitman arm, tie-rods, steering arms, or steering linkage ball studs | 4. Tighten to specified torque, or replace if necessary. |
| | 5. Steering gear thrust bearings loosely adjusted | 5. Adjust preload to specifications. |
| | 6. Excessive over-center lash in steering gear | 6. Adjust preload to specifications as per shop manual. |
| | 7. Worn intermediate rod or tie-rod sockets | 7. Replace worn part. |
| Excessive looseness in tie-rod or intermediate rod pivots, or excessive vertical lash in idler support | Seal damage and leakage resulting in loss of lubricant, corrosion, and excessive wear | Replace damaged parts as necessary. Properly position upon reassembly. |
| Hard steering—excessive effort required at steering wheel | 1. Low or uneven tire pressure | 1. Inflate to specified pressures. |
| | 2. Steering linkage or bolt joints need lubrication | 2. Lube with specified lubricant. |
| | 3. Tight or frozen intermediate rod, tie-rod, or idler socket | 3. Lube, replace, or reposition as necessary. |
| | 4. Steering gear-to-column misalignment | 4. Align column. |
| | 5. Steering gear adjusted too tightly | 5. Adjust over-center and thrust bearing preload to specification. |
| | 6. Front wheel alignment incorrect (manual gear) | 6. Check alignment and correct as necessary. |
| Poor returnability | 1. Steering linkage or ball joints need lubrication | 1. Lube with specified lubricant. |
| | 2. Steering gear adjusted too tightly | 2. Adjust over-center and thrust bearing preload to specifications. |

**Table 9-4 STEERING LINKAGE DIAGNOSIS (continued)**

| Condition | Possible Cause | Correction |
|---|---|---|
| | 3. Steering gear-to-column misalignment | 3. Align columns. |
| | 4. Front wheel aligment incorrect (caster) | 4. Check alignment and correct as necessary. |

## Guidelines for Servicing Steering Columns and Steering Linkage Mechanisms

1. Prior to servicing an air bag system component, the technician must disconnect the negative battery cable and wait one minute.
2. Only the vehicle manufacturer's recommended tools should be used when servicing an air bag system or the electrical system on an air-bag-equipped vehicle. Do not use such tools as a 12-V test lamp or a self-powered test lamp.
3. Always store an air bag deployment module facing upward on the work bench.
4. Do not hammer on the top of the steering shaft to remove a steering wheel.
5. Do not pull on the steering wheel to remove it from the steering shaft.
6. Prior to installing a clock spring electrical connector, or spiral cable, the front wheels must be in the straight-ahead position and the clock spring must be centered.
7. When a steering wheel is installed, the marks on the steering wheel and the steering shaft must be aligned.
8. When a collapsible steering column is checked for collision damage, the column jacket, gearshift tube, and steering shaft should be checked for sheared injected plastic.
9. To check for collision damage in a collapsible steering column, the distance should be measured from the end of the bearing assembly to the lower edge of the upper column jacket.
10. Some ignition key cylinders must be removed and replaced with the key in the ACC position.
11. Some vehicle manufacturers recommend coating all rubbing steering column components with molybdenum disulphide lithium-based grease.
12. When a flexible coupling is installed, there should be a specified clearance between the coupling clamp and the steering gear adjusting plug.
13. When a tie rod end stud is installed in the steering arm opening, only the threaded part of the stud should appear above the knuckle surface.
14. The slots in the tie rod sleeves must be positioned away from the openings in the sleeve clamps.
15. The front wheel toe should be checked after steering linkage components are replaced.
16. After tie rod end nuts are tightened to the specified torque, they must never be loosened to install the cotter key.
17. Never straighten steering linkage components.
18. An idler arm with excessive vertical movement must be replaced.
19. Some idler arms with steel bushings require an adjustment to set the clearance between the idler arm bracket flange and the idler arm surface.
20. If there is any indication of oil dripping from a steering damper, this component must be replaced.
21. The distance may be measured from the center of each outer tie rod end stud to the inside edge of each rim to check for a bent steering arm.

## CASE STUDY

A customer complained about excessive steering effort and poor steering wheel returnability after a turn on a 1993 Lincoln Town Car with power steering. The technician road tested the car and found the customer's description of the problems to be accurate except for one point. The steering continually required excessive steering effort and the steering wheel did not return properly after a turn. However, the customer did not mention, or possibly did not notice, that a squawking and creaking noise was sometimes heard during a turn.

The technician checked the power steering fluid level and condition, and found this fluid to be in good condition and at the proper level. Next, the technician checked the power steering belt condition and the power steering pump pressure. The belt tension and condition were satisfactory, and the power steering pump pressure was normal. A check of the power steering pump mounting bolts and brackets indicated they were in good condition. The technician disconnected the center link from the pitman arm and rotated the steering wheel with the engine running. Under this condition, the steering wheel turned very easily. Therefore, the technician concluded the excessive steering effort and poor returnability problems were not in the steering column or steering gear. Next, the technician disconnected the idler arm from the center link. When the technician attempted to move the idler arm back and forth, it was discovered that the idler arm had a severe binding problem. After an idler arm replacement, a road test revealed the excessive steering effort and poor returnability problems had disappeared.

## Terms to Know

Air bag deployment module
Backup power supply
Accidental air bag deployment
Clock spring electrical connector
Spiral cable
Collapsible steering column
Sheared injected plastic
Tapered-head bolts
Moiybdenum disulphide lithium-based grease
Steering wheel freeplay

## ASE Style Review Questions

**1.** While discussing steering column service on an air-bag-equipped vehicle:
*Technician A* says service personnel should disconnect the negative battery cable and wait one minute prior to servicing an air bag system component.
*Technician B* says a 12-V test lamp may be used to diagnose an air bag system.
Who is correct?
**A.** A only
**B.** B only
**C.** Both A and B
**D.** Neither A nor B

**2.** While discussing steering column service on an air-bag-equipped vehicle:
*Technician A* says the backup power supply provides power to deploy the air bag when a frontal collision occurs.
*Technician B* says the backup power supply provides power to deploy the air bag if the battery is disconnected during a frontal collision.
Who is correct?
**A.** A only
**B.** B only
**C.** Both A and B
**D.** Neither A nor B

3. While discussing steering wheel and clock spring electrical connector removal and replacement on an air-bag-equipped vehicle:
*Technician A* says the clock spring electrical connector may be installed in any position with the front wheels straight ahead.
*Technician B* says when the clock spring electrical connector is installed, the front wheels should be straight ahead and the clock spring should be turned fully clockwise.
Who is correct?
**A.** A only
**B.** B only
**C.** Both A and B
**D.** Neither A nor B

4. While discussing steering wheel removal:
*Technician A* says a steering wheel puller must be used to remove the steering wheel from the shaft.
*Technician B* says the steering wheel should be grasped firmly with both hands and pulled from the shaft.
Who is correct?
**A.** A only
**B.** B only
**C.** Both A and B
**D.** Neither A nor B

5. While discussing collapsible steering column damage:
*Technician A* says if the vehicle has been in a frontal collision, the injected plastic in the column jacket, gearshift tube, and steering shaft may be sheared.
*Technician B* says if the injected plastic is sheared in the column jacket, this jacket must be replaced.
Who is correct?
**A.** A only
**B.** B only
**C.** Both A and B
**D.** Neither A nor B

6. While discussing steering column service:
*Technician A* says some ignition switches must be removed and replaced with the key in the run position.
*Technician B* says some ignition switch cylinders must be removed and replaced with the key in the accessory (ACC) position.
Who is correct?
**A.** A only
**B.** B only
**C.** Both A and B
**D.** Neither A nor B

7. While discussing steering linkage service:
*Technician A* says a rattling noise may be caused by a worn universal joint in the steering shaft.
*Technician B* says a rattling noise may be caused by a bent center link.
Who is correct?
**A.** A only
**B.** B only
**C.** Both A and B
**D.** Neither A nor B

8. While discussing tie rod end service:
*Technician A* says a worn tie rod end causes excessive front wheel toe change and tread wear on the front tires.
*Technician B* says the slot in the tie rod sleeve must be positioned away from the opening in the sleeve clamp.
Who is correct?
**A.** A only
**B.** B only
**C.** Both A and B
**D.** Neither A nor B

9. While discussing tie rod end service:
*Technician A* says the machined part of the tie rod stud should be visible above the surface of the steering knuckle when the tie rod stud is installed in the knuckle opening.
*Technician B* says the tie rod nut may be loosened to install the cotter key in the nut and stud openings.
Who is correct?
**A.** A only
**B.** B only
**C.** Both A and B
**D.** Neither A nor B

10. While discussing idler arm service:
*Technician A* says the normal vertical idler arm movement should not exceed 1/2 inch with a 25-pound vertical load applied to the idler arm.
*Technician B* says the some idler arms require an adjustment between the top of the bracket and the lower end of the bushing.
Who is correct?
**A.** A only
**B.** B only
**C.** Both A and B
**D.** Neither A nor B

## Table 9-5 ASE TASK

Diagnose steering column noises, looseness, and binding problems (including tilt mechanisms); determine needed repairs.

| Problem Area | Symptoms | Possible Causes | Classroom Manual | Shop Manual |
|---|---|---|---|---|
| NOISES | Rattling on road irregularities | Loose, worn coupling or universal joints | 198 | 204 |
| | Clunk while tilting | 1. Worn universal joint, or spherical bearing | 198 | 204 |
| | | 2. Worn tilt bumper | 204 | 196 |
| BINDING | Excessive steering effort | 1. Column misaligned | 198 | 208 |
| | | 2. Improperly installed or defective dust shield | 198 | 208 |
| | | 3. Worn, defective column bearings | 203 | 208 |
| LOOSENESS | Excessive steering wheel freeplay | 1. Worn universal joints or flexible coupling | 203 | 208 |
| | | 2. Worn tilt spherical bearing, or universal joint | 203 | 208 |
| | | 3. Defective column bearings | 203 | 208 |

## Table 9-6 ASE TASK

Inspect and replace steering shaft universal joint(s), flexible coupling(s), collapsible columns, steering wheels, including steering wheels with air bags.

| Problem Area | Symptoms | Possible Causes | Classroom Manual | Shop Manual |
|---|---|---|---|---|
| NOISE | Rattling on road irregularities | Worn, loose flexible coupling, or universal joint | 198 | 204 |
| STEERING CONTROL | Excessive steering wheel freeplay | Worn, loose flexible coupling, or universal joint | 198 | 204 |
| | Improper steering wheel position | Collapsed column | 199 | 195 |
| | Damaged steering wheel | Frontal collision | 198 | 194 |
| AIR BAG DEPLOYMENT | Failure to deploy | Defective clock spring | 200 | 193 |

## Table 9-7 ASE TASK

Inspect and adjust (where applicable) front and rear steering linkage geometry, including attitude and parallelism.

| Problem Area | Symptoms | Possible Causes | Classroom Manual | Shop Manual |
|---|---|---|---|---|
| TIRE LIFE | Excessive front tire wear | 1. Bent pitman arm | 207 | 211 |
| | | 2. Worn or misadjusted idler arm | 208 | 213 |

## Table 9-8 ASE TASK

Inspect and replace pitman arm.

| Problem Area | Symptoms | Possible Causes | Classroom Manual | Shop Manual |
|---|---|---|---|---|
| TIRE LIFE | Excessive front tire wear | Bent pitman arm | 207 | 211 |
| LOOSENESS | Excessive steering wheel freeplay | Worn ball socket in pitman arm | 207 | 211 |

## Table 9-9 ASE TASK

Inspect and replace relay (center link/intermediate) rod.

| Problem Area | Symptoms | Possible Causes | Classroom Manual | Shop Manual |
|---|---|---|---|---|
| LOOSENESS | Excessive steering wheel freeplay | 1. Worn ball sockets in center link | 208 | 213 |
| | | 2. Worn stud openings in center link | 208 | 213 |

## Table 9-10 ASE TASK

Inspect, adjust (where applicable), and replace idler arm and mountings.

| Problem Area | Symptoms | Possible Causes | Classroom Manual | Shop Manual |
|---|---|---|---|---|
| NOISE | Rattling on road irregularities | Worn idler arm | 208 | 213 |
| STEERING CONTROL | Excessive steering wheel freeplay | Worn idler arm or loose mounting | 208 | 213 |
| | Binding, excessive steering effort | Binding idler arm | 208 | 213 |
| TIRE LIFE | Excessive front tire wear | Worn, misadjusted idler arm | 208 | 213 |

## Table 9-11 ASE TASK

Inspect, replace, and adjust tie rods, tie rod sleeves, clamps, and tie rod ends (sockets).

| Problem Area | Symptoms | Possible Causes | Classroom Manual | Shop Manual |
|---|---|---|---|---|
| NOISE | Rattling noise on road irregularities | Loose tie rod ends | 206 | 211 |
| TIRE LIFE | Excessive front tire wear | Loose or improperly adjusted tie rod ends | 206 | 211 |

## Table 9-12 ASE TASK

Inspect and replace steering linkage damper.

| Problem Area | Symptoms | Possible Causes | Classroom Manual | Shop Manual |
|---|---|---|---|---|
| STEERING CONTROL | Shock, vibration on steering wheel | Worn or loose steering damper | 212 | 214 |

# Power Steering Pump Diagnosis and Service

CHAPTER 10

Upon completion and review of this chapter, you should be able to:

- ❑ Check power steering belt condition and adjust belt tension.
- ❑ Diagnose power steering belt problems.
- ❑ Check power steering fluid level, and add fluid as required.
- ❑ Drain and flush power steering system.
- ❑ Bleed air from power steering system.
- ❑ Perform power steering pump pressure test.
- ❑ Check power steering pump fluid leaks.
- ❑ Remove and replace power steering pumps, and inspect pump mounts.
- ❑ Remove and replace power steering pump pulleys.
- ❑ Remove and replace power steering pump integral reservoirs.
- ❑ Remove, replace, and check flow control valve and pressure relief valve.
- ❑ Remove, replace, and check power steering pump rotating components.
- ❑ Remove and replace power steering pump seals and O-rings.
- ❑ Check, remove, and replace power steering lines.

## Power Steering Pump Service and Diagnosis

### Checking Belt Condition and Tension

Power steering belt condition and tension are extremely important for satisfactory power steering pump operation. A loose belt causes low pump pressure and hard steering. The steering wheel may jerk and surge during a turn if the power steering pump belt is loose. A loose, dry, or worn belt may cause squealing and chirping noises, especially during engine acceleration and cornering.

The power steering pump belt should be checked for tension, cracks, oil soaking, worn or glazed edges, tears, and splits (Figure 10-1). If any of these conditions are present, belt replacement is necessary.

Since the friction surfaces are on the sides of a V-type belt, wear occurs in this area. If the belt edges are worn, the belt may be rubbing on the bottom of the pulley. This condition requires belt replacement. Belt tension may be checked by measuring the belt deflection. Press on the belt

**Basic Tools**

Basic technician's tool set

Service manual

Pry bar

Floor jack

Jack stands

Oil drain pan

Crocus cloth

**Special Tools**

Belt tension gauge

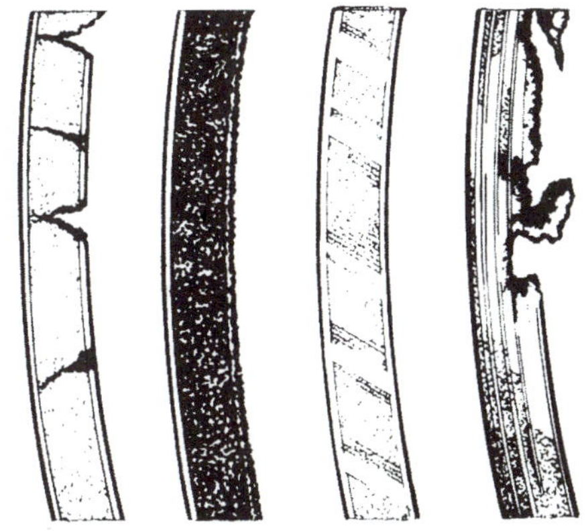


Figure 10-1 Defective belt conditions (Courtesy of Chrysler Corporation)

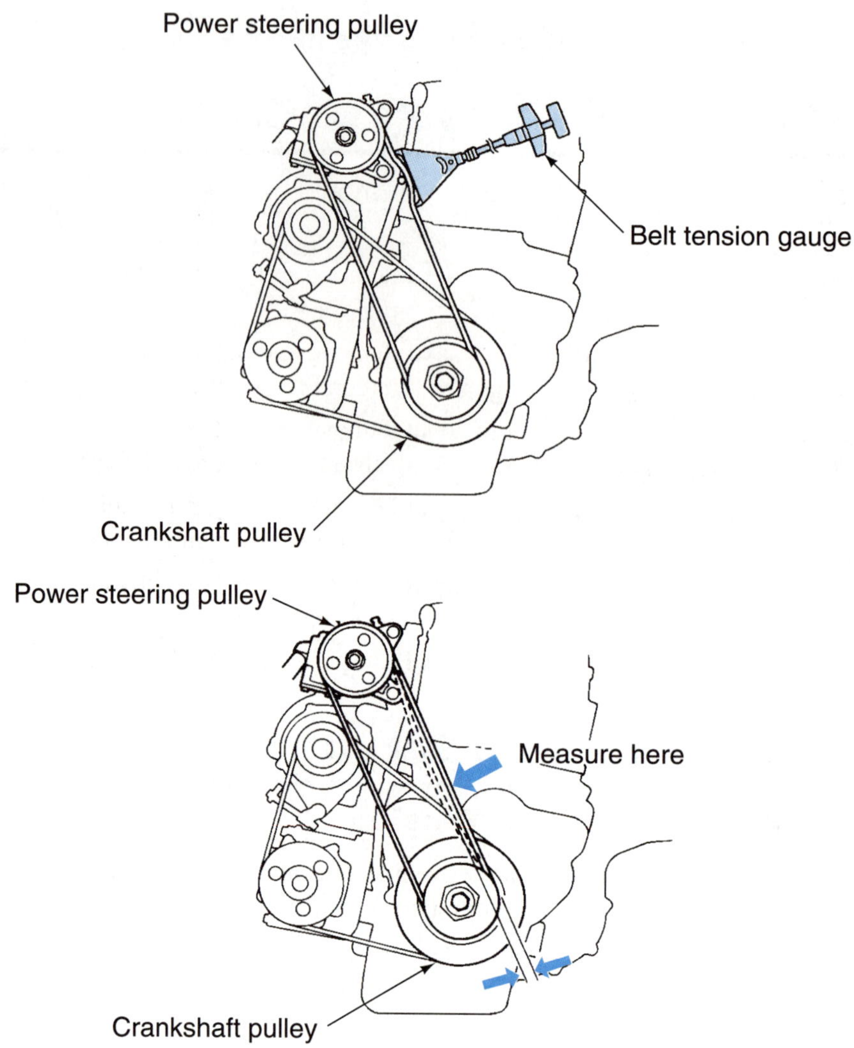


**Figure 10-2** Methods of checking belt tension (Courtesy of American Honda Motor Co., Inc.)

with the engine stopped to measure the belt deflection, which should be 1/2 inch per foot of free span. The belt tension may be checked with a belt tension gauge placed over the belt (Figure 10-2). The tension on the gauge should equal the vehicle manufacturer's specifications.

If the belt requires tightening, follow this procedure:

**WARNING:** Do not pry on the pump reservoir with a pry bar. This action may damage the reservoir.

1. Loosen the power steering pump bracket, or tension adjusting, bolt.
2. Loosen the power steering pump mounting bolts.
3. Check the bracket and pump mounting bolts for wear. If these bolts or bolt openings in the bracket or pump housing are worn, replacement is necessary.
4. Pry against the pump ear and hub with a pry bar to tighten the belt. Some pump brackets have a 1/2-inch square opening in which a breaker bar may be installed to move the pump and tighten the belt.
5. Hold the pump in the position described in step 4, and tighten the bracket, or tension adjusting, bolt.

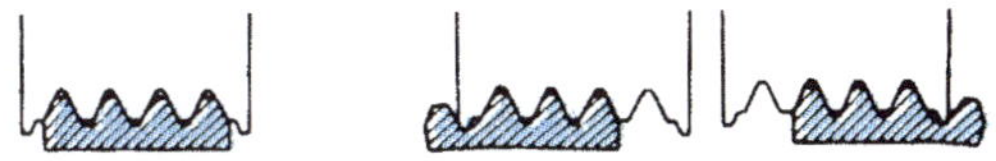


**Figure 10-3** Proper and improper installation of ribbed V-belt on a pulley (Courtesy of Toyota Motor Corporation)

6. Recheck the belt tension with the tension gauge. If the belt does not have the specified tension, repeat steps 1 through 5.
7. Tighten the tension adjusting bolt and the mounting bolts to the manufacturer's specified torque.

Some power steering pumps have a ribbed V-belt. Many ribbed V-belts have an automatic tensioning pulley; therefore, a tension adjustment is not required. The ribbed V-belt should be checked to make sure it is installed properly on each pulley in the belt drive system (Figure 10-3). The tension on a ribbed V-belt may be checked with a belt tension gauge in the same way as the tension on a V-belt.

A ribbed V-belt may be referred to as a serpentine belt.

Many ribbed V-belts have a spring-loaded tensioner pulley which automatically maintains belt tension. As the belt wears, or stretches, the spring moves the tensioner pulley to maintain the belt tension. Some of these tensioners have a belt length scale that indicates new belt range and used belt range (Figure 10-4). If the indicator on the tensioner is out of the used belt length range, belt replacement is required. Many belt tensioners have a 1/2-inch drive opening in which a ratchet or flex handle may be installed to move the tensioner pulley off the belt during belt replacement (Figure 10-5).

Belt pulleys must be properly aligned to minimize belt wear. The edges of the pulleys must be in line when a straight edge is placed on the pulleys (Figure 10-6).

**Classroom Manual**
Chapter 10, page 217

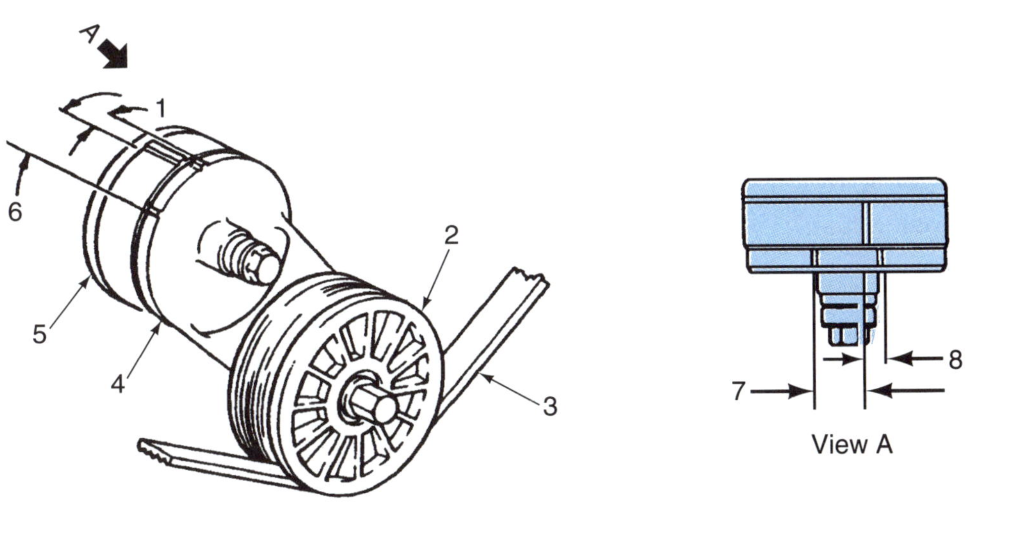


1. New belt range
2. Pulley
3. Belt
4. Arm
5. Spindle
6. Used belt acceptable wear range
7. Used belt length range
8. New belt length range

**Figure 10-4** Belt tensioner scale (Courtesy of Chevrolet Motor Division, General Motors Corporation)

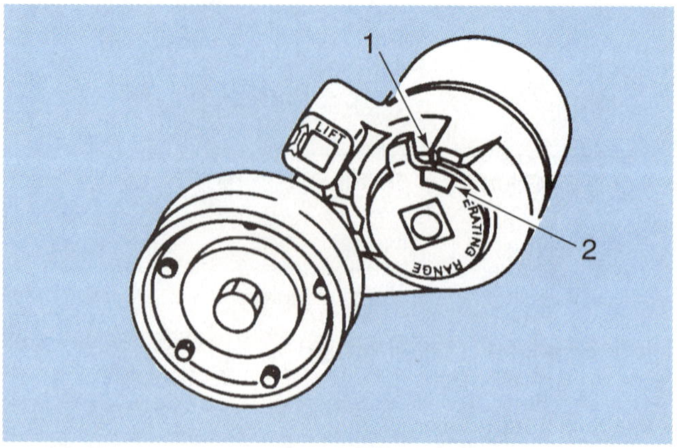


1. Cast arrow
2. Range indicator

**Figure 10-5** One-half inch drive opening in the tensioner pulley (Courtesy of Cadillac Motor Car Division, General Motors Corporation)

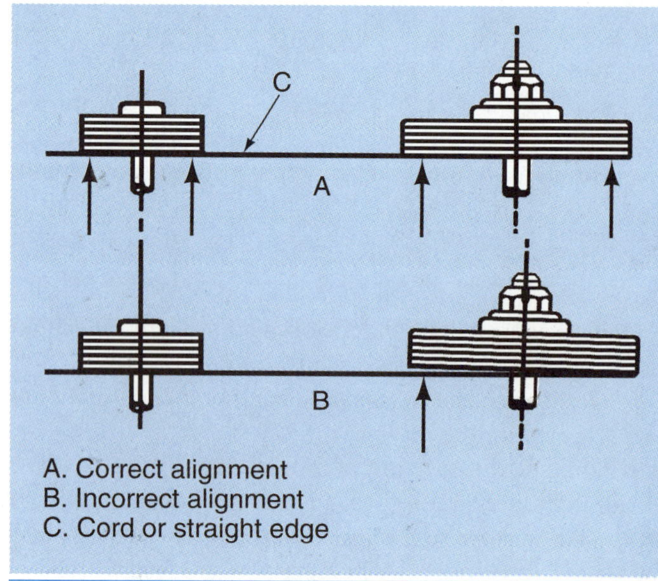


**Figure 10-6** Checking pulley alignment (Courtesy of Chevrolet Motor Division, General Motors Corporation)

**CUSTOMER CARE:** When servicing vehicles, always promote preventative maintenance to the customer. For example, if you are performing some underhood service such as replacing a battery and starting motor, always make a quick check of the belt, or belts, and cooling system hoses. If you see a ribbed V-belt with chunks out of it, recommend a belt replacement to the customer. The customer will usually appreciate this service, and authorize the belt replacement. If you ignore the belt with chunks out of it, and the belt breaks the next day while the customer is driving on the freeway, the customer is inconvenienced and has to pay a tow bill. Although you did a satisfactory battery and starting motor replacement, the customer will not be happy if this problem occurs the day after you worked on the vehicle. The next time this unhappy customer requires automotive service, he or she may take the vehicle to another repair shop.

## Fluid Level Checking

Most car manufacturers recommend power steering fluid or automatic transmission fluid in power steering systems. Always use the type of power steering fluid recommended in the vehicle manufacturer's service manual. If the power steering fluid level is low, steering effort is increased and may be erratic. A low fluid level may cause a growling noise in the power steering pump. Some car manufacturers now recommend checking the power steering pump fluid level with the fluid at an ambient temperature of 176°F (80°C). Follow these steps to check the power steering fluid level:

1. With the engine idling at 1,000 rpm or less, turn the steering wheel slowly and completely in each direction several times to boost the fluid temperature (Figure 10-7).
2. If the vehicle has a remote power steering fluid reservoir, check for foaming in the reservoir, which indicates low fluid level or air in the system.
3. Observe the fluid level in the remote reservoir. This level should be at the hot full mark. Shut off the engine and remove dirt from the neck of the reservoir with a shop towel. If the power steering pump has an integral reservoir, the level should be at the hot level on the dipstick. When an external reservoir is used, the dipstick is located in the external reservoir (Figure 10-8).

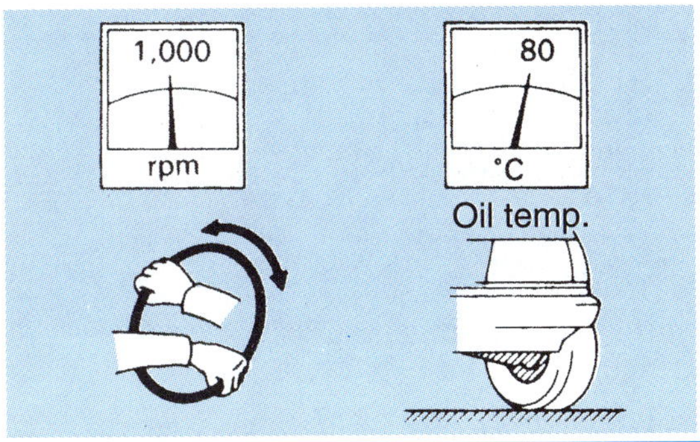


**Figure 10-7** Boosting fluid temperature (Courtesy of Toyota Motor Corporation)

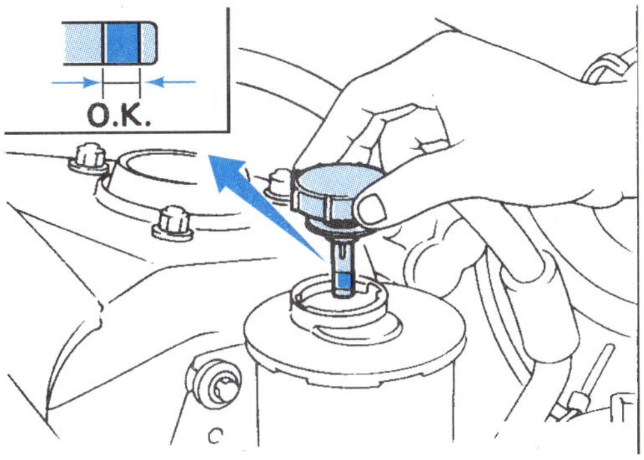


**Figure 10-8** Power steering pump dipstick (Courtesy of Nissan Motor Co., Ltd.)

4. Pour the required amount of the car manufacturer's recommended power steering fluid into the reservoir to bring the fluid level to the hot full mark on the reservoir or dipstick with the engine idling.

## Power Steering System Draining and Flushing

If the power steering fluid is contaminated with moisture, dirt, or metal particles, the system must be drained and new fluid installed. Follow these steps to drain and flush the power steering system:

1. Lift the front of the vehicle with a floor jack and install jack stands under the suspension. Lower the vehicle onto the jack stands and remove the floor jack.
2. Remove the return hose from the remote reservoir that is connected to the steering gear. Place a plug on the reservoir outlet, and position the return hose in an empty drain pan (Figure 10-9).
3. With the engine idling, turn the steering wheel fully in each direction, and stop the engine (Figure 10-10).
4. Fill the reservoir to the hot full mark with the manufacturer's recommended fluid.
5. Start the engine and run the engine at 1,000 rpm while observing the return hose in the drain pan. When fluid begins to discharge from the return hose, shut the engine off.
6. Repeat steps 4 and 5 until there is no air in the fluid discharging from the return hose.
7. Remove the plug from the reservoir and reconnect the return hose. Bleed the power steering system.

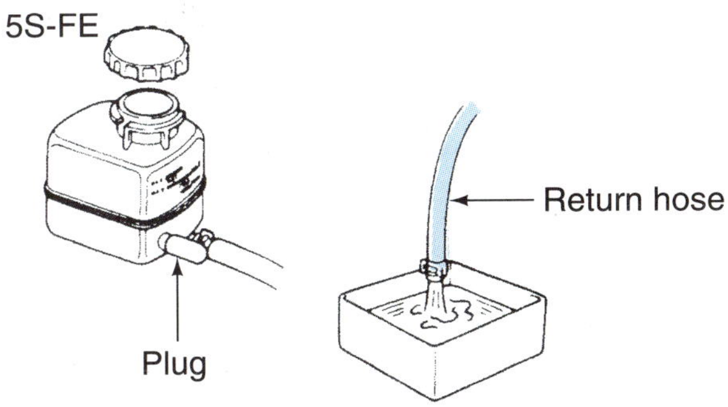


**Figure 10-9** Return hose installed in drain pan for power steering draining and flushing (Courtesy of Toyota Motor Corporation)

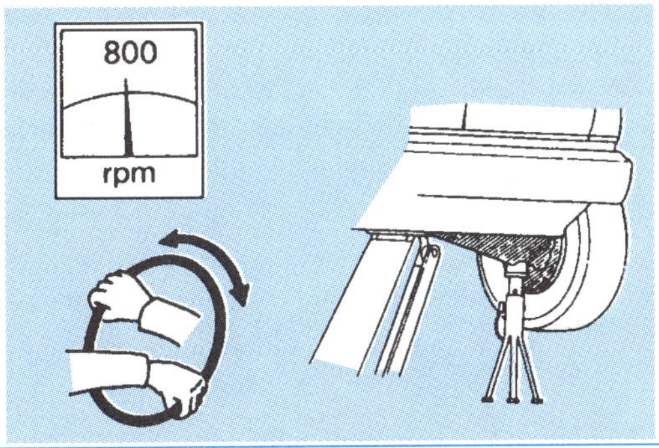


**Figure 10-10** Draining fluid from the power steering system (Courtesy of Toyota Motor Corporation)

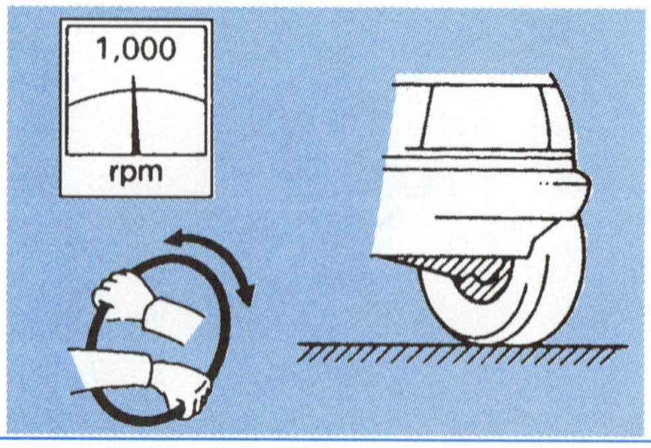


**Figure 10-11** Turning steering wheel slowly to bleed air from the power steering system (Courtesy of Toyota Motor Corporation)

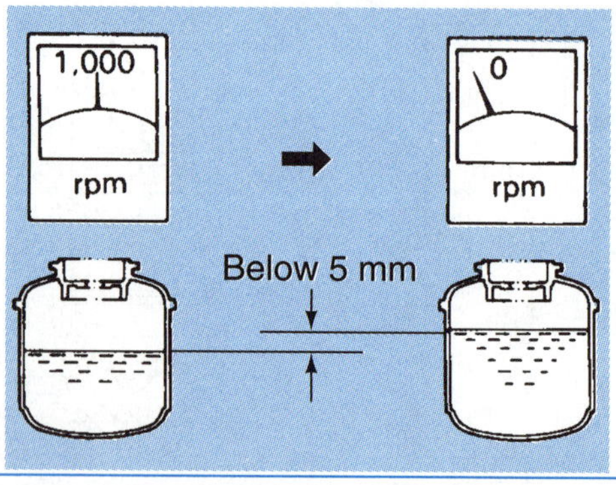


**Figure 10-12** Power steering fluid level after bleeding (Courtesy of Toyota Motor Corporation)

## Bleeding Air From the Power Steering System

When air is present in the power steering fluid, a growling noise may be heard in the pump, and steering effort may be increased or erratic. When a power steering system has been drained and refilled, follow this procedure to remove air from the system:

1. Fill the power steering pump reservoir as outlined previously.
2. With the engine running at 1,000 rpm, turn the steering wheel fully in each direction three or four times (Figure 10-11). Each time the steering wheel is turned fully to the right or left, hold it there for 2 to 3 seconds before turning it in the other direction.
3. Check for foaming of the fluid in the reservoir. When foaming is present, repeat steps 1 and 2.
4. Check the fluid level with the engine running, and be sure it is at the hot full mark. Shut off the engine and make sure the fluid level does not increase more than 0.020 in (5 mm) (Figure 10-12).

## Power Steering Pump Pressure Test

Since there are some variations in power steering pump pressure test procedures and pressure specifications, the vehicle manufacturer's test procedures and specifications must be used. If the power steering pump pressure is low, steering effort is increased. Erratic power steering pump pressure causes variations in steering effort, and the steering wheel may jerk as it is turned. Since a power steering pump will never develop the specified pressure if the belt is slipping, the belt tension must be checked and adjusted, if necessary, prior to a pump pressure test. Following is a typical power steering pressure test procedure:

**Special Tools**

Power steering pressure test gauge

1. With the engine stopped, disconnect the pressure line from the power steering pump, and connect the gauge side of the pressure gauge to the pump outlet fitting. Connect the valve side of the gauge to the pressure line (Figure 10-13).
2. Start the engine and turn the steering wheel fully in each direction two or three times to bleed air from the system (Figure 10-14). Be sure the fluid level is correct and the fluid temperature is at least 176°F (80°C). A thermometer may be inserted in the pump reservoir fluid to measure the fluid temperature.

**CAUTION:** During the power steering pump pressure test with the pressure gauge valve closed, if this valve is closed for more than 10 seconds, excessive pump pressure may cause power steering hoses to rupture, resulting in personal injury.

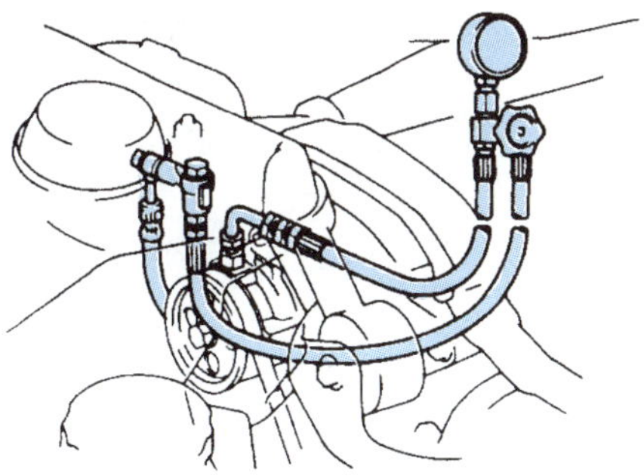

Figure 10-13 Pressure gauge connections to power steering pump (Courtesy of Toyota Motor Corporation)

Figure 10-14 Bleeding air from the power steering pump system and checking fluid temperature (Courtesy of Toyota Motor Corporation)

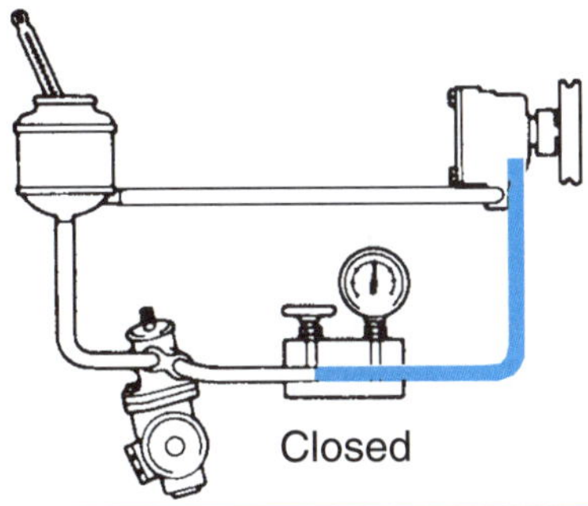


Figure 10-15 Power steering pump pressure test with pressure gauge valve closed (Courtesy of Toyota Motor Corporation)

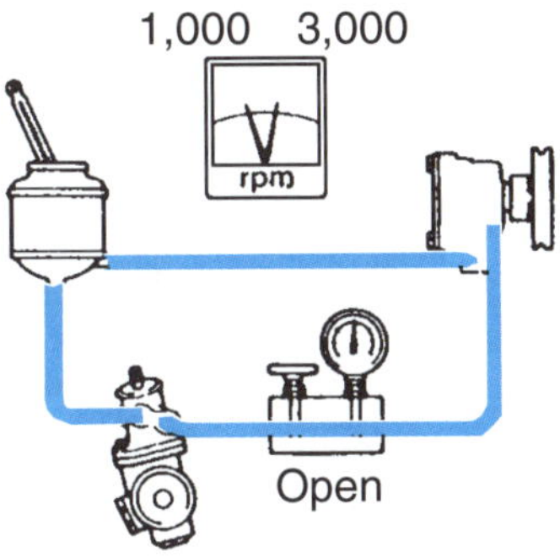


Figure 10-16 Power steering pressure test at 1,000 rpm and 3,000 rpm (Courtesy of Toyota Motor Corporation)

**CAUTION:** Do not allow the fluid to become too hot during the power steering pump pressure test. Excessively high fluid temperature reduces pump pressure. Wear protective gloves and always shut the engine off before disconnecting gauge fittings, because the hot fluid may cause burns.

3. With the engine idling, close the pressure gauge valve for no more than 10 seconds, and observe the pressure gauge reading (Figure 10-15). Turn the pressure gauge valve to the fully open position. If the pressure gauge reading does not equal the vehicle manufacturer's specifications, repair or replace the power steering pump.
4. Check the power steering pump pressure with the engine running at 1,000 rpm and 3,000 rpm, and record the pressure difference between the two readings (Figure 10-16). If the pressure difference between the pressure readings at 1,000 rpm and 3,000 rpm does not equal the vehicle manufacturer's specifications, repair or replace the flow control valve in the power steering pump.
5. With the engine running, turn the steering wheel fully in one direction, and observe the steering pump pressure while holding the steering wheel in this position (Figure 10-17). If the pump pressure is less than the vehicle manufacturer's specifications, the steering gear housing has an internal leak and should be repaired or replaced.
6. Be sure the front tire pressure is correct, and center the steering wheel with the engine idling. Connect a spring scale to the steering wheel and measure the steering effort in

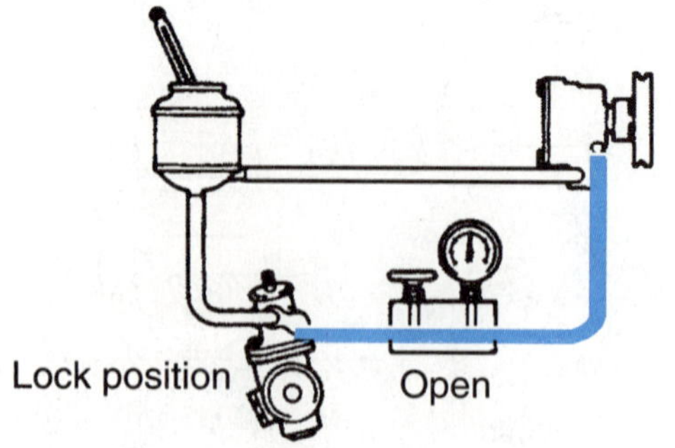


Figure 10-17 Power steering pump pressure test with the front wheels turned fully in one direction (Courtesy of Toyota Motor Corporation)

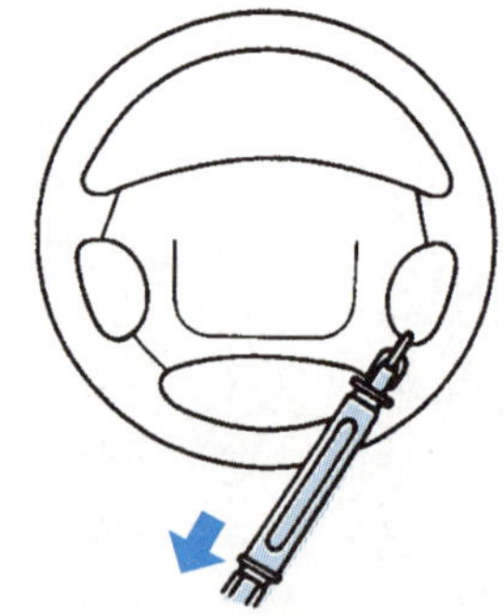

Figure 10-18 Steering effort measurement (Courtesy of Toyota Motor Corporation)

**Special Tools**

Spring scale

both directions (Figure 10-18). If the power steering pump pressure is satisfactory and the steering effort is more than the vehicle manufacturer's specifications, the power steering gear should be repaired. Photo Sequence 8 shows a typical procedure for pressure testing a power steering pump.

## Power Steering Pump Oil Leak Diagnosis

The possible sources of power steering pump oil leaks are the drive shaft seal, reservoir O-ring seal, high-pressure outlet fitting, and the dipstick cap. If leaks occur at any of the seal locations, seal replacement is necessary. When a leak is present at the high-pressure outlet fitting, tighten this fitting to the specified torque (Figure 10-19). If the leak still occurs, replace the O-ring seal on the fitting and retighten the fitting.

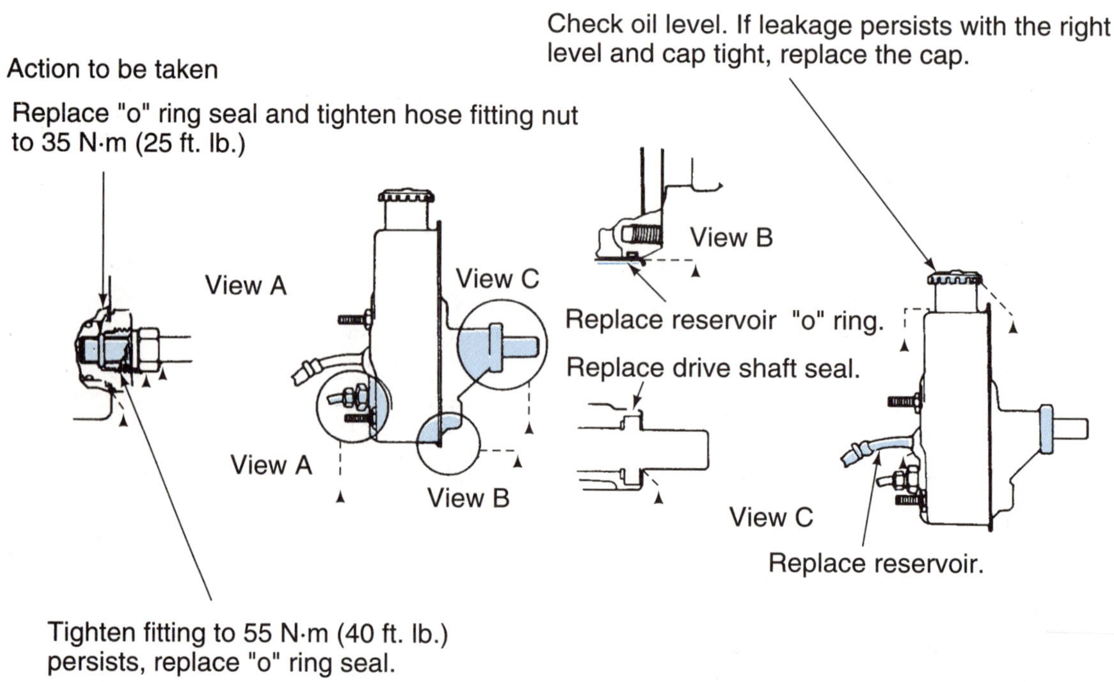


Figure 10-19 Power steering pump oil leak diagnosis (Courtesy of Chrysler Corporation)

## Photo Sequence 8
## Typical Procedure for Pressure Testing a Power Steering Pump

P8-1 Connect the pressure gauge to the power steering pump.

P8-2 Connect a tachometer to the ignition system.

P8-3 Look up the vehicle manufacturer's specified power steering pump pressure in the service manual.

P8-4 Start the engine and turn the steering wheel from lock to lock three times to bleed air from the system and heat the power steering fluid.

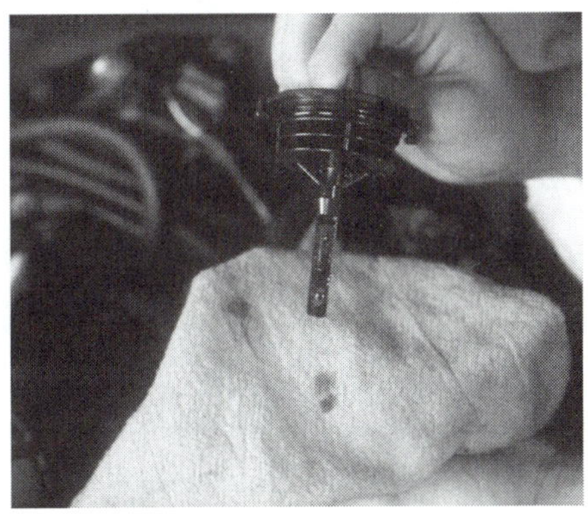

P8-5 Check the power steering fluid level, and add fluid as required.

P8-6 Place a thermometer in the power steering fluid reservoir, and be sure the fluid temperature is at least 176°F (80°C).

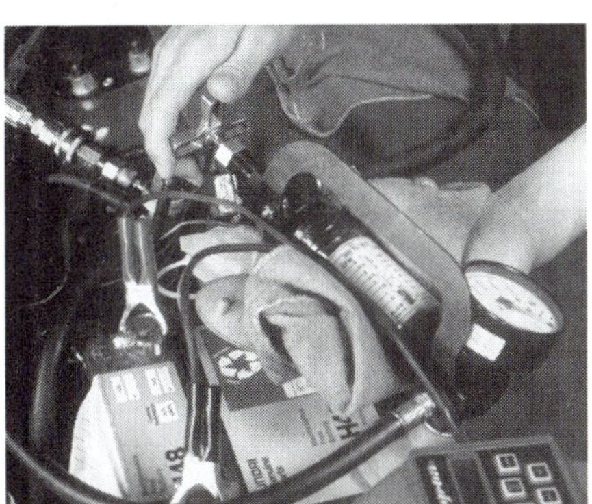

P8-7 With the engine idling, close the pressure gauge valve for no more than 10 seconds. Observe and record the pressure gauge reading.

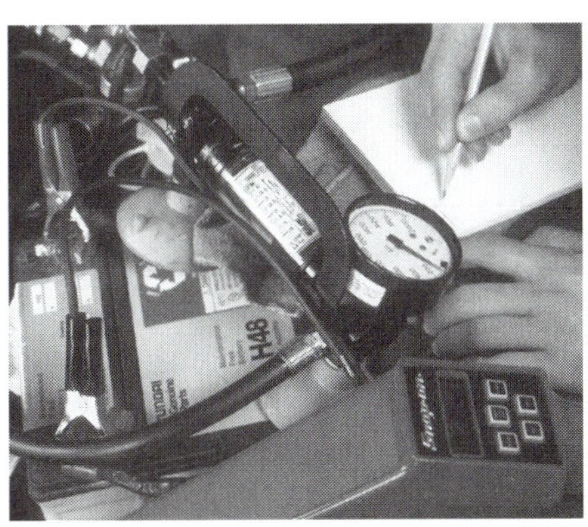

P8-8 Check and record the power steering pump pressure at 1,000 rpm.

P8-9 Check and record the power steering pump pressure at 3,000 rpm.

## Photo Sequence 8
## Typical Procedure for Pressure Testing a Power Steering Pump (continued)

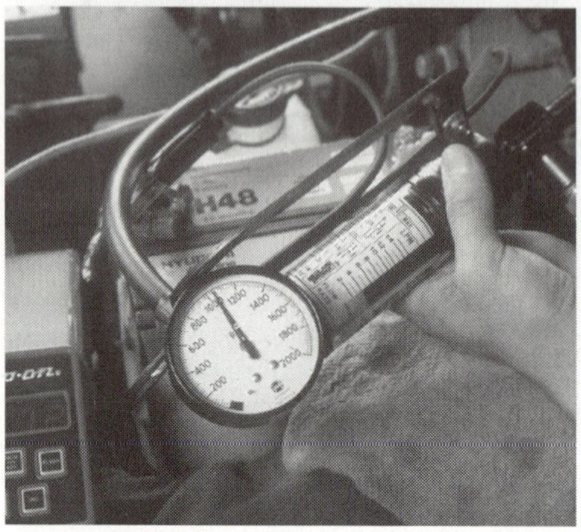

**P8-10** Check and record the power steering pump pressure with the steering wheel turned fully in one direction and the engine idling.

### Power Steering Pump Replacement

**CAUTION:** If the vehicle has been driven recently, the pump, hoses, and fluid could be extremely hot. Use caution when handling components to avoid burns.

If a growling noise is present in the power steering pump after the fluid level is checked and air has been bled from the system, the pump bearings or other components are defective and pump replacement or repair is required. When the power steering pump pressure is lower than specified, pump replacement or repair is necessary. When the power steering pump is replaced, proceed as follows:

1. Disconnect the power steering return hose from the remote reservoir or pump, and allow the fluid to drain from this hose into a drain pan. Discard the used fluid.
2. Loosen the bracket, or belt tension adjusting bolt, and the pump mounting bolt.
3. Loosen the belt tension until the belt can be removed. On some cars, it is necessary to lift the vehicle on a hoist and gain access to the power steering pump from underneath the vehicle.
4. Remove the hoses from the pump, and cap the pump fittings and hoses.
5. Remove the belt tension adjusting bolt and the mounting bolt, and remove the pump.
6. Check the pump mounting bolts and bolt holes for wear. Worn bolts must be replaced. If the bolt mounting holes in the pump are worn, pump replacement is necessary.
7. Reverse steps 1 through 5 to install the power steering pump. Tighten the belt as described previously, and tighten the pump mounting and bracket bolts to the manufacturer's specifications. If O-rings are used on the pressure hose, replace the O-rings. Be sure the hoses are not contacting the exhaust manifold, catalytic converter, or exhaust pipe during or after pump replacement.

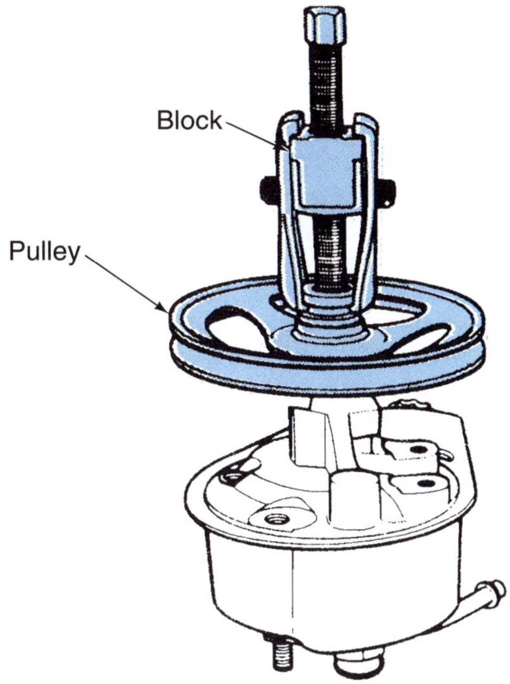


**Figure 10-20** Press-on type power steering pump pulley removal (Courtesy of Chrysler Corporation)

8. Fill the pump reservoir with the manufacturer's recommended power steering fluid, and bleed air from the power steering system as described previously.

## Power Steering Pump Pulley Replacement

If the pulley wobbles while it is rotating, the pulley is probably bent, and pulley replacement is necessary. Worn pulley grooves also require pulley replacement. Always check the pulley for cracks. If this condition is present, pulley replacement is essential. A pulley that is loose on the pump shaft must be replaced. Never hammer on the pump drive shaft during pulley removal or replacement. This action will damage internal pump components. If the pulley is pressed onto the pump shaft, a special puller is required to remove the pulley (Figure 10-20), and a pulley installation tool is used to install the pulley (Figure 10-21).

If the power steering pump pulley is retained with a nut, mount the pump in a vise. Always tighten the vise on one of the pump mounting bolt surfaces. Do not tighten the vise with excessive force. Use a special holding tool to keep the pulley from turning, and loosen the pulley nut with a box end wrench (Figure 10-22). Remove the nut, pulley, and woodruff key. Inspect the pulley, shaft, and woodruff key for wear. Be sure the key slots in the shaft and pulley are not worn. Replace all worn components.

**Special Tools**

Puller for power steering pump pulley, press-on pulley

**Special Tools**

Power steering pump pulley installing tool, press-on pulley

A woodruff key is a half-moon shaped key that fits snugly in a shaft opening, and the top edge of the key extends out of the shaft. The top edge of the key fits snugly in a groove cut inside the pulley hub. When the pulley is installed on the shaft, the pulley groove slides over the key to prevent pulley rotation on the shaft.

## Remove and Replace the Power Steering Pump Reservoir

Power steering pump service procedures vary depending on the type of pump. Always follow the vehicle manufacturer's recommended service procedures in the service manual. The following service procedures are for a Chrysler power steering pump with an integral reservoir. When the power steering pump has an integral reservoir, follow these steps for reservoir removal and replacement:

1. Remove the filler cap and drain the oil from the reservoir.
2. Remove the two studs and the fitting from the rear of the reservoir.
3. Rock the reservoir by hand or tap it gently with a soft hammer to remove it from the housing.

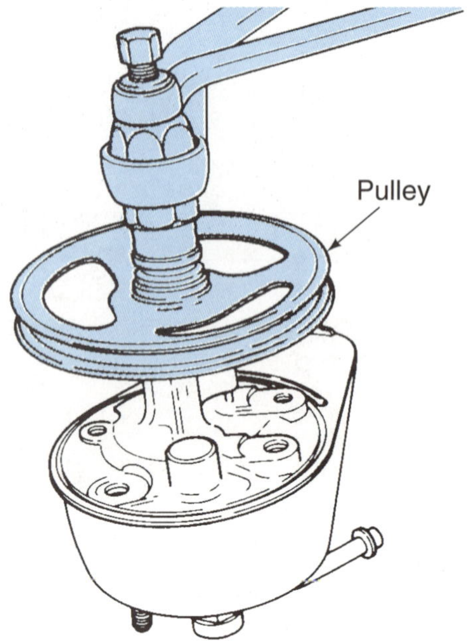


**Figure 10-21** Press-on type power steering pump pulley installation (Courtesy of Chrysler Corporation)

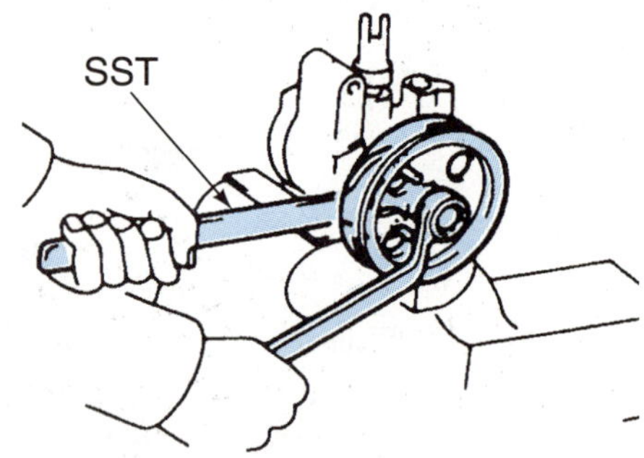


**Figure 10-22** Removing power steering pump pulley and nut (Courtesy of Toyota Motor Corporation)

4. Clean all the parts and discard O-rings and seals. Inspect all O-ring and seal surfaces for damage.
5. Lubricate the new seals and O-rings with the manufacturer's recommended power steering fluid and install them (Figure 10-23). Be sure the pressure union fitting O-ring is installed in the groove nearest the fitting hex head.

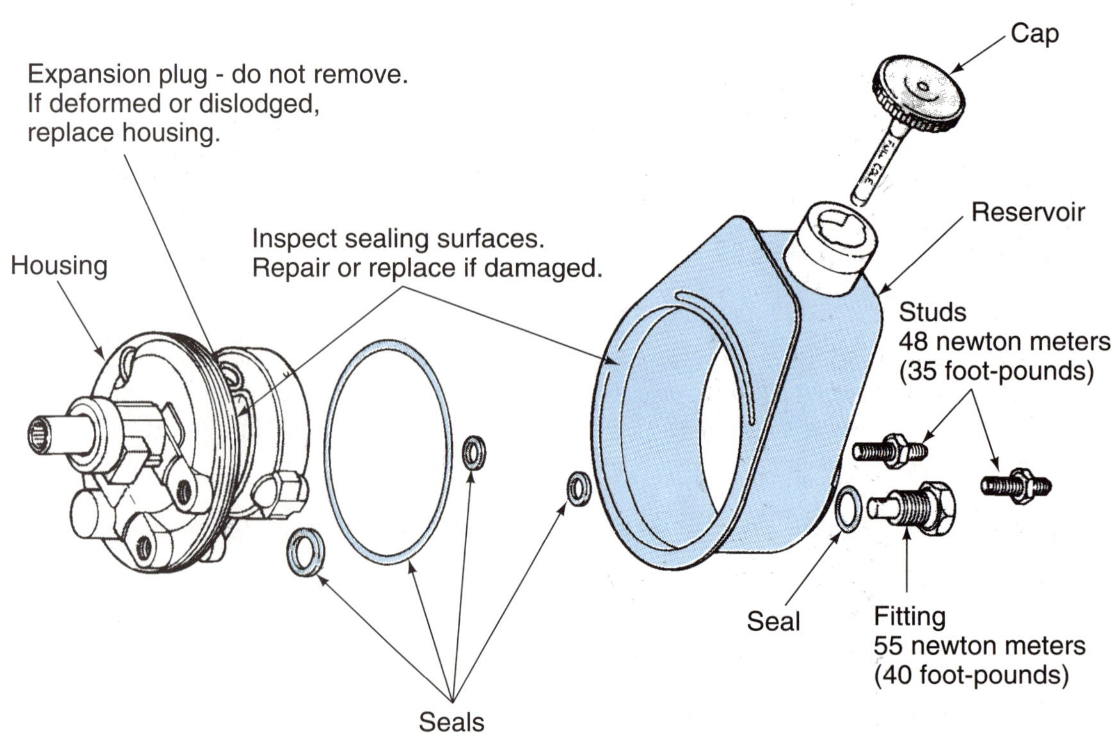


**Figure 10-23** Integral power steering pump reservoir seal and O-ring replacement (Courtesy of Chrysler Corporation)

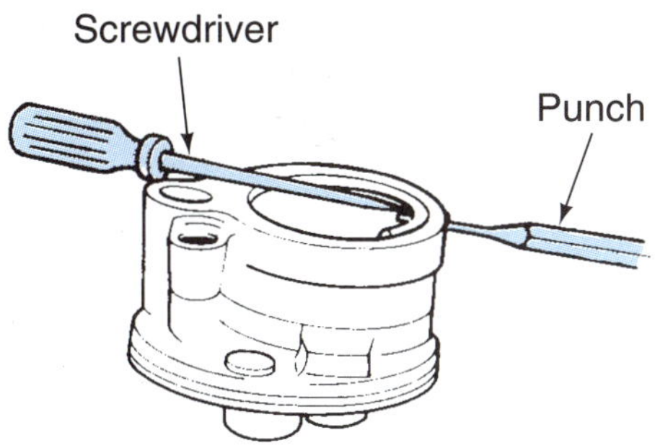


Figure 10-24 Retaining ring removal (Courtesy of Chrysler Corporation)

## Remove and Replace Flow Control Valve and End Cover

When the flow control valve and end plate are serviced, follow these steps:

1. Remove the retaining ring with a slotted screwdriver and a punch (Figure 10-24).
2. Remove the flow control valve, end cover, spring, and magnet. Check the flow control valve for burrs. Remove minor burrs with crocus cloth, and clean the flow control valve in solvent. Damaged or worn flow control valves must be replaced. Inspect the end cover sealing surface for damage. Check the pump drive shaft for corrosion and damage. Remove corrosion with crocus cloth (Figure 10-25). Clean the magnet with a shop towel.

Crocus cloth is fine poli… emery p…

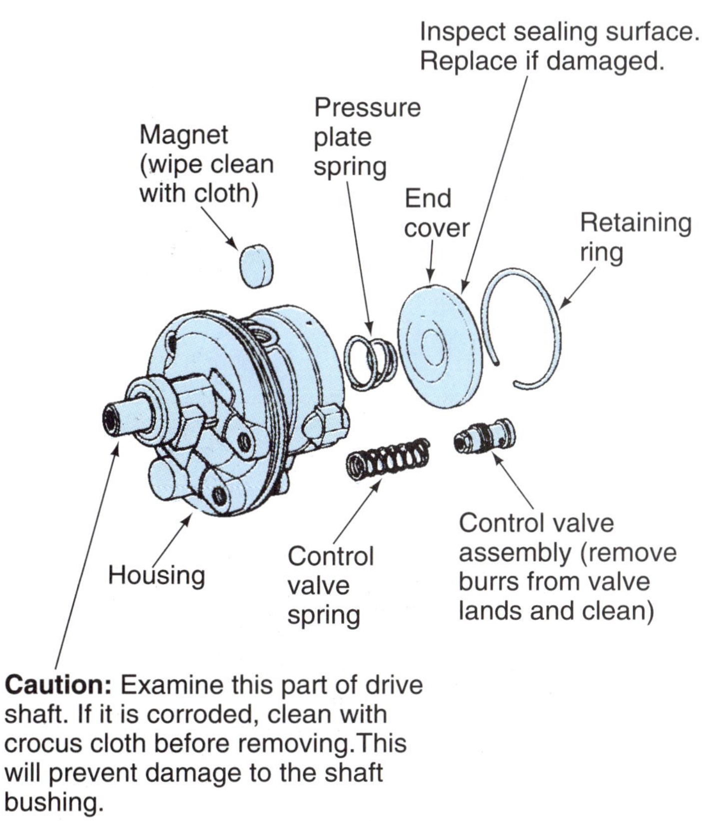


Figure 10-25 End cover and related component service (Courtesy of Chrysler Corporation)

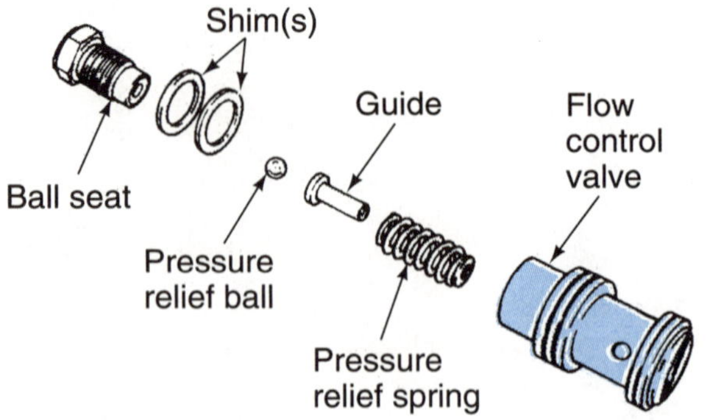


**Figure 10-26** Pressure relief valve removal (Courtesy of Chrysler Corporation)

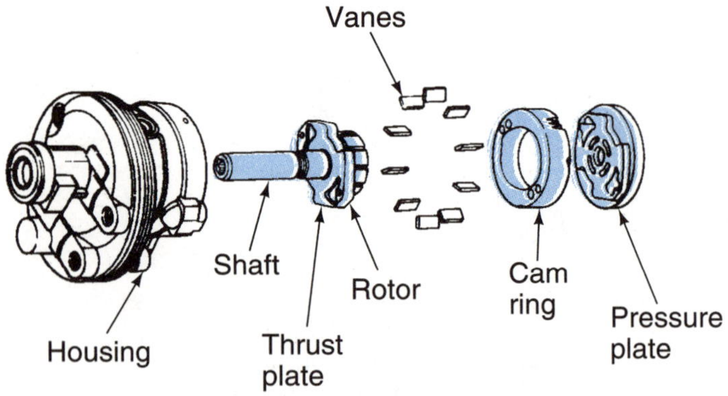


**Figure 10-27** Removing pressure plate, pump ring, rotor, and shaft assembly (Courtesy of Chrysler Corporation)

3. Clean all parts and lubricate the end cover with power steering fluid.
4. Install the end cover, retaining ring, and related components.

## Remove and Replace Pressure Relief Valve

Follow this procedure to service the pressure relief valve:

1. Wrap a shop towel around the land end of the flow control valve and clamp this end in a soft-jawed vice. Be very careful not to mark valve lands.
2. Remove the hex-head ball seat (Figure 10-26). Clean the components in solvent. A worn or damaged pressure relief ball, spring, guide, or seat must be replaced.

## Servicing Power Steering Pump Rotating Components

1. After the end cover and spring have been removed, pull the pressure plate, pump ring, rotor, and shaft out of the pump housing (Figure 10-27).
2. Remove and discard the rotor lock ring (Figure 10-28).
3. Remove the rotor and thrust plate from the shaft (Figure 10-29).
4. Examine all the wear surfaces, and replace parts that are worn, scuffed, or scored. Check vanes and vane slots in the rotor for wear.

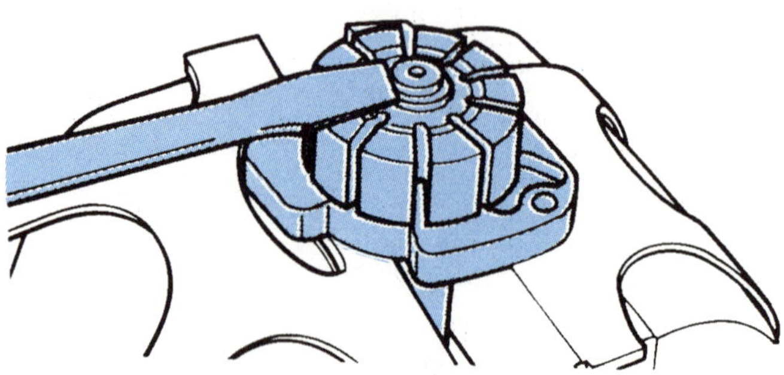

**Figure 10-28** Rotor lock ring removal (Courtesy of Chrysler Corporation)

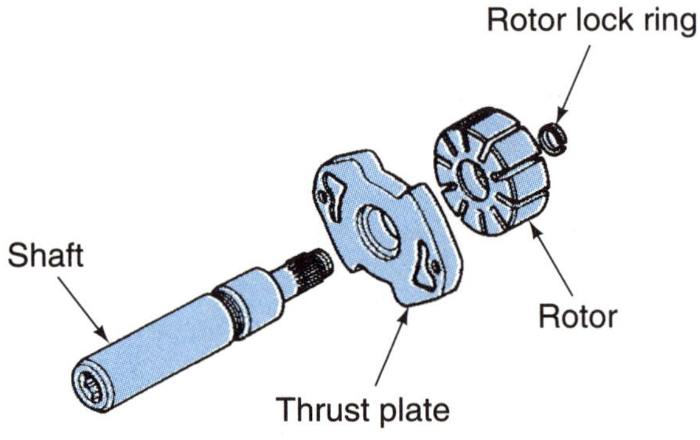


**Figure 10-29** Pump rotor and thrust plate removal (Courtesy of Chrysler Corporation)

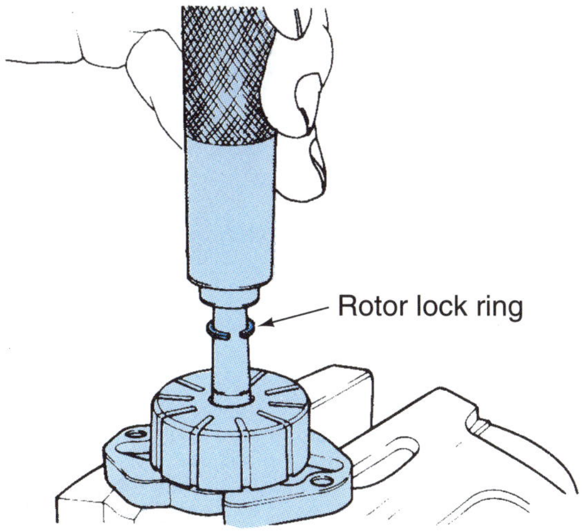


Figure 10-30 Thrust plate, rotor, and lock ring installation (Courtesy of Chrysler Corporation)

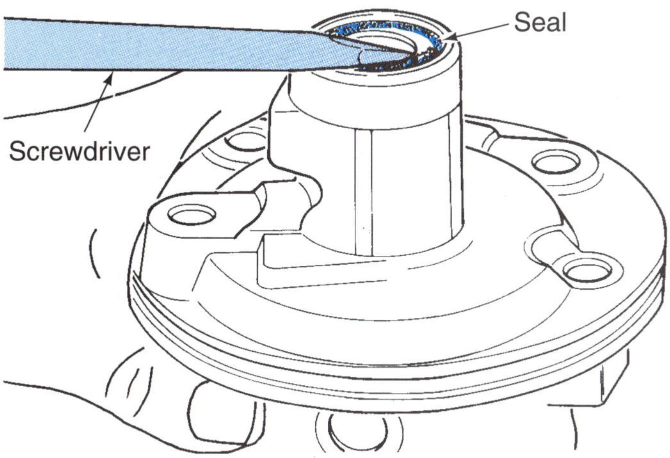


Figure 10-31 Drive shaft seal removal (Courtesy of Chrysler Corporation)

5. Clean all the components and install the inner thrust plate, rotor, and a new lock ring on the shaft (Figure 10-30).
6. Remove the drive shaft seal from the pump housing with a slotted screwdriver (Figure 10-31), and install a new seal with the proper driver (Figure 10-32). Lubricate the new seal with the manufacturer's recommended fluid.
7. Lubricate the new housing O-rings with the recommended fluid and position the O-rings in the housing.
8. Install the thrust plate, shaft, and pump rotor assembly in the housing, and install the pump ring (Figure 10-33).
9. Install the vanes in the rotor and lubricate all parts with power steering fluid.
10. Install and seat the pressure plate (Figure 10-34).
11. Install the pressure plate spring, end cover, and retaining ring. Reassemble the magnet and flow control valve and install the pump reservoir with a new O-ring and seals.

**Special Tools**

Seal drivers

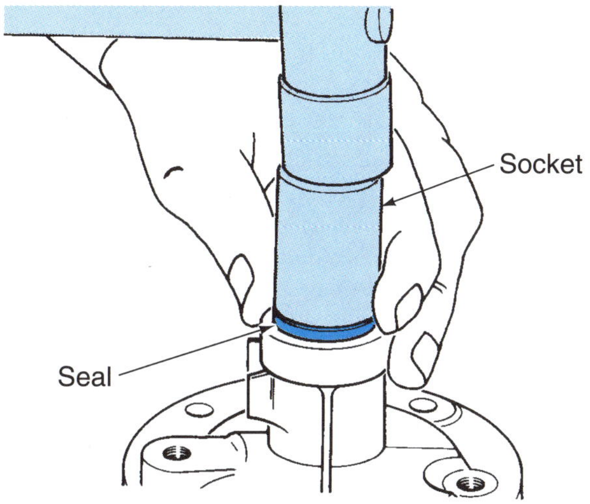


Figure 10-32 Installation of drive shaft seal (Courtesy of Chrysler Corporation)

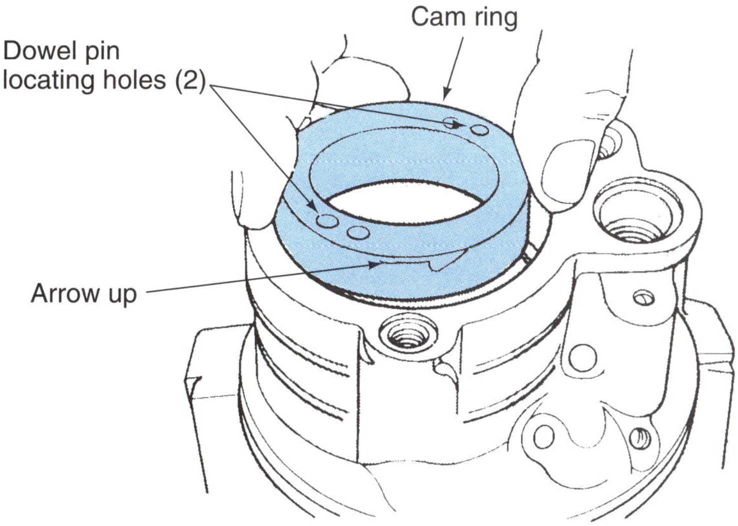


Figure 10-33 Pump ring installation (Courtesy of Chrysler Corporation)

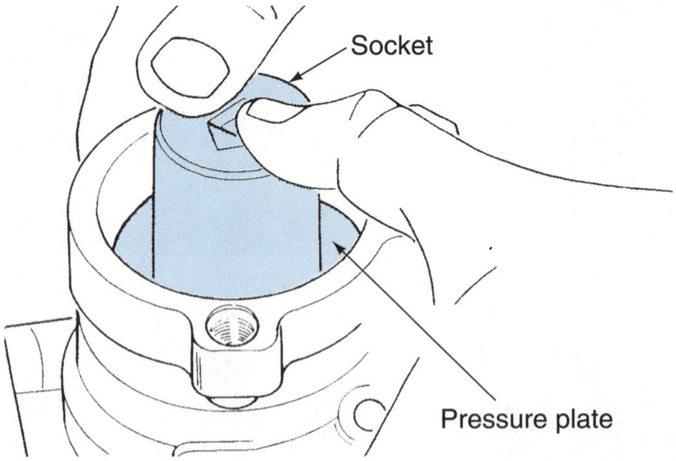


**Figure 10-34** Pressure plate installation and seating (Courtesy of Chrysler Corporation)

## Checking and Servicing Power Steering Lines and Hoses

Power steering lines should be checked for leaks, dents, sharp bends, cracks, and contact with other components. Lines and hoses must not rub against other components. This action could wear a hole in the line or hose. Many high-pressure power steering lines are made from high-pressure steel-braided hose with molded steel fittings on each end (Figure 10-35).

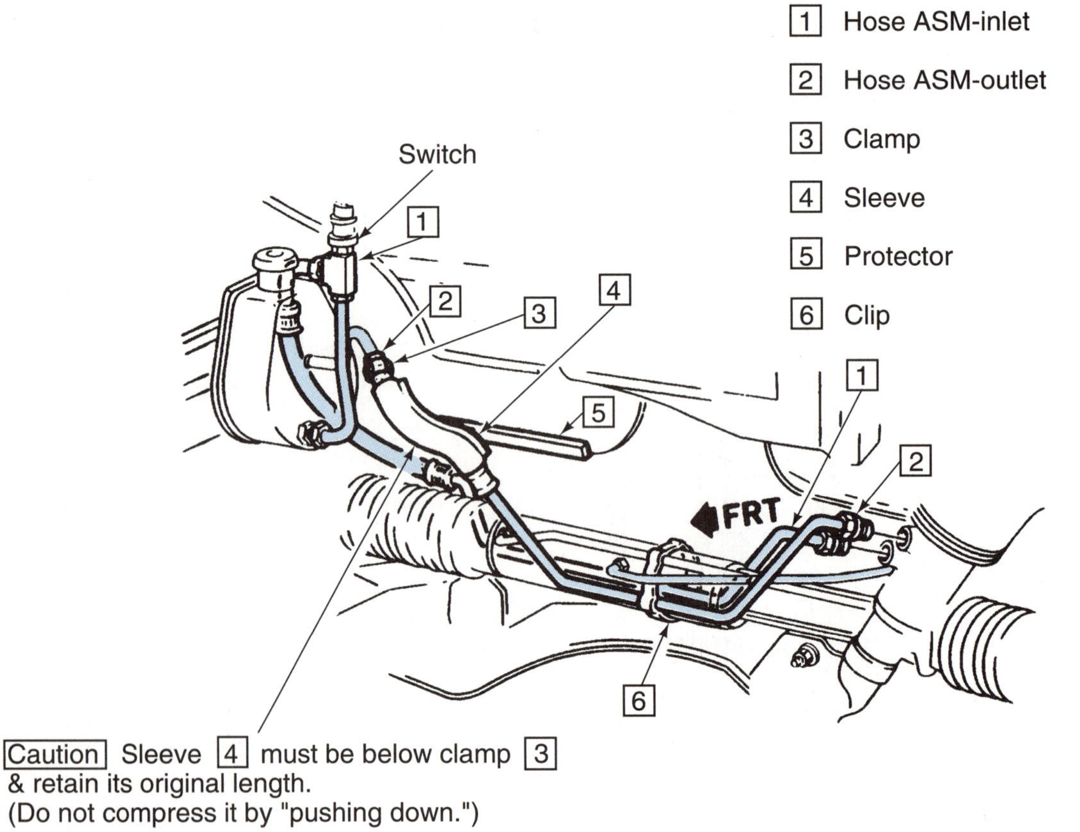


**Figure 10-35** Power steering lines (Courtesy of Chevrolet Motor Division, General Motors Corporation)

## Power Steering Hose Replacement

**CAUTION:** If the engine has been running, the power steering hoses, components, and fluid may be extremely hot. Wear protective gloves and use caution to avoid burns.

When power steering hose replacement is required, follow these steps:

1. With the engine stopped, remove the return hose at the power steering gear, and allow the fluid to drain from this hose into a drain pan. On some cars, the vehicle must be lifted on a hoist to replace the power steering hoses.
2. Loosen and remove all hose fittings from the pump and steering gear.
3. Remove all hose-to-chassis clips.
4. Remove the hoses from the chassis, and cap the pump and steering gear fittings.
5. If O-rings are used on the hose ends, install new O-rings. TRW steering gear cylinder lines have gaskets that are serviced as a unit. The old gasket must be pried out of the fitting in the housing before the new lines and gaskets are installed. Lubricate O-rings with power steering fluid.
6. Reverse steps 1 through 4 to install the power steering hoses. Tighten all fittings to the manufacturer's specified torque. Be sure all hose-to-chassis clips are in place. Do not position hoses where they rub on other components.
7. Fill the pump reservoir to the full mark with the manufacturer's recommended fluid. Bleed air from the power steering system as mentioned previously in this chapter. Check the fluid level in the reservoir and add fluid as required.

**Classroom Manual**
Chapter 10, page 218

## Guidelines for Servicing Power Steering Pumps

1. A loose power steering belt causes low pump pressure, hard steering, and belt chirping, or squealing, especially on acceleration.
2. Power steering belt tension may be measured with belt tension gauge or by checking the amount of belt deflection.
3. When prying on a power steering pump to tighten the belt, always pry on the pump ear and hub. Do not pry on the reservoir.
4. Most car manufacturers recommend power steering fluid or automatic transmission fluid in the power steering system.
5. Many car manufacturers recommend checking the power steering pump fluid level when the fluid is warmed up to 176°F (80°C).
6. If the power steering fluid is contaminated with moisture, dirt, or metal particles, the system should be drained and flushed, and new fluid should be installed.
7. Air may be bled from the power steering system by turning the steering wheel fully in each direction several times with the engine running. Each time the wheel is turned fully to the right or left, it should be held in that position for 2 to 3 seconds.
8. A pressure gauge and manual valve are connected in series in the power steering pump pressure hose to check pump pressure. Maximum pump pressure is checked by closing the manual valve for less than 10 seconds.
9. The power steering pump pressure may be checked at 1,000 rpm and 3,000 rpm to check the operation of the pump flow control valve.
10. After the power steering pump pressure is tested and proven to be satisfactory, steering gear leakage may be tested by turning the steering wheel fully in one direction and observing the pressure reading. If the power steering pump pressure is less than specified, the steering gear has an internal leak.

11. With the engine idling, a spring scale may be attached at the outer end of the steering wheel cross bar to measure steering effort.
12. The possible sources of power steering pump leaks are the drive shaft seal, reservoir O-ring seals, high-pressure outlet fitting, dipstick cap, and low-pressure inlet pipe.
13. When a power steering pump is replaced, the mounting bolts and bolt holes must be checked for wear. If excessive wear is present on these components, pump and belt misalignment may occur.
14. Power steering pump pulleys may be pressed onto the pump shaft or retained on the shaft with a woodruff key and nut.
15. Never tap on the pump shaft while removing or replacing the pump pulley.
16. A sticking flow control valve causes erratic power steering pump pressure. This valve may be cleaned and polished with crocus cloth.
17. A defective power steering pump may be overhauled or replaced.
18. Power steering pump lines should be checked for leaks, dents, sharp bends, cracks, and contact with other components.
19. When power steering pump components are serviced, always use caution because these components may be very hot if the engine has been running.
20. If it is necessary to drain the power steering fluid, disconnect the return hose at the remote reservoir or pump and drain the fluid from this hose into a drain pan.

## CASE STUDY

The owner of a Dodge Spirit with a 2.5 L turbocharged engine requested a price on a power steering gear replacement. The service writer questioned the owner regarding the reason for the steering gear replacement, and discovered that the problem was intermittent hard steering. Further discussion with the owner also revealed that the car had been taken to another automotive service center, and the owner was informed that her car required a steering gear replacement. The owner was now looking for a second opinion. The technician road tested the vehicle and found no indication of hard steering. The owner was asked about the exact driving conditions when the hard steering was experienced, and the answer to this question provided the solution to the problem. The owner indicated that the hard steering was usually experienced while cornering and always on a rainy day. Once this information was revealed, the technician suspected a problem with the power steering belt, because a wet belt slips more easily. A check of the power steering belt indicated the belt was oil soaked from engine oil leaks and slightly loose. The power steering belt was replaced and adjusted, and the owner was informed that the engine oil leaks should be corrected or the belt would become oil soaked again. A short while later, after a rain, the owner brought the car to the shop to have the engine oil leaks repaired. The owner reported that the belt corrected the hard steering problem. In this situation, a brief discussion about the customer's complaint lead to accurate diagnosis of the problem, and because of this accurate diagnosis the shop gained a steady customer.

## Terms to Know

V-belt
Ribbed V-belt
Serpentine belt
Steering effort
Woodruff key
Crocus cloth
Belt tension gauge
Power steering pump pressure gauge
Air bleeding
Integral reservoir
Remote reservoir
Flow control valve
Pressure relief valve

## ASE Style Review Questions

1. While discussing power steering belt adjustment:
*Technician A* says the tension on a power steering pump V-belt should be checked with a tension gauge.
*Technician B* says that some ribbed V-belts have an automatic tensioning pulley that eliminates the need for tension adjustments.
Who is correct?
**A.** A only **C.** Both A and B
**B.** B only **D.** Neither A nor B

2. While discussing power steering belt adjustment:
*Technician A* says that a power steering pump V-belt should be tightened with a pry bar installed against the pump reservoir.
*Technician B* says that the power steering pump V-belt should be tightened with a pry bar against the pump ear and hub.
Who is correct?
**A.** A only **C.** Both A and B
**B.** B only **D.** Neither A nor B

3. While discussing power steering fluid level checking:
*Technician A* says foaming in the remote reservoir may indicate air in the power steering system.
*Technician B* says many car manufacturers recommend checking the power steering fluid level with the fluid warmed up.
Who is correct?
**A.** A only **C.** Both A and B
**B.** B only **D.** Neither A nor B

4. While discussing draining of the power steering fluid:
*Technician A* says the return hose from the gear to the remote reservoir should be disconnected at the remote reservoir to drain the fluid.
*Technician B* says the pressure hose should be disconnected from the steering gear to drain the fluid.
Who is correct?
**A.** A only **C.** Both A and B
**B.** B only **D.** Neither A nor B

5. While discussing refilling of the power steering system:
*Technician* A says when the power steering system has been drained and refilled, the steering wheel should be turned fully in both directions with the engine running to bleed air from the system.
*Technician B* says that the low-pressure return hose should be loosened with the engine running to bleed air from the system.
Who is correct?
**A.** A only **C.** Both A and B
**B.** B only **D.** Neither A nor B

6. While discussing the steering pump pressure test:
*Technician A* says the pressure gauge and valve should be connected in the power steering pump return hose to check pump pressure.
*Technician B* says the pressure gauge valve should be closed for 30 seconds during the power steering pump pressure test.
Who is correct?
**A.** A only **C.** Both A and B
**B.** B only **D.** Neither A nor B

7. While discussing power steering pump pressure:
*Technician A* says if the maximum pump pressure is satisfactory with the pressure gauge valve closed, but the pressure is lower than specified with the steering wheel turned fully in one direction, the steering gear may have an internal leak.
*Technician B* says if the pump pressure is satisfactory and the front tires are properly inflated, if the steering effort is higher than specified, the steering gear may require replacing or repairing.
Who is correct?
**A.** A only **C.** Both A and B
**B.** B only **D.** Neither A nor B

8. While discussing steering effort measurement:
*Technician A* says steering effort may be checked with a dial indicator.
*Technician B* says steering effort may be checked with a spring scale.
Who is correct?
**A.** A only **C.** Both A and B
**B.** B only **D.** Neither A nor B

9. While discussing power steering pump pulley replacement:
*Technician A* says when a nut retains the pulley on the shaft, a woodruff key prevents pulley rotation on the shaft.
*Technician B* says a soft hammer may be used to tap on the shaft while removing the pulley.
Who is correct?
**A.** A only **C.** Both A and B
**B.** B only **D.** Neither A nor B

10. While discussing power steering pump diagnosis:
*Technician A* says a sticking flow control valve may cause erratic steering effort.
*Technician B* says if a V-belt is rubbing on the bottom of the pulley, the belt may slip.
Who is correct?
**A.** A only **C.** Both A and B
**B.** B only **D.** Neither A nor B

## Table 10-1 ASE TASK

Inspect power steering fluid level and condition; adjust level in accordance with vehicle manufacturer's recommendations.

| Problem Area | Symptoms | Possible Causes | Classroom Manual | Shop Manual |
|---|---|---|---|---|
| STEERING EFFORT | Hard, erratic steering | 1. Low fluid level | 218 | 226 |
| | | 2. Contaminated fluid | 218 | 227 |

## Table 10-2 ASE TASK

Inspect, adjust tension and alignment, and replace power steering belt(s).

| Problem Area | Symptoms | Possible Causes | Classroom Manual | Shop Manual |
|---|---|---|---|---|
| NOISE | Squealing while accelerating or cornering | Loose belt | 217 | 223 |
| STEERING EFFORT | Hard, erratic steering | Loose belt | 217 | 223 |

## Table 10-3 ASE TASK

Remove and replace power steering pump; inspect pump mounts.

| Problem Area | Symptoms | Possible Causes | Classroom Manual | Shop Manual |
|---|---|---|---|---|
| NOISE | Growling noise from pump with engine running | 1. Low fluid level | 223 | 226 |
| | | 2. Defective pump bearing or components | 223 | 228 |
| STEERING EFFORT | Hard, erratic steering | Defective pump | 223 | 228 |

## Table 10-4 ASE TASK

Inspect and replace power steering pump seals and gaskets.

| Problem Area | Symptoms | Possible Causes | Classroom Manual | Shop Manual |
|---|---|---|---|---|
| FLUID LOSS | Fluid leaks | Worn pump shaft seal, O-rings | 223 | 230 |

## Table 10-5 ASE TASK

Inspect and replace power steering pump pulley.

| Problem Area | Symptoms | Possible Causes | Classroom Manual | Shop Manual |
|---|---|---|---|---|
| NOISE | Squealing while accelerating or cornering | Worn pump pulley | 223 | 233 |
| BELT REPLACEMENT | Rapid belt wear | Bent pulley | 223 | 233 |

## Table 10-6 ASE TASK

Perform power steering pressure test; determine needed repairs.

| Problem Area | Symptoms | Possible Causes | Classroom Manual | Shop Manual |
|---|---|---|---|---|
| STEERING EFFORT | Hard, erratic steering | 1. Low pump pressure | 226 | 228 |
| | | 2. Sticking, worn flow control valve | 226 | 228 |
| | | 3. Defective steering gear | 226 | 228 |

## Table 10-7 ASE TASK

Inspect and replace power steering hoses, fittings, and O-rings.

| Problem Area | Symptoms | Possible Causes | Classroom Manual | Shop Manual |
|---|---|---|---|---|
| STEERING EFFORT | Hard steering, loss of fluid | 1. Leaking hoses | 218 | 238 |
| | | 2. Restricted hoses | 218 | 238 |

## Table 10-8 ASE TASK

Flush, fill, and bleed power steering system.

| Problem Area | Symptoms | Possible Causes | Classroom Manual | Shop Manual |
|---|---|---|---|---|
| NOISE | Growling noise from pump | Air in power steering system | 227 | 226 |
| STEERING EFFORT | Hard, erratic steering | Contaminated fluid | 227 | 226 |

CHAPTER 11

# Recirculating Ball Steering Gear Diagnosis and Service

Upon completion and review of this chapter, you should be able to:

- ❑ Diagnose manual recirculating ball steering gear problems.
- ❑ Remove and replace manual recirculating ball steering gears.
- ❑ Adjust worm shaft bearing preload on manual recirculating ball steering gears.
- ❑ Adjust sector lash on manual recirculating ball steering gears.
- ❑ Repair oil leaks in manual recirculating ball steering gears.
- ❑ Disassemble, repair, and reassemble manual recirculating ball steering gears.
- ❑ Diagnose power recirculating ball steering gear problems.
- ❑ Remove and replace power recirculating ball steering gears.
- ❑ Adjust worm shaft thrust bearing preload in power recirculating ball steering gears.
- ❑ Adjust sector lash in power recirculating ball steering gears.
- ❑ Diagnose and repair oil leaks in power recirculating ball steering gears.
- ❑ Disassemble, repair, and reassemble power recirculating ball steering gears.

## Manual Recirculating Ball Steering Gear Diagnosis

**Basic Tools**

Basic technician's tool set

Service manual

Feeler gauge set

Masking tape

**CAUTION:** Recirculating ball steering gears are often mounted near the exhaust manifold, which may be extremely hot. Wear protective gloves and use caution when inspecting, adjusting, and servicing the steering gear.

When the oil-level plug is removed from the steering gear, the lubricant should be level with the bottom of the plug opening. If the oil level is low, fill the steering gear with the manufacturer's specified steering gear lubricant. A steering gear provides binding and hard uneven steering effort if the oil level is very low. If the oil level is low, visually check the sector shaft seal and the worm shaft seal area for leaks. Leaking seals must be replaced. Hard steering effort may also be caused by wheel alignment problems. Defective worm shaft bearings cause uneven turning effort and steering gear noise. Excessive steering effort may be caused by worn steering gears. A rattling noise from the steering gear may be caused by loose mounting bolts or worn steering shaft U-joints. Excessive steering wheel free-play may be caused by these items:

1. loose worm shaft bearing preload adjustment
2. loose sector shaft lash adjustment
3. worn steering gears
4. loose steering gear mounting bolts
5. worn steering shaft U-joints or flexible coupling

## Manual Recirculating Ball Steering Gear Replacement

**WARNING:** Hard steel bolts may be used for steering gear mounting. Bolt hardness is indicated by ribs on the bolt head. Harder bolts have five, six, or seven ribs on the bolt heads. Never substitute softer steel bolts in place of the original harder bolts, because these softer bolts may break, allowing the steering box to detach from the frame, which results in loss of steering control.

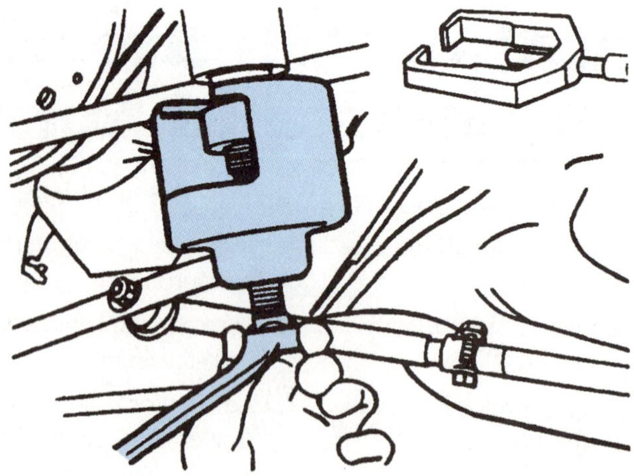

Figure 11-1 Disconnecting pitman shaft from the steering gear (Courtesy of Chevrolet Motor Division, General Motors Corporation)

**WARNING:** When the steering linkage is disconnected from the gear, do not turn the steering wheel hard against the stops. This action may damage the ball guides in the steering gear.

Follow these preliminary steps prior to steering gear removal or adjustment:

1. Disconnect the battery ground cable.
2. Raise the vehicle with the front wheels in the straight-ahead position. If the vehicle is lifted with a floor jack, place jack stands under the chassis, or suspension, and lower the vehicle onto the jack stands.
3. Remove the pitman arm nut and washer. Mark the pitman arm position in relation to the pitman shaft with a center punch, and use a puller to remove the pitman arm (Figure 11-1).
4. Loosen the sector shaft backlash adjuster lock nut, and back the adjuster screw off one-quarter turn.
5. Turn the steering wheel gently in one direction until it is stopped by the gear, and then turn the steering wheel back one-half turn toward the center position.
6. Remove the center steering wheel cover, and place a socket and inch-pound (in-lb) torque wrench on the steering wheel nut (Figure 11-2). Do not use a torque wrench with a maximum scale reading above 50 in. lb.
7. Rotate the steering wheel through a 90° arc and record the turning torque, which indicates the worm shaft bearing preload.

**Special Tools**

Pitman arm puller

**Special Tools**

Inch-pound torque wrench

Figure 11-2 Measuring worm shaft bearing preload on vehicle (Courtesy of Buick Motor Division, General Motors Corporation)

Follow these steps for manual steering gear removal and replacement:

1. Disconnect the flexible coupling from the worm shaft.
2. Remove the steering-gear-to-frame mounting bolts.
3. Remove the steering gear from the chassis. The steering gear may be cleaned externally with solvent or a parts washer.
4. Reverse steps 1 through 3 to reinstall the steering gear. All bolts must be tightened to the specified torque. Be sure the steering gear is filled with the manufacturer's specified steering gear lubricant.

**Special Tools**

Foot-pound torque wrench

# Manual Recirculating Ball Steering Gear Adjustments

## Worm Shaft Bearing Preload Adjustment

If the worm shaft bearing preload adjustment is loose, steering wheel free-play is excessive. This results in steering wander when the vehicle is driven straight ahead. Steering effort is increased if the worm shaft bearing preload adjustment is too tight. When the worm shaft bearing preload is adjusted, use the following procedure:

When a bearing has preload, all end-play is removed and there is a slight tension on the bearing.

1. Follow the preliminary steps listed previously.
2. Loosen the worm shaft adjuster plug lock nut with a brass punch and a hammer, and tighten the adjuster plug until all the worm shaft end-play is removed. Then loosen the plug one-quarter turn.

**WARNING:** Applying force to the worm shaft at either of the stops may damage the steering gear.

3. Turn the worm shaft fully to the right with a socket and an inch-pound torque wrench, and then turn the worm shaft one-half turn toward the center position.
4. Tighten the adjuster plug until the specified bearing preload is indicated on the torque wrench as the worm shaft is rotated (Figure 11-3). The specification on some steering gears is 5 to 8 in-lb (0.56 to 0.896 N·m). Always use the vehicle manufacturer's specified preload.
5. Tighten the adjuster plug lock nut to 85 ft-lb (114 N·m).

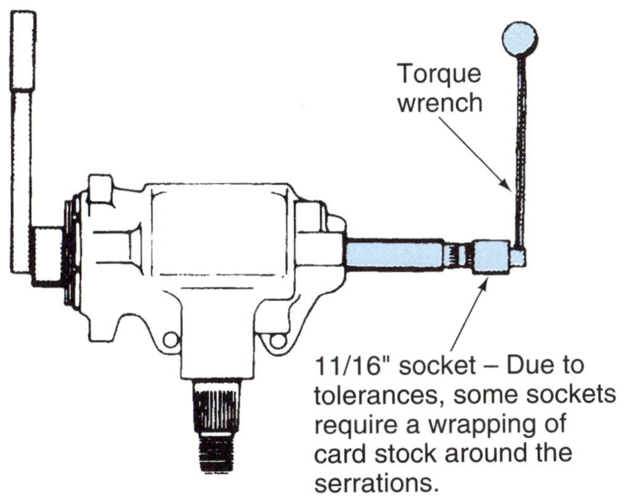


**Figure 11-3** Adjusting worm shaft bearing preload, manual recirculating ball steering gear (Courtesy of Chevrolet Motor Division, General Motors Corporation)

Steering wheel free-play refers to the amount of steering wheel rotation before the front wheels begin to turn right or left.

Vehicle wander is the tendency of a vehicle to steer right or left as it is driven straight ahead.

Road feel is experienced by a driver when the steering wheel is turned, and the driver has a positive feeling that the front wheels are turning in the intended direction.

Gear tooth backlash refers to movement between gear teeth that are meshed with each other.

### Sector Lash Adjustment

When the sector shaft lash adjustment is too loose, steering wheel free-play is excessive and vehicle wander occurs when the vehicle is driven straight ahead. A loose sector shaft lash adjustment decreases driver road feel. If the sector lash adjustment is too tight, steering effort is increased, especially with the front wheels in the straight-ahead position.

The following procedure may be used when the pitman shaft sector lash is adjusted:

1. Turn the pitman backlash adjuster screw outward until it stops, and then turn it in one turn.
2. Rotate the worm shaft fully from one stop to the other stop, and carefully count the number of shaft rotations.
3. Turn the worm shaft back exactly one-half the total number of turns from one of the stops.
4. With the steering gear positioned as it was in step 3, connect an inch-pound torque wrench and socket to the worm shaft, and note the steering gear turning torque while rotating the worm shaft 45° in each direction.
5. Turn the pitman backlash adjuster screw until the torque wrench reading is 6 to 10 in-lb (0.44 to 1.12 Nm) more than the worm shaft bearing preload torque in step 4 (Figure 11-4).
6. Tighten the pitman backlash adjuster screw lock nut to the specified torque.

## Manual Recirculating Ball Steering Gear Disassembly

After the steering gear is removed from the chassis, clean the housing externally with solvent and place the gear in a vise. The worm shaft seal may be changed without disassembling the steering

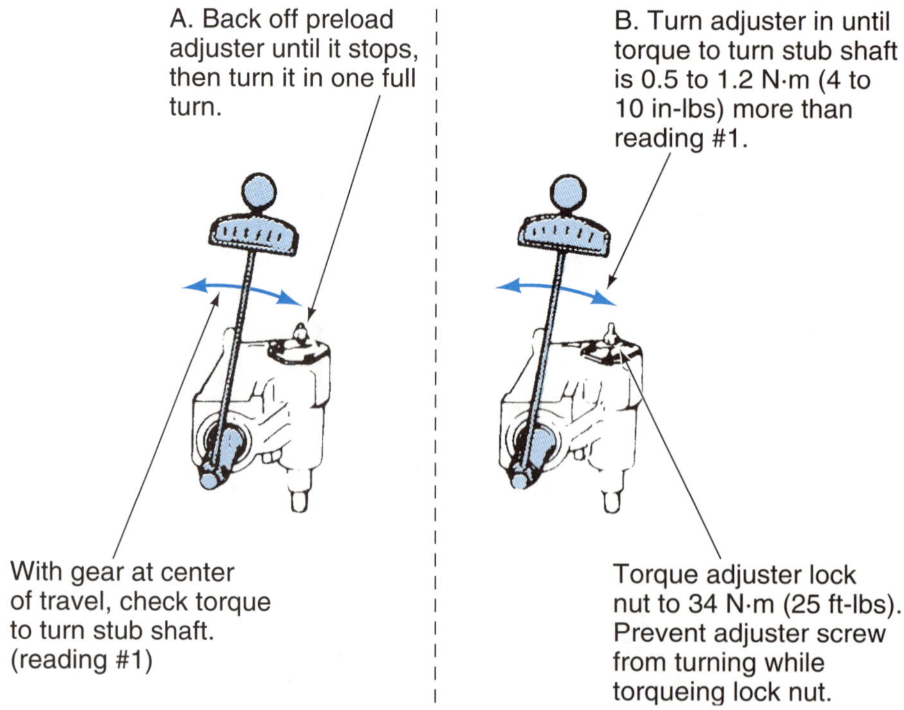


**Figure 11-4** Adjusting pitman backlash adjuster screw, manual recirculating ball steering gear (Courtesy of Chevrolet Motor Division, General Motors Corporation)

gear. Do not damage the shaft or housing during the seal replacement procedure. Follow these steps for steering gear disassembly:

1. Rotate the worm shaft until it is in the centered position, and use a center punch to mark the worm shaft in relation to the housing.
2. Remove the sector lash adjusting screw lock nut and the cover attaching screws. Turn the lash adjusting screw clockwise down through the cover to remove the cover and gasket.
3. Remove the lash adjusting screw and shim from the end of the pitman shaft, and remove this shaft from the housing. Place all steering gear components on clean shop towels on the work bench.
4. Use a brass punch and a hammer to loosen the worm shaft bearing adjuster lock nut, and remove the adjuster and bearing. Remove the bearing retainer with a slotted screwdriver.
5. Remove the worm shaft and ball nut assembly. Remove the screws to release the ball nut return guide, and remove this guide. Turn the ball nut over and rotate the worm shaft from side to side to remove all the steel balls. Remove the ball nut from the worm shaft.
6. Clean all internal parts in solvent, and inspect all components for wear, scoring, or pitting.
7. Pry the pitman shaft seal from the housing with a slotted screwdriver.
8. Check for movement between the pitman shaft and the bearing in the housing. If movement is present, use the proper bearing driver to remove the bearing (Figure 11-5). Install the new bearing with the proper bearing driver (Figure 11-6). If the pitman shaft is worn or scored in the bearing contact area, shaft replacement is necessary.
9. Remove the steering shaft seal with a slotted screwdriver, and use the proper seal driver to tap the new seal into the housing.
10. If the upper and lower worm shaft bearing cups in the housing or adjuster nut are scored, pitted, or worn, use a slide-hammer-type puller to remove these cups. Install the new bearing cups with the proper cup driver.

**Special Tools**

Pitman arm bearing puller

**Special Tools**

Seal drivers

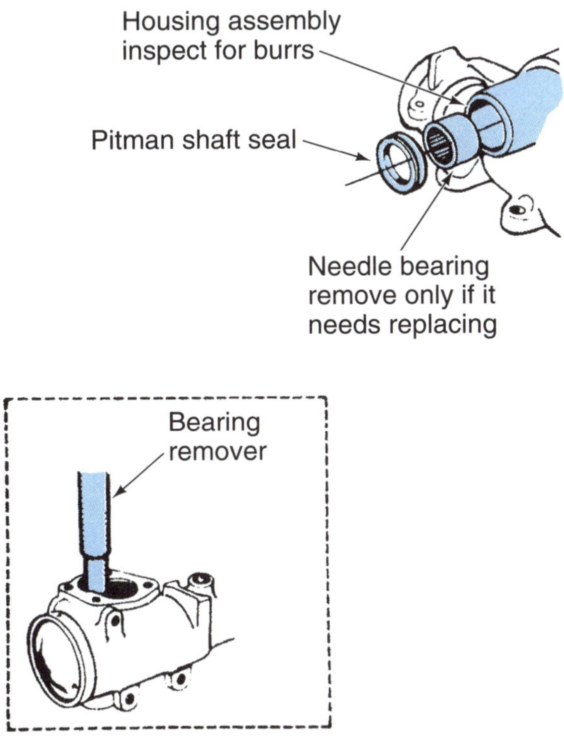


**Figure 11-5** Removing pitman shaft bearing with proper driving tool (Courtesy of Chevrolet Motor Division, General Motors Corporation)

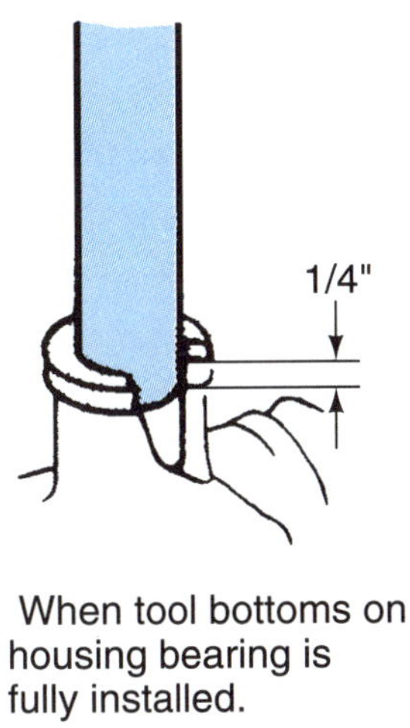


**Figure 11-6** Installing pitman shaft bearing with the proper driving tool (Courtesy of Chevrolet Motor Division, General Motors Corporation)

## Manual Recirculating Ball Steering Gear Reassembly

1. Lubricate all seals, bushings, and bearings with the manufacturer's recommended steering gear lubricant.
2. Position the ball nut on the worm shaft and install the balls in the return guides and the ball nut. Place an equal number of balls in each ball nut circuit. Install the return guide clamp and tighten the screws.
3. Install the upper bearing over the worm shaft and center the ball nut on the worm gear. Slide the worm shaft assembly into the gear housing. Be careful not to damage the upper worm shaft seal.
4. Place the lower bearing on the worm shaft adjuster plug with the bearing retainer, and install the adjuster and lock nut in the housing. Tighten the adjuster enough to hold the bearing in place.
5. Install the pitman shaft adjusting screw and shim in the pitman shaft. Measure the end-play between the screw and the slot. If this measurement exceeds 0.002 inch, replace the selective shim. These shims are available in four thicknesses: 0.063, 0.065, 0.067, and 0.069 inch.
6. Install the pitman shaft and adjusting screw in the housing. The sector teeth must mesh properly with the centered ball nut teeth.
7. Install the sector cover and gasket on the adjusting screw, and turn the adjusting screw counterclockwise to thread the screw up through the cover. Continue turning the adjusting screw until the cover and gasket contact the housing and the screw extends 5/8 to 3/4 inch above the cover. Install, but do not tighten, the adjusting screw lock nut. Install and tighten the cover screws.
8. Tighten the sector lash adjusting screw until the sector teeth and the ball nut teeth are meshed, with some movement between the teeth. Final adjustment is made later.
9. Wrap masking tape around the pitman shaft splines to protect the seal, and install the pitman shaft seal with the proper driver.
10. Fill the steering gear housing with the manufacturer's recommended steering lubricant, and turn the worm shaft fully in each direction several times to check for a binding condition. Do not apply force to the worm shaft when it reaches the end of its travel in either direction.
11. Perform the worm shaft bearing preload and sector lash adjustments as mentioned previously.

**Classroom Manual**
Chapter 11, page 233

## Power Recirculating Ball Steering Gear Diagnosis

If the steering gear is noisy, check these items:

1. Loose pitman shaft lash adjustment— may cause a rattling noise when the steering wheel is turned.
2. Cut or worn dampener O-ring on the valve spool—when this defect is present, a squawking noise is heard during a turn.
3. Loose steering gear mounting bolts.
4. Loose or worn flexible coupling or steering shaft U-joints.

A hissing noise from the power steering gear is normal if the steering wheel is at the end of its travel, or when the steering wheel is rotated with the vehicle standing still. If the steering wheel jerks or surges when the steering wheel is turned with the engine running, check the power steering pump belt condition and tension. When excessive kick-back is felt on the steering wheel, check the poppet valve in the steering gear.

When the steering is loose, check these defects:

1. Air in the power steering system. To remove the air, fill the power steering pump reservoir and rotate the steering wheel fully in each direction several times.
2. Loose pitman lash adjustment.
3. Loose worm shaft thrust bearing preload adjustment.
4. Worn flexible coupling or universal joint.
5. Loose steering gear mounting bolts.
6. Worn steering gears.

A complaint of hard steering while parking could be caused by one of these defects:

1. Loose or worn power steering pump belt.
2. Low oil level in the power steering pump.
3. Excessively tight steering gear adjustments.
4. Defective power steering pump with low-pressure output.
5. Restricted power steering hoses.
6. Defects in the steering gear such as:
   (a) pressure loss in the cylinder because of scored cylinder, worn piston ring, or damaged backup O-ring.
   (b) excessively loose spool in the valve body.
   (c) defective or improperly installed gear check poppet valve.

## Power Recirculating Ball Steering Gear Replacement

When the power steering gear is replaced, proceed as follows:

1. Disconnect the hoses from the steering gear and cap the lines and fittings to prevent dirt from entering the system.
2. Remove the pitman arm nut and washer and mark the pitman arm in relation to the shaft with a center punch. Use a puller to remove the pitman arm.
3. Disconnect the steering shaft from the worn shaft.
4. Remove the steering gear mounting bolts, and remove the steering gear from the chassis (Figure 11-7).
5. Reverse steps 1 through 4 to install the steering gear. All bolts must be tightened to the specified torque. Be sure the pitman arm is installed in the original position.

## Power Recirculating Ball Steering Gear Adjustments

### Worm Shaft Thrust Bearing Preload Adjustment

A loose worm shaft thrust bearing preload adjustment or sector lash adjustment causes excessive steering free-play and steering wander. The power recirculating ball steering gear adjustment procedures may vary depending on the vehicle make and model year. Always follow the vehicle manufacturer's recommended procedure in the service manual. When the worm shaft thrust bearing preload adjustment is performed, use this procedure:

1. Remove the worm shaft thrust bearing adjuster plug lock nut with a hammer and brass punch (Figure 11-8).
2. Turn this adjuster plug inward, or clockwise, until it bottoms, and tighten the plug to 20 ft-lbs (27 Nm).
3. Place an index mark on the steering gear housing next to one of the holes in the adjuster plug (Figure 11-9).

**Special Tools**

Worm shaft adjuster plug rotating tool

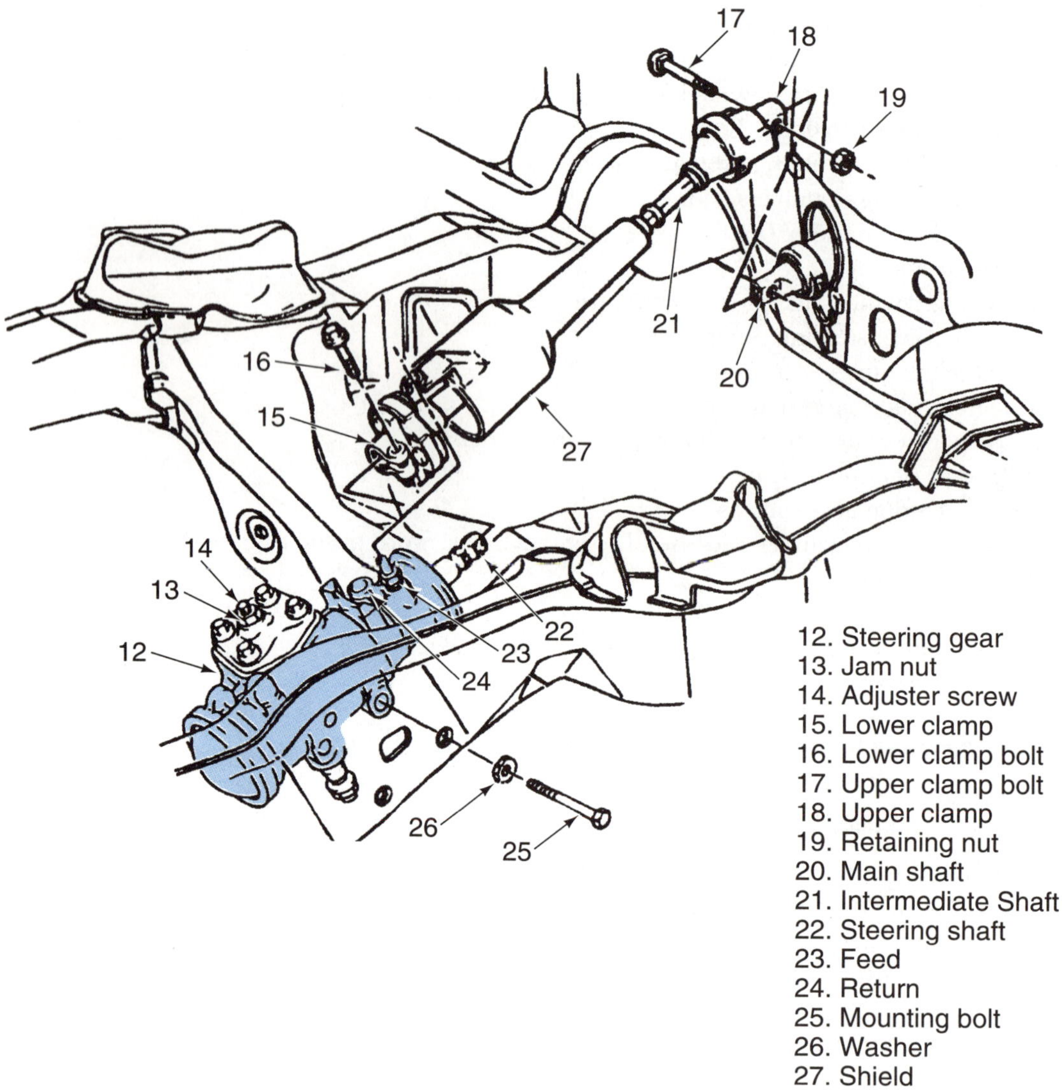


**Figure 11-7** Removing steering gear from the chassis (Courtesy of Chevrolet Motor Division, General Motors Corporation)

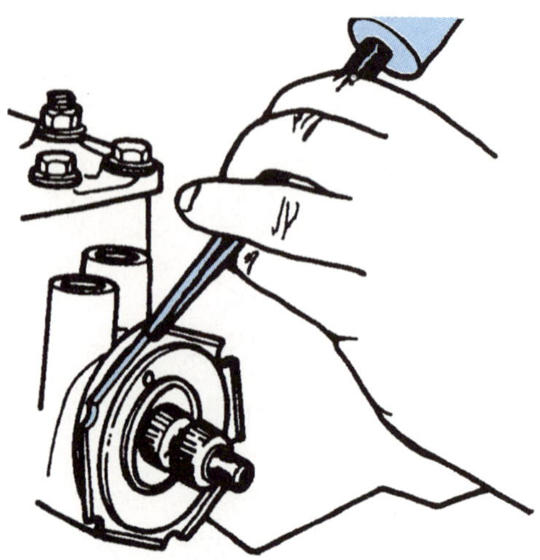

**Figure 11-8** Removing worm shaft thrust bearing adjuster plug lock nut (Courtesy of Chevrolet Motor Division, General Motors Corporation)

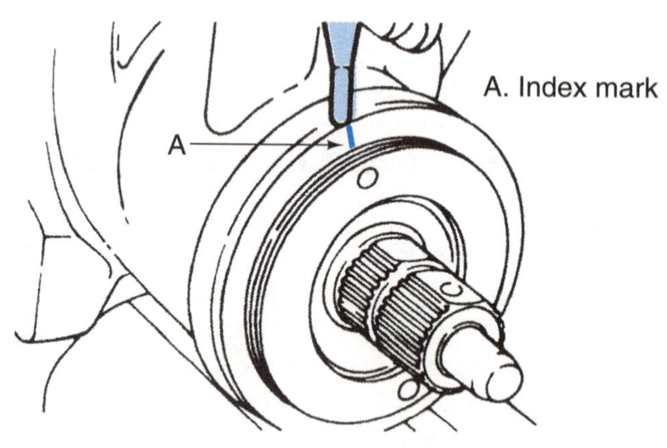


**Figure 11-9** Placing index mark on steering gear housing opposite one of the adjuster plug holes (Courtesy of Chevrolet Motor Division, General Motors Corporation)

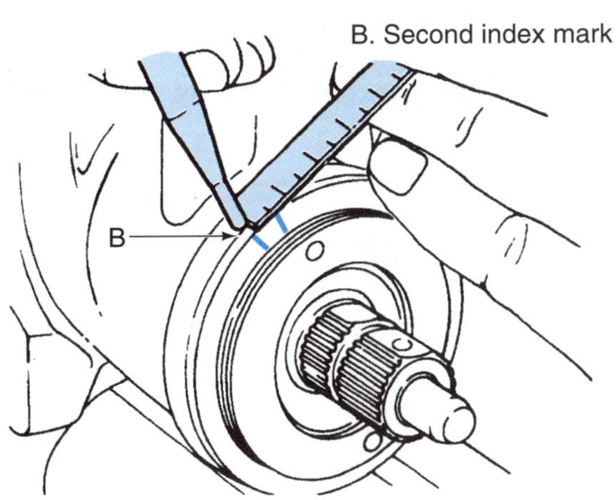


Figure 11-10 Measuring 0.50 in (13 mm) counterclockwise from the index mark on the steering gear housing (Courtesy of Chevrolet Motor Division, General Motors Corporation)

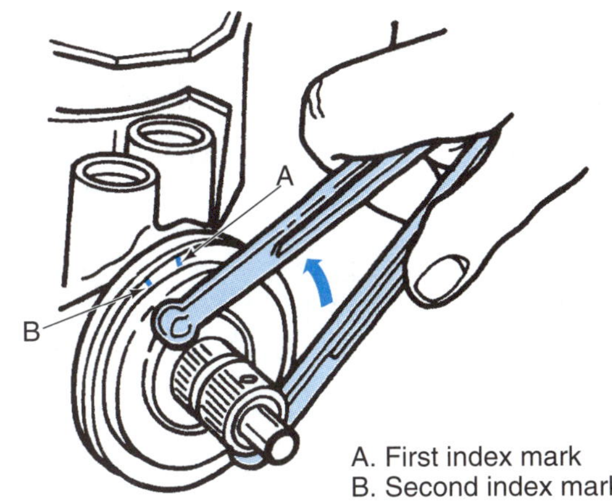


Figure 11-11 Aligning adjuster plug hole with second index mark placed on steering gear housing (Courtesy of Chevrolet Motor Division, General Motors Corporation)

4. Measure 0.50 in (13 mm) counterclockwise from the index mark, and place a second index mark at this position (Figure 11-10).
5. Rotate the adjuster plug counterclockwise until the hole in the adjuster plug is aligned with the second index mark placed on the housing (Figure 11-11).
6. Install and tighten the adjuster plug lock nut to the specified torque.

Photo Sequence 9 shows a typical procedure for performing a worm shaft bearing preload adjustment.

## Pitman Sector Shaft Lash Adjustment

When the pitman sector shaft lash adjustment is performed, proceed as follows:

1. Rotate the stub shaft from stop to stop and count the number of turns.
2. Starting at either stop, turn the stub shaft back two-thirds of the total number of turns. In this position, the flat on the stub shaft should be facing upward (Figure 11-12), and the master spline on the pitman shaft should be aligned with the pitman shaft backlash adjuster screw (Figure 11-13).

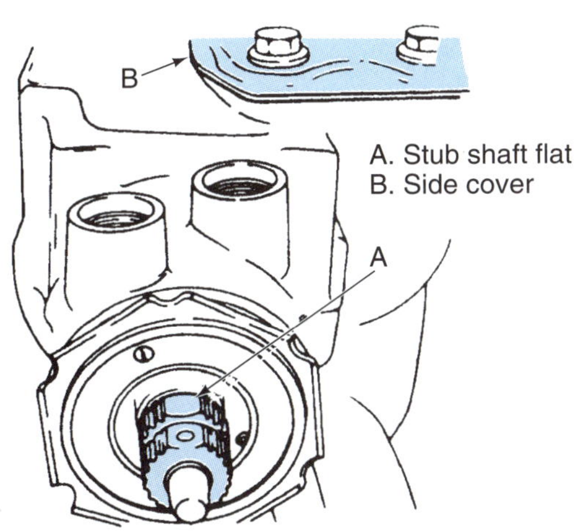


Figure 11-12 Pitman shaft flat facing upward and parallel with the side cover (Courtesy of Chevrolet Motor Division, General Motors Corporation)

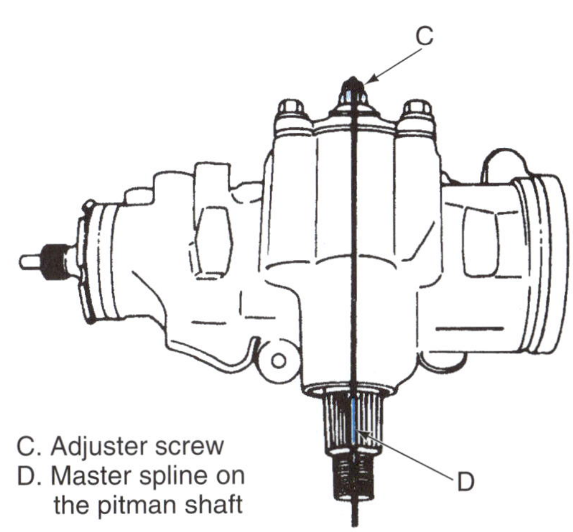


Figure 11-13 Pitman shaft master spline aligned with the pitman backlash adjuster screw (Courtesy of Chevrolet Motor Division, General Motors Corporation)

## Photo Sequence 9

# Typical Procedure for Performing a Worm Shaft Bearing Preload Adjustment

**P9-1** Retain the steering gear in a vise.

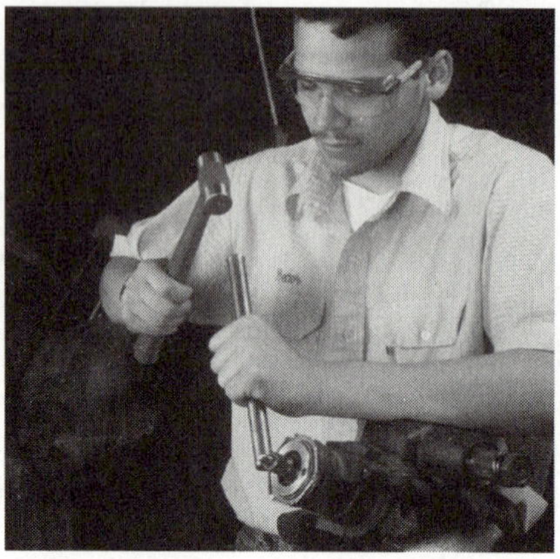

**P9-2** Remove the worm shaft bearing lock nut with a hammer and brass punch.

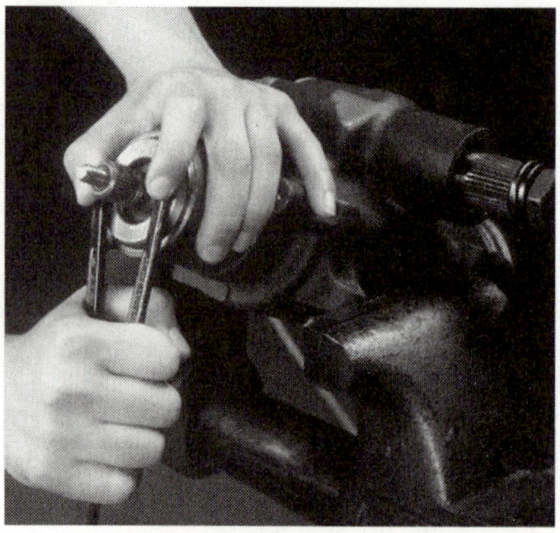

**P9-3** Use the proper tool to turn the worm shaft adjuster plug clockwise until it bottoms, and tighten this plug to 20 ft-lbs (27 Nm).

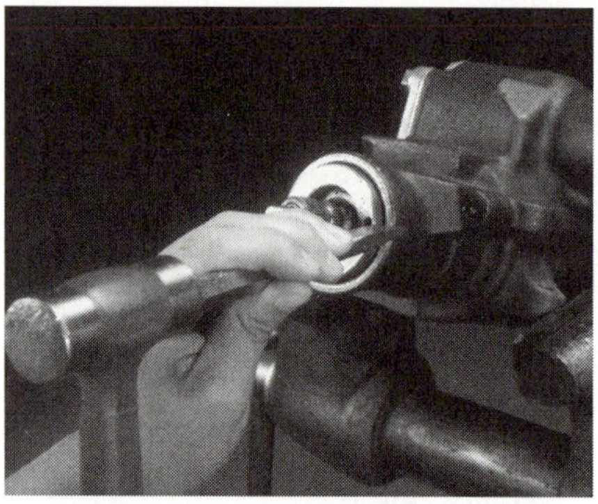

**P9-4** Place an index mark on the steering gear housing next to one of the adjuster plug holes.

**P9-5** Measure 0.50 in (13 mm) counterclockwise from the mark placed on the housing in step 4, and place a second mark on the housing.

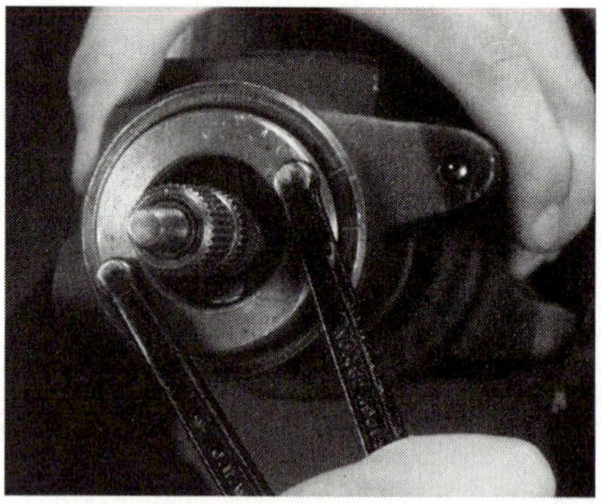

**P9-6** Rotate the worm shaft adjuster plug counterclockwise until the adjuster plug hole is aligned with the second mark placed on the housing.

**P9-7** Install and tighten the worm shaft adjuster plug lock nut.

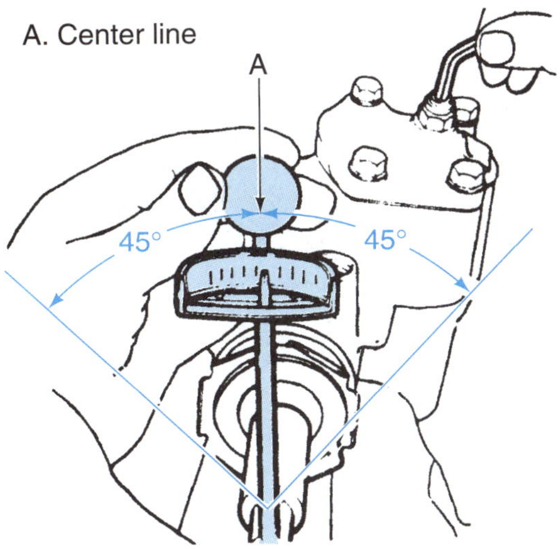


Figure 11-14 Measuring worm shaft turning torque to adjust pitman backlash adjuster screw (Courtesy of Chevrolet Motor Division, General Motors Corporation)

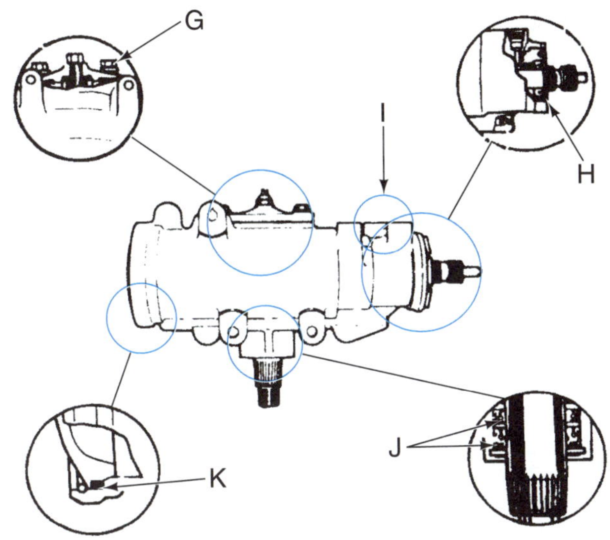


Figure 11-15 Power recirculating ball steering gear oil leak locations (Courtesy of Chevrolet Motor Division, General Motors Corporation)

3. Turn the pitman shaft backlash adjuster screw fully counterclockwise, and then turn it clockwise one turn.
4. Use an inch-pound torque wrench to turn the stub shaft through a 45° arc on each side of the position in step 2. Read the over-center torque as the stub shaft turns through the center position (Figure 11-14).
5. Continue to adjust the pitman shaft adjuster screw until the torque is 6 to 10 in-lbs (0.6 to 1.2 Nm) more than the torque in step 4.
6. Hold the pitman shaft adjuster screw in this position and tighten the lock nut to the specified torque.

## Power Recirculating Ball Steering Gear Oil Leak Diagnosis

Five locations where oil leaks may occur in a power steering gear are the following:

1. side cover O-ring seal (Figure 11-15G)
2. adjuster plug seal (H)
3. pressure line fitting (I)
4. pitman shaft oil seals (J)
5. top cover seal (K)

If an oil leak is present at any of these areas, complete or partial steering gear disassembly, and seal or O-ring replacement is necessary.

## Power Recirculating Ball Steering Gear Seal Replacement

### Side Cover O-Ring Replacement

Prior to any disassembly procedure, clean the steering gear with solvent or a parts washer. The steering gear service procedures vary depending on the make of gear. Always follow the vehicle

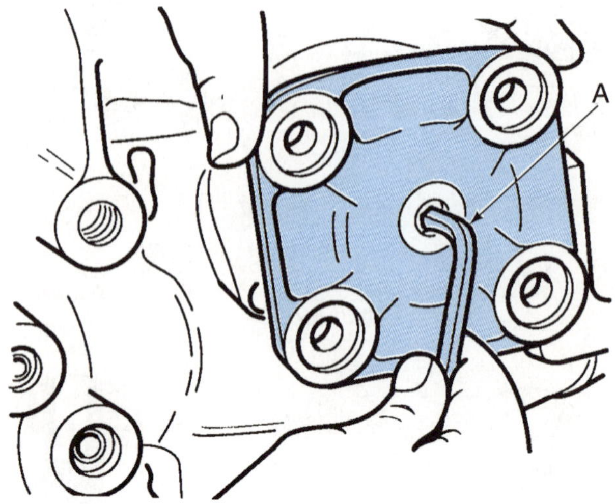


Figure 11-16 Removing steering gear side cover (Courtesy of Chevrolet Motor Division, General Motors Corporation)

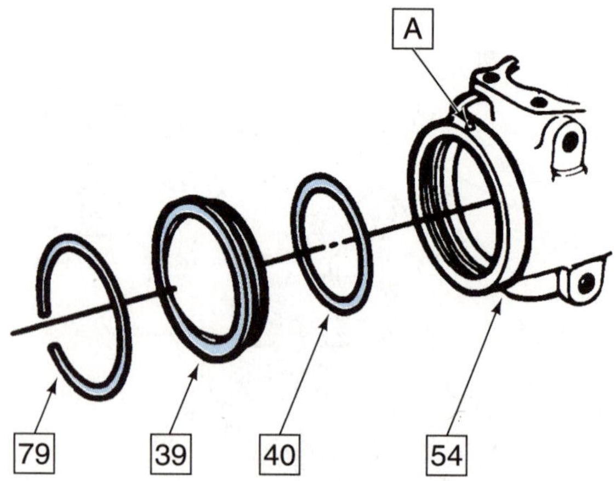


Figure 11-17 Removing steering gear end plug, retaining ring, and seal (Courtesy of Chevrolet Motor Division, General Motors Corporation)

manufacturer's recommended procedure in the service manual. Following is a typical side-cover O-ring replacement procedure:

1. Loosen the pitman backlash adjuster screw lock nut and remove the side-cover bolts. Rotate the pitman backlash adjuster screw clockwise to remove the cover from the screw (Figure 11-16).
2. Discard the O-ring and inspect the side-cover matching surfaces for metal burrs and scratches.
3. Lubricate a new O-ring with the vehicle manufacturer's recommended power steering fluid and install the O-ring.
4. Rotate the pitman backlash adjuster screw counterclockwise into the side cover until the side cover is properly positioned on the gear housing. Turn this adjuster screw fully counterclockwise, and then one turn clockwise. Install and tighten the side-cover bolts to the specified torque. Adjust the pitman sector shaft lash as explained previously.

## End Plug Seal Replacement

Follow these steps for end plug seal replacement:

1. Insert a punch into the access hole in the steering gear housing to unseat the retaining ring, and remove the ring (Figure 11-17).
2. Remove the end plug and seal.
3. Clean the end plug and seal contact area in the housing with a shop towel.
4. Lubricate a new seal with the vehicle manufacturer's recommended power steering fluid, and install the seal.
5. Install the end plug and retaining ring.

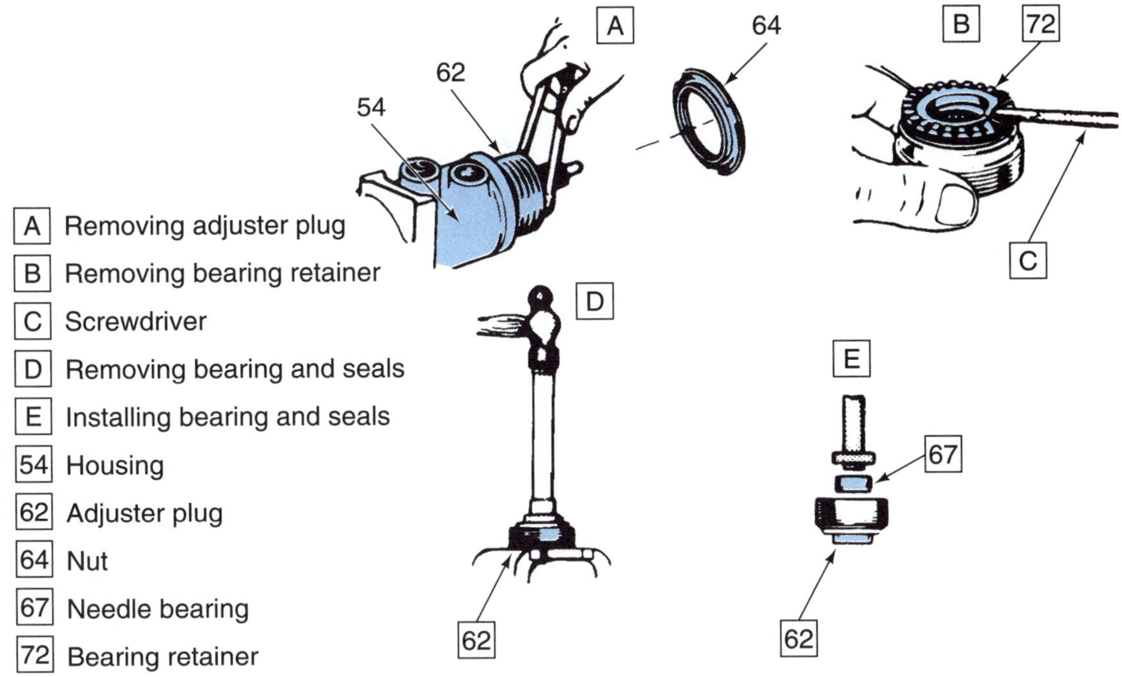


**Figure 11-18** Removing and replacing worm shaft adjuster plug, bearing, and seal (Courtesy of Chevrolet Motor Division, General Motors Corporation)

## Worm Shaft Bearing Adjuster Plug Seal and Bearing Replacement

Follow these steps for worm shaft bearing adjuster plug seal and bearing service:

1. Remove the adjuster plug lock nut, and use a special tool to remove the adjuster plug.
2. Use a screwdriver to pry at the raised area of the bearing retainer to remove this retainer from the adjuster plug (Figure 11-18).
3. Place the adjuster plug face down on a suitable support, and use the proper driver to remove the needle bearing, dust seal, and lip seal.

**WARNING:** The bearing identification number must face the driving tool to prevent bearing damage during installation.

4. Place the adjuster plug outside face up on a suitable support, and use the proper driver to install the needle bearing dust seal and lip seal.
5. Install the bearing retainer in the adjuster plug, and lubricate the bearing and seal with the vehicle manufacturer's recommended power steering fluid.
6. Install the adjuster plug and lock nut, and adjust the worm shaft bearing preload as discussed previously.

**Classroom Manual**
Chapter 11, page 237

**CUSTOMER CARE:** As an automotive technician, you should be familiar with the maintenance schedules recommended by various vehicle manufacturers. Of course, it is impossible to memorize all the maintenance schedules on different makes of vehicles, but maintenance schedule books are available. This maintenance schedule information is available in the owner's manual, but the vehicle owner may not take time to read this manual. If you advise the customer that his or her vehicle requires some service, such as a cooling system flush, according to the vehicle manufacturer's maintenance schedule, the customer will often have the service performed. The customer will usually appreciate your interest in his or her vehicle, and the shop will benefit from the increased service work.

## Guidelines for Servicing Manual and Power Recirculating Ball Steering Gears

1. Oil leaks may occur at the pitman shaft seal or worm shaft seal in a manual recirculating ball steering gear.
2. A rattling noise and excessive steering free-play may be caused by loose gear mounting bolts, worn steering shaft U-joints, or a worn flexible coupling in a recirculating ball steering gear system.
3. A loose worm shaft bearing preload adjustment or sector lash adjustment on a manual or power recirculating ball steering gear causes excessive steering wheel free-play, steering wander, and reduced feel of the road.
4. If the worm shaft bearing preload adjustment or sector lash adjustment is tighter than specified on a manual or power recirculating ball steering gear, steering effort is higher than normal.
5. On a manual recirculating ball steering gear, the worm shaft bearing preload adjuster plug is rotated until the correct worm shaft turning torque is obtained with the worm shaft positioned one-half turn from the full-right or full-left position.
6. On a manual recirculating ball steering gear, the sector lash adjusting screw is tightened until the correct worm shaft turning torque is obtained as the worm shaft is rotated through the center position.
7. Oil leaks in a power recirculating ball steering gear may occur at the end plug O-ring seal, side-cover O-ring seal, pitman shaft seal, or worm shaft thrust bearing seal.
8. On some power recirculating ball steering gears, the worm shaft thrust bearing adjuster plug is tightened until the correct worm shaft turning torque is obtained with the worm shaft positioned one-half turn from the full-right or full-left position.
9. On a power recirculating ball steering gear, the sector lash screw is tightened until the correct worm shaft turning torque is obtained as the worm shaft is rotated through the center position.

### CASE STUDY

The owner of a rear wheel drive Oldsmobile 98 complained about increased and somewhat erratic steering effort. The service writer asked the customer about the conditions when this problem occurred, and the customer said that the condition was always present. During a road test, the technician discovered that the customer's description of the problem was accurate.

The technician checked the power steering fluid level, and made a careful check of the belt tension and condition without finding any problems. Next, the technician checked the power steering pump pressure, and found it to be normal. A check of the power steering hoses did not reveal any hose restrictions.

The technician removed and disassembled the steering gear, and found a severely scored cylinder bore in the gear housing. The ball nut piston ring was also worn and scored. A replacement steering gear was installed, and the system filled with the manufacturer's recommended power steering fluid. A road test indicated a new gear worked satisfactorily.

## Terms to Know

Bearing preload
Steering free-play
Vehicle wander
Gear backlash, or lash
Road feel
Selective shim

## ASE Style Review Questions

1. While discussing manual recirculating ball steering gear diagnosis:
*Technician A* says excessive steering wheel free-play may be caused by a worm shaft bearing preload adjustment that is tighter than normal.
*Technician B* says excessive steering wheel free-play may be caused by loose steering gear mounting bolts.
Who is correct?
**A.** A only
**B.** B only
**C.** Both A and B
**D.** Neither A nor B

2. While discussing manual recirculating ball steering gear diagnosis:
*Technician A* says the worm shaft bearing preload adjustment is performed with the sector lash adjuster screw backed off one-quarter turn.
*Technician B* says the worm shaft bearing preload adjustment is performed with the worm shaft turned fully to the right or left.
Who is correct?
**A.** A only
**B.** B only
**C.** Both A and B
**D.** Neither A nor B

3. While discussing manual recirculating ball steering gear service:
*Technician A* says steering wander may be caused by a loose sector lash adjustment.
*Technician B* says a loose sector lash adjustment may cause reduced feel of the road.
Who is correct?
**A.** A only
**B.** B only
**C.** Both A and B
**D.** Neither A nor B

4. While discussing manual recirculating steering gear diagnosis:
*Technician A* says the sector lash adjustment is performed with the worm shaft half-way between the centered position and the full-right position.
*Technician B* says the worm shaft bearing preload adjustment is performed before the sector lash adjustment.
Who is correct?
**A.** A only
**B.** B only
**C.** Both A and B
**D.** Neither A nor B

5. While discussing manual recirculating ball steering gear service:
*Technician A* says if the end-play between the sector lash screw and the slot is excessive, selective shims are available to correct this end-play.
*Technician B* says if the end-play between the sector lash screw and the slot is excessive, the screw or pitman shaft must be replaced.
Who is correct?
**A.** A only
**B.** B only
**C.** Both A and B
**D.** Neither A nor B

6. While discussing power recirculating ball steering gear diagnosis:
*Technician A* says if the steering gear has excessive kick-back, the sector lash adjustment may be loose.
*Technician B* says if the steering gear has excessive kick-back, the worm shaft thrust bearing preload adjustment may be tighter than specified.
Who is correct?
**A.** A only
**B.** B only
**C.** Both A and B
**D.** Neither A nor B

7. While discussing power recirculating ball steering gear diagnosis:
*Technician A* says excessive steering effort may be caused by a defective power steering pump.
*Technician B* says excessive steering effort may be caused by a scored steering gear cylinder.
Who is correct?
**A.** A only
**B.** B only
**C.** Both A and B
**D.** Neither A nor B

8. While discussing power recirculating ball steering gear adjustment:
*Technician A* says the worm shaft bearing adjuster plug should be bottomed and then backed off until the specified worm shaft turning torque is obtained.
*Technician B* says the sector lash adjustment screw should be tightened one turn before the worm shaft thrust bearing preload is adjusted.
Who is correct?
**A.** A only
**B.** B only
**C.** Both A and B
**D.** Neither A nor B

9. While discussing power recirculating ball steering gear adjustment:
*Technician A* says the sector lash adjustment screw should be tightened until the specified worm shaft turning torque is obtained as the worm shaft is rotated fully to the right.
*Technician B* says the sector lash adjustment screw should be tightened until the specified worm shaft turning torque is obtained as the worm shaft is rotated fully to the left.
Who is correct?
**A.** A only
**B.** B only
**C.** Both A and B
**D.** Neither A nor B

10. While discussing power recirculating ball steering gear diagnosis:
*Technician A* says a hissing noise from the steering system is normal if the steering wheel is at the end of its travel.
*Technician B* says if the steering wheel jerks when it is turned, the power steering belt may be slipping.
Who is correct?
**A.** A only
**B.** B only
**C.** Both A and B
**D.** Neither A nor B

# Table 11-1 ASE TASK

Diagnose manual steering gear (non-rack and pinion type noises, binding, uneven turning effort, looseness, hard steering, and lubricant leakage problems; determine needed repairs.

| Problem Area | Symptoms | Possible Causes | Classroom Manual | Shop Manual |
|---|---|---|---|---|
| **LOSS OF LUBRICANT** | Fluid leaks | 1. Worn pitman shaft seal | 234 | 245 |
| | | 2. Worn worm shaft seal | 234 | 245 |
| **NOISE** | Rattling on road irregularities | 1. Loose steering shaft U-joints, flexible coupling | 234 | 245 |
| | | 2. Loose steering gear mounting bolts | 236 | 245 |
| **LOOSENESS** | Excessive steering wheel free-play | 1. Loose worm bearing preload adjustment | 234 | 245 |
| | | 2. Loose sector lash adjustment | 234 | 245 |
| | | 3. Worn steering gears | 234 | 245 |
| | | 4. Loose stearing gear mounting bolts | 236 | 245 |
| **STEERING EFFORT** | Excessive steering | 1. Tight worm bearing preload adjustment | 234 | 245 |
| | | 2. Tight sector lash adjustment | 235 | 245 |
| | | 3. Low steering gear lubricant | 234 | 245 |
| | | 4. Worn steering gears or bearings | 234 | 245 |
| | Uneven steering effort | Worn steering gears or bearings | 234 | 245 |

## Table 11-2 ASE TASK

Remove and replace manual steering gear (non-rack and pinion type).

| Problem Area | Symptoms | Possible Causes | Classroom Manual | Shop Manual |
|---|---|---|---|---|
| **STEERING EFFORT** | Excessive steering effort, binding | 1. Worn steering gears or bearings | 234 | 245 |
| | | 2. Tight steering gear adjustments | 234 | 245 |
| | | 3. Low steering gear lubricant | 234 | 245 |
| **LOSS OF LUBRICANT** | Fluid leaks | 1. Worn pitman shaft seal | 234 | 245 |
| | | 2. Worn worm shaft seal | 234 | 245 |

## Table 11-3 ASE TASK

Adjust manual steering gear (non-rack and pinion type) worm bearing preload and sector lash.

| Problem Area | Symptoms | Possible Causes | Classroom Manual | Shop Manual |
|---|---|---|---|---|
| **LOOSENESS** | Excessive steering wheel free-play | 1. Loose worm bearing preload adjustment | 234 | 247 |
| | | 2. Loose sector lash adjustment | 235 | 248 |
| **STEERING EFFORT** | Excessive steering effort | 1. Tight worm bearing preload adjustment | 234 | 247 |
| | | 2. Tight sector lash adjustment | 235 | 248 |

## Table 11-4 ASE TASK

Diagnose power steering gear (non-rack and pinion type) noises, binding, uneven turning effort, looseness, hard steering, and fluid leakage problems; determine needed repairs.

| Problem Area | Symptoms | Possible Causes | Classroom Manual | Shop Manual |
|---|---|---|---|---|
| **LOSS OF LUBRICANT** | Fluid leaks | 1. Worn pitman shaft seal | 237 | 255 |
| | | 2. Worn end-plug O-ring seal | 237 | 255 |
| | | 3. Worn side-cover O-ring seal | 237 | 255 |
| | | 4. Worn thrust-bearing O-ring seal | 237 | 255 |
| **NOISE** | Rattling on road irregularities | 1. Loose steering shaft U-joints, flexible coupling | 238 | 250 |
| | | 2. Loose steering gear mounting bolts | 238 | 250 |
| **LOOSENESS** | Excessive steering wheel free-play | 1. Loose thrust bearing preload adjustment | 239 | 251 |
| | | 2. Loose sector lash adjustment | 239 | 253 |
| | | 3. Loose steering gear mounting bolts | 239 | 250 |
| | | 4. Worn steering gears or components | 239 | 251 |
| **STEERING EFFORT** | Excessive steering effort | 1. Tight sector lash adjustment | 239 | 253 |
| | | 2. Tight thrust bearing adjustment | 239 | 251 |
| | | 3. Low fluid level | 239 | 251 |
| | | 4. Worn steering gears, bearings, or cylinder | 239 | 251 |
| | Uneven steering | 1. Low fluid level | 239 | 251 |
| | | 2. Worn steering gears, bearings, or components | 239 | 251 |

## Table 11-5 ASE TASK

Remove and replace power steering gear (non-rack and pinion type).

| Problem Area | Symptoms | Possible Causes | Classroom Manual | Shop Manual |
|---|---|---|---|---|
| **STEERING EFFORT** | Excessive steering effort | 1. Worn steering gears bearings, or components | 234 | 251 |
| | | 2. Tight steering gear adjustments | 234 | 251 |
| | | 3. Low fluid level | 234 | 251 |
| **LOSS OF FLUID** | Fluid leaks | 1. Worn pitman shaft seal | 238 | 253 |
| | | 2. Worn thrust-bearing O-ring seal | 238 | 255 |
| | | 3. Worn side-cover O-ring | 238 | 255 |
| | | 4. Worn end-plug O-ring | 238 | 255 |

## Table 11-6 ASE TASK

Adjust power steering gear (non-rack and pinion type) worm bearing preload and sector lash.

| Problem Area | Symptoms | Possible Causes | Classroom Manual | Shop Manual |
|---|---|---|---|---|
| **LOOSENESS** | Excessive steering wheel free-play | 1. Loose thrust bearing preload adjustment | 239 | 251 |
| | | 2. Loose sector lash adjustment | 239 | 253 |
| **STEERING EFFORT** | Excessive steering effort | 1. Tight thrust bearing preload adjustment | 239 | 251 |
| | | 2. Tight sector lash adjustment | 239 | 253 |

## Table 11-7 ASE TASK

Inspect and replace power steering gear (non-rack and pinion type) seals and gaskets.

| Problem Area | Symptoms | Possible Causes | Classroom Manual | Shop Manual |
|---|---|---|---|---|
| LOSS OF FLUID | Fluid leaks | 1. Worn pitman shaft seal | 239 | 253 |
| | | 2. Worn end-plug O-ring seal | 239 | 255 |
| | | 3. Worn thrust-bearing O-ring | 239 | 255 |
| | | 4. Worn side-cover O-ring seal | 239 | 255 |

# Rack and Pinion Steering Gear Diagnosis and Service

CHAPTER 12

Upon completion and review of this chapter, you should be able to:

- ❑ Perform a manual or power rack and pinion steering gear inspection.
- ❑ Remove and replace manual or power rack and pinion steering gears.
- ❑ Disassemble, inspect, repair, and reassemble manual rack and pinion steering gears.
- ❑ Adjust manual rack and pinion steering gears.
- ❑ Diagnose manual rack and pinion steering systems.
- ❑ Diagnose oil leaks in power rack and pinion steering gears.
- ❑ Disassemble, inspect, and repair power rack and pinion steering gears.
- ❑ Adjust power rack and pinion steering gears.

## Manual or Power Rack and Pinion Steering Gear On-Car Inspection

**Basic Tools**

Basic technician's tool set

Service manual

Floor jack

Jack stands

Machinist's rule

The wear points are reduced to four in a rack and pinion steering gear. These wear points are the inner and outer tie rod ends on both sides of the rack and pinion assembly (Figure 12-1).

The first step in manual or power rack and pinion steering gear diagnosis is a very thorough inspection of the complete steering system. During this inspection, all steering system components such as the inner and outer tie rod ends, bellows boots, mounting bushings, couplings or universal joints, ball joints, tires, and steering wheel free-play must be checked. Follow these steps for manual or power rack and pinion steering gear inspection:

Steering wheel free-play is the distance that the steering wheel moves before the front wheels begin to turn when the steering wheel is rocked back and forth with light finger pressure.

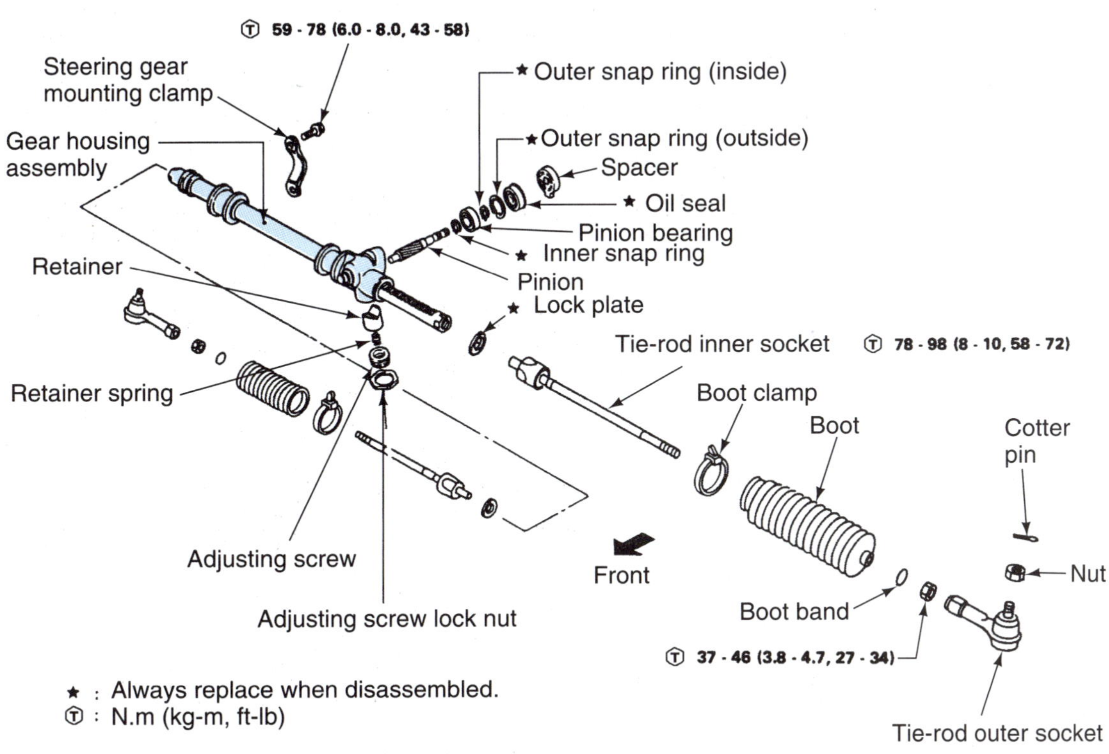


Figure 12-1 Wear points at the inner and outer tie rod ends in a rack and pinion steering gear (Courtesy of Nissan Motor Co., Ltd.)

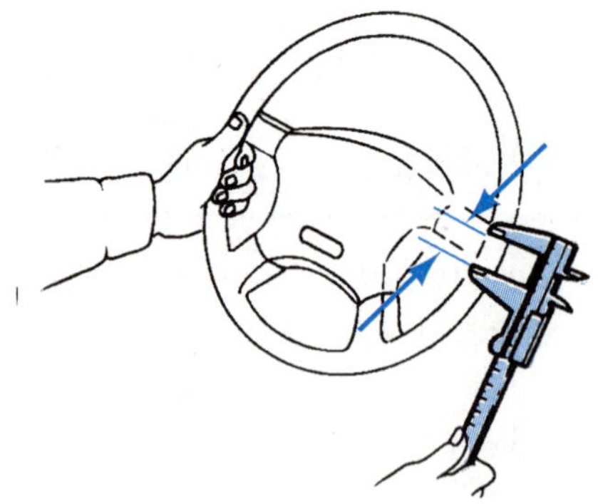

**Figure 12-2** Checking steering wheel free-play (Courtesy of Nissan Motor Co., Ltd.)

1. With the front wheels straight ahead and the engine stopped, rock the steering wheel gently back and forth with light finger pressure (Figure 12-2). Measure the maximum steering wheel free-play. The maximum specified steering wheel free-play on some vehicles is 1.18 in (30 mm). Always refer to the vehicle manufacturer's specifications in the service manual. Excessive steering wheel free-play indicates worn steering components.
2. With the vehicle sitting on the shop floor and the front wheels straight ahead, have an assistant turn the steering wheel about 1/4 turn in both directions. Watch for looseness in the flexible coupling or universal joints in the steering shaft. If looseness is observed, replace the coupling or universal joint.
3. While an assistant turns the steering wheel about 1/2 turn in both directions, watch for movement of the steering gear housing in the mounting bushings. If there is any movement of the housing in these bushings, replace the bushings. The steering gear mounting bushings may be deteriorated by oil soaking, heat, or age.
4. Grasp the pinion shaft extending from the steering gear and attempt to move it vertically. If there is steering shaft vertical movement, a pinion bearing preload adjustment may be required. When the steering gear does not have a pinion bearing preload adjustment, replace the necessary steering gear components.
5. Road test the vehicle and check for excessive steering effort. A bent steering rack, tight rack bearing adjustment, or damaged front drive axle joints in a front wheel drive car may cause excessive steering effort.
6. Visually inspect the bellows boots for cracks, splits, leaks, and proper clamp installation. Replace any boot that indicates any of these conditions. If the boot clamps are loose or improperly installed, tighten or replace the clamps as necessary. Since the bellows boots protect the inner tie rod ends and the rack from contamination, boot condition is extremely important. Boots should be inspected each time undercar service such as oil and filter change or chassis lubrication is performed.
7. Loosen the inner bellows boot clamps and move each boot toward the outer tie rod end until the inner tie rod end is visible. Push outward and inward on each front tire and watch for movement in the inner tie rod end. If any movement or looseness is present, replace the inner tie rod end. An alternate method of checking the inner tie rod ends is to squeeze the bellows boots and grasp the inner tie rod end socket. Movement in the inner tie rod end is then felt as the front wheel is moved inward and outward. Hard plastic bellows boots may be found on some applications. With this type of bellows boot, remove the ignition key from the switch to lock the steering column and push inward and outward on the front tire while observing any lateral movement in the tie rod. When lateral movement is observed, replace the inner tie rod end.

**WARNING:** Bent steering components must be replaced. Never straighten steering components, because this action may weaken the metal and result in sudden component failure, serious personal injury, and vehicle damage.

8. Grasp each outer tie rod end and check for vertical movement. While an assistant turns the steering wheel 1/4 turn in each direction, watch for looseness in the outer tie rod ends. If any looseness or vertical movement is present, replace the tie rod end. Check the outer tie rod end seals for cracks and proper installation of the nuts and cotter keys. Cracked seals must be replaced. Inspect the tie rods for a bent condition. Bent tie rods or other steering components must be replaced. Do not attempt to straighten these components.

## Manual or Power Rack and Pinion Steering Gear Removal and Replacement

The replacement procedure is similar for manual or power rack and pinion steering gears. This removal and replacement procedure varies depending on the vehicle. On some vehicles, the front crossmember, or engine support cradle, must be lowered to remove the rack and pinion steering gear. Always follow the vehicle manufacturer's recommended procedure in the service manual. Following is a typical rack and pinion steering gear removal and replacement procedure:

1. Place the front wheels in the straight-ahead position, and remove the ignition key from the ignition switch to lock the steering wheel. Place the driver's seat belt through the steering wheel to prevent wheel rotation if the ignition switch is turned on (Figure 12-3). This action maintains the clock spring electrical connector, or spiral cable, in the centered position on air-bag-equipped vehicles.
2. Lift the front end with a floor jack, and place jack stands under the vehicle chassis. Lower the vehicle onto the jack stands. Remove the left and right fender apron seals (Figure 12-4).
3. Place punch marks on the lower universal joint and the steering gear pinion shaft so they may be reassembled in the same position (Figure 12-5). Loosen the upper universal joint bolt and remove the lower universal joint bolt, and disconnect this joint.
4. Remove the cotter keys from the outer tie rod ends. Loosen, but do not remove, the tie rod end nuts. Use a tie rod end puller to loosen the outer tie rod ends in the steering arms (Figure 12-6). Remove the tie rod end nuts and remove the tie rod ends from the arms.
5. Use the proper wrenches to disconnect the high-pressure hose and return hose from the steering gear (Figure 12-7). This step is not required on a manual steering gear.
6. Remove the four stabilizer bar mounting bolts (Figure 12-8).

**Special Tools**

Tie rod end puller

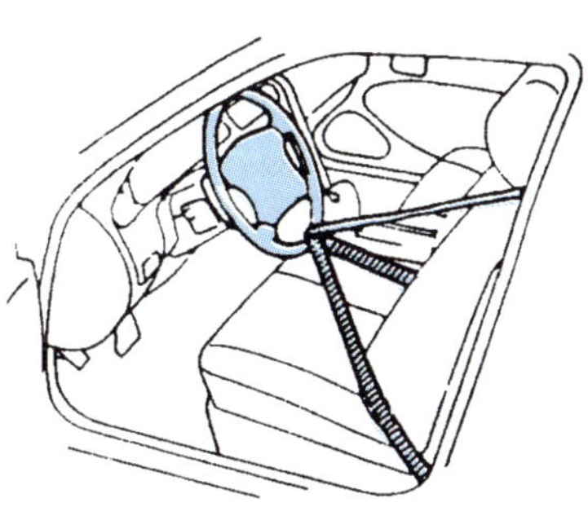

Figure 12-3 Driver's seat belt wrapped around the steering wheel to prevent wheel rotation (Courtesy of Toyota Motor Corporation)

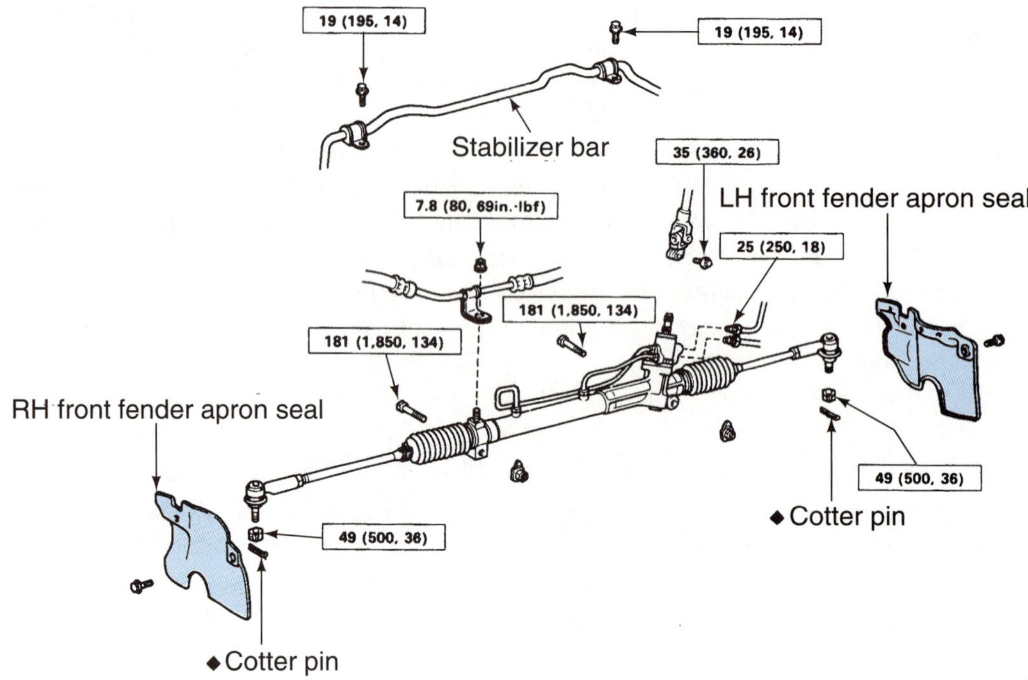


**Figure 12-4** Left and right fender aprons (Courtesy of Toyota Motor Corporation)

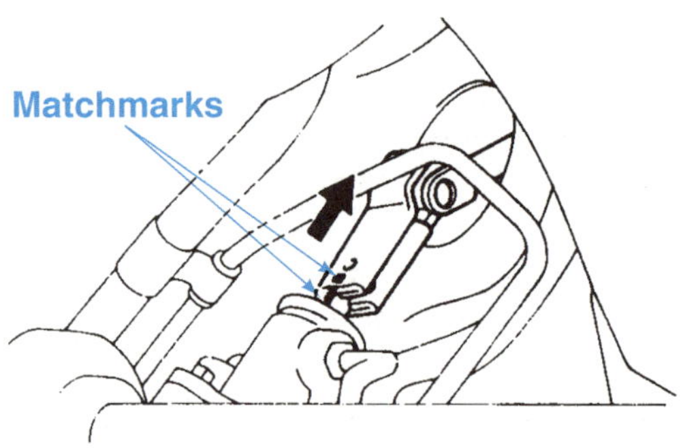


**Figure 12-5** Punch marks on universal joint and pinion shaft (Courtesy of Toyota Motor Corporation)

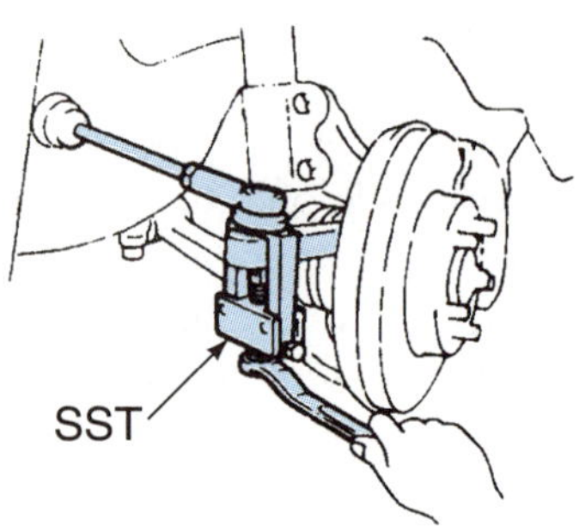


**Figure 12-6** Removing outer tie rod ends (Courtesy of Toyota Motor Corporation)

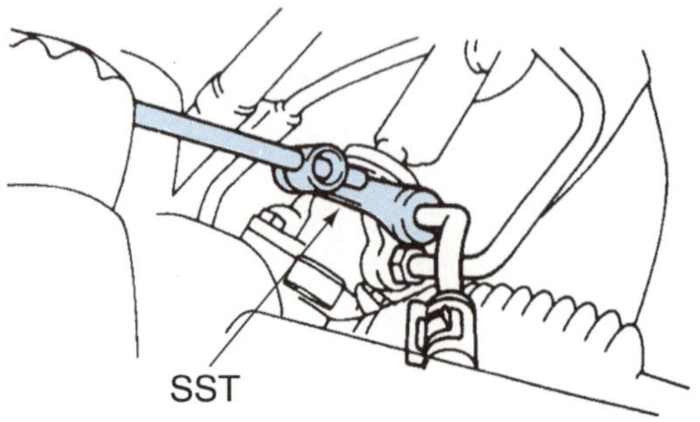


**Figure 12-7** Removing high-pressure and return hoses from the steering gear (Courtesy of Toyota Motor Corporation)

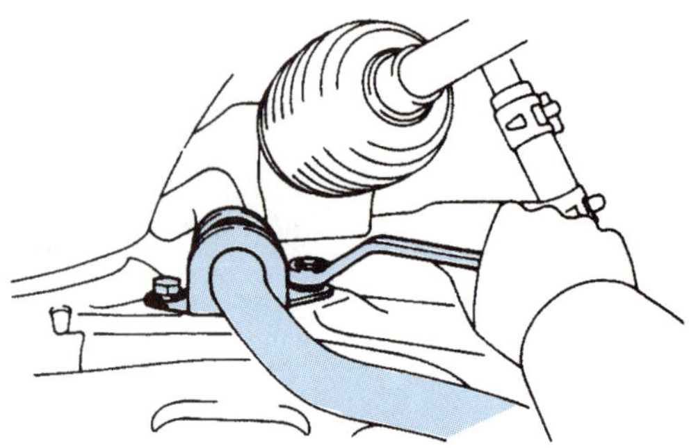

**Figure 12-8** Removing stabilizer bar mounting bolts (Courtesy of Toyota Motor Corporation)

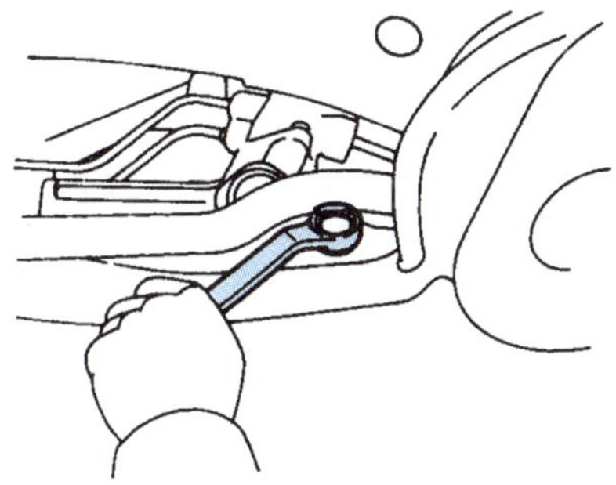

Figure 12-9 Removing steering gear mounting bolts (Courtesy of Toyota Motor Corporation)

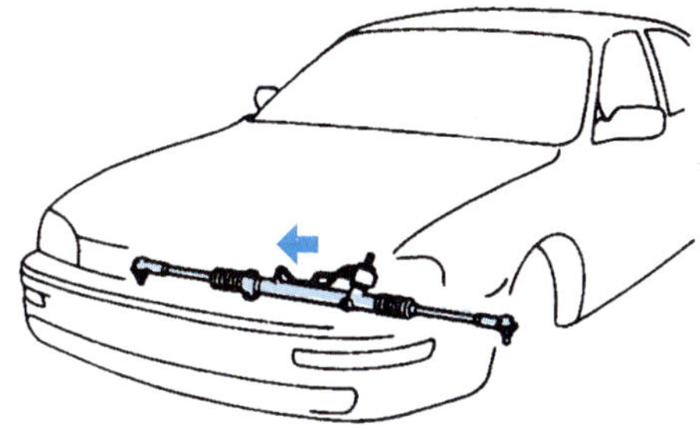

Figure 12-10 Removing steering gear from the car (Courtesy of Toyota Motor Corporation)

7. Remove steering gear mounting bolts (Figure 12-9).
8. Remove the steering gear assembly from the right side of the car (Figure 12-10).
9. Position the right and left tie rods the specified distance from the steering gear housing (Figure 12-11). Install the steering gear through the right fender apron.
10. Install the pinion shaft in the universal joint with the punch marks aligned. Tighten the upper and lower universal joint bolts to the specified torque.
11. Install the steering gear mounting bolts and tighten these bolts to the specified torque.
12. Install the stabilizer bar mounting bolts and torque these bolts to specifications.
13. Install and tighten the high-pressure and return hoses to the specified torque. This step is not required on a manual rack and pinion steering gear.

**Special Tools**

Foot-pound torque wrench

**WARNING:** Do not loosen the tie rod end nuts to align cotter key holes. This action causes improper torquing of these nuts.

14. Install the outer tie rod ends in the steering knuckles, and tighten the nuts to the specified torque. Install the cotter keys in the nuts.
15. Check the front wheel toe, and adjust as necessary. Tighten the outer tie rod end jam nuts to the specified torque, and tighten the outer bellows boot clamps.
16. Install the left and right fender apron seals, and lower the vehicle with a floor jack.

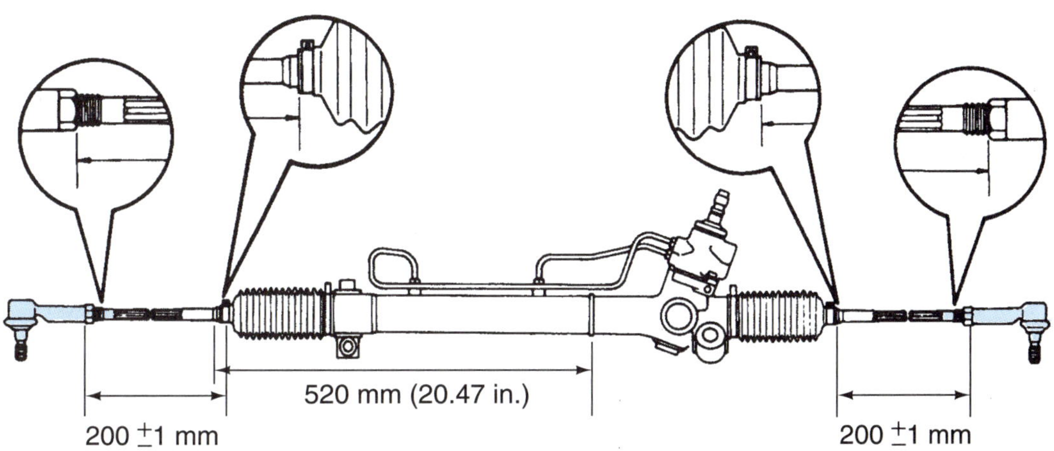


Figure 12-11 Right and left tie rod position in relation to the steering gear housing prior to installation (Courtesy of Toyota Motor Corporation)

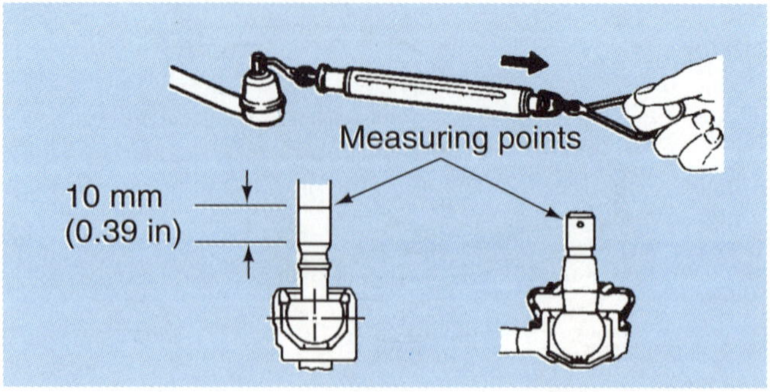


Figure 12-12 Measuring inner tie rod end articulation effort (Courtesy of Nissan Motor Co., Ltd.)

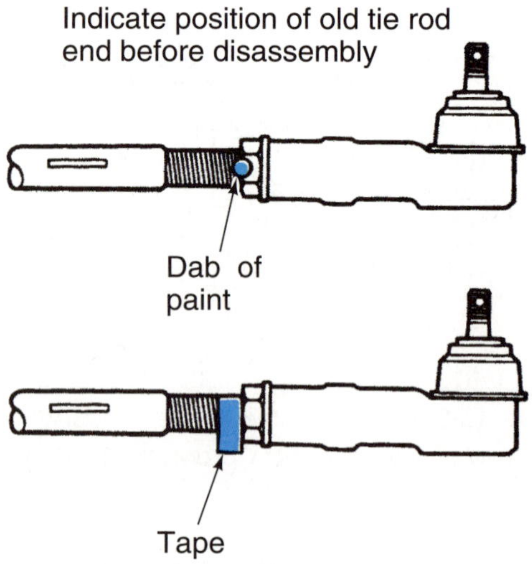


Figure 12-13 Marking outer tie rod end and jam nut position (Courtesy of Chrysler Corporation)

17. Fill the power steering pump reservoir with the vehicle manufacturer's recommended power steering fluid, and bleed air from the power steering system. (Refer to Chapter 10 for these tasks.) This step is not required on a manual rack and pinion steering gear.
18. Road test the vehicle and check for proper steering gear operation and steering control.

# Manual Rack and Pinion Steering Gear Diagnosis and Service

## Manual Rack and Pinion Steering Gear Disassembly

Follow these steps for manual rack and pinion steering gear disassembly:

1. Clamp the center of the steering gear housing in a soft-jaw vise. Do not apply excessive force to the vise.
2. Insert a short piece of wire through each outer tie rod end cotter key hole and connect a pull scale to this wire. Pull upward on the scale to check the tie rod articulation effort (Figure 12-12). If this effort is not within specifications, replace the inner tie rod end.
3. Mark the outer tie rod and jam nut position with masking tape wrapped around the tie rod threads next to the jam nut, or place a dab of paint on the jam nut and tie rod threads (Figure 12-13). Loosen the jam nuts and remove the outer tie rod ends.
4. Remove the inner and outer bellows boot clamps and pull the bellows boots from the tie rods.
5. Hold the rack in a soft-jaw vise, and straighten the lock plate where it is bent over the inner tie rod end (Figure 12-14). Hold the rack with a wrench, and use the proper size wrench to remove the inner tie rod from the rack (Figure 12-15). A jam nut is used in place of the lock ring on some inner tie rod ends, and a roll pin retains some inner tie rod ends to the rack. Various inner tie rod socket removal procedures are required depending on the socket design.
6. Rotate the pinion shaft until the specified distance is obtained between the end of the rack and the steering gear housing. Mark the pinion shaft position in relation to the housing with a center punch. If the specified distance from the end of the rack to the housing is not available, rotate the pinion shaft fully clockwise. Rotate the pinion shaft

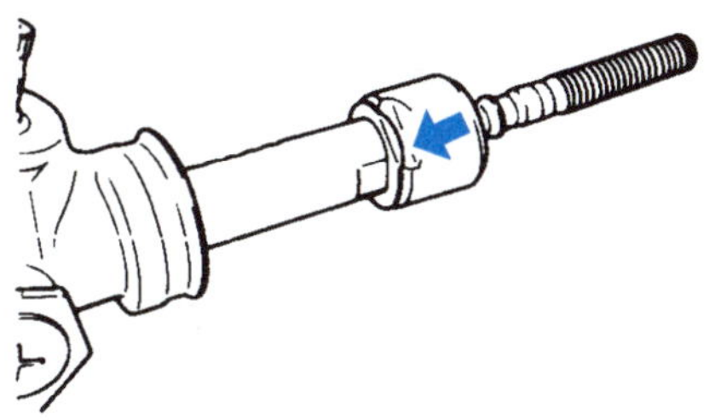

Figure 12-14 Straightening lock plate where it is bent over the inner tie rod end (Courtesy of Nissan Motor Co., Ltd.)

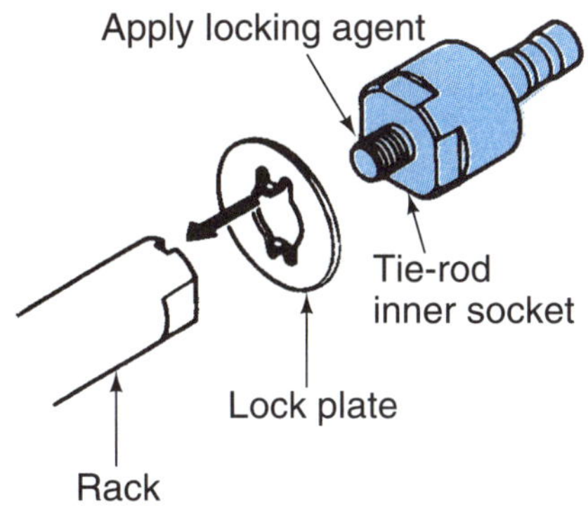


Figure 12-15 Removing inner tie rod end from the rack (Courtesy of Nissan Motor Co., Ltd.)

fully counterclockwise and count the number of turns. Turn the pinion shaft clockwise one-half this number of turns. Measure the distance from each end of the rack to the housing and record this measurement. Use a center punch to mark the pinion shaft position in relation to the housing.

7. Loosen the adjuster plug lock nut, and remove the adjuster plug and spring.
8. Remove the rack bearing through the adjuster plug opening.
9. Clean the surface around the pinion shaft seal, and pierce the pinion shaft seal at one of the two round areas on the seal surface. Pry this seal from the housing.
10. Remove the pinion shaft retaining ring with snap ring pliers.
11. Place the end of the pinion shaft in a soft-jaw vise, and tap the gear housing with a soft hammer to remove the pinion shaft and bearing from the housing.
12. Remove the rack from the housing.
13. Use an approved solvent to clean all steering gear components except the inner tie rod assemblies, and blow dry these parts with compressed air.

## Inspection of Manual Rack and Steering Gear Components

**Inner Tie Rod Sockets.** After the steering gear components are removed and cleaned, they must be visually inspected. If the inner tie rod ends were loose during the on-car inspection or the inner tie rod end articulation effort did not meet specifications, replace these ends.

**Pinion and Bearing Assembly.** If the pinion bearing is loose on the pinion shaft, replace the pinion and bearing assembly. A pinion shaft with worn or chipped teeth must be replaced. Inspect the pilot bearing contact area on the pinion shaft. Wear, pitting, or scoring in this area indicates that a new pinion shaft is required. If the pinion bearing is rough, loose, or noisy, replace the pinion shaft and bearing assembly.

**Pilot Bearing.** Check the pilot bearing in the steering gear housing. If this bearing is worn or scored, replace the pilot bearing and the pinion shaft and bearing assembly. Use an appropriate bearing driver to remove and replace the pilot bearing in the steering gear housing.

**Rack Bushing.** If the rack bushing is worn, bushing replacement is necessary. Remove the bushing retaining ring prior to bushing removal. Position the puller fingers behind the bushing, and operate the slide hammer on the puller to remove the bushing. If a puller is not available, an appropriate bushing driver or socket may be used to drive the bushing out of the housing.

**Mounting Bushings.** If the mounting bushings are loose during the on-car inspection, replace the bushings. Always replace the bushings in pairs. If the bushings are in satisfactory condition, do not disturb them.

## Manual Rack and Pinion Steering Gear Assembly

Proceed as follows to reassemble the manual rack and pinion steering gear:

1. Install the rack in the housing with the teeth parallel to the pinion shaft. Move the rack until the specified distance is obtained between the end of the rack and the housing. If this specification is not available, position the rack ends at the distance from the housing recorded during the disassembly procedure.
2. Use an anhydrous calcium grease to lubricate the pilot bearing and the pilot bearing contact area on the pinion shaft.
3. Install the pinion shaft with the punch marks aligned on the shaft and housing.
4. Install the pinion shaft retaining ring with the beveled edge facing upward.
5. Place a generous coating of anhydrous calcium grease on top of the pinion shaft bearing.
6. Use the proper seal driver to install the pinion shaft seal.
7. Apply a coating of lithium-based grease to the rack bearing, and install this bearing.
8. Apply lithium-based grease to the ends of the preload spring and the adjuster plug threads, and install the preload spring and adjuster plug.
9. Turn the adjuster plug until it bottoms, and then rotate the plug 45° to 60° counterclockwise.
10. Rotate the pinion shaft with a socket and a torque wrench, and observe the turning torque on the torque wrench. If necessary, rotate the adjuster plug to obtain the specified turning torque. Tighten the adjuster plug lock nut to the specified torque.
11. Coat both ends of the rack with lithium-based grease, and fill the rack teeth with the same lubricant. Rotate the pinion shaft fully in each direction several times, and apply more grease to the rack teeth after each complete pinion shaft rotation.
12. Rotate the inner tie rod ends onto the rack until they bottom. Install the inner tie rod pins, or stake these ends as required. Always use a wooden block to support the opposite side of the rack and inner tie rod end while staking. If jam nuts are located on the inner tie rod ends, be sure they are tightened to the specified torque.
13. Place a large bellows boot clamp over each end of the gear housing. Install the bellows boots, and be sure the boots are seated in the housing and the tie rod undercuts. Install and tighten the large inner boot clamps.
14. Install the outer bellows boot clamps on the tie rods, but do not install these clamps on the boots until the steering gear is installed and the toe is adjusted.
15. Install the jam nuts and the outer tie rod ends. Align the marks placed on these components during the disassembly procedure. Leave the jam nuts loose until the steering gear is installed and the toe is adjusted.
16. Install the steering gear in the vehicle and check the front wheel toe. Tighten the outer bellows boot clamps, and the outer tie rod end jam nuts.

## Diagnosis of Manual Steering and Suspension Systems

**Classroom Manual**
Chapter 12, page 245

It is sometimes difficult to separate the diagnosis of manual steering gears, steering columns, and suspension systems, because some noises and symptoms may be caused by problems in any of these components or systems. Therefore, we have attempted to provide a complete diagnosis that includes manual recirculating ball steering gears, manual rack and pinion steering gears, and some of the common steering column and suspension system complaints (Table 12-1).

**Table 12-1 MANUAL STEERING GEARS, SUSPENSION SYSTEMS, AND STEERING COLUMNS DIAGNOSIS**

| Condition | Possible Cause | Correction |
|---|---|---|
| Loose steering | 1. Steering linkage connections loose | 1. Tighten as required. |
| | 2. Steering linkage ball stud rubber deteriorated | 2. Replace as required. |
| | 3. Pitman arm-to-steering gear attaching nut loose | 3. Tighten to specifications. |
| | 4. Flex coupling-to-steering gear attaching bolt loose | 4. Tighten the bolt to specifications. |
| | 5. Intermediate shaft-to-steering column shaft attaching bolt loose | 5. Tighten the bolt to specifications. |
| | 6. Flex coupling damaged or worn | 6. Replace as required. |
| | 7. Front and/or rear suspension components loose, damaged, or worn | 7. Tighten or replace as required. |
| | 8. Steering gear adjustment set too low | 8. Check the steering gear adjustment and readjust as required. |
| Noise, knock, clunk, rapping, squeaking noise when turning | 1. Steering wheel-to-steering column shroud interference | 1. Adjust or replace as required. |
| | 2. Lack of lubrication where speed control brush contacts steering wheel pad | 2. Lubricate as required. |
| | 3. Steering column mounting bolts loose | 3. Tighten the bolt to specifications. |
| | 4. Intermediate shaft-to-steering column shaft attaching bolt loose | 4. Tighten the bolt to specifications. |
| | 5. Flex coupling-to-steering gear attaching bolt loose | 5. Tighten the bolt to specifications. |
| | 6. Steering gear mounting bolts loose | 6. Tighten the bolt to specifications. |
| | 7. Tires rubbing or grounding out against body or chassis | 7. Adjust/replace as required. |
| | 8. Front suspension components loose, worn, or damaged | 8. Tighten or replace as required. |
| | 9. Steering linkage ball stud rubber deteriorated | 9. Replace as required. |
| | 10. Normal lash between sector gear teeth and ball nut gear teeth when the steering gear is in the off-center position permits gear tooth chuckle noise while turning on rough road surfaces | 10. This is a normal condition and cannot be eliminated. |

**Table 12-1 MANUAL STEERING GEARS, SUSPENSION SYSTEMS, AND STEERING COLUMNS DIAGNOSIS (continued)**

| Condition | Possible Cause | Correction |
|---|---|---|
| Vehicle pulls/drifts to one side *Note:* This condition cannot be caused by the manual steering gear. | 1. Vehicle overloaded or unevenly loaded | 1. Correct as required. |
| | 2. Improper tire pressure | 2. Adjust air pressure as required. |
| | 3. Mismatched tires and wheels | 3. Install correct tire and wheel combination. |
| | 4. Unevenly worn tires | 4. Replace as required (check for cause). |
| | 5. Loose, worn, or damaged steering linkage | 5. Tighten or replace as required. |
| | 6. Steering linkage stud not centered within the socket | 6. Repair or replace as required. |
| | 7. Bent spindle or spindle arm | 7. Repair or replace as required. |
| | 8. Broken and/or sagging front and/or rear suspension springs | 8. Replace as required. |
| | 9. Loose, bent, or damaged suspension components | 9. Tighten or replace as required. |
| | 10. Bent rear axle housing | 10. Replace as required. |
| | 11. Excessive camber/caster split (Excessive side-to-side variance in the caster/camber settings) | 11. Adjust camber/caster split as required. |
| | 12. Improper toe setting | 12. Adjust as required. |
| | 13. Front-wheel bearings out of adjustment | 13. Adjust as required. |
| | 14. Investigate tire variance (conicity for radial tires and unequal circumference for bias-ply belted bias-ply tires) | 14. Repair or replace as required. |
| High efforts/poor returnability | 1. Improper tire/wheel size | 1. Install correct tire and wheel combinations. |
| | 2. Underinflated tires | 2. Adjust tire pressure as required. |
| | 3. Misaligned steering column and/or distorted flex coupling | 3. Align/replace as required. |
| | 4. Steering wheel-to-column interference | 4. Remove interference. |
| | 5. Steering gear with insufficient grease | 5. Check steering gear seals and replace, if required. Add lubricant as required. |
| | 6. Improper steering column indexing and/or improper toe setting | 6. Set front wheels in straight-ahead position and ensure the following:<br>— Steering wheel hub and steering column shaft index |

**Table 12-1** MANUAL STEERING GEARS, SUSPENSION SYSTEMS, AND STEERING COLUMNS DIAGNOSIS (continued)

| Condition | Possible Cause | Correction |
|---|---|---|
| | | marks must be aligned.<br>— The index marks must be in 12 o'clock position.<br>— Steering gear input shaft flat must be in 12 o'clock position. If the indexing is improper, reindex the steering column. If the steering gear input shaft flat is not in 12 o'clock position, reset the toe. |
| | 7. Steering gear preload/meshload set too high | 7. Check the steering gear preload/meshload and adjust as required. |
| | 8. Steering linkage and rubber captured ball stud binding | 8. Repair/replace as required. |
| | 9. Spindle ball joint torque too high. | 9. Repair/replace as required. |
| Vehicle wanders (side to side)—Constant corrective steering is required at highway speeds. | 1. Vehicle overloaded or unevenly loaded | 1. Correct as required. |
| | 2. Improper tire/wheel size | 2. Install correct tire and wheel combination. |
| | 3. Steering gear mounting bolts loose | 3. Tighten the bolts to specifications. |
| | 4. Flex coupling-to-steering gear attaching bolt loose | 4. Tighten the bolt to specifications. |
| | 5. Flex coupling damaged or worn | 5. Replace as required. |
| | 6. Improper steering column and/or improper toe setting | 6. Set front wheels in straight-ahead position and ensure the following:<br>— Steering wheel hub and steering column shaft index marks must be aligned.<br>— The index marks must be in 12 o'clock position.<br>— Steering gear input shaft flat must be in 12 o'clock position. If the indexing is improper, reindex the steering column. If the steering gear input shaft flat is not in 12 o'clock position, reset the toe. |
| | 7. Steering linkage loose, worn, or damaged | 7. Correct as required. |
| | 8. Broken or sagging suspension springs | 8. Replace as required. |

**Table 12-1** **MANUAL STEERING GEARS, SUSPENSION SYSTEMS, AND STEERING COLUMNS DIAGNOSIS (continued)**

| Condition | Possible Cause | Correction |
|---|---|---|
| | 9. Steering gear adjustment set too low | 9. Check the steering gear adjustment and adjust as required. |
| | 10. Improper caster/camber | 10. Adjust caster/camber. |
| | 11. Front-wheel bearing misadjusted | 11. Adjust as required. |
| Uneven efforts | Steering gear input shaft not centered due to improper toe setting | Center gear and adjust toe as required. |
| Pointing, darting, sticky, or down-the-road steering feel is not smooth; seems to be sticky and cannot make minor corrections with ease | 1. Steering column misaligned | 1. Align as required. |
| | 2. Flex coupling distorted | 2. Replace as required. |
| | 3. Steering gear grease leaking from input and/or output shaft seal | 3. Replace the steering gear seals. |
| | 4. Steering gear input shaft not centered due to incorrect toe settings | 4. Center gear and adjust toe as required. |
| | 5. Steering gear adjustment set too high | 5. Check the steering gear adjustment and adjust as required to specifications. |
| | 6. Excessive caster split (excessive side-to-side variance in the caster setting) | 6. Adjust caster split as required. |
| | 7. Spindle ball joint torque too high | 7. Repair/replace as required. |
| Rough, lumpy, binding, steering—Rugged or catchy steering feel while turning in either direction | 1. Steering column misaligned | 1. Align as required. |
| | 2. Flex coupling distorted | 2. Replace as required. |
| | 3. Steering linkage components damaged due to impact | 3. Replace as required. |
| | 4. Excessive steering gear roughness | 4. Check steering gear adjustment and readjust as required. If, during the check, torque wrench indicates excessive roughness, repair the gear. |
| Poor center feel/poor groove feel—Increase in steering efforts when the steering wheel is rotated off-center during straight-ahead highway driving. Vehicle with poor groove feel might require repeated small corrections to maintain a desired straight-ahead path. | 1. Improper steering column indexing and/or improper toe setting | 1. Set front wheels in straight-ahead position and ensure the following:<br>—Steering wheel hub and steering column shaft index marks must be aligned.<br>—The index marks must be in 12 o'clock position.<br>—Steering gear input shaft flat must be in 12 o'clock |

**Table 12-1** **MANUAL STEERING GEARS, SUSPENSION SYSTEMS, AND STEERING COLUMNS DIAGNOSIS (continued)**

| Condition | Possible Cause | Correction |
|---|---|---|
| | | position. If the indexing is improper, reindex the steering column. If the steering gear input shaft flat is not in 12 o'clock position, reset the toe. |
| | 2. Steering gear adjustment set too high | 2. Check the steering gear adjustment and adjust as required. |
| | 3. Excessive caster split (side-to-side variance is too large) | 3. Adjust as required. |
| Shimmy—A consistent side-to-side wobble in the steering wheel or road wheels usually felt at over 40 mph and amplified when driven over pot holes or bumps in the road surface. *Note:* This condition cannot be caused by the manual steering gear. | 1. Wheels out of balance | 1. Check/correct as required. |
| | 2. Improper tire pressure | 2. Adjust as required. |
| | 3. Excessive tire sidewall deflection | 3. Inspect, adjust air pressure, and replace as required. |
| | 4. Irregular tire wear | 4. Replace as required. |
| | 5. Loose wheel lug nuts | 5. Tighten to specifications. |
| | 6. Radius arm bushings deteriorated | 6. Replace as required. |
| | 7. Loose or leaky shock absorber | 7. Tighten or replace as required. |
| | 8. Broken or sagging suspension spring | 8. Replace as required. |
| | 9. Loose, worn, or damaged steering linkage | 9. Tighten/replace as required. |
| | 10. Steering gear mounting bolts loose | 10. Tighten the bolts. |
| | 11. Front-wheel bearings out of adjustment | 11. Adjust as required. |
| | 12. Out-of-rounds wheels | 12. Replace as required. |
| | 13. Out-of-round tires | 13. Replace as required. |
| | 14. Wheel and tire lateral runout not within specification | 14. Check/replace as required. |
| | 15. Worn spindle ball joints | 15. Replace as required. |
| | 16. Steering linkage ball stud rubber deteriorated | 16. Replace as required. |
| Rattle or chuck in steering gear | 1. Insufficient or improper lubricant in steering gear | 1. Add lube specified. |
| | 2. Pitman arm loose on shaft or steering gear mounting bolt loose | 2. Tighten to specified torque. |
| | 3. Loose or worn steering shaft bearing | 3. Replace steering shaft bearing. |

**Table 12-1 MANUAL STEERING GEARS, SUSPENSION SYSTEMS, AND STEERING COLUMNS DIAGNOSIS (continued)**

| Condition | Possible Cause | Correction |
|---|---|---|
| | 4. Excessive over-center lash or worm thrust bearing adjustment too loose<br>*Note:* On turns, a slight rattle might occur, due to the increased lash between ball nut and pitman shaft as gear moves off the center of high point position. This is normal—lash must not be reduced to eliminate the slight rattle. | 4. Adjust steering gear to specified preloads. |
| Poor returnability | 1. Steering column misaligned<br>2. Insufficient or improper lubricant in steering gear or front suspension<br>3. Steering gear adjusted too tightly<br>4. Front wheel alignment incorrect (caster) | 1. Align column.<br>2. Lubricate as specified.<br>3. Adjust over-center and thrust bearing preload to specifications.<br>4. Adjust to specifications. |
| Excessive play or looseness in steering system | 1. Front wheel bearings loosely adjusted<br>2. Worn upper ball joints<br>3. Steering wheel loose on shaft, loose pitman arm, tie-rods, steering arms, or steering linkage ball nuts<br>4. Excessive over-center lash<br>5. Worm thrust bearings loosely adjusted | 1. Adjust to obtain proper end-play.<br>2. Check and replace ball joints if necessary.<br>3. Tighten to specifications; replace if worn or damaged.<br>4. Adjust over-center preload to specifications.<br>5. Adjust worm thrust bearing preload to specifications. |
| Hard steering—Excessive effort required at steering wheel | 1. Low or uneven tire pressure<br>2. Insufficient or improper lubricant in steering gear or front suspension<br>3. Steering shaft flexible coupling misaligned<br>4. Steering gear adjusted too tightly<br>5. Front wheel alignment incorrect (manual gear) | 1. Inflate to specified pressures.<br>2. Lubricate as specified. Relubricate at specified intervals.<br>3. Align column and coupling.<br>4. Adjust over-center and thrust bearing preload to specifications.<br>5. Adjust to specifications. |

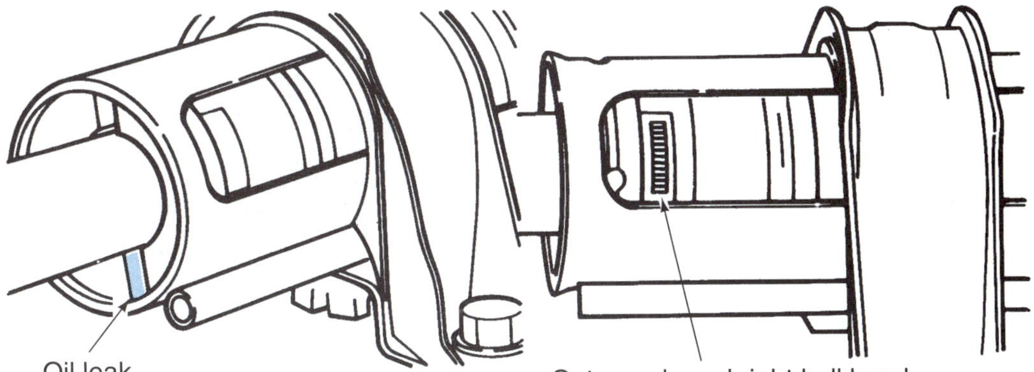


**Figure 12-16** Oil leak diagnosis at outer rack seal (Courtesy of Chrysler Corporation)

# Power Rack and Pinion Steering Gear Diagnosis and Service

## Oil Leak Diagnosis

**CAUTION:** If the engine has been running for a length of time, power steering gears, pumps, lines, and fluid may be very hot. Wear eye protection and protective gloves when servicing these components.

If power steering fluid leaks from the cylinder end of the power steering gear, the outer rack seal is leaking (Figure 12-16).

The inner rack seal is defective if oil leaks from the pinion end of the housing when the rack reaches the left internal stop (Figure 12-17). An oil leak at one rack seal may result in oil leaks from both boots, because the oil may travel through the breather tube between the boots.

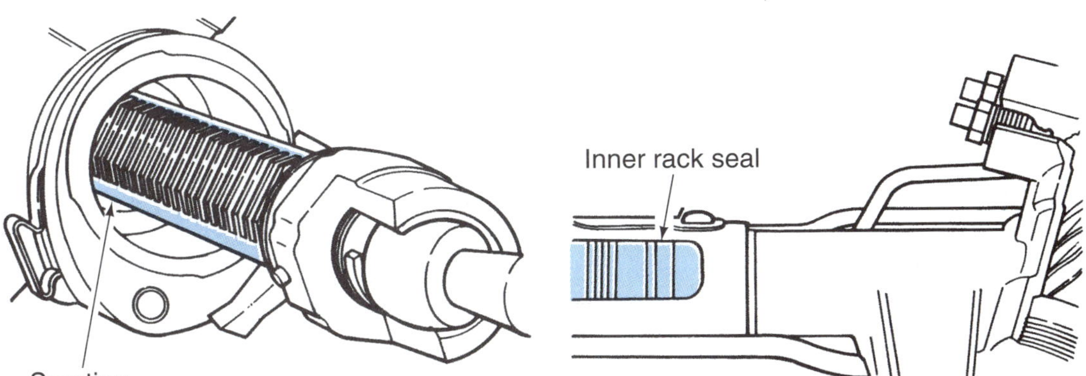


**Figure 12-17** Inner rack seal leak diagnosis (Courtesy of Chrysler Corporation)

If you detect a leak at the pinion end of the housing and it is not influenced by the direction of the turn, the origin of the leak is the pinion seal.

**Figure 12-18** Pinion seal leak diagnosis (Courtesy of Chrysler Corporation)

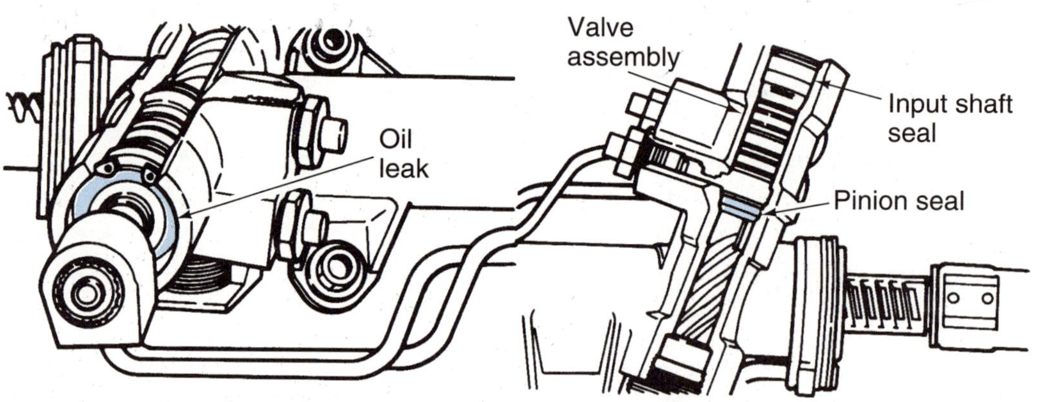


If you discover a leak at the pinion coupling area, you'll have to replace both the input shaft seal and the lower pinion seal.

**Figure 12-19** Oil leak diagnosis in pinion coupling area (Courtesy of Chrysler Corporation)

If an oil leak occurs at the pinion end of the housing, and this leak is not influenced when the steering wheel is turned, the pinion seal is defective (Figure 12-18).

If an oil leak occurs in the pinion coupling area, the input shaft seal is leaking (Figure 12-19). This seal and the pinion seal will require replacement, because the pinion seal must be replaced if the pinion is removed.

When the rack is removed, the inner and outer rack seals and the pinion seal must be replaced (Figure 12-20). If oil leaks occur at fittings, these fittings must be torqued to manufacturer's specifications. If the leak is still present, the line and fitting should be replaced. Leaks in the lines or hoses require line or hose replacement.

## Turning Imbalance Diagnosis

**SERVICE TIP:** To provide equal turning effort in both directions, the rack must be centered with the front wheels straight ahead.

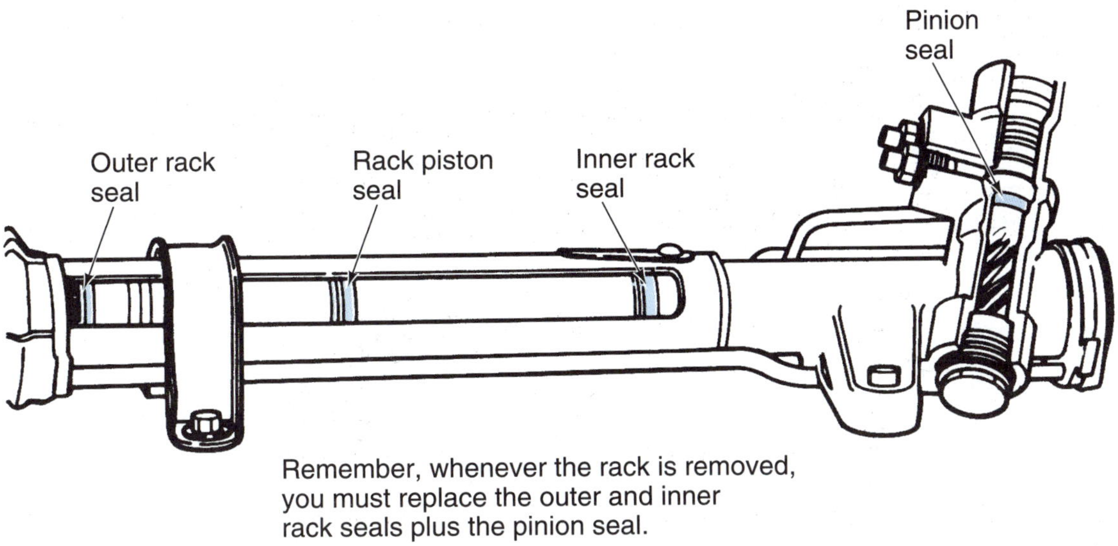


**Figure 12-20** Rack seals and pinion seal (Courtesy of Chrysler Corporation)

The same amount of effort should be required to turn the steering wheel in either direction. A pressure gauge connected to the high-pressure hose should indicate the same pressure when the steering wheel is turned in each direction. Steering effort imbalance, or lower power assist in each direction, may be caused by defective rack seals (Figure 12-21). Steering effort imbalance, or low power assist in both directions, may also be caused by defective rotary valve rings and seals or restricted hoses and lines (Figure 12-22).

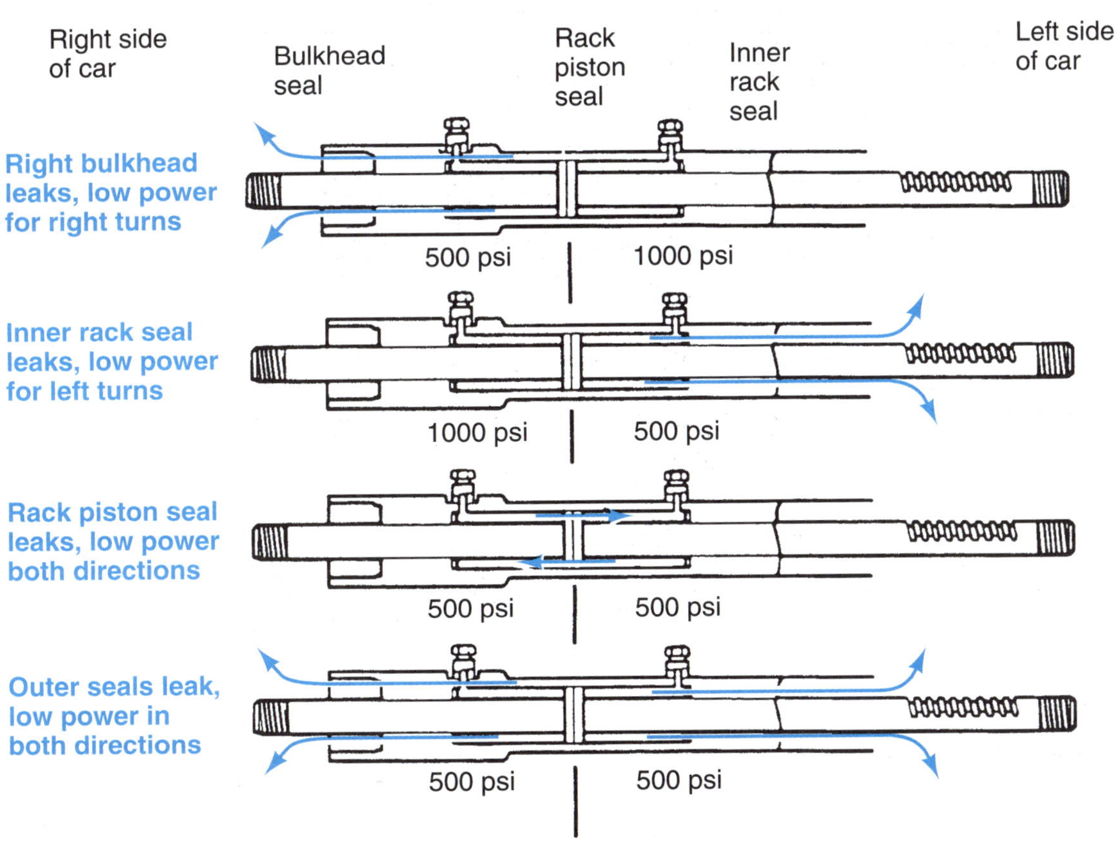


**Figure 12-21** Effect of defective rack seals on steering effort imbalance and low power assist (Courtesy of Chrysler Corporation)

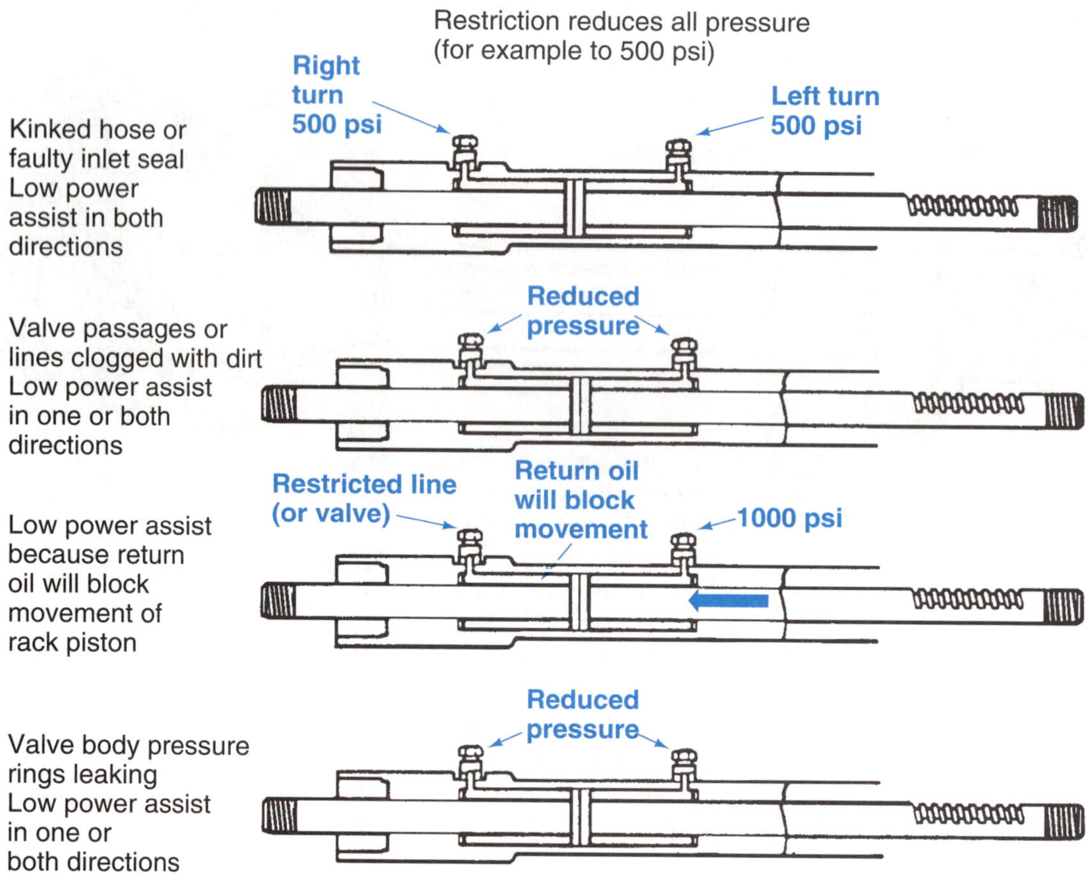


**Figure 12-22** Effect of worn rotary valve rings and seals, or restricted lines on steering effort (Courtesy of Chrysler Corporation)

## Power Rack and Pinion Steering Gear Disassembly

**Special Tools**

Steering gear holding tool

The power rack and pinion steering gear disassembly procedure varies depending on the vehicle and steering gear manufacturer. Always follow the vehicle manufacturer's recommended procedure in the service manual. Following is a typical power rack and pinion steering gear disassembly procedure:

1. Install a holding tool on the steering gear housing, and clamp this tool in a vise (Figure 12-23).
2. Use the proper wrench to disconnect the left and right turn tubes (Figure 12-24). Remove the O-rings from the turn tubes.

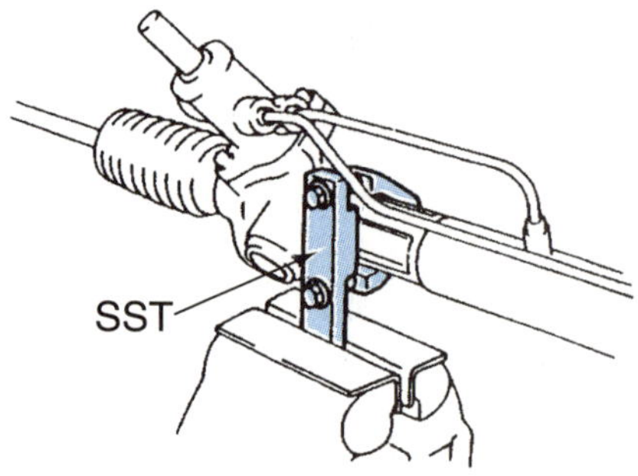


**Figure 12-23** Steering gear holding tool (Courtesy of Toyota Motor Corporation)

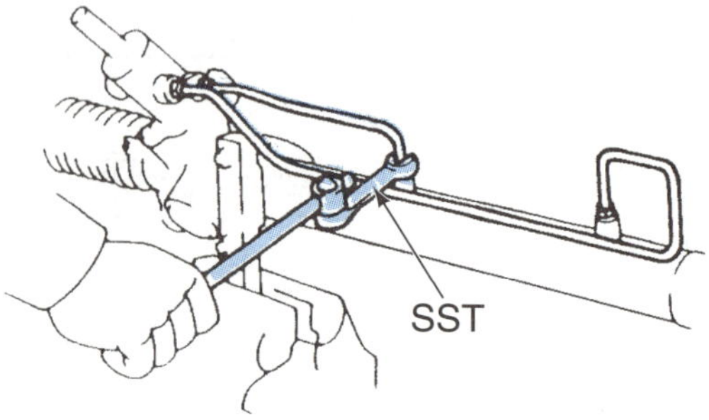


**Figure 12-24** Removing left and right turn tubes (Courtesy of Toyota Motor Corporation)

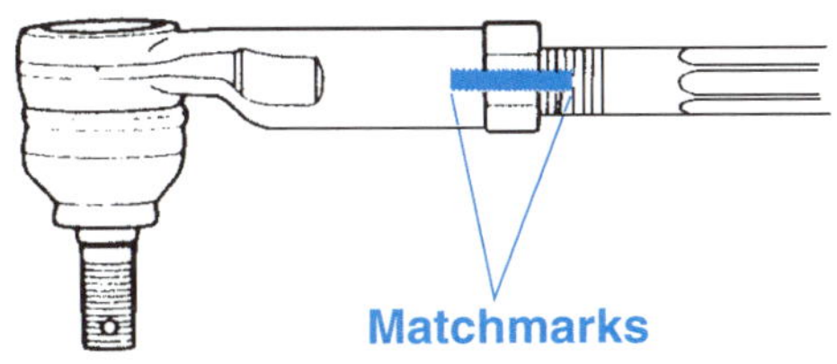


**Figure 12-25** Marking outer tie rod ends, jam nuts, and tie rods prior to outer tie rod end removal (Courtesy of Toyota Motor Corporation)

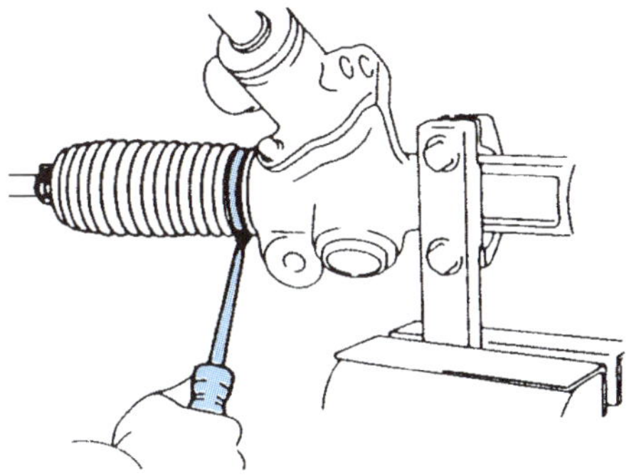

**Figure 12-26** Removing bellows boot clamps (Courtesy of Toyota Motor Corporation)

3. Mark the outer tie rod ends, jam nuts, and tie rods (Figure 12-25). Loosen the jam nuts and remove the tie rod ends and jam nuts.
4. Loosen the inner and outer clamps on both bellows boots, and remove these boots (Figure 12-26).
5. Clamp the inner tie rod socket lightly in a soft-jaw vise, and use a hammer and chisel to unstake the claw washers on the inner tie rod sockets (Figure 12-27).
6. Hold the rack with an adjustable wrench and use the proper tool to loosen the inner tie rod ends (Figure 12-28). Remove the inner tie rod ends and the claw washers.
7. Rotate the pinion shaft until the specified distance is obtained between the end of the rack and the steering gear housing. Mark the pinion shaft position in relation to the housing with a center punch. If the specified distance from the end of the rack to the housing is not available, rotate the pinion shaft fully clockwise. Rotate the pinion shaft fully counterclockwise and count the number of turns. Turn the pinion shaft clockwise one-half this number of turns. Measure the distance from each end of the rack to the housing and record this measurement. Use a center punch to mark the pinion shaft position in relation to the housing.
8. Use the proper tool to loosen the adjuster plug lock nut (Figure 12-29).
9. Remove the adjuster plug with the proper tool (Figure 12-30).
10. Remove the spring, rack guide, and the seat from the adjuster plug opening (Figure 12-31).
11. Use the appropriate socket and breaker bar to remove the pinion housing cap (Figure 12-32).

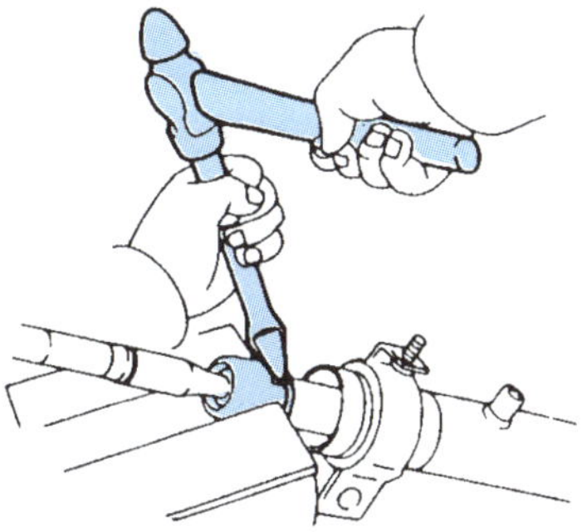

**Figure 12-27** Removing claw washers from inner tie rod sockets (Courtesy of Toyota Motor Corporation)

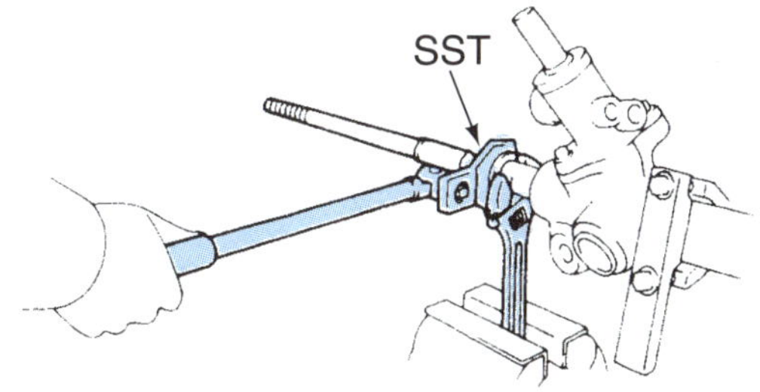


**Figure 12-28** Loosening inner tie rod ends (Courtesy of Toyota Motor Corporation)

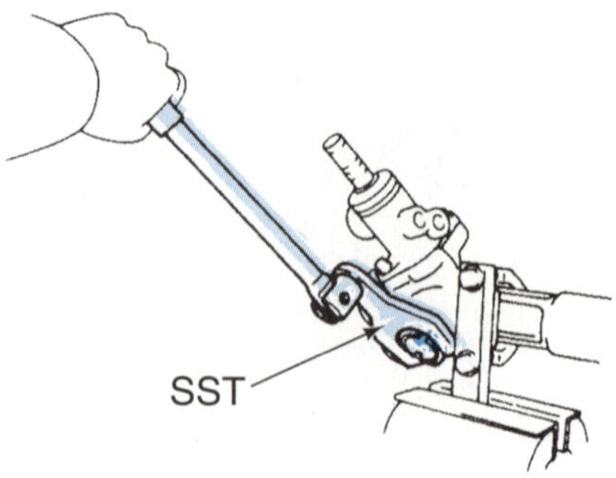


**Figure 12-29** Loosening adjuster plug lock nut (Courtesy of Toyota Motor Corporation)

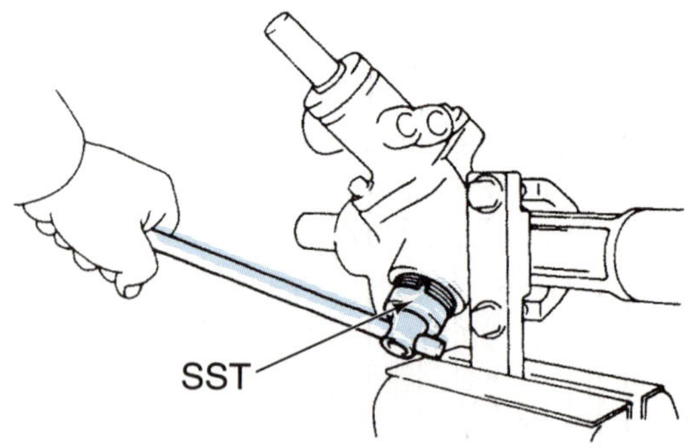


**Figure 12-30** Removing adjuster plug (Courtesy of Toyota Motor Corporation)

**Figure 12-31** Removing spring, rack guide, and seat (Courtesy of Toyota Motor Corporation)

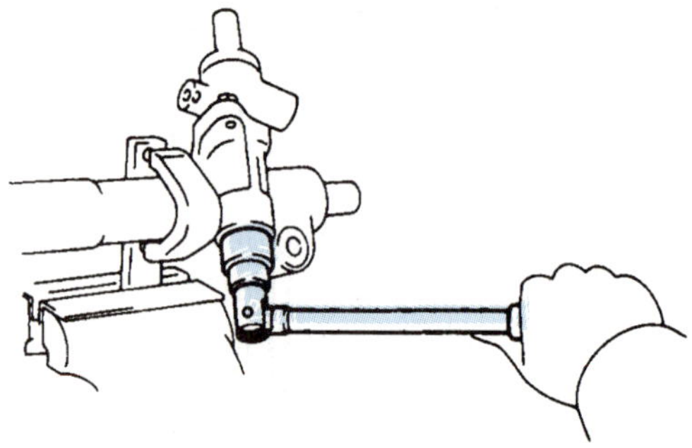

**Figure 12-32** Removing pinion housing cap (Courtesy of Toyota Motor Corporation)

12. Hold the top of the pinion shaft with a socket and breaker bar, and loosen the self-locking nut on the end of the pinion shaft (Figure 12-33). Remove this self-locking nut.
13. Remove the dust cover from the control valve housing and mark this housing in relation to the steering gear housing (Figure 12-34).
14. Remove the control valve housing retaining bolts, and remove this housing from the steering gear housing. Use a soft hammer to tap the pinion shaft from the control valve housing (Figure 12-35).
15. Place matching marks on the number 2 bracket and steering gear housing (Figure 12-36).

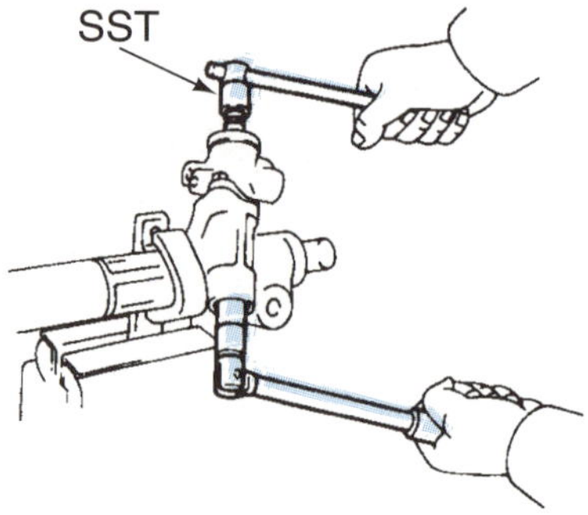


**Figure 12-33** Loosening pinion shaft self-locking nut (Courtesy of Toyota Motor Corporation)

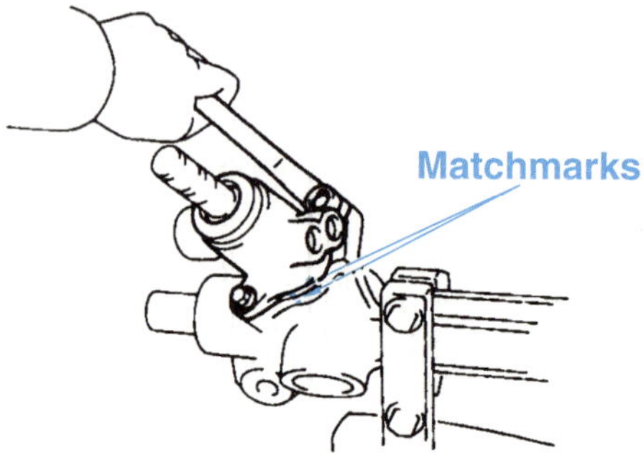


**Figure 12-34** Marking control valve housing in relation to steering gear housing (Courtesy of Toyota Motor Corporation)

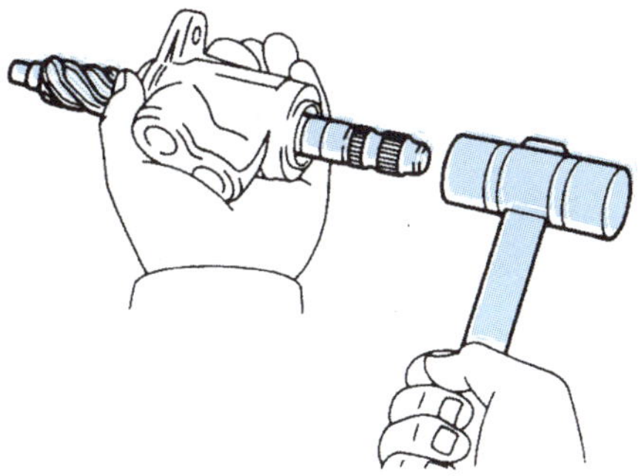

Figure 12-35 Removing pinion shaft from control valve housing (Courtesy of Toyota Motor Corporation)

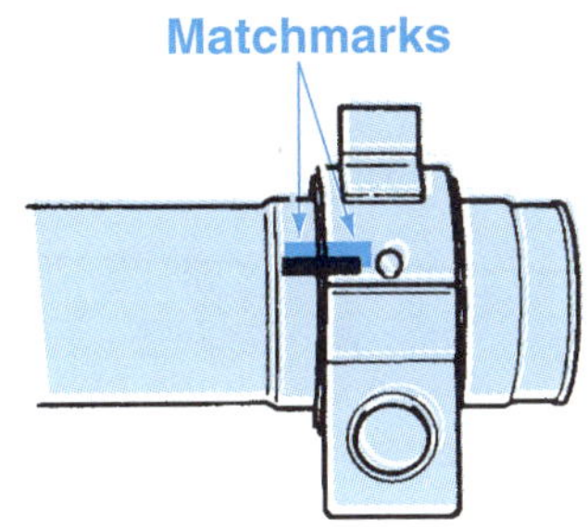


Figure 12-36 Matching marks placed on the bracket and steering gear housing (Courtesy of Toyota Motor Corporation)

Use a slotted screwdriver to pry the clasp open on the housing bracket (Figure 12-37). Remove the bushing and bracket from the housing, and disconnect the bushing from the bracket.

**16.** Use an end stopper removal tool to rotate the cylinder end stopper clockwise until the wire end comes through the hole in the gear housing (Figure 12-38). Rotate the end stopper counterclockwise with the same tool and remove the wire through the gear housing opening.

**Special Tools**

End stopper removal tool

**17.** Use a brass bar and a hammer to tap the rack and rack bushing from the housing (Figure 12-39). Remove the O-ring from the housing.

**18.** Drive the cylinder side oil seal and spacer out of the housing with the appropriate driving tool and a brass bar (Figure 12-40).

**Special Tools**

Seal driving tools

## Power Rack and Pinion Steering Gear Parts Inspection

Clean all steering gear components, except the inner tie rod ends and any sealed bearings, with an approved solvent. Blow dry components with compressed air. The power rack and pinion steering gear inspection and parts replacement procedure may vary depending on the vehicle and steering

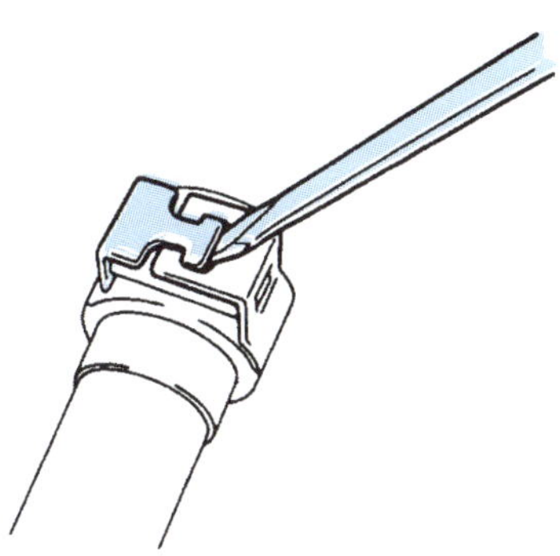

Figure 12-37 Prying the clasp open on the steering gear mounting bracket (Courtesy of Toyota Motor Corporation)

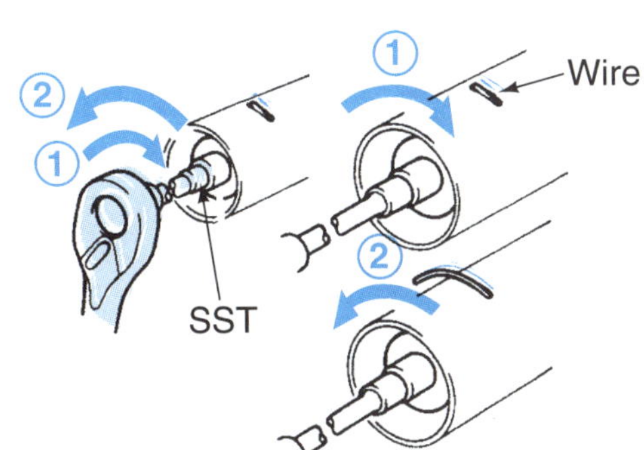


Figure 12-38 Removing wire from the end stopper (Courtesy of Toyota Motor Corporation)

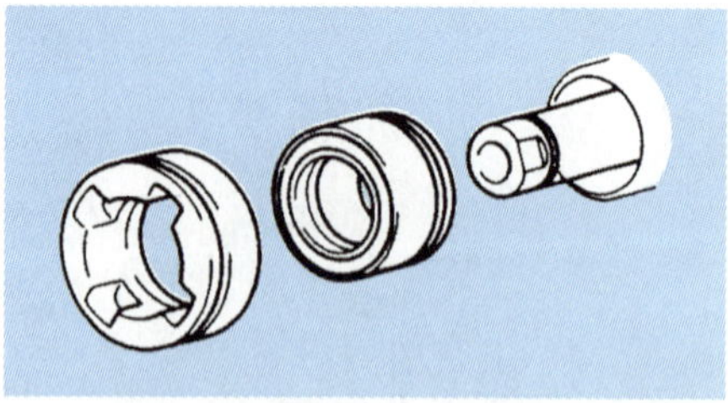

**Figure 12-39** Removing rack and rack bushing (Courtesy of Toyota Motor Corporation)

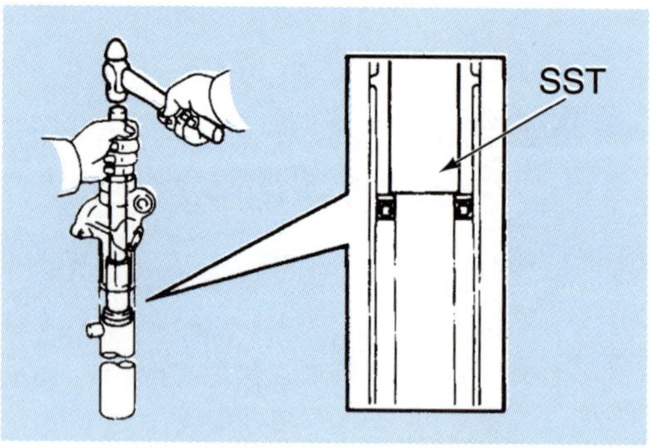


**Figure 12-40** Removing cylinder side oil seal and spacer (Courtesy of Toyota Motor Corporation)

gear manufacturer. Always follow the procedure in the vehicle manufacturer's service manual. Following is a typical inspection and parts replacement procedure:

**Special Tools**

V-blocks

Dial indicator

1. Inspect the rack for tooth wear or damage. If these conditions are present, replace the rack or complete steering gear. Place the rack on V-blocks and slowly rotate the rack with a dial indicator stem positioned near the center of the rack (Figure 12-41). If the rack runout exceeds specifications, replace the rack or complete steering gear.
2. If the outer circumference of the rack piston is scored or worn, replace the rack. Visually check the rack piston contact area in the housing for scoring, pitting, and wear. When any of these conditions are present, replace the housing or complete steering gear.

**Special Tools**

Bearing driving tools

3. Use the proper driving tool to drive the oil seal and upper bearing out of the control valve housing (Figure 12-42). If the control valve housing is scored, pitted, or worn in the ring contact area, replace the housing.
4. Drive the new seal into the control valve housing with the proper driving tool (Figure 12-43). With the new seal in place, use the proper driving tool to install the new upper bearing (Figure 12-44).
5. Check the lower bearing and center bearing. If the bearings are scored or worn, replacement is necessary. Visually check the pinion shaft in the bearing contact areas. If the pinion shaft is worn, scored, or pitted in these areas, replace the shaft. When the

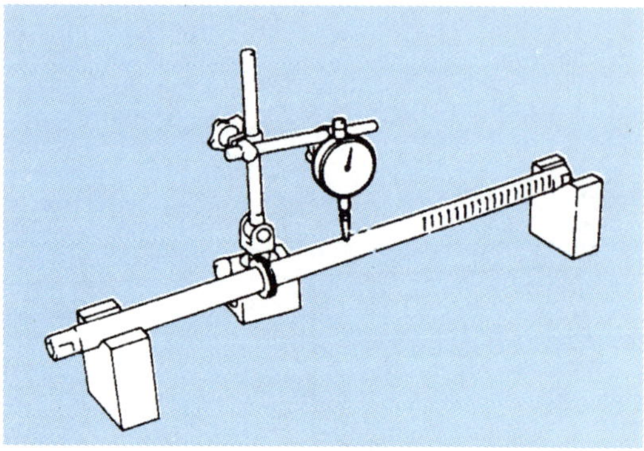

**Figure 12-41** Measuring rack runout (Courtesy of Toyota Motor Corporation)

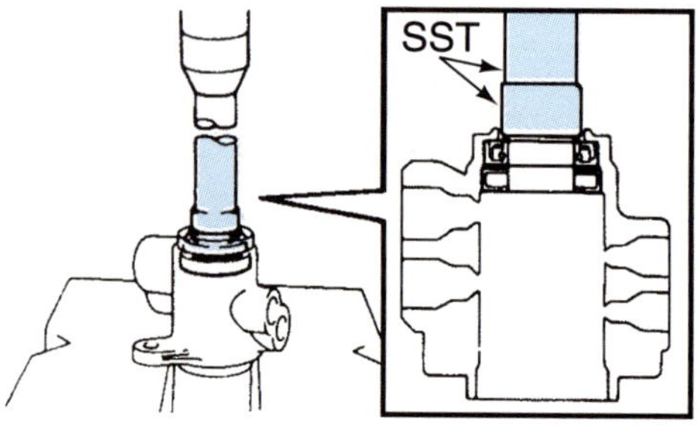


**Figure 12-42** Removing upper bearing and seal in the control valve housing (Courtesy of Toyota Motor Corporation)

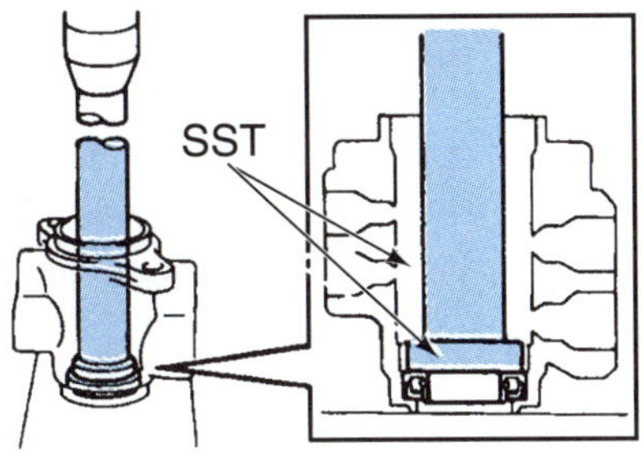


**Figure 12-43** Installing control valve housing seal (Courtesy of Toyota Motor Corporation)

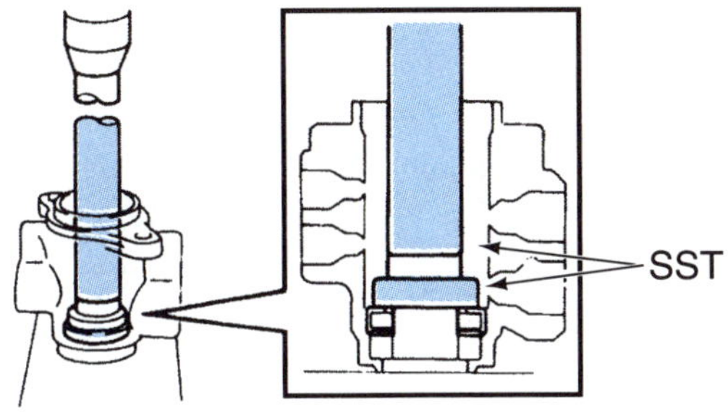


**Figure 12-44** Installing upper bearing in the control valve housing (Courtesy of Toyota Motor Corporation)

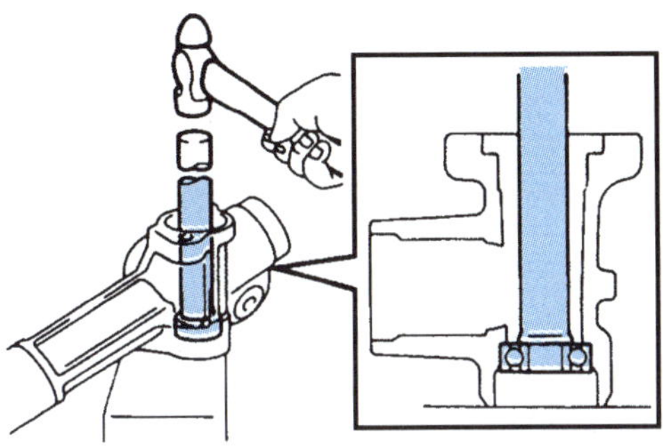

**Figure 12-45** Removing lower control valve bearing (Courtesy of Toyota Motor Corporation)

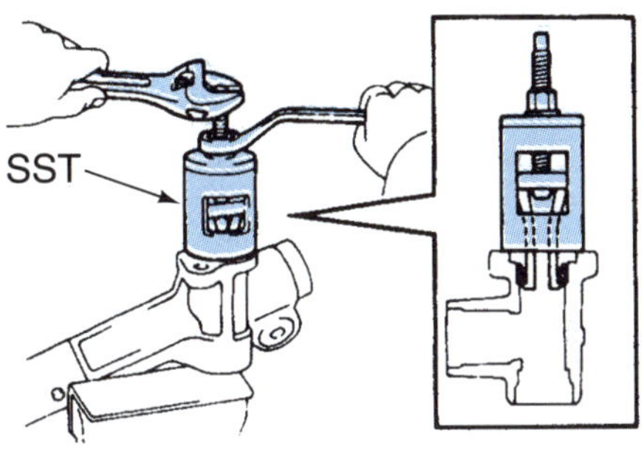


**Figure 12-46** Removing center control valve bearing (Courtesy of Toyota Motor Corporation)

pinion shaft teeth are worn, chipped, or pitted, replace the pinion shaft. If lower and center bearing replacement is required, drive the lower bearing out of the housing with a hammer and brass bar (Figure 12-45). Use the appropriate bearing puller to remove the center bearing (Figure 12-46).

6. Use the proper driving tool to press the new center bearing into the housing (Figure 12-47).
7. Press the new lower bearing into the steering gear housing with the proper driving tool (Figure 12-48).
8. Pull the rack bushing oil seal from the housing with a seal puller (Figure 12-49). Coat the new seal with power steering fluid and use the correct driving tool to install the seal (Figure 12-50).

**Special Tools**

Seal puller

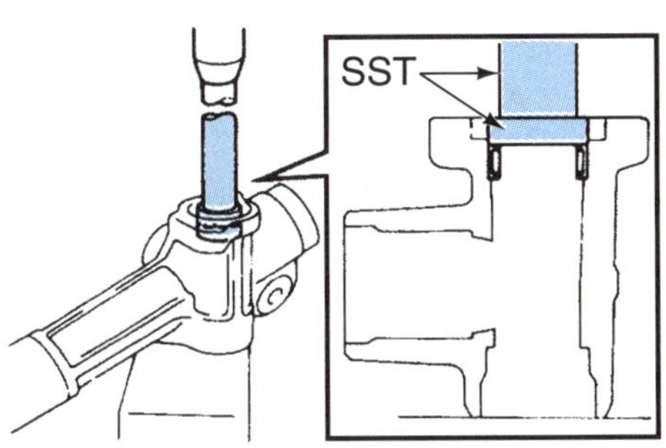


**Figure 12-47** Installing center bearing (Courtesy of Toyota Motor Corporation)

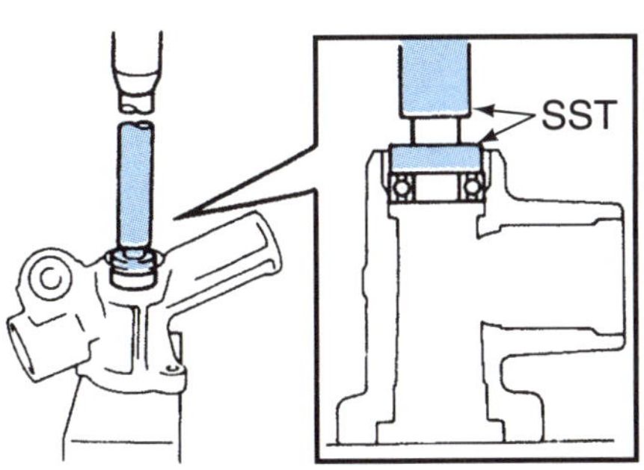


**Figure 12-48** Installing lower control valve bearing (Courtesy of Toyota Motor Corporation)

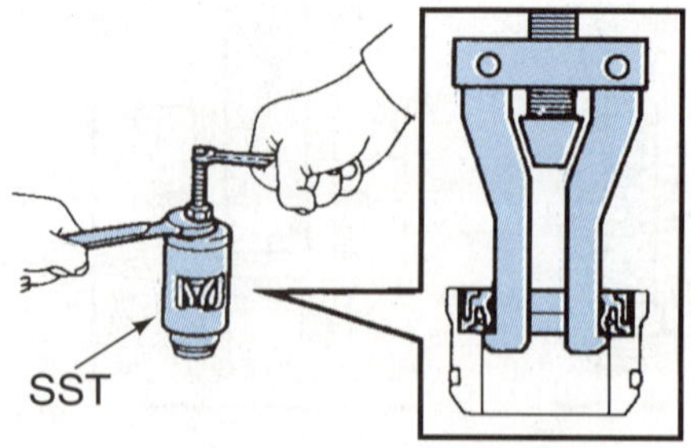


**Figure 12-49** Removing rack bushing seal (Courtesy of Toyota Motor Corporation)

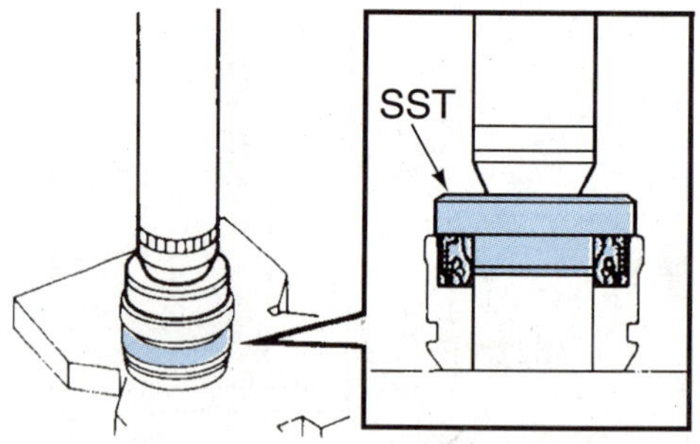


**Figure 12-50** Installing rack bushing seal (Courtesy of Toyota Motor Corporation)

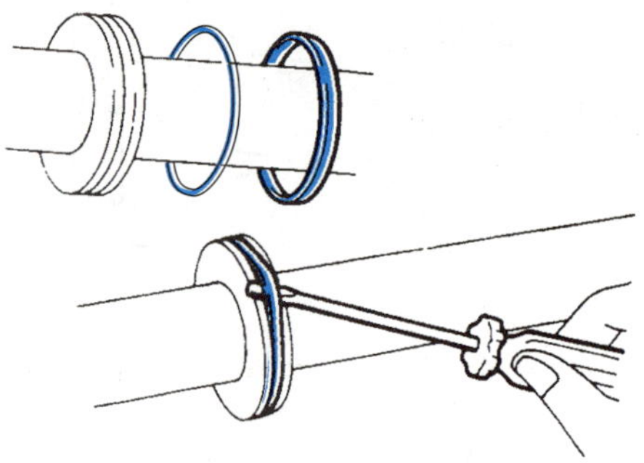

**Figure 12-51** Removing Teflon ring and O-ring from the rack piston (Courtesy of Toyota Motor Corporation)

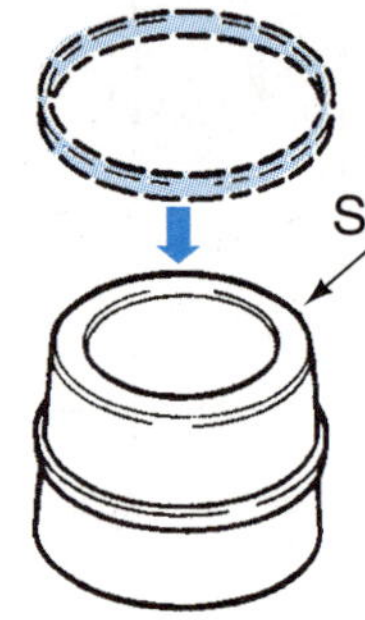


**Figure 12-52** Teflon ring expanding tool (Courtesy of Toyota Motor Corporation)

**WARNING:** Be careful not to score, or mark, the rack piston while removing and replacing the piston ring or O-ring. A scored rack piston must be replaced.

**Special Tools**

Teflon ring expanding tool

**9.** Pry the Teflon ring and O-ring from the rack piston with a small slotted screwdriver (Figure 12-51). Be careful not to mark the rack piston.

**10.** Coat the new Teflon ring with power steering fluid and install this ring on the expanding tool (Figure 12-52). Some steering gear manufacturers recommend placing the Teflon rings in boiling water for 10 minutes to expand the rings prior to ring installation.

**11.** Remove the Teflon ring from the expanding tool and install it on the rack piston. Wrap your hand around the Teflon ring, and squeeze it down into the rack piston groove (Figure 12-53).

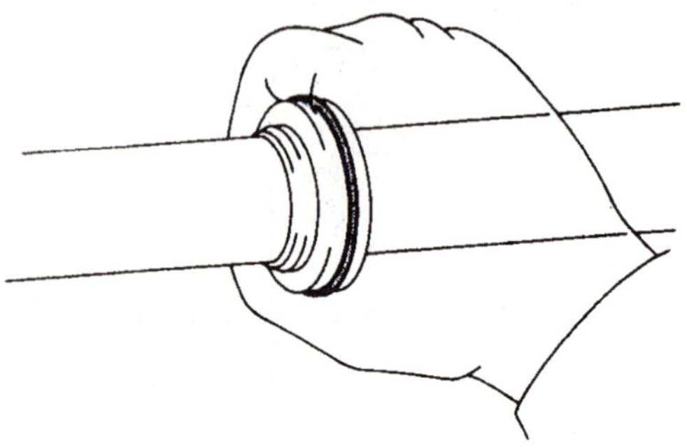

**Figure 12-53** Installing the Teflon ring in the rack piston groove (Courtesy of Toyota Motor Corporation)

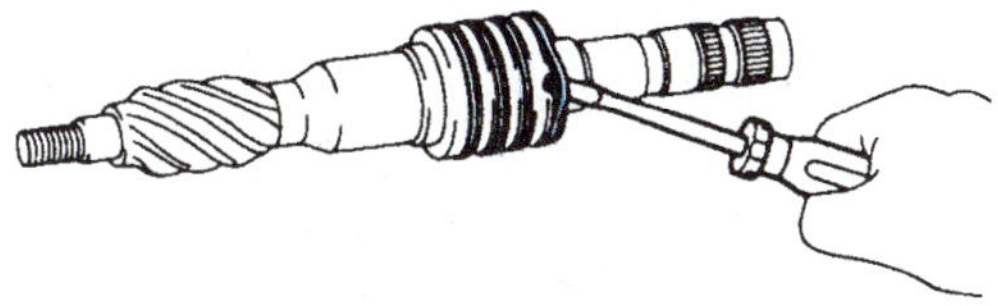

Figure 12-54 Removing Teflon rings from the control valve (Courtesy of Toyota Motor Corporation)

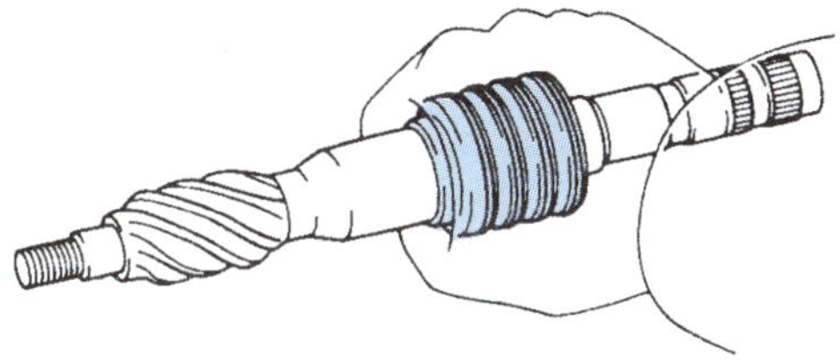

Figure 12-55 Installing rings on control valve (Courtesy of Toyota Motor Corporation)

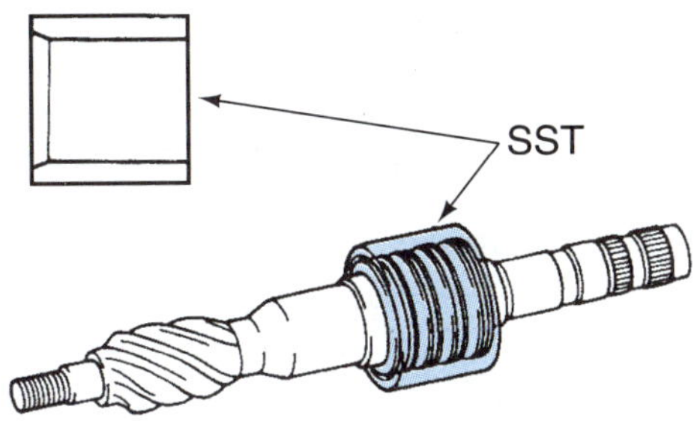


Figure 12-56 Compressing tool installed on Teflon rings (Courtesy of Toyota Motor Corporation)

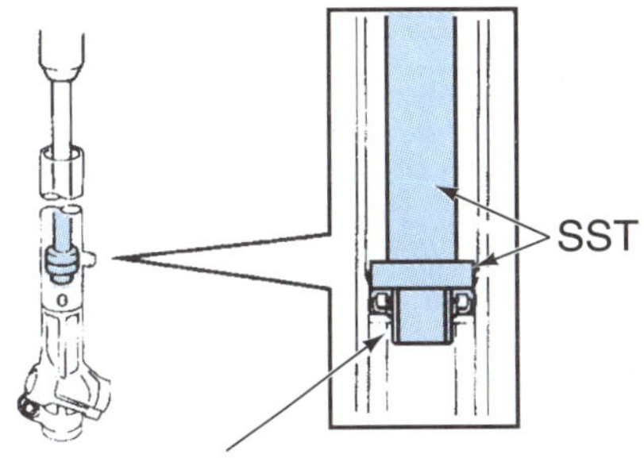


Figure 12-57 Installing cylinder side oil seal and spacer (Courtesy of Toyota Motor Corporation)

**WARNING:** Do not score, or mark, the control valve while removing the Teflon rings. Scored control valves must be replaced.

12. Pry the Teflon rings from the control valve with a small slotted screwdriver (Figure 12-54). Be careful not to mark the control valve.
13. Coat one of the control valve Teflon rings with power steering fluid and install it on the expanding tool. Remove the Teflon ring from the expanding tool and install it in a control valve groove. Wrap your hand around the Teflon ring and squeeze it into the groove (Figure 12-55). Follow the same procedure for each Teflon ring. When all the Teflon rings are installed on the control valve, coat them with power steering fluid and install the compressing tool over the rings (Figure 12-56).

**Special Tools**

Teflon ring compressing tool

## Power Rack and Pinion Steering Gear Assembly

1. Coat the new cylinder side oil seal with power steering fluid. Place one layer of vinyl tape on the seal driving tool extension, and install the seal and spacer on the tool. Press the cylinder side oil seal and spacer into the housing with the driving tool (Figure 12-57).
2. Check the rack surface and rack teeth for small metal burrs. Remove any burrs with crocus cloth, and wipe the rack with a clean shop towel. Place the rack installation tool over the rack and coat the tool with power steering fluid. Slide the rack and installation tool into the housing (Figure 12-58). The installation tool prevents seal damage. Remove the installation tool.
3. Coat a new rack bushing O-ring with power steering fluid, and install the O-ring in the bushing. Push the rack bushing and cylinder end stopper into the housing until the wire installation hole appears (Figure 12-59).
4. Insert the end of a new wire into the wire installation hole. Use the proper tool and a ratchet to rotate the cylinder end stopper clockwise until the outer wire end disappears in the wire installation hole (Figure 12-60).

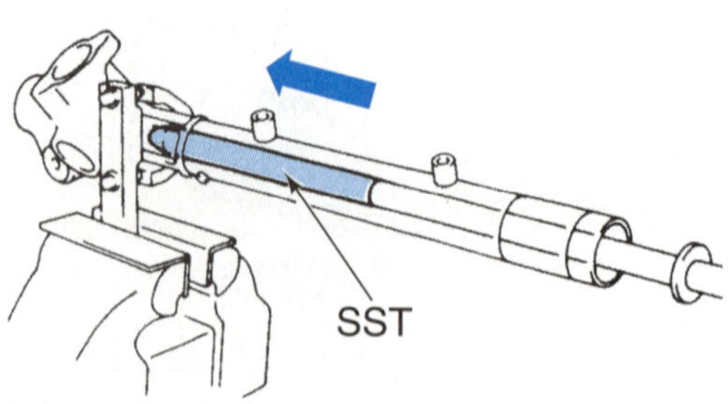


**Figure 12-58** Rack installation (Courtesy of Toyota Motor Corporation)

**Figure 12-59** Installing rack bushing and cylinder end stopper (Courtesy of Toyota Motor Corporation)

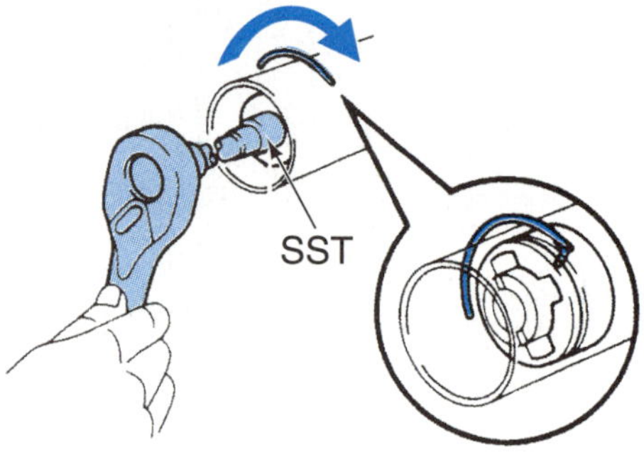


**Figure 12-60** Installing retaining wire (Courtesy of Toyota Motor Corporation)

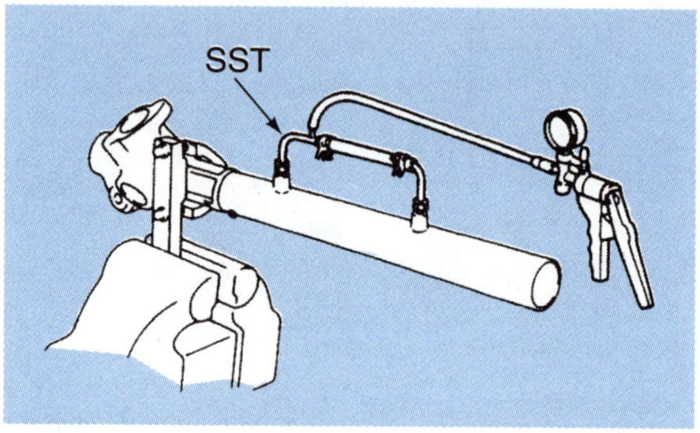


**Figure 12-61** Leak testing rack cylinders (Courtesy of Toyota Motor Corporation)

**Special Tools**

Leak test hose, power rack and pinion steering gear

Hand vacuum pump

5. Connect a leak test hose between the left and right turn ports on the steering gear housing. Attach a vacuum pump hose to the leak test hose, and operate the hand pump until 15.75 in Hg (53.3 kPa) is obtained on the vacuum gauge (Figure 12-61). After 30 seconds, observe the vacuum gauge. If the gauge reading decreased, the rack housing seal or rack bushing seal must be leaking and the rack should be removed to re-check these seals.
6. If the bushing and bracket was removed from the rack housing, install a new bushing in the bracket. Be sure the projection on the bushing is aligned with the bracket hole (Figure 12-62). Install the bracket and bushing on the housing, and align the marks placed on the housing and bracket during disassembly (Figure 12-63).

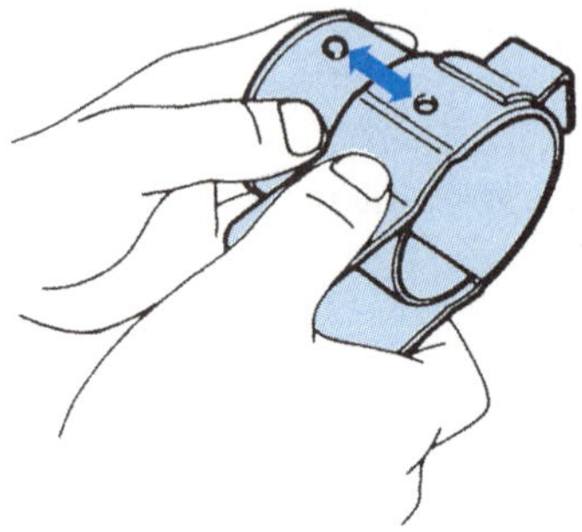

**Figure 12-62** Installing bushing in housing bracket (Courtesy of Toyota Motor Corporation)

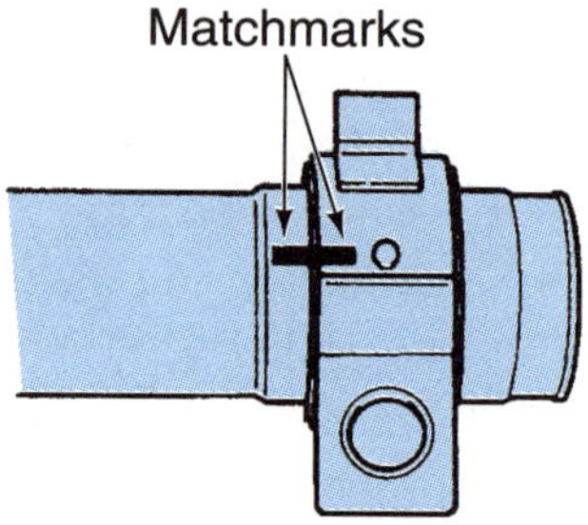


**Figure 12-63** Installing bracket and bushing on steering gear housing (Courtesy of Toyota Motor Corporation)

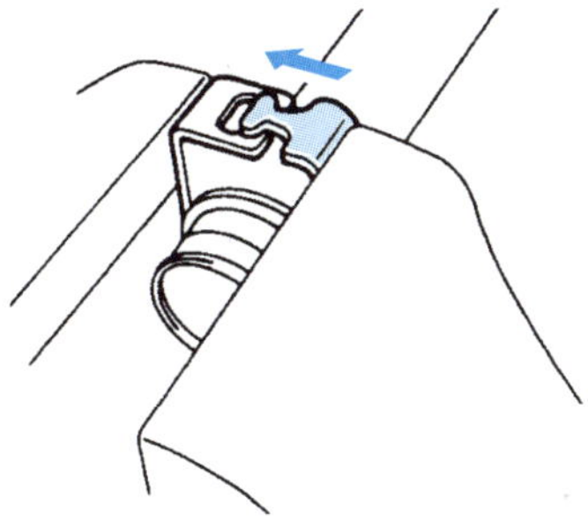

Figure 12-64 Fastening the bracket clasp (Courtesy of Toyota Motor Corporation)

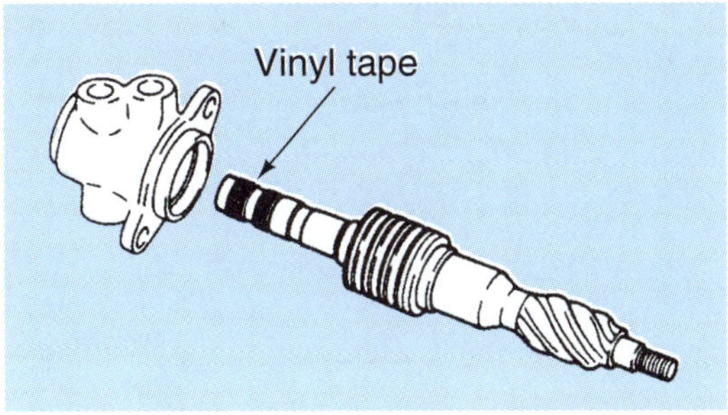


Figure 12-65 Installing control valve and pinion shaft assembly in control valve housing (Courtesy of Toyota Motor Corporation)

7. Place the housing bracket in a vise, and tighten the vise to fasten the bracket clasp (Figure 12-64).
8. Wind a layer of vinyl tape over the serrations on the upper end of the pinion shaft. Coat the Teflon rings with power steering fluid, and push the control valve and pinion shaft assembly into the housing (Figure 12-65). Coat a new oil seal with power steering fluid and press the seal into the lower side of the control valve housing with the appropriate seal driving tool (Figure 12-66).
9. Position the rack ends at the specified distance from the housing or at the distance from the housing measured during the disassembly procedure.
10. Install a new control valve housing gasket on the steering gear housing. Align the marks on the control valve and steering gear housings, and align the marks on the pinion shaft and housing. Install the control valve housing, and tighten the two housing bolts to the specified torque (Figure 12-67). Install the housing dust cover.
11. Install a new self-locking nut on the lower end of the pinion shaft. Use the proper tool to hold the upper end of the pinion shaft, and tighten the self-locking nut to the specified torque with a socket and foot-pound torque wrench (Figure 12-68).
12. Coat the first two or three threads on the housing cap with Loctite 242 sealant or its equivalent. Install the housing cap and tighten it to the specified torque (Figure 12-69). Use a center punch to stake the steering gear housing in two locations (Figure 12-70).
13. Coat the rack guide seat and rack guide with power steering fluid. Install the rack guide seat, rack guide, and spring (Figure 12-71). Apply Loctite 242 sealant or its equivalent to the first two or three threads on the rack spring cap. Install the rack spring cap and tighten it to the specified torque (Figure 12-72).

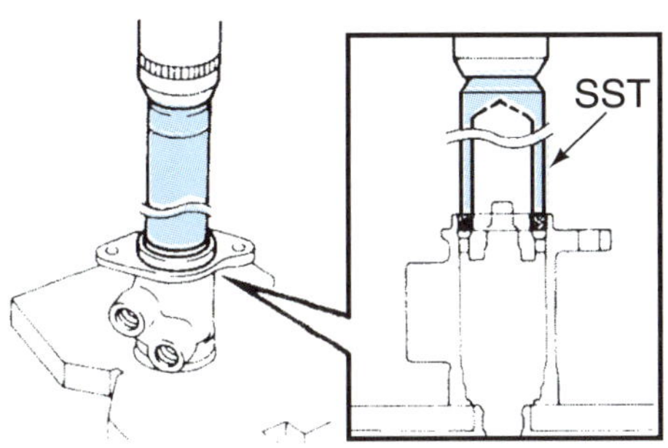


Figure 12-66 Installing oil seal in lower side of control valve housing (Courtesy of Toyota Motor Corporation)

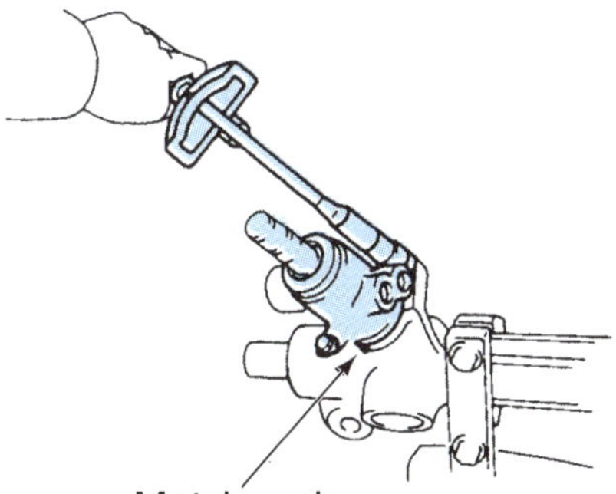


Figure 12-67 Installing control valve housing (Courtesy of Toyota Motor Corporation)

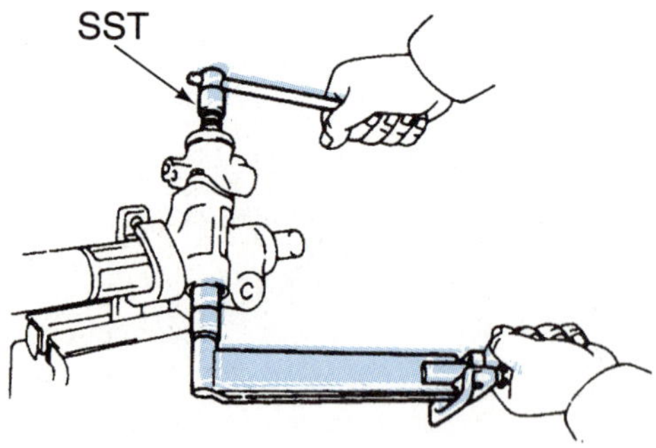


**Figure 12-68** Tightening self-locking nut on pinion shaft (Courtesy of Toyota Motor Corporation)

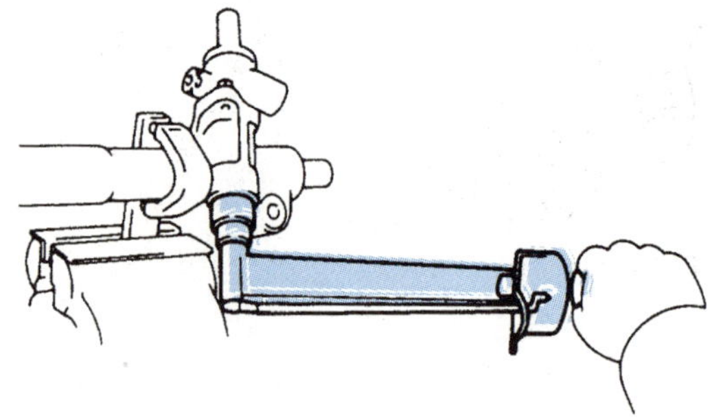

**Figure 12-69** Installing steering gear housing cap (Courtesy of Toyota Motor Corporation)

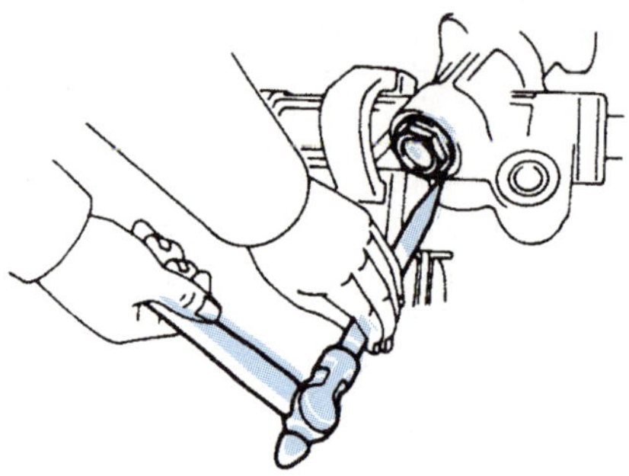

**Figure 12-70** Staking steering gear housing (Courtesy of Toyota Motor Corporation)

**Figure 12-71** Installing rack guide seat, rack guide, and spring (Courtesy of Toyota Motor Corporation)

**14.** Use the proper tool to rotate the rack spring cap 12° counterclockwise. Turn the pinion shaft fully in each direction, and repeat this action. Loosen the rack spring cap until there is no tension on the rack guide spring. Place the proper turning tool and a foot-pound torque wrench on top of the pinion shaft. Tighten the rack spring cap while rotating the pinion shaft back and forth (Figure 12-73). Continue tightening the rack spring cap until the specified turning torque is indicated on the torque wrench.

**15.** Apply Loctite 242 sealant or its equivalent to the first two or three threads on the lock nut. Install the lock nut on the rack spring cap. Use the proper tool to hold the rack spring cap, and tighten the lock nut to the specified torque (Figure 12-74).

**16.** Install a new claw washer on the inner tie rod end, and install the inner tie rod end on the rack. Hold the rack with an adjustable wrench, and use a torque wrench with the correct tool to tighten the inner tie rod end to the specified torque (Figure 12-75).

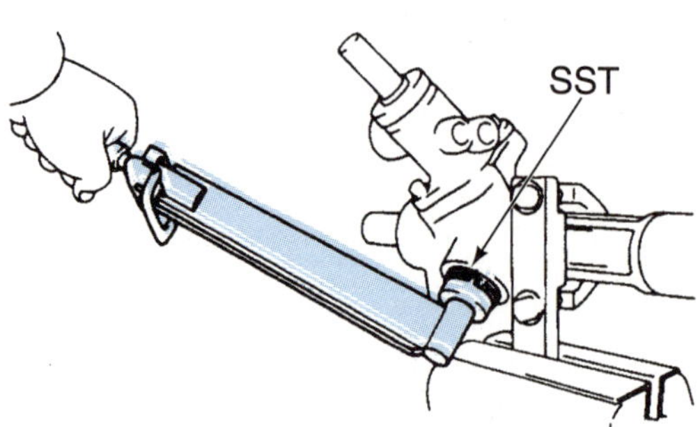


**Figure 12-72** Installing rack spring cap (Courtesy of Toyota Motor Corporation)

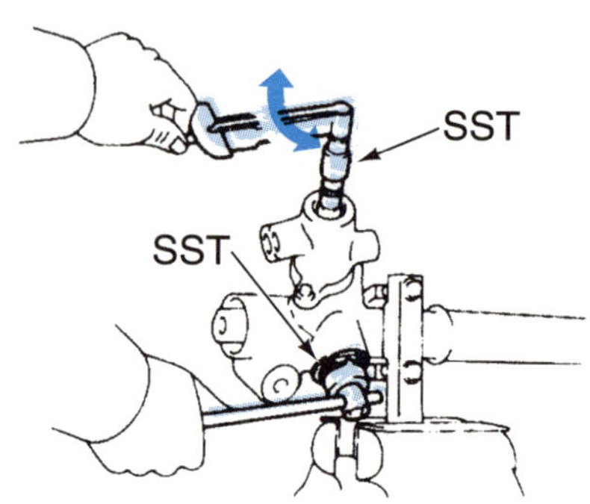


**Figure 12-73** Adjusting pinion shaft turning torque (Courtesy of Toyota Motor Corporation)

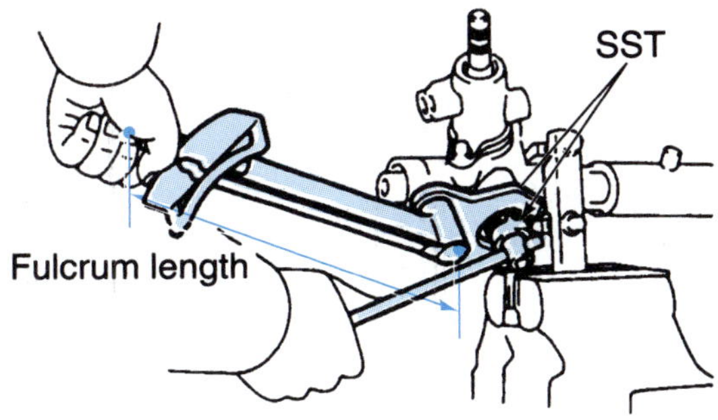


Figure 12-74 Tightening lock nut on rack spring cap (Courtesy of Toyota Motor Corporation)

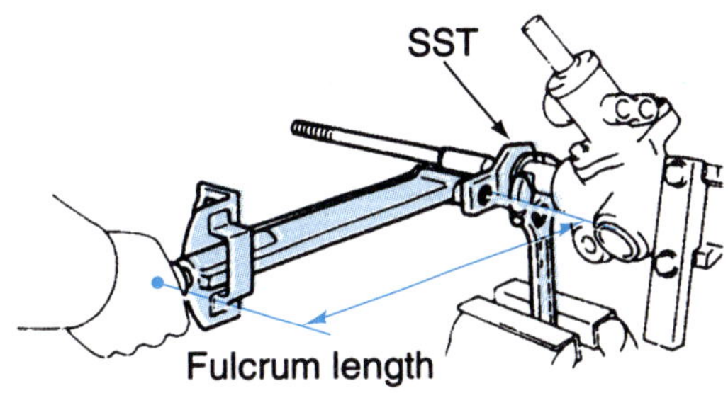


Figure 12-75 Tightening inner tie rod end on the rack (Courtesy of Toyota Motor Corporation)

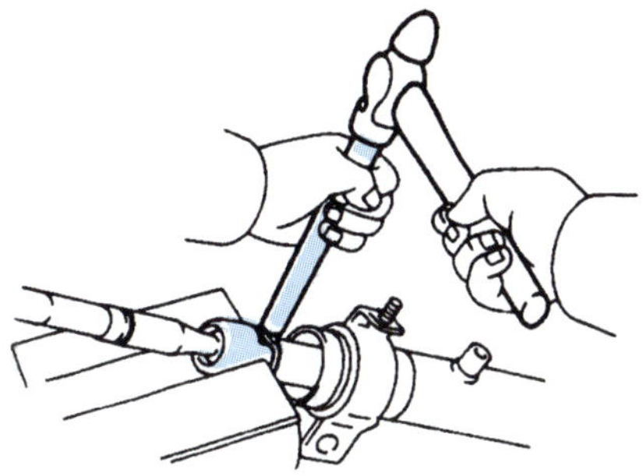

Figure 12-76 Staking claw washer on the inner tie rod end (Courtesy of Toyota Motor Corporation)

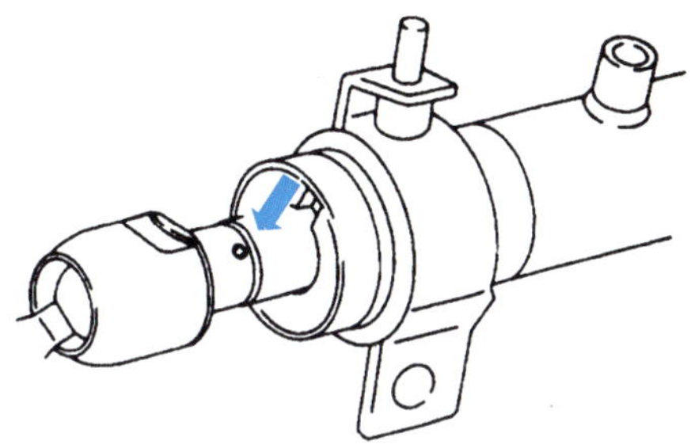

Figure 12-77 Rack vent holes must be open (Courtesy of Toyota Motor Corporation)

Clamp the inner tie rod end in a soft-jaw vise and stake the claw washer with a brass bar and a hammer (Figure 12-76). Repeat this procedure on both inner tie rod ends.

**17.** Check the vent hole in each end of the rack and be sure they are not clogged with grease or other material (Figure 12-77). These vent holes allow air flow from one boot to the other during a turn.

**18.** Install the bellows boots, clamps, and clips. When the inner boot clamps are tightened, there must be a minimum of 0.20 in (0.08 mm) between the ends of the clamps (Figure 12-78).

**19.** Install the outer tie rod ends and jam nuts. Align the marks placed on the tie rod ends, jam nuts, and tie rods during disassembly (Figure 12-79). Leave the jam nuts loose until the steering gear is installed and the front wheel toe is adjusted.

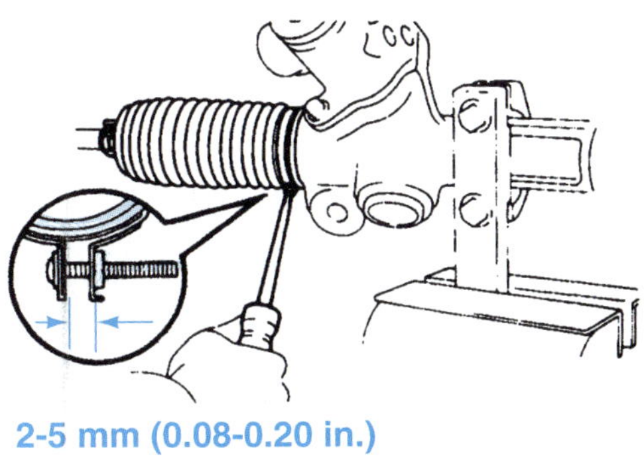


Figure 12-78 Minimum clearance between boot clamp ends (Courtesy of Toyota Motor Corporation)

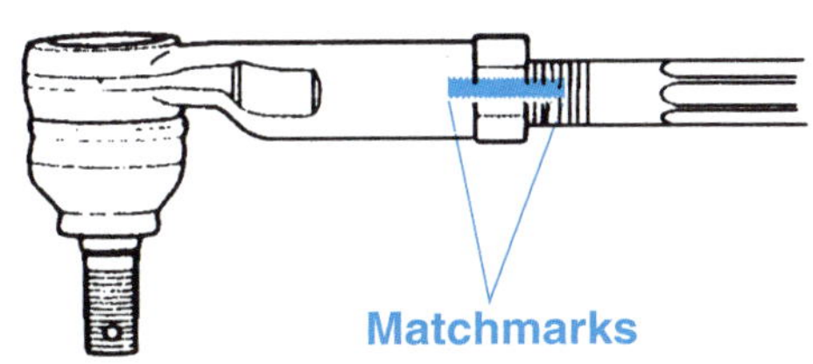


Figure 12-79 Aligning marks on outer tie rod ends, jam nuts, and tie rods (Courtesy of Toyota Motor Corporation)

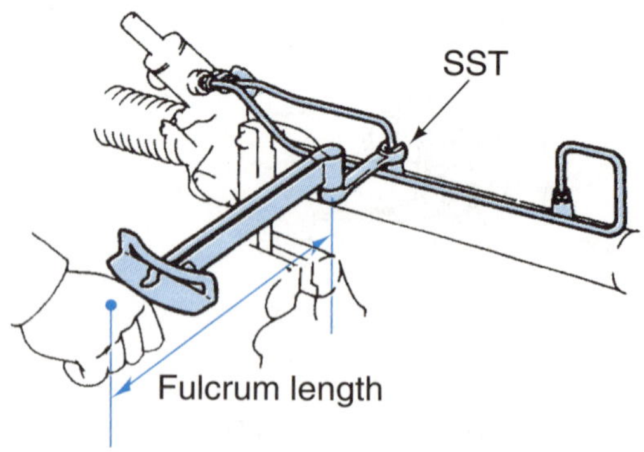


**Figure 12-80** Tightening turn tube fittings (Courtesy of Toyota Motor Corporation)

20. Coat the new turn tube O-rings with power steering fluid and install the O-rings on the right and left turn tubes. Tighten the left and right turn tube fittings to the specified torque with the proper wrench attached to a torque wrench (Figure 12-80).
21. Follow the steering gear installation procedure described previously in this chapter to install the steering gear in the vehicle. Tighten all fasteners to the specified torque. (See Chapter 10 for power steering pump filling and bleeding procedure.) Check the front wheel toe and tighten the outer tie rod end jam nuts to the specified torque. Tighten the outer bellows boot clamps. Road test the vehicle and check for proper steering operation and control.

Photo Sequence 10 shows a typical procedure for removing and replacing inner tie rod end, power rack and pinion steering gear.

## Photo Sequence 10
## Typical Procedure for Removing and Replacing Inner Tie Rod End, Power Rack and Pinion Steering Gear

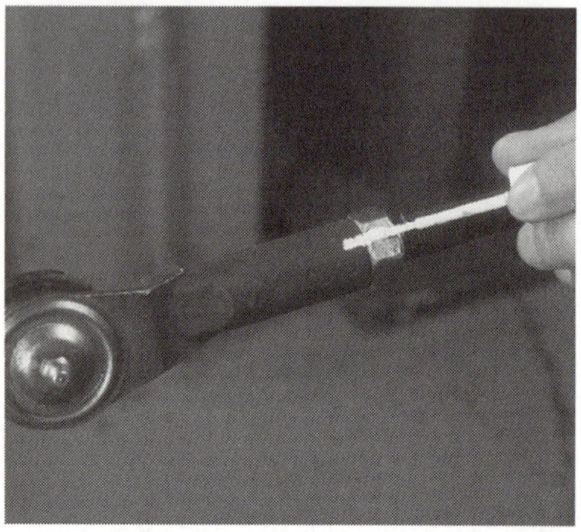

**P10-1** Place an index mark on the outer tie rod, jam nut, and tie rod.

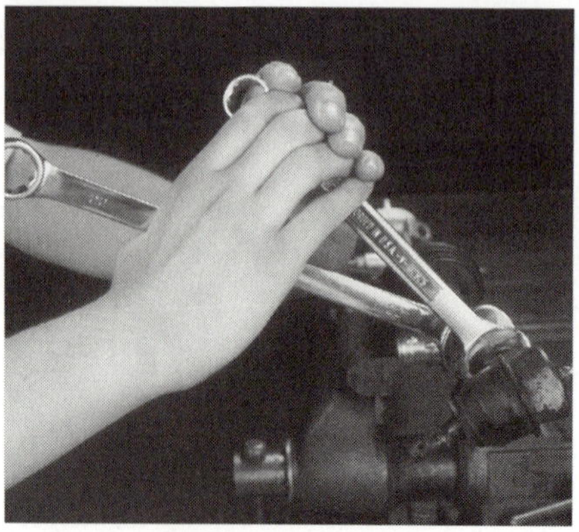

**P10-2** Loosen the jam nut, and remove the outer tie rod end and jam nut.

**P10-3** Remove the inner and outer boot clamps.

# Photo Sequence 10
## Typical Procedure for Removing and Replacing Inner Tie Rod End, Power Rack and Pinion Steering Gear (cont.)

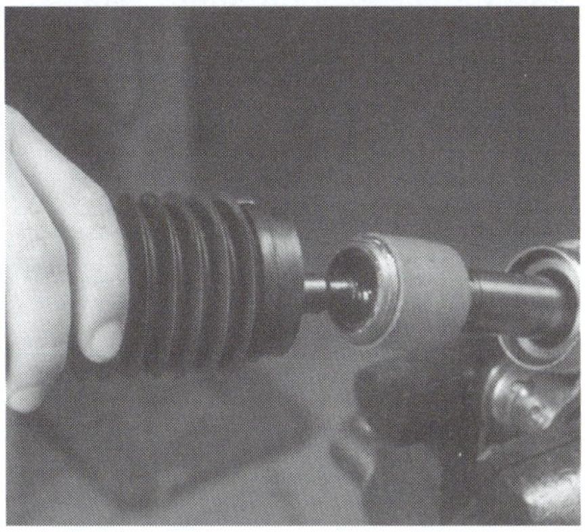

**P10-4** Remove the bellows boot from the tie rod.

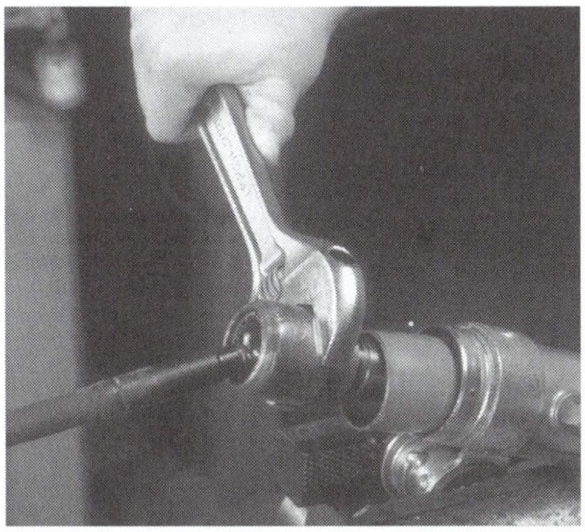

**P10-5** Hold the rack with the proper size wrench, and loosen the inner tie rod end with the proper wrench.

**P10-6** Remove the inner tie rod from the rack.

**P10-7** Be sure the shock damper ring is in place on the rack.

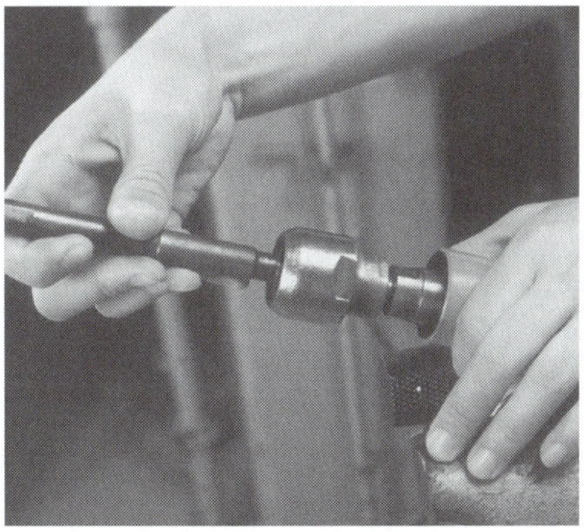

**P10-8** Install the inner tie rod on the rack and tighten the tie rod to the specified torque while holding the rack with the proper size wrench.

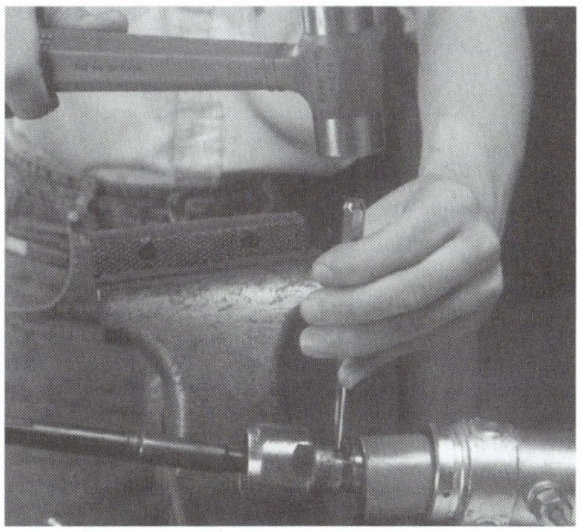

**P10-9** Support the inner tie rod on a vise and stake both sides of the inner tie rod joint with a hammer and punch.

**P10-10** Use a feeler gauge to measure the clearance between the rack and the inner tie rod joint housing stake.

**P10-11** Install the bellows boot and new clamps.

**P10-12** Install the jam nut and outer tie rod end with the index marks aligned.

## Diagnosis of Power Steering and Suspension Systems

Some power steering problems may be caused by defects in the steering gear, pump, or column. For example, heavy steering effort may be caused by a defective power steering pump, damaged rack piston and ring, or misalignment of the steering column. Therefore, we have combined the diagnoses of these components (Table 12-2).

**Table 12-2 POWER STEERING SYSTEM AND STEERING COLUMN DIAGNOSES**

| Condition | Possible Cause | Correction |
|---|---|---|
| Wander—Vehicle wander is a condition where the vehicle wanders side to side on the roadway when it is driven straight ahead, while the steering wheel is held in a firm position. Evaluation should be conducted on a level road (little road crown). | 1. Loose/worn tie rod ends or ball socket | 1. Replace tie rod end or tie rod assembly. |
| | 2. Inner ball housing loose or worn | 2. Replace tie rod assemblies. |
| | 3. Gear assembly loose crossmember | 3. Tighten the two mounting nuts to specification. |
| | 4. Excessive yoke clearance | 4. Adjust yoke clearance. |
| | 5. Loose suspension struts or ball joints | 5. Tighten or replace as required. |
| | 6. Column intermediate shaft connecting bolts loose | 6. Tighten bolts to specification at gear and column. |
| | 7. Column intermediate shaft joints loose or worn | 7. Replace intermediate shaft assembly. |
| | 8. Improper wheel alignment | 8. Set alignment to specification. |
| | 9. Tire size and pressure | 9. Check tire sizes and adjust tire pressure. |
| | 10. Vehicle unevenly loaded or overloaded | 10. Adjust load. |
| | 11. Steering gear mounting insulators and/or attachment bolts loose or damaged | 11. Replace insulators and/or attachment nuts and bolts. Tighten to specification. |
| | 12. Steering gear adjustments | 12. Refer to steering gear section of shop manual. |
| | 13. Front end misaligned | 13. Check and align to specifications. |
| | 14. Worn front-end parts or wheel bearings | 14. Inspect and replace affected parts. |
| | 15. Unbalanced or badly worn steering gear control valve | 15. Inspect and replace affected parts. |
| Pulls to one side—A condition in which the vehicle tends to pull to one side when driven on a level surface. | 1. Improper tire pressure | 1. Adjust tire pressure. |
| | 2. Improper tire size or different type | 2. Replace as required. |
| | 3. Vehicle unevenly or excessively loaded | 3. Adjust load. |
| | 4. Improper wheel alignment | 4. Adjust as required. |

**Table 12-2 POWER STEERING SYSTEM AND STEERING COLUMN DIAGNOSES (continued)**

| Condition | Possible Cause | Correction |
|---|---|---|
| | 5. Damaged front-suspension components | 5. Refer to front-suspension section of shop manual. |
| | 6. Damaged rear-suspension components | 6. Refer to rear-suspension section of shop manual. |
| | 7. Steering gear valve effort out of balance | 7. Place transmission in NEUTRAL while driving and turn engine off (coasting).<br>• If vehicle does not pull with the engine off, replace the steering gear valve assembly. Refer to steering gear section of shop manual.<br>• If vehicle does drift with engine off:<br>— Cross switch front tire wheel assemblies.<br>— If vehicle pulls to opposite side, cross switch tire/wheel assemblies that were on the rear to the same on the front.<br>— If vehicle pull direction is not changed, check front-suspension components and wheel alignment. |
| | 8. Front and rear brakes operation | 8. Adjust if necessary. |
| | 9. Bent rear axle housing, or damaged or sagging springs in the front and/or rear suspension | 9. Replace if necessary. |
| | 10. Rear suspension for loose/worn shock absorber or suspension attaching fasteners | 10. Tighten all attaching fasteners to specification. |
| Feedback—(Rattle, chuckle, knocking noises in the steering gear.) Feedback is a condition where roughness is felt in the steering wheel by the driver when the vehicle is driven over rough pavement. | 1. Column U-joints loose | 1. Replace if bad. |
| | 2. Loose tie rod ends | 2. Replace tie rod end assemblies. |
| | 3. Loose or worn tie rod ball | 3. Replace tie rod assemblies. |
| | 4. Gear assembly loose on cross-member | 4. Tighten the two mounting nuts to specification. |
| | 5. Loose pinion bearing cap | 5. Adjust cap to specification. |
| | 6. Loose yoke plug or lock nut | 6. Adjust yoke preload to specification, and tighten lock nut to specification. |

**Table 12-2 POWER STEERING SYSTEM AND STEERING COLUMN DIAGNOSES (continued)**

| Condition | Possible Cause | Correction |
|---|---|---|
| | 7. Loose pinion bearing lock nut | 7. Tighten lock nut to specification. |
| | 8. Piston disengaged or loose on rack | 8. Replace rack assembly. |
| | 9. Oversize pinion shaft bushing | 9. Replace gear housing. |
| | 10. Steering gear yoke worn | 10. Replace yoke assembly. |
| | 11. Column intermediate shaft connecting bolts loose | 11. Tighten bolts to specification. |
| | 12. Loose suspension bushings/ fasteners or ball joints | 12. Adjust or replace as necessary. |
| | 13. Column intermediate shaft joints loose/worn | 13. Replace intermediate shaft assembly. |
| | 14. Steering gear mounting insulators and/or attaching bolts loose or damaged | 14. Replace insulators and/or tighten attachment nuts to specification. |
| Poor steering wheel returnability—(Sticky feel.) Poor returnability is noticed when the steering fails to return to center following a turn without manual effort from the driver. In addition, when the driver returns the steering to center, it might have a sticky or catchy feel. | 1. Misaligned steering column or column flange rubbing steering wheel and/or flange | 1. Align column. Refer to steering column section of shop manual. |
| | 2. Check rotational torque of intermediate shaft joints. | 2. If binding, replace intermediate shaft. |
| | 3. Yoke plug too tight | 3. Adjust yoke preload to specification. |
| | 4. Tight inner toe rod ball joints | 4. Replace tie rod as required. |
| | 5. Tight tie rod end ball studs | 5. Replace tie rod end assemblies. |
| | 6. Undersized pinion shaft bushing in the housing | 6. Replace gear housing assemblies. |
| | 7. Binding in valve assembly | 7. Replace input shaft valve assembly. |
| | 8. Bent or damaged rack | 8. Replace rack assembly. |
| | 9. Bent or damaged crossmember | 9. Replace as necessary. |
| | 10. Column bearing binding | 10. Replace bearing. |
| | 11. Tight suspension struts or lower control arm ball joints | 11. Adjust or replace as required. |
| | 12. Improper wheel alignment | 12. Set to specification. Adjust toe as required. |
| | 13. Contamination in system | 13. Flush power steering system. |
| | 14. Deformed engine mounts | 14. Replace as required. |
| | 15. Improper tire pressure | 15. Adjust tire pressures. |
| | 16. Improper tire size or different type | 16. Replace as required. |
| | 17. Column intermediate shaft universal joints binding | 17. Replace intermediate shaft assembly. |

## Table 12-2 POWER STEERING SYSTEM AND STEERING COLUMN DIAGNOSES (continued)

| Condition | Possible Cause | Correction |
|---|---|---|
| | 18. Boot tears and/or binding or damage to tie rod ends or ball joints | 18. Replace as necessary. |
| | 19. Damaged/worn front suspension components | 19. Inspect struts and lower control arm ball joints. |
| | 20. Steering gear adjustments | 20. Refer to steering gear section of shop manual. |
| | 21. Front end misaligned | 21. Check and align to specifications. |
| | 22. Lack of lubrication in steering linkage or ball joints | 22. Inspect, lubricate, or replace affected parts. |
| | 23. Sticky or plugged steering gear spool valve | 23. Clean spool valve or replace. |
| | 24. Internal leakage in steering gear | 24. Inspect seals, replace as required, or overhaul steering gear. |
| Excessive steering wheel return or loose steering | 1. Air in the system | 1. Add oil to reservoir and bleed the system. Check all connections. |
| | 2. Excessive over-center lash | 2. Adjust to specifications. |
| | 3. Loose steering gear mounting | 3. Retorque mounting bolts to specifications. |
| | 4. Steering linkage worn or damaged | 4. Inspect and replace affected parts. |
| | 5. Steering coupler loose or damaged | 5. Inspect, tighten, or replace as required. |
| | 6. Loose thrust bearing preload adjustment | 6. Adjust to specifications. |
| | 7. Loose or damaged front wheel bearings | 7. Inspect, adjust, or replace as required. |
| Steering wheel—Surges or jerks with engine running, especially at slow speeds. | 1. Pump belt loose | 1. Tighten or replace belt. |
| | 2. Low oil level in reservoir | 2. Check level; fill as needed. |
| | 3. Engine idle too low. Air in the system | 3. Raise idle as required, check all hose connections and pump for leaks. |
| | 4. Insufficient pump pressure | 4. Check pump pressure. Replace affected parts or replace pump. |
| | 5. Flow control valve sticks | 5. Check valve for dirt or burrs. Replace as needed. |
| | 6. Steering linkage hitting obstruction | 6. Locate obstruction and repair. |
| | 7. Sticky flow control valve | 7. Inspect for varnish or damage, replace if necessary. |
| Momentary increase in steering wheel effort when turned rapidly | 1. Fluid level low in reservoir | 1. Check level; fill as needed. |
| | 2. Pump belt slipping | 2. Tighten or replace belt. |
| | 3. High internal leakage | 3. Check pump pressure. Replace affected parts or replace pump. |

**Table 12-2** **POWER STEERING SYSTEM AND STEERING COLUMN DIAGNOSES (continued)**

| Condition | Possible Cause | Correction |
|---|---|---|
| Heavy or hard steering effort—(Poor or loss of assist.) A heavy effort and poor assist condition is recognized by the driver while turning corners and especially while parking. A road test will verify this condition. | 1. Leakage or loss of fluid | 1. Refer to external diagnosis for service. |
| | 2. Low pump fluid | 2. Fill as necessary. |
| | 3. Pump flow/pressure not to specifications | 3. Refer to pump section of shop manual. |
| | 4. Valve plastic ring cut or twisted | 4. Replace ring. |
| | 5. Damaged or worn plastic piston ring | 5. Replace ring. |
| | 6. Loose or missing rubber backup piston O-ring | 6. Replace or install O-ring. |
| | 7. Loose rack piston | 7. Replace rack assembly. |
| | 8. Gear assembly oil passages restricted | 8. Clear or service as required. |
| | 9. Bent or damaged rack assembly | 9. Replace rack assembly. |
| | 10. Valve assembly internal leakage | 10. Replace valve assembly. |
| | 11. Improper drive belt tension | 11. Readjust belt tension. |
| | 12. Hose or cooler external leakage | 12. Replace as necessary. |
| | 13. Improper engine idle speed | 13. Readjust idle. |
| | 14. Pulley loose or warped | 14. Replace pulley. |
| | 15. Hose or cooler line restrictions | 15. Clear or replace as required. |
| | 16. System contamination | 16. Inspect system for foreign objects, kinked hose, etc.<br>—Flush system.<br>—Refer to power steering pump section of shop manual. |
| | 17. Plugged spool valve screen | 17. Prior to rebuilding a pump, examine the spool valve screen for contamination. Replace all valves that have plugged or contaminated valve screens. |
| | 18. Steering gear adjustments | 18. Refer to steering gear section of shop manual. |
| | 19. Low tire pressure | 19. Inflate to correct pressure. |
| | 20. Improper front suspension alignment | 20. Realign to specifications. |
| | 21. Low oil level in reservoir | 21. Add P/S oil to proper level. |
| | 22. Pump belt loose or glazed | 22. Tighten or replace belt. |
| | 23. Lack of lubricant in suspension or ball joints. Over-center adjustment too tight | 23. Lubricate or replace parts as needed. Readjust to specifications. |
| | 24. Steering gear coupler to column misaligned | 24. Align steering column. |

## Table 12-2 POWER STEERING SYSTEM AND STEERING COLUMN DIAGNOSES (continued)

| Condition | Possible Cause | Correction |
|---|---|---|
| Fluid leakage | 1. Overfilled system<br>2. Component leakage | 1. Correct fluid level as required.<br>2. Locate suspected component, and refer to appropriate section for service. |
| Power steering pump leaks | 1. Excessive fluid fill | 1. Adjust fluid to proper level. |
| | 2. Dipstick missing, loose, or damaged or missing O-ring | 2. Service or replace if required. |
| | 3. Broken/cracked reservoir | 3. Replace reservoir. |
| | 4. Loose or damaged hose fittings | 4. Service or replace. |
| | 5. Leakage at shaft seal | 5. Service shaft seal. |
| | — Shaft seal not pressed in flush with housing surface | — Use an appropriate seal installation tool to correct installation; if not possible or still leaking, replace seal. |
| | — Seal damage | — Replace seal. |
| | — Rotor shaft damage, such as helical grooving, or the OD has an axial scratch | — Replace shaft and seal. |
| | — Worn shaft bushing | — Replace plate and bushing assembly. |
| | — Plugged drainback hole | — Disassemble pump and clean drainback hole. If hole is not drilled through, replace plate and bushing assembly. |
| | 6. Damaged or missing reservoir O-ring | 6. Replace O-ring. |
| | 7. Damaged or missing outlet fitting O-rings | 7. Replace O-rings. |
| | 8. Loose outlet fitting | 8. Tighten outlet fitting. |
| | 9. Excessive pump assembly bracket vibration | 9. Correct bracket alignment and tighten bracket bolts to specification. |
| | 10. Plate and bushing reservoir seal groove porosity/damage, metal chips, or foreign material on seal or in seal groove | 10. Clean groove; replace plate and bushing assembly if damaged. |
| | 11. Outlet fitting damage | 11. Replace outlet fitting. |
| Noise—In system: chirp or squeal when steering wheel is cycled lock to lock/Pump: whine or moan | 1. Loose or worn pump belt | 1. Adjust to specification, or replace as required. |
| | 2. Fluid aeration | 2. Purge system of air. Refer to purging procedures in shop manual. |
| | 3. Low fluid | 3. Check fluid level. Correct as required. |

**Table 12-2** **POWER STEERING SYSTEM AND STEERING COLUMN DIAGNOSES (continued)**

| Condition | Possible Cause | Correction |
| --- | --- | --- |
| | 4. Pump brackets loose or misaligned | 4. Check bracket(s), bolt torques, and bracket alignment. Correct as required. |
| | 5. Power steering hose(s) grounded | 5. Check for component grounding, and correct as required. |
| | 6. Fluid level low in reservoir | 6. Check level and fill as needed. |
| | 7. Air in the system | 7. Check hose connections and bleed the system. |
| | 8. Pump shaft bushing or bearing is scored or worn | 8. Inspect pump shaft, bushing, or bearing. Replace affected parts. |
| Noise/rattle—Steering column | 1. Loose bolts/attaching brackets | 1. Tighten to specification. |
| | 2. Looseness of ball bearings or insufficient lube | 2. Lube or replace bearings. |
| | 3. Steering shaft insulators cracked or dry | 3. Replace or lube insulators as required. |
| | 4. Flex coupling compressed or extended | 4. Reposition shaft assembly to flatten flex coupling. |
| | 5. Pressure hose coming in contact with other parts of the car | 5. Reposition power-steering pressure hose. |
| | 6. Loose pump or steering gear mounting | 6. Retorque to specifications. |
| | 7. Loose steering linkage | 7. Inspect steering linkage for excessive play; replace affected parts as necessary. |
| | 8. High point on steering gear improperly adjusted | 8. Readjust to proper setting. |
| | 9. Improperly installed pump vanes | 9. Remove burrs and varnish from rotor and vanes. Replace as necessary. |
| | 10. Pump vanes sticking in rotor slots | 10. Remove burrs and varnish from rotor and vanes. Replace as necessary. |
| Noise/squeak or cracks—Steering column | 1. Dry bushings | 1. Lube shaft seal and shift tube seal. |
| | 2. Loose or mispositioned shrouds | 2. Tighten or reposition shrouds as required. |
| | 3. Steering wheel rubbing against shrouds | 3. Replace shroud(s) or wheel or reposition shrouds as required. |
| | 4. Dry shift lever grommets | 4. Lube grommets. |
| | 5. Insufficient lube on speed control slip ring | 5. Lube slip ring. |
| | 6. Upper or lower bearing sleeve out of position | 6. Reposition bearing sleeve. |

## Table 12-2 POWER STEERING SYSTEM AND STEERING COLUMN DIAGNOSES (continued)

| Condition | Possible Cause | Correction |
|---|---|---|
| Gear squawk | Cut damper O-ring on spool valve | Replace the O-ring. |
| Growling | 1. Restriction in the power-steering system<br>2. Extreme wear on cam ring<br>3. Scored pump pressure plate, thrust plate, or rotor | 1. Locate restriction and correct. Replace parts as necessary.<br>2. Replace affected parts.<br>3. Replace affected parts. |
| Click | A noise caused by the pump slippers being too long, broken slipper springs, excessive wear, nicked rotors, and damaged cam contour | Check for excessive wear, or slipper, rotor, or cam damage. Replace cam rotor assembly if necessary. |
| Power steering pump—Noise, moan, or whine | 1. Fluid aeration<br>2. Low fluid<br>3. Hose grounded<br>4. Column grounded<br>5. Valve cover O-ring missing/damaged<br>6. Valve cover baffle missing/damaged<br>7. Interference of components in pumping elements<br>8. Loose or poor bracket alignment | 1. Purge the power steering system to reduce aeration noise. Refer to general steering service section purging procedure of shop manual.<br>2. Check fluid level.<br>3. Check for hose being grounded.<br>4. Check column alignment.<br>5. Check valve cover O-ring damage. Replace if necessary.<br>6. Check valve cover for baffle missing/damage. Replace if necessary.<br>7. Check rotor assembly and pressure plates for indication of interference. Replace if necessary.<br>8. Check bracket alignment and bolt torque. Correct if required. |
| Swish | 1. Noise created by the flow of excessive fluid into the bypass port of the pump valve housing (with temperature below 130°). The shearing effect of the cooler (heavier) oil is not detrimental to pump operation.<br>2. Defective pump control valve | 1. A normal condition. Noise will diminish with fluid temperature increase.<br>2. Inspect valve for burrs or nicks. Replace as necessary. |
| Gear hiss—Some hissing noise is normal in all power-steering systems. Under certain conditions, such as tight parking maneuvers or | 1. Column intermediate shaft alignment | 1. Adjust column and/or intermediate shaft. Refer to steering column section of shop manual. |

## Table 12-2 POWER STEERING SYSTEM AND STEERING COLUMN DIAGNOSES (continued)

| Condition | Possible Cause | Correction |
|---|---|---|
| turning the wheel from stop to stop, a hissing sound is most noticeable. There is no direct relationship between the hissing noise and performance of the steering system. | 2. Grounded or loose boot at dash panel<br>3. Input shaft and control valve assembly | 2. Align boot or tighten fasteners as required. Refer to steering column section of shop manual.<br>3. Replace the pump control valve only if the hissing noise is extremely objectionable. Check the coupler on the steering column for any metal-to-metal contact that may transmit the sound to the passenger compartment. |

**CUSTOMER CARE:** Diagnosing is an extremely important part of a technician's job on today's high-tech vehicles. Always take time to diagnose a customer's vehicle accurately. Fast, inaccurate diagnosis of automotive problems leads to unnecessary, expensive repairs, and unhappy customers who may take their business to another shop. Accurate diagnosis may take more time, but in the long term it will improve customer relations, and bring customers back to the shop.

# Guidelines for Servicing Manual and Power Rack and Pinion Steering Gears

1. During a manual or power rack and pinion steering gear on-car inspection, components to be checked include the bellows boots, inner and outer tie rod ends, tie rods, mounting bushings, flexible coupling or universal joint, and tires.
2. Since bellows boots protect the inner tie rod ends from contamination, boot condition should be checked each time undercar service such as oil and filter change or chassis lubrication is performed.
3. Bellows boots that are cracked, split, leaking, or deteriorated must be replaced.
4. Do not straighten bent steering components such as tie rods.
5. If there is any movement, or looseness, in the inner tie rod ends, these ends must be replaced.
6. Outer tie rod ends with any looseness or cracked seals must be replaced.
7. Prior to steering gear removal, the steering wheel should be centered and the ignition key removed from the ignition switch. This action locks the steering column and maintains the clock spring, or spiral cable, in the centered position on an air-bag-equipped vehicle. The driver's seat belt may be wrapped through the steering wheel to maintain the steering wheel centered position if the ignition switch is turned on.
8. The flexible coupling or lower universal joint and the steering gear pinion shaft should be punch marked prior to steering gear removal.
9. Prior to outer tie rod end removal, a dab of paint may be used to mark the jam nut position on the tie rods. An alternate method of marking this position is to wrap masking tape around the tie rod directly behind the jam nut.
10. Always hold the rack while loosening the inner tie rod ends.

11. Prior to steering gear disassembly, the ends of the rack should be positioned at the specified distance from the steering gear housing and the pinion shaft position should be marked in relation to the housing. If this rack specification is not available, center the steering gear and measure the distance from the ends of the rack to the housing. After recording this distance, mark the pinion shaft in relation to the housing.
12. The rack adjuster plug must be adjusted until the correct turning torque is obtained on the pinion shaft.
13. The outer bellows boot clamps and the outer tie rod end jam nuts should not be tightened until the steering gear is installed in the vehicle and the front wheel toe checked.

## CASE STUDY

A customer complains about excessive steering effort on a 1993 Olds 88. The technician road tested the car and found no evidence of hard steering. Further questioning of the customer revealed that the problem only occurred when the engine was first started in the morning, after the car had been parked all night. The service writer informed the customer that the car would have to be parked overnight in the shop, and then diagnosed the following morning. The customer complied with this request.

Prior to starting the engine the next morning, the technician connected a power steering pressure test gauge in series with the high-pressure hose. When the engine was started, the technician discovered the steering effort was very high, and the steering felt much like manual steering while the steering wheel was turned. However, the power steering pump pressure on the pressure gauge was higher than specified. Once the steering wheel was turned in one direction, the steering effort quickly decreased to normal. Since the power steering pump pressure was higher than specified, the technician concluded the pump and drive belt must be in satisfactory condition. The technician also reasoned that other causes of excessive steering effort, such as binding in the steering column or flexible coupling, would be constant. Therefore, the technician decided that the problem must be in the steering gear.

After receiving the customer's approval to remove and inspect the steering gear, the gear was removed and disassembled. The technician found the rack cylinder was severely ridged and scored in the center area, and the Teflon ring on the rack piston was badly worn. The customer was advised that the steering gear required replacement. A replacement steering gear was installed, and the system was flushed and refilled with power steering fluid. When the engine was started, the steering effort and the power steering pump pressure were normal.

## Terms to Know

Steering wheel free-play
Bellows boots
Pinion and bearing assembly
Pilot bearing
Rack bushing
Anhydrous calcium grease
Lithium-based grease
Turning imbalance
Soft-jaw vise
Self-locking nut
Cylinder end stopper
Teflon ring expander
Teflon ring compressing tool
Claw washer

## ASE Style Review Questions

1. While discussing steering wheel free-play:
*Technician A* says excessive steering wheel free-play may be caused by a worn outer tie rod end.
*Technician B* says excessive steering wheel free-play may be caused by loose steering gear mounting bushings.
Who is correct?
**A.** A only
**B.** B only
**C.** Both A and B
**D.** Neither A nor B

2. While discussing steering effort:
*Technician A* says a tight rack bearing adjustment may cause excessive steering effort.
*Technician B* says a loose inner tie rod end may cause excessive steering effort.
Who is correct?
**A.** A only
**B.** B only
**C.** Both A and B
**D.** Neither A nor B

3. While discussing steering gear removal and replacement on air-bag-equipped vehicles:
*Technician A* says the steering column should be locked in the centered position to prevent air bag deployment.
*Technician B* says the steering column should be locked in the centered position to keep the clock spring in the centered position.
Who is correct?
**A.** A only
**B.** B only
**C.** Both A and B
**D.** Neither A nor B

4. While discussing power rack and pinion steering gear adjustment:
*Technician A* says the self-locking nut on the pinion shaft should be rotated to adjust the pinion shaft turning torque.
*Technician B* says the rack spring cap should be rotated to obtain the specified turning torque on the pinion shaft.
Who is correct?
**A.** A only
**B.** B only
**C.** Both A and B
**D.** Neither A nor B

5. While discussing manual rack and pinion steering gear lubrication:
*Technician A* says the steering gear should be filled with SAE 90 lubricant.
*Technician B* says the rack and rack teeth should be coated with lithium-based grease.
Who is correct?
**A.** A only
**B.** B only
**C.** Both A and B
**D.** Neither A nor B

6. While discussing power rack and pinion steering gear oil leaks:
*Technician A* says the inner rack seal is defective if oil leaks from the pinion end of the housing when the rack contacts the left internal stop.
*Technician B* says the pinion seal is leaking if oil leaks from the pinion end of the housing and the leak is not affected when the steering wheel is rotated.
Who is correct?
**A.** A only
**B.** B only
**C.** Both A and B
**D.** Neither A nor B

7. While discussing turning imbalance on a power rack and pinion steering gear:
*Technician A* says a leaking rack piston seal may cause steering effort imbalance.
*Technician B* says worn rings on the rotary valve may cause low power assist while turning the steering wheel in either direction.
Who is correct?
**A.** A only
**B.** B only
**C.** Both A and B
**D.** Neither A nor B

8. While discussing power rack and pinion steering gear disassembly:
*Technician A* says the rack should be held while the inner tie rod ends are loosened.
*Technician B* says claw washers retain the outer tie rod ends to the tie rods in some steering gears.
Who is correct?
**A.** A only
**B.** B only
**C.** Both A and B
**D.** Neither A nor B

9. While discussing the diagnosis of a bent rack in a power rack and pinion steering gear:
*Technician A* says a straight edge may be placed against the rack to check for a bent condition.
*Technician B* says a bent rack may cause excessive steering wheel free-play.
Who is correct?
**A.** A only
**B.** B only
**C.** Both A and B
**D.** Neither A nor B

10. While discussing power rack and pinion steering gear service:
*Technician A* says Teflon control valve rings may be expanded with an expansion tool.
*Technician B* says Teflon control valve rings should be heated with a propane torch to expand them.
Who is correct?
**A.** A only
**B.** B only
**C.** Both A and B
**D.** Neither A nor B

## Table 12-3 ASE TASK

Diagnose manual rack and pinion steering gear noises, vibration, looseness, and hard steering problems; determine needed reapairs.

| Problem Area | Symptoms | Possible Causes | Classroom Manual | Shop Manual |
|---|---|---|---|---|
| NOISES | Knocks, clunks, rattles | 1. Loose steering gear mounting bolts | 245 | 267 |
| | | 2. Flexible coupling loose, or worn | 245 | 268 |
| | | 3. Universal joint loose or worn | 245 | 268 |
| | | 4. Loose tie rod ends | 246 | 269 |
| | | 5. Lack of lubrication on horn bushing contact | 246 | 275 |
| STEERING CONTROL | Looseness, wander | 1. Worn inner or outer tie rod ends | 246 | 268 |
| | | 2. Worn steering gear mounting bushings | 247 | 268 |
| | | 3. Improper steering gear adjustment | 245 | 274 |
| | | 4. Worn flexible coupling or universal joint | 245 | 268 |
| STEERING EFFORT | Hard steering | 1. Tight steering gear adjustment | 245 | 274 |
| | | 2. Lack of lubrication in steering gear | 245 | 274 |

## Table 12-4 ASE TASK

Remove and replace manual rack and pinion steering gear.

| Problem Area | Symptoms | Possible Causes | Classroom Manual | Shop Manual |
|---|---|---|---|---|
| STEERING CONTROL | Lack of steering control, wander | 1. Worn inner tie rod ends | 245 | 275 |
| | | 2. Worn rack or pinion teeth | 245 | 268 |
| | | 3. Improper steering gear adjustment | 245 | 273 |
| | Binding, uneven turning effort | Worn rack or pinion teeth | 245 | 273 |

## Table 12-5 ASE TASK

Adjust manual rack and pinion steering gear.

| Problem Area | Symptoms | Possible Causes | Classroom Manual | Shop Manual |
|---|---|---|---|---|
| STEERING CONTROL | Looseness | Loose steering gear adjustment | 246 | 274 |
| | Hard steering | Tight steering gear adjustment | 246 | 274 |

## Table 12-6 ASE TASK

Inspect and replace manual rack and pinion steering gear inner tie rod ends (sockets) and bellows boots.

| Problem Area | Symptoms | Possible Causes | Classroom Manual | Shop Manual |
|---|---|---|---|---|
| STEERING CONTROL | Looseness | Worn inner tie rod ends | 246 | 274 |
| PREMATURE INNER TIE ROD END FAILURE | Looseness | Cracked or split bellows boots | 246 | 274 |

## Table 12-7 ASE TASK

Inspect and replace rack and pinion steering gear mounting bushings and brackets.

| Problem Area | Symptoms | Possible Causes | Classroom Manual | Shop Manual |
|---|---|---|---|---|
| STEERING CONTROL | Looseness | Worn steering gear mounting bushings | 247 | 268 |

# Table 12-8 ASE TASK

Diagnose power rack and pinion steering gear noises, vibration, looseness, hard steering, and fluid leakage problems; determine needed repairs.

| Problem Area | Symptoms | Possible Causes | Classroom Manual | Shop Manual |
|---|---|---|---|---|
| **NOISES** | Rattles, clunks, knocks, squeaks | 1. Loose steering gear mounting bolts | 248 | 300 |
| | | 2. Worn steering gear mounting bushings | 248 | 300 |
| | | 3. Loose tie rod ends | 251 | 299 |
| | | 4. Worn flexible coupling or universal joints | 251 | 299 |
| | | 5. Lack of lubrication on horn bushing contact | 251 | 304 |
| **STEERING CONTROL** | Looseness | 1. Worn inner or outer tie rod ends | 251 | 298 |
| | | 2. Worn steering gear mounting bushings | 251 | 298 |
| | | 3. Worn flexible coupling or universal joint | 251 | 301 |
| | | 4. Loose steering gear adjustment | 251 | 301 |
| | Hard steering | 1. Tight steering gear adjustment | 251 | 302 |
| | | 2. Low fluid level | 251 | 302 |
| **LOW FLUID LEVEL** | Fluid leaks | 1. Cylinder end rack seal | 251 | 281 |
| | | 2. Pinion end rack seal | 252 | 281 |
| | | 3. Pinion seal, upper or lower | 255 | 282 |

## Table 12-9 ASE TASK

Remove and replace power rack and pinion steering gear; inspect and replace mounting bushings and brackets.

| Problem Area | Symptoms | Possible Causes | Classroom Manual | Shop Manual |
|---|---|---|---|---|
| STEERING CONTROL | Looseness, wander | 1. Worn inner tie rod ends | 251 | 298 |
| | | 2. Improper steering gear adjustment | 251 | 298 |
| | | 3. Worn rack or pinion teeth | 251 | 298 |
| | | 4. Worn steering gear mounting bushings | 251 | 298 |
| | Uneven steering effort | Worn rack or pinion teeth | 251 | 281 |
| | Steering effort imbalance | 1. Leaking rack seals | 248 | 283 |
| | | 2. Worn control valve Teflon rings | 248 | 284 |

## Table 12-10 ASE TASK

Adjust power rack and pinion steering gear.

| Problem Area | Symptoms | Possible Causes | Classroom Manual | Shop Manual |
|---|---|---|---|---|
| STEERING CONTROL | Looseness | 1. Loose steering gear adjustment | 252 | 294 |
| | Hard steering | 1. Tight steering gear adjustment | 253 | 294 |

## Table 12-11 ASE TASK

Inspect and replace power rack and pinion steering gear inner tie rod ends (sockets), seals, gaskets, O-rings, and bellows boots.

| Problem Area | Symptoms | Possible Causes | Classroom Manual | Shop Manual |
|---|---|---|---|---|
| **STEERING CONTROL** | Looseness | Worn inner tie rod ends | 251 | 295 |
| **LOW FLUID LEVEL** | Fluid leaks | 1. Worn rack or pinion seals | 248 | 283 |
| | | 2. Defective control valve housing gasket | 255 | 286 |
| **PREMATURE INNER TIE ROD END FAILURE** | Looseness | Cracked or split bellows boots | 255 | 295 |

CHAPTER 13

# Electronic Four Wheel Steering Service and Diagnosis

Upon completion and review of this chapter, you should be able to:

- ❑ Perform a preliminary inspection on a four wheel steering (4WS) system.
- ❑ Perform a trouble code diagnosis on a 4WS system with the ignition switch on.
- ❑ Perform a trouble code diagnosis on a 4WS system with the engine running.
- ❑ Remove and replace the rear steering actuator.
- ❑ Remove and replace the tie rod ends on the rear steering actuator.
- ❑ Remove and replace tie rods and boots on the rear steering actuator.
- ❑ Remove and replace the rear main steering angle sensor.
- ❑ Remove and replace the rear sub steering angle sensor.
- ❑ Perform an electronic neutral check on a 4WS system.
- ❑ Adjust the front main steering angle sensor.
- ❑ Adjust the front sub steering angle sensor.
- ❑ Adjust the rear sub steering angle sensor.

**Basic Tools**

Basic technician's tool set

Service manual

Jumper wire

Floor jack

Jack stands

Wax marker

Silicone grease

Chassis lubricant

Cotter keys

Length of stiff wire

## Preliminary Inspection

Prior to any four wheel steering diagnosis, these concerns should be considered:

1. Have any suspension modifications been made that would affect steering?
2. Are the tire sizes the same as specified by the vehicle manufacturer?
3. Are the tires inflated to the pressure specified by the vehicle manufacturer?
4. Is the power steering belt adjusted to the vehicle manufacturer's specified tension?
5. Is the power steering pump reservoir filled to the proper level with the type of fluid specified by the vehicle manufacturer?
6. Is the engine idling at the speed specified by the vehicle manufacturer, and is the idle speed steady?
7. Is the steering wheel original equipment?

The fail-safe mode is entered by the four wheel steering control unit if a defect occurs in the system. In this mode, the rear wheels move to the centered position.

## Fail-Safe Function

If the four wheel steering (4WS) control unit senses a failure in the system, the control unit switches to a fail-safe mode. In this mode, the control unit stores a trouble code, or codes, and illuminates the 4WS indicator light to inform the driver that a problem exists in the system. When this mode is entered, the 4WS control unit shuts off voltage to the rear steering unit, and the rear wheels remain in the straight-ahead position.

The Society of Automotive Engineers (SAE) J1930 terminology is an attempt to standardize electronics terminology in the automotive industry.

## Damper Control

In the SAE J1930 terminology, the term malfunction indicator light (MIL) replaces other terms for computer system indicator lights.

When the 4WS control unit enters the fail-safe mode, a quick return of the rear wheels to the straight-ahead position would adversely affect steering under certain steering wheel and rear wheel positions. To prevent this action, the 4WS control unit energizes the damper relay when it enters the fail-safe mode. The rear steering actuator motor is spun by the steering shaft movement as this shaft is moved to the centered position by centering spring force. This action causes the motor armature to act as a voltage generator, and the voltage generated by the armature is fed back through the damper relay to the motor armature. Under this condition, the motor rotation is slowed, and the return spring slowly moves the rear steering shaft to the straight-ahead position. Without the action of the damper relay, the return spring would move the rear steering shaft quickly to the straight-ahead position.

## Trouble Code Diagnosis

### Road Test

**CUSTOMER CARE:** While discussing the customers' automotive problems, always remain polite, and never make statements that make customers feel uninformed about their vehicles.

The 4WS control unit stores a fault code and illuminates the 4WS indicator light if a defect occurs in the system, even if the defect is temporary. Always ask the customer about the conditions that caused the 4WS indicator light to come on, and duplicate this condition during a road test. If the 4WS light is not illuminated during the road test, the system is satisfactory electronically and does not require further electronic diagnosis. The troubleshooting procedures in the vehicle manufacturer's service manual assume that the problem is present at the time of diagnosis.

### Trouble Code Display with Ignition Switch On

**WARNING:** When diagnosing a computer system, never connect or ground any terminals, unless instructed to do so in the vehicle manufacturer's service manual.

**WARNING:** When diagnosing a computer system, never disconnect or connect any computer system component with the ignition switch on unless instructed to do so in the vehicle manufacturer's service manual. This action may damage the computer or system components.

**WARNING:** When performing electronic diagnosis on a vehicle equipped with an air bag, most vehicle manufacturers recommend turning the ignition switch off, disconnecting the negative battery cable, and waiting one minute before proceeding with electronic component diagnosis or service.

**CAUTION:** When performing electronic diagnosis on a vehicle equipped with an air bag, follow all the service precautions recommended in the vehicle manufacturer's service manual.

Always follow the exact 4WS service and diagnostic procedures in the vehicle manufacturer's service manual. These procedures vary depending on the make and year of the vehicle. Following are typical procedures for a Honda Prelude. The following procedures should be avoided until after the diagnosis is complete, because any of these procedures will erase trouble codes.

1. Disconnect the battery terminals.
2. Disconnect the 4WS control unit connector.
3. Remove the number 43 clock-radio 10-A fuse from the underhood fuse/relay box.

Follow these steps to obtain the trouble codes:

1. Remove the dual-terminal service check connector located behind the center console, and connect the two terminals in this connector with a jumper wire (Figure 13-1).

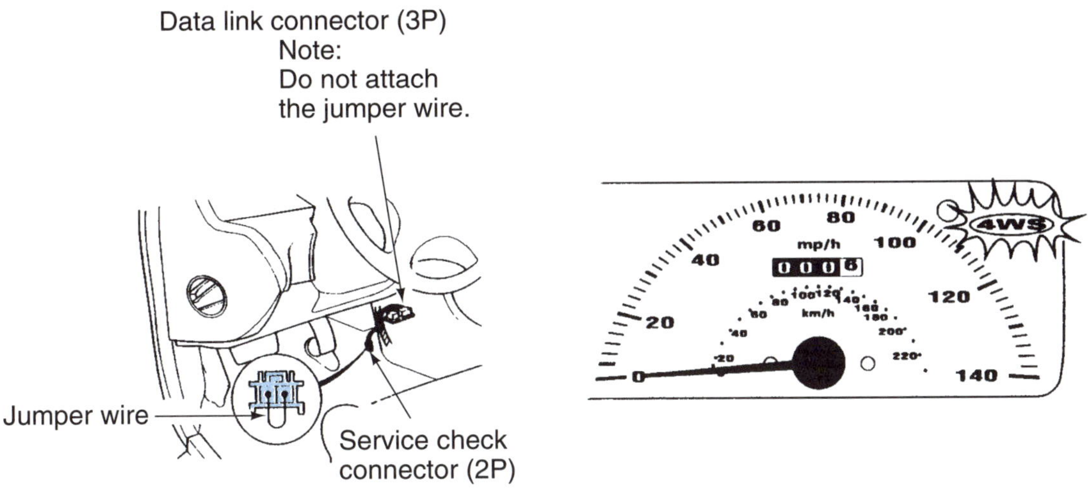


Figure 13-1 Dual-terminal service check connector positioned behind the center console (Courtesy of American Honda Motor Co., Inc.)

2. Turn on the ignition switch, but do not start the engine.
3. Observe the 4WS indicator light to read the trouble codes. Three longer flashes followed by a brief pause and one quicker flash indicates code 31. The codes are given in numerical order.
4. Record the fault codes.

## Trouble Code Display with Engine Running

The 4WS control unit actually contains two processing units, which are referred to as the main and sub processing units. Each processing unit can store a maximum of 10 trouble codes. If the trouble diagnosis is performed with the engine running, the code display indicates whether the codes are stored in the main or sub processor. When the service connector terminals are connected with a jumper wire and the engine is started, the 4WS indicator light follows this sequence if there are trouble codes in the main and sub processors:

1. blinks quickly once when the ignition switch is turned on
2. pauses for 3 seconds
3. displays codes stored in the main processor
4. pauses for 1.6 seconds
5. blinks quickly for 3 seconds to indicate a separation between the main and sub processor codes
6. pauses for 1.6 seconds
7. displays codes stored in the sub processor
8. pauses for 3 seconds and then repeats the cycle (Figure 13-2)

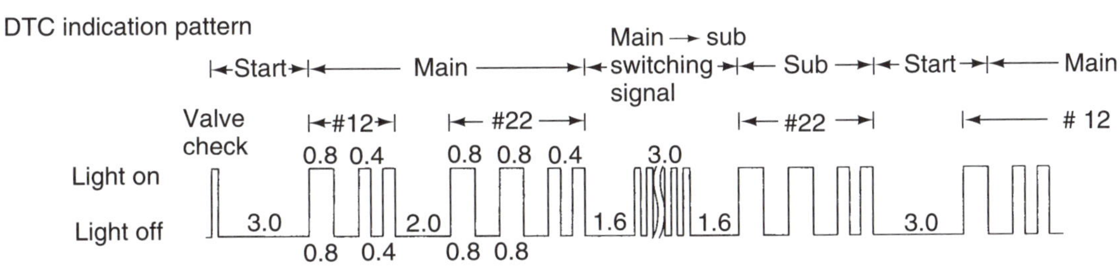


Figure 13-2 Trouble codes in main and sub processors obtained with the engine running (Courtesy of American Honda Motor Co., Inc.)

### Main Steering Angle Sensor Trouble Code

If a defect occurs in the main steering angle sensor system, the clock-radio 10-A fuse must be disconnected to cancel the 4WS indicator light. When defects occur in other parts of the electronic 4WS system, the 4WS indicator light is cancelled when the ignition switch is turned off. However, the 4WS indicator light is illuminated again when the ignition switch is turned on, and the 4WS control unit detects the problem again.

**Classroom Manual**
Chapter 13, page 267

### Trouble Codes Representing Temporary Driving Conditions

Codes 70, 71, 73, and 74 represent problems resulting from abnormal or harsh driving conditions (Figure 13-3). When the 4WS control unit detects one of these problems, it does not illuminate the 4WS indicator light, but these codes are flashed during the diagnostic procedure.

## Rear Steering Actuator Service

### Rear Steering Actuator Removal

**CAUTION:** Many steering service and diagnostic procedures require the installation of the rear steering center lock pin in the rear steering actuator to lock this unit in the centered position.

**CAUTION:** Do not start the engine with the rear steering center lock pin in place. This action may damage the lock pin and rear steering actuator.

**CAUTION:** Do not attempt to disassemble the rear steering actuator other than tie rods, tie rod ends, and sensors. This actuator is serviced as an assembly.

**CAUTION:** Do not damage the tie rod boot when using the tie rod end removal tool.

Follow these steps to remove the rear steering actuator:

1. Raise the vehicle on a hoist or lift the rear of the vehicle with a floor jack and support the chassis with jack stands placed in the vehicle manufacturer's recommended locations.
2. Remove the cotter key and nut from each tie rod end.
3. Install a 12 millimeter (mm) nut on each tie rod end until the nuts are flush with the tie rod stud.

| DTC | Operation | 4WS indicator light |
|---|---|---|
| 70 | The ignition is turned from OFF to ON while driving. | ON |
| 71 | The car is driven aggressively with the driver and three passengers on board, or the steering wheel is turned with a rear wheel blocked by the curb, etc. | ——— |
| 73 | The engine is started while quick-charging the battery. | ——— |
| 74 | Driving the car with the parking brake ON | ON 5 minutes after detection |

**Figure 13-3** Codes that represent problems caused by abnormal or harsh driving conditions (Courtesy of American Honda Motor Co., Inc.)

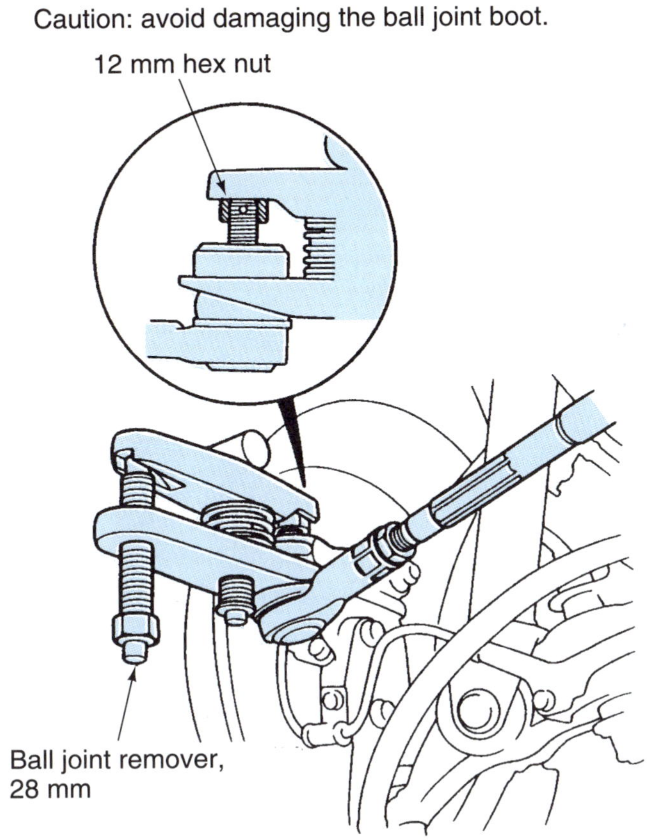


**Figure 13-4** Removing a tie rod end with a special tool (Courtesy of American Honda Motor Co., Inc.)

4. Install the special tool on the tie rod end, and with the tool arms parallel, tighten the screw on the tool to loosen the tie rod end (Figure 13-4). Repeat the procedure on both tie rod ends.
5. Remove the nuts from the tie rods, and remove the tie rods from the steering arms.
6. Remove the rear steering actuator cover (Figure 13-5).

**Special Tools**

Tie rod end removal tool

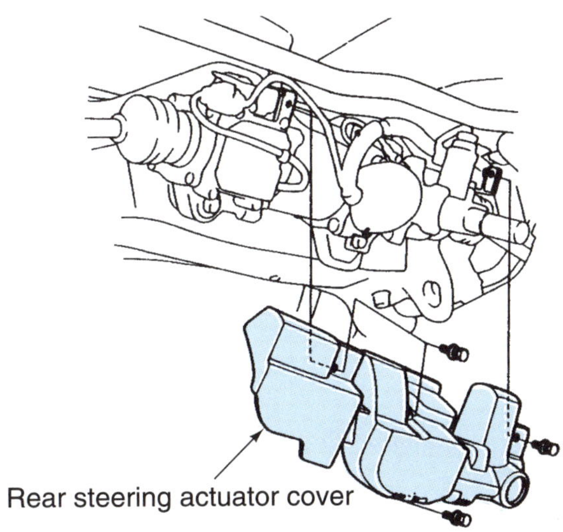


**Figure 13-5** Removing the rear steering actuator cover (Courtesy of American Honda Motor Co., Inc.)

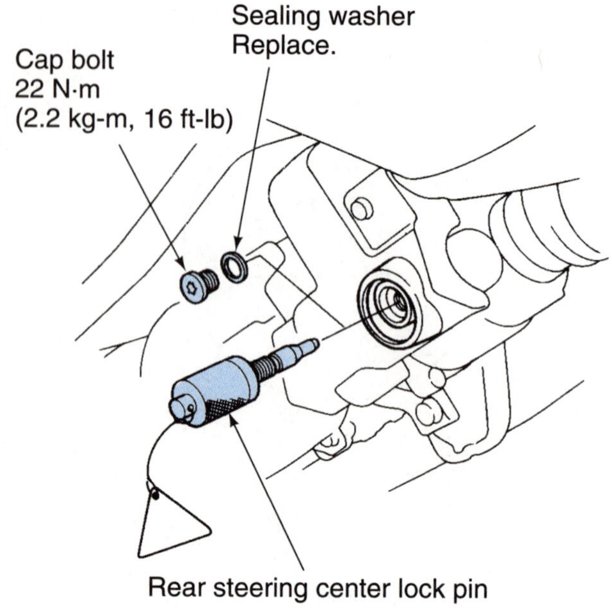


**Figure 13-6** Removing cap bolt and washer, and installing rear steering center lock pin (Courtesy of Honda Motor Co., Inc.)

7. Remove the cap bolt and washer and install the rear steering center lock pin (Figure 13-6).
8. Remove the ground cable connector and all wiring harness connectors on the rear steering actuator (Figure 13-7).
9. Remove the four mounting bolts and bracket, and remove the rear steering actuator (Figure 13-8).

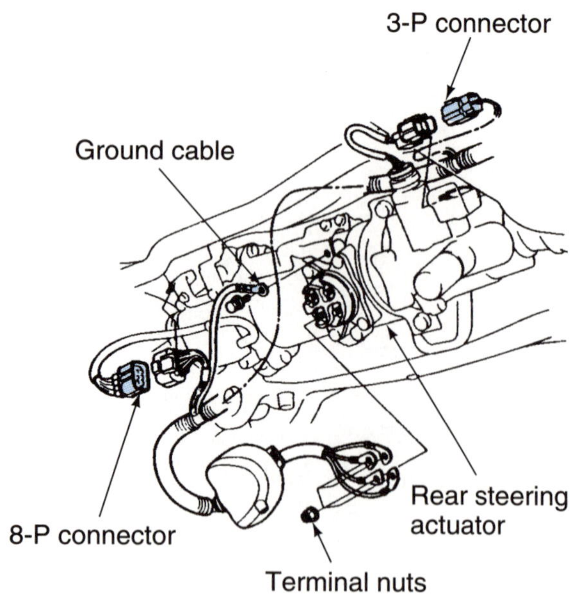


**Figure 13-7** Removing ground cable connector and all wiring harness connectors on the rear steering actuator (Courtesy of American Honda Motor Co., Inc.)

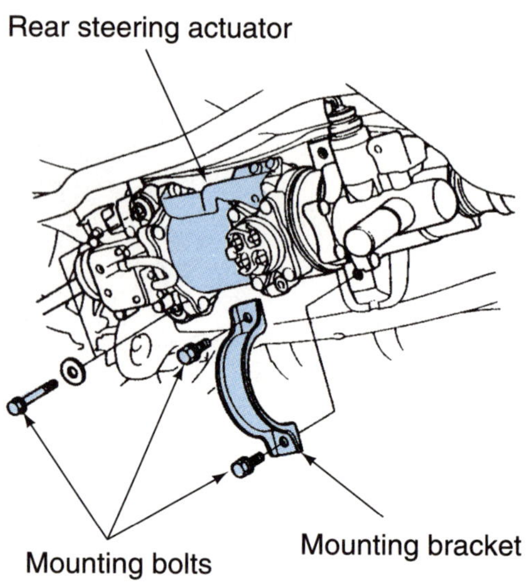


**Figure 13-8** Removing four mounting bolts, bracket, and rear steering actuator (Courtesy of American Honda Co., Inc.)

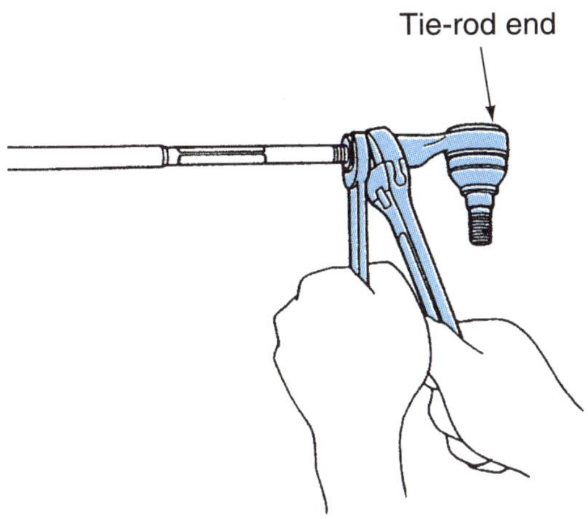


Figure 13-9 Loosening the tie rod lock nut (Courtesy of American Honda Motor Co., Inc.)

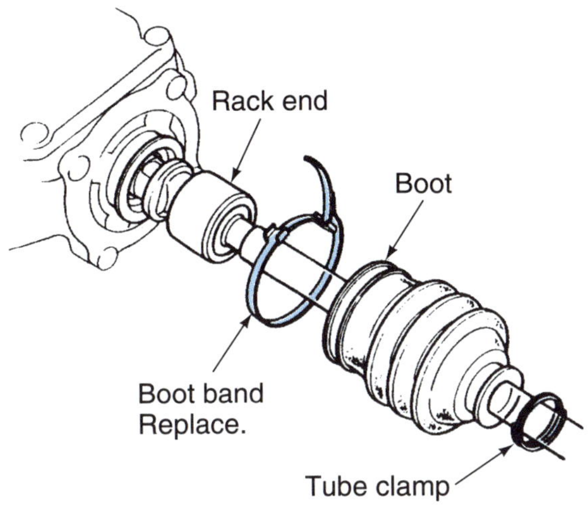


Figure 13-10 Removing boot bands and clamps from the inner tie rod end (Courtesy of American Honda Motor Co., Inc.)

## Tie Rod and Tie Rod End Removal

Follow these steps for tie rod end and tie rod removal:

1. Mark the relative position of the tie rod end, lock nut, and tie rod with a wax marker.
2. Hold the tie rod end with a wrench and loosen the lock nut (Figure 13-9).
3. Remove the tie rod end.
4. Remove the boot bands and clamps from the inner tie rod ends (Figure 13-10).
5. Place the flat side of the rack holding tool toward the actuator housing and drive the special rack holding tool between the actuator housing and the stop washer with a soft hammer (Figure 13-11).
6. Straighten the tabs on the tie rod lock washer.

**Special Tools**

Rack holding tool

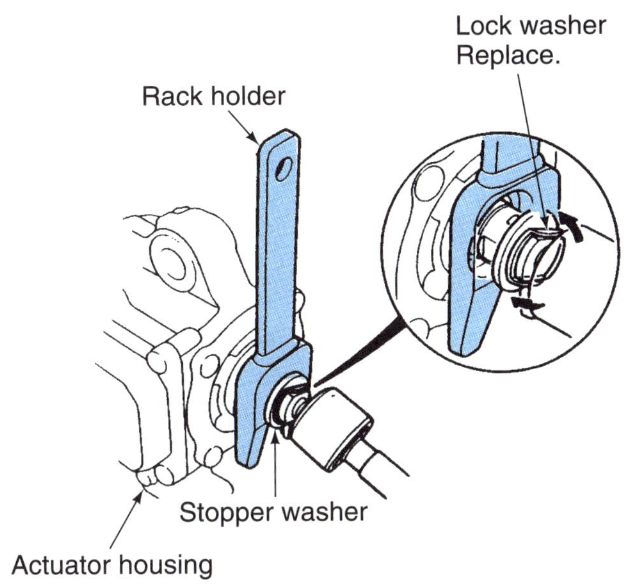


Figure 13-11 Installing special rack holding tool between the actuator housing and the stop washer (Courtesy of American Honda Motor Co., Inc.)

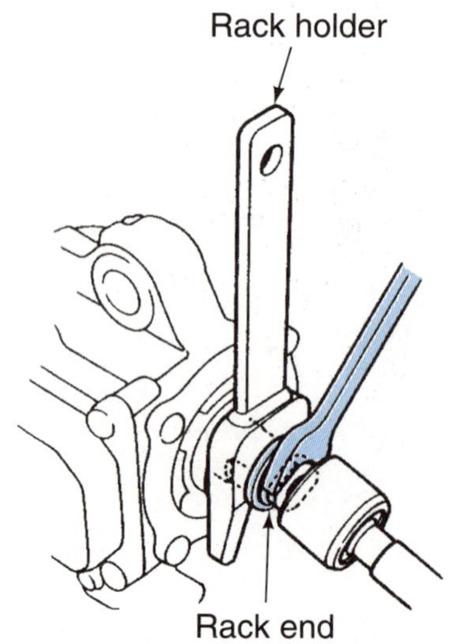


**Figure 13-12** Removing the tie rod from the shaft screw in the rear steering actuator (Courtesy of Honda Motor Co., Inc.)

**SERVICE TIP:** Hold the special holding tool firmly while loosening the tie rod to avoid applying rotational force to the shaft screw in the actuator.

7. Hold the shaft screw with the holding tool and loosen the tie rod with a wrench (Figure 13-12).
8. Thread the tie rod off the shaft screw and repeat this procedure on each tie rod end.

## Tie Rod End Boot Removal and Replacement

Tie rod end boots must be replaced if they are cracked, split, deteriorated, or loose. Follow these steps to remove and replace the tie rod end boots:

**CAUTION:** Do not put grease on the boot installation shoulder and tapered section of the ball pin in the tie rod end.

**CAUTION:** Do not allow dust, dirt, or foreign material to enter the tie rod end ball joint or boot.

**Special Tools**

Tie rod boot driving tool

1. Use a large screwdriver to pry the old boot from the tie rod end.
2. Pack the interior of the new boot with the vehicle manufacturer's recommended grease, and place a light coating of grease on the boot lip.
3. Wipe the grease off the sliding surfaces of the ball pin with a shop towel, and then pack the lower area around the ball pin and body with grease (Figure 13-13).
4. Use the special driving tool to install the new boot on the tie rod end (Figure 13-14).
5. Wipe any grease from the tapered section of the ball pin with a shop towel, and apply sealant around the lower edge of the boot and tie rod body.

## Installation of Tie Rods and Tie Rod Ends

**CAUTION:** Never apply axial impact or rotational force to the shaft screw in the rear steering actuator. Either of these actions may cause internal actuator damage.

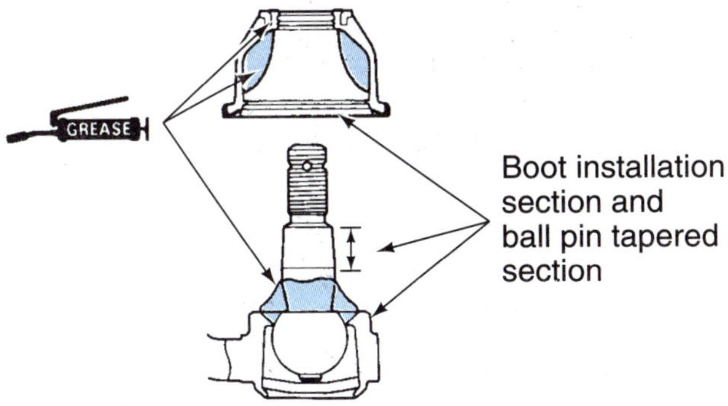


**Figure 13-13** Packing boot and tie rod end with grease (Courtesy of Honda Motor Co., Inc.)

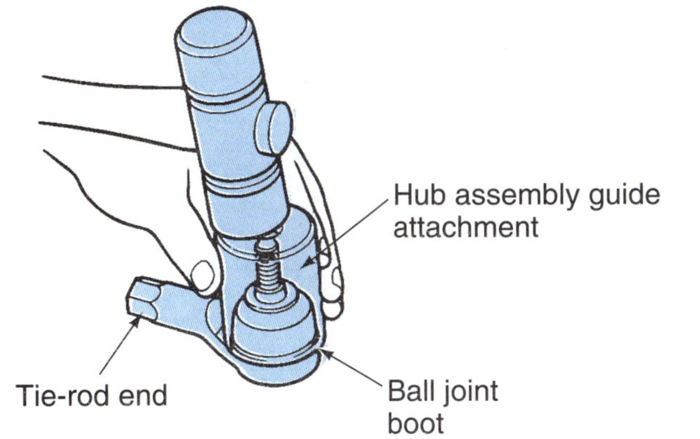


**Figure 13-14** Driving the boot onto the tie rod end with a special driving tool (Courtesy of American Honda Motor Co., Inc.)

Inner tie rod boots must be replaced if they are cracked, split, deteriorated, or damaged. Follow these steps for tie rod end and tie rod installation:

1. Install the tie rod ends so the marks on the tie rod ends, nuts, and tie rods are aligned, and tighten the tie rod nuts to the specified torque.
2. Screw each inner tie rod onto the shaft screw while holding the lock washer so its tabs are in the inner tie rod end. The stop washer must be installed on the shaft screw with the chamfered side facing outward (Figure 13-15).
3. Drive the special holding tool between the actuator housing and the stop washer with a soft hammer (Figure 13-16).
4. Hold the shaft screw with the holding tool, and tighten the inner tie rod end to the specified torque.
5. Bend the lock washer tabs against the flat on the inner tie rod end.

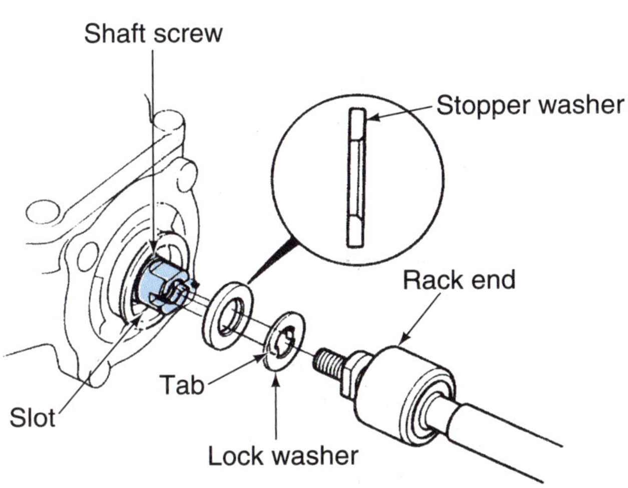


**Figure 13-15** Installing the inner tie rod end on the shaft screw (Courtesy of American Honda Motor Co., Inc.)

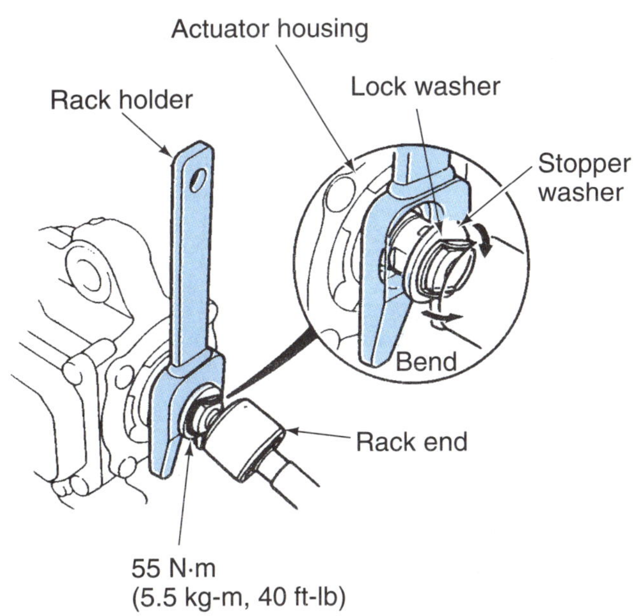


**Figure 13-16** Installing special tool to hold the shaft screw while tightening the inner tie rod end (Courtesy of American Honda Motor Co., Inc.)

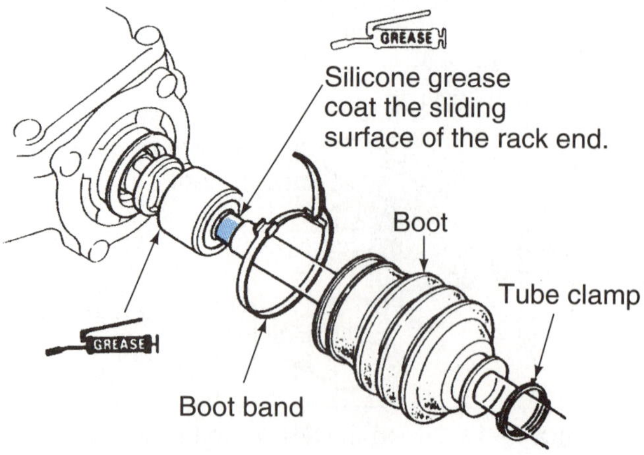


**Figure 13-17** Lubrication of the inner tie rod joint housing (Courtesy of American Honda Motor Co., Inc.)

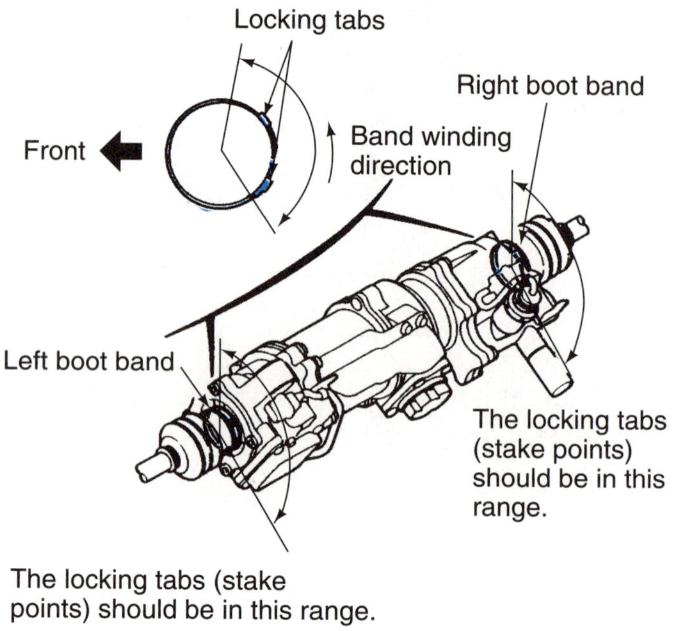


**Figure 13-18** Proper boot band position in relation to the actuator housing (Courtesy of American Honda Motor Co., Inc.)

6. Remove the special holding tool, and apply silicone grease to the sliding surface of the tie rod (Figure 13-17). Place a light coating of silicone grease inside the tie rod boot.
7. Apply the vehicle manufacturer's recommended grease to the circumference of the inner tie rod joint housing.
8. Install the boots on the actuator housing, and install the boot bands with the locking tabs properly positioned in relation to the actuator housing (Figure 13-18).

**CAUTION:** While staking the boot clamps, be careful not to damage the inner tie rod boots.

9. Tighten the boot bands and bend both sets of locking tabs over the band (Figure 13-19). Tap lightly on the doubled over portion of the band to reduce its height, and stake the locking tabs firmly.

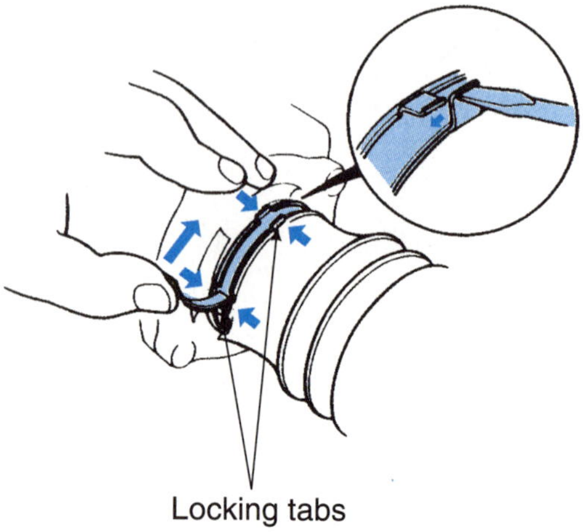


**Figure 13-19** Tightening and staking the inner tie rod boot clamps (Courtesy of American Honda Motor Co., Inc.)

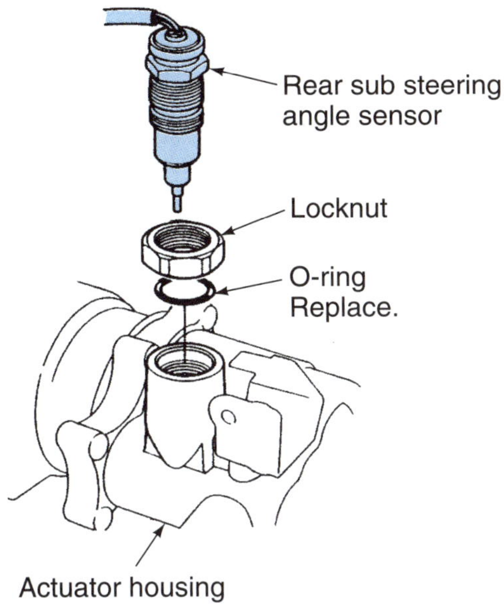


**Figure 13-20** Removing the rear sub steering angle sensor (Courtesy of American Honda Motor Co., Inc.)

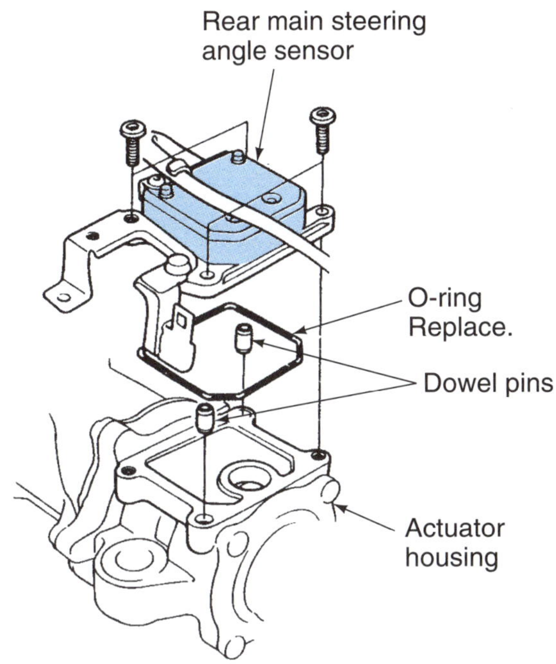


**Figure 13-21** Removing the rear main steering angle sensor (Courtesy of American Honda Motor Co., Inc.)

## Remove and Replace Rear Steering Actuator Sensors

Follow these steps to remove and replace the rear sub steering angle sensor and the rear main steering angle sensor:

1. Loosen the rear sub steering angle sensor lock nut, and rotate the sensor to thread it out of the housing (Figure 13-20). Discard the sensor O-ring.
2. Remove the two mounting bolts in the rear main steering angle sensor, and remove the sensor from the actuator housing (Figure 13-21). Note the position of the dowel pins, and discard the O-ring.

**CAUTION:** Cover the rear main steering sensor and rear sub steering sensor openings in the actuator housing with masking tape or its equivalent to keep dirt and foreign material out of the actuator housing.

3. Install the lock nut and a new O-ring on the rear sub steering angle sensor.
4. Place a light coating of grease on the O-ring, and install the sensor in the actuator housing.
5. Rotate the sensor until it touches the tapered shaft, and back it out one-half turn. Tighten the lock nut finger tight. Final adjustment of the rear sub steering angle sensor is completed with the actuator installed in the vehicle.
6. Place a light coating of grease on the rear main steering angle sensor O-ring and install this O-ring on the sensor.
7. Install the rear main steering angle sensor and O-ring in the actuator housing with the dowel pins properly positioned, and tighten the mounting bolts to the specified torque.

## Installing Rear Steering Actuator

Follow these steps for rear steering actuator installation:

1. Install the rear steering actuator and the four mounting bolts and bracket. The arrow on the bracket must face upward (Figure 13-22).
2. Tighten the rear steering actuator mounting bolts to the specified torque.

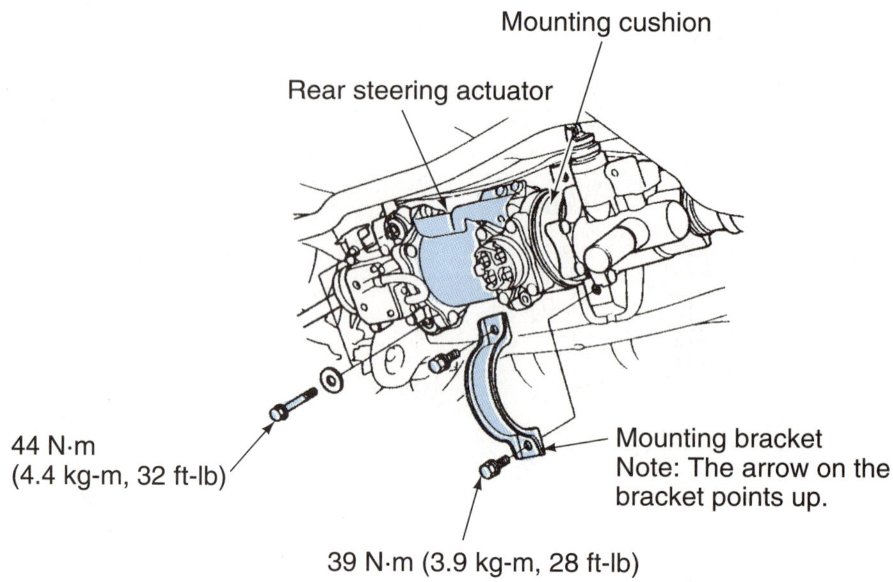


**Figure 13-22** Installing rear steering actuator, mounting screws, and bracket (Courtesy of American Honda Motor Co., Inc.)

**CAUTION:** Tighten the castle nut on the tie rod ends to the specified torque, and then tighten these nuts enough to align the slots in the nut with the hole in the tie rod pin. Do not loosen the nut to align the nut slots with the tie rod pin hole.

3. Reconnect the tie rod ends to the steering arms, and tighten the castle nut to the specified torque. If necessary, tighten the nut slightly to align the nut slots with the tie rod pin hole.
4. Install the cotter key in the nut and tie rod end pin openings, and bend one leg of the cotter key downward over the nut. Bend the other cotter key leg upward over the top of the tie rod end pin (Figure 13-23).

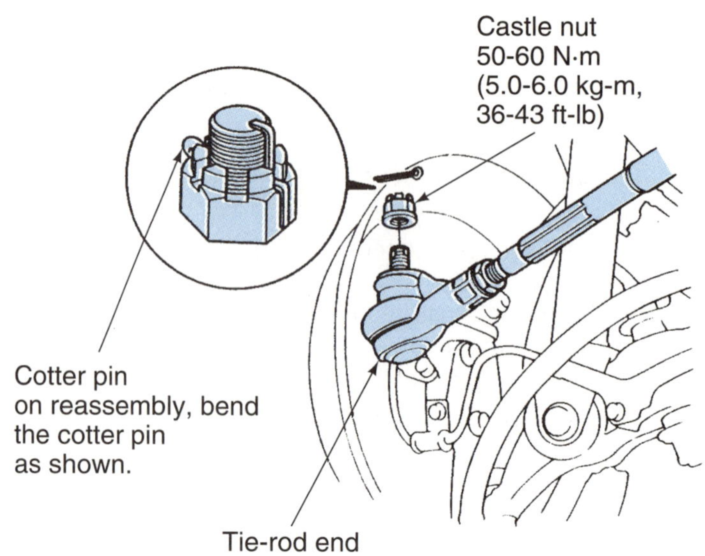


**Figure 13-23** Proper installation of cotter key in the tie rod end (Courtesy of American Honda Motor Co., Inc.)

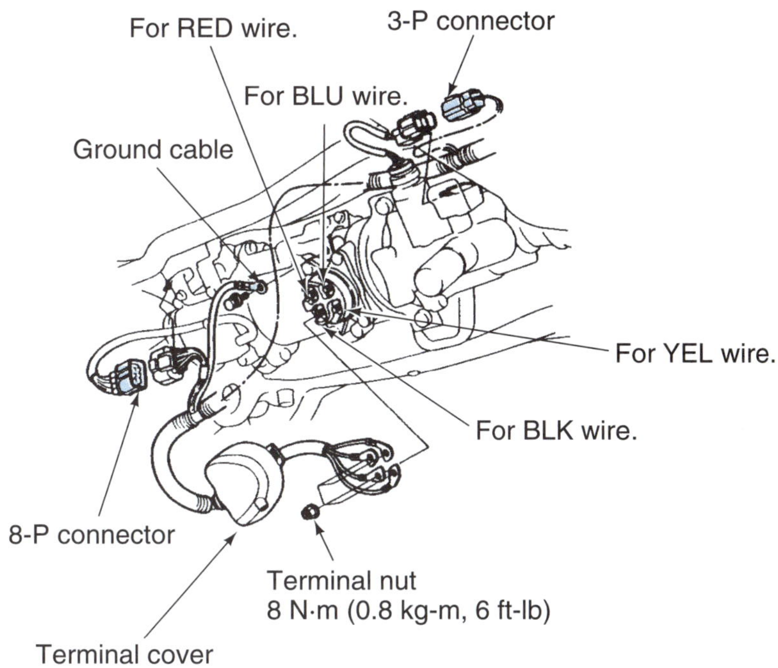


**Figure 13-24** Installing wiring connectors on the rear steering actuator (Courtesy of American Honda Motor Co., Inc.)

5. Check all the wiring connectors for contamination, and clean as necessary. Install all the wiring connectors on the rear steering actuator, and tighten all the terminal nuts to the specified torque (Figure 13-24).
6. Install the terminal cover on the rear main steering sensor terminals. Remove the rear steering lock pin, and install the cap bolt and washer. Leave the steering actuator cover removed until after the final rear steering actuator adjustments.

# Rear Steering Actuator Adjustment

## Electronic Neutral Check

**Preliminary Checks.** If the power to the 4WS control unit has been shut down for any of these operations, start the engine and turn the steering wheel fully right and left.

**WARNING:** Do not start the engine with the rear steering actuator lock pin in place. This action may damage the lock pin and rear steering actuator.

1. Battery cables have been disconnected.
2. The 4WS control unit connector has been disconnected.
3. The number 43 clock-radio fuse has been disconnected.

Prior to the electronic neutral check, be sure the steering wheel spoke is at the designated angle while driving straight ahead. Be sure the rear wheels are in the straight-ahead driving position before the electronic neutral check.

**Steering Wheel Marking and Diagnostic Mode Entry.** Follow these steps to mark the steering wheel and enter the front steering sensor diagnostic mode:

**Figure 13-25** All four wheels must be on radius gauge turn tables with the wheels straight ahead and the turn tables in the zero position (Courtesy of American Honda Motor Co., Inc.)

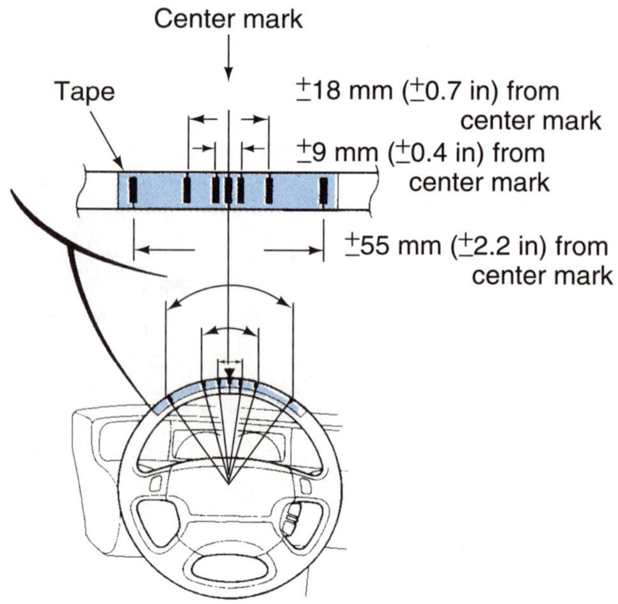


**Figure 13-26** Steering wheel marking for electronic neutral check (Courtesy of American Honda Motor Co., Inc.)

**Special Tools**

Turning radius gauge turn tables

1. Drive the vehicle on an alignment rack, and place all four wheels on turning radius gauge turn tables. Be sure the wheels are in the center of the turn tables with the wheels straight ahead and the turn tables in the zero-degree position (Figure 13-25).
2. Place a piece of masking tape 300 mm (12.0 in) long on top of the steering wheel. Mark the tape at each of these locations on the outer circumference of the wheel (Figure 13-26):
   - center, highest point on the wheel
   - 9 mm (0.4 in) right and left of the center mark
   - 18 mm (0.7 in) right and left of the center mark
   - 55 mm (2.2 in) right and left of the center mark
3. Bend a stiff piece of wire so it can be taped to the top or the dash with the outer end of the wire positioned over the steering wheel marks. Be sure the front wheels are straight ahead with the tip of the wire over the center mark on the steering wheel (Figure 13-27). Be sure the wire is securely taped to the top of the instrument panel.

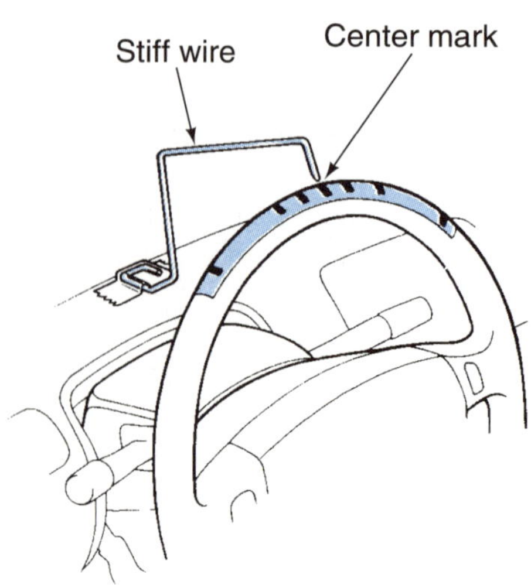


**Figure 13-27** Front wheels straight ahead, and wire pointer positioned over the center mark on the steering wheel (Courtesy of American Honda Motor Co., Inc.)

4. Connect a jumper wire to the 4WS service check connector terminals. This is the same connection for obtaining 4WS control unit trouble codes. Check and verify the trouble codes prior to the electronic neutral check. The 4WS indicator light will not indicate the electronic neutral check and trouble codes at the same time.
5. Pull the parking brake fully on until the parking brake warning light is on, and turn on the ignition switch to set the front steering sensor test mode.

**Front Sensor Inspection, Electronic Neutral Check.** Follow these steps to check the front main steering sensor:

1. With the ignition switch on, turn the steering wheel slowly to the left and slowly to the right until the 4WS indicator light comes on. Repeat this step several times to find the exact steering wheel position where the 4WS indicator light is illuminated for more than two seconds.
2. The 4WS indicator light should be illuminated when the steering wheel is 9 mm (0.4 in) to the left and right of the center mark on the steering wheel. If the 4WS indicator light comes on at a point outside of this specified range, the 4WS system requires adjustment.

Follow this procedure to check the front sub steering sensor:

1. Slowly turn the steering wheel to the left and right of the center position until the 4WS indicator light blinks at intervals of 0.2 seconds. Repeat this procedure several times to locate the exact steering wheel position where the indicator light begins blinking. The light should begin blinking within 55 mm (2.2 in) to the left or right of the center mark on the steering wheel.
2. If the 4WS indicator light does not begin flashing within this specified range, a 4WS system adjustment is necessary. After adjusting the front sub steering sensor, the 4WS indicator light should begin flashing when the steering wheel is turned 18 mm (0.7 in) to the left or right of the center mark on the steering wheel.

**Rear Sensor Inspection, Electronic Neutral Check.** Complete this procedure to complete the rear sensor inspection:

1. Release the parking brake, and be sure the parking brake warning light is off. This causes the 4WS control unit to enter the rear steering sensor inspection mode.
2. Remove the rear cap bolt and sealing washer from the rear steering actuator, and install the rear steering center lock pin until it bottoms in the actuator (Figure 13-28).

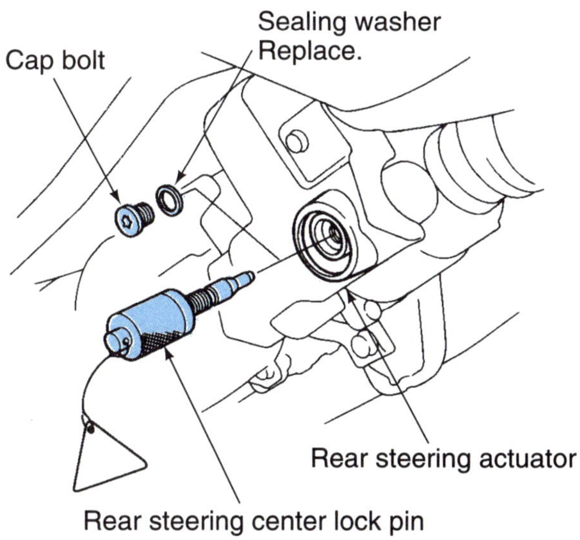


**Figure 13-28** Removing rear cap bolt and sealing washer and installing the rear steering center lock pin (Courtesy of American Honda Motor Co., Inc.)

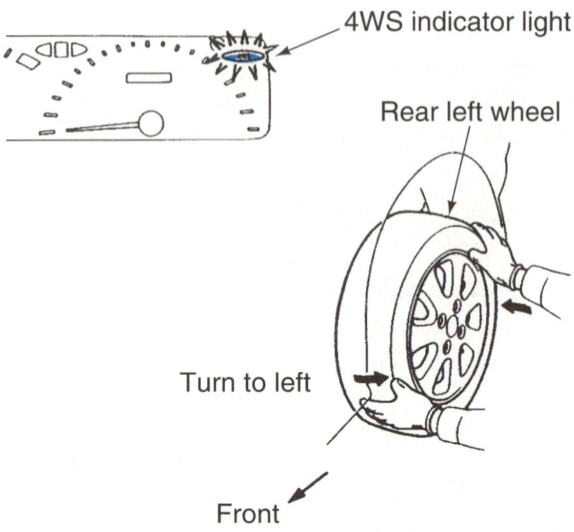


**Figure 13-29** The 4WS indicator light should flash when the left rear wheel is pushed to the left if the rear sub steering angle sensor is properly adjusted. (Courtesy of American Honda Motor Co., Inc.)

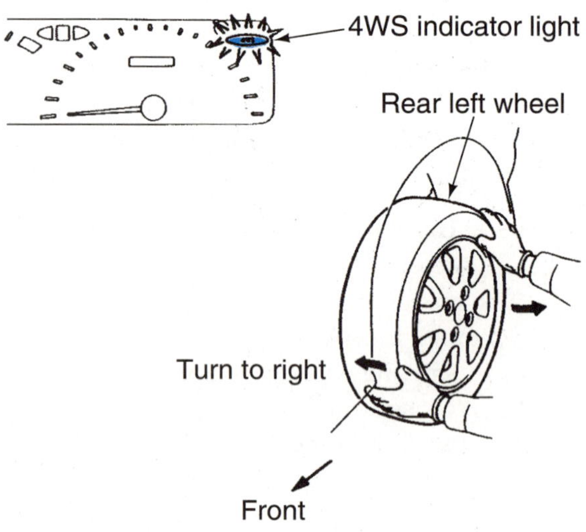


**Figure 13-30** Pushing the left rear wheel to the right and observing the 4WS indicator light checks the rear main steering angle sensor (Courtesy of American Honda Motor Co., Inc.)

3. Position the front wheels in the straight-ahead position to prevent the rear wheels from steering if the engine is started by mistake.
4. Turn the ignition switch on, and push the left rear wheel fully to the right by hand. Then push this wheel fully to the left by hand while a co-worker observes the 4WS indicator light (Figure 13-29). The 4WS indicator light should begin to flash at 0.2 second intervals when the left rear wheel is pushed to the left a small amount. If the 4WS indicator light does not flash, adjust the rear sub steering angle sensor.
5. With the ignition switch on, push the left rear wheel fully to the left by hand, and then slowly push it to the right. The 4WS indicator light should be illuminated for more than two seconds when the left rear wheel is pushed to the right (Figure 13-30). If the 4WS indicator light is not illuminated, remove the rear main steering angle sensor and check it for damage.
6. Turn off the ignition switch.
7. Remove the rear steering center lock pin, and install the cap bolt and washer. Tighten the cap bolt to the specified torque.
8. Remove the jumper wire from the service check connector.
9. Install the rear steering actuator cover.

## Front Main Steering Angle Sensor Adjustment

Proceed as follows for the front main steering angle sensor adjustment:

1. Place the car on an alignment rack with each wheel on a turning radius gauge turn table. Turn the steering wheel fully to the right and then fully to the left, and count the number of turns from fully right to fully left.
2. Turn the steering wheel back from full left exactly one-half the number of turns from fully right to fully left. This action centers the front steering rack, and the steering wheel spoke should be within the vehicle manufacturer's specified number of degrees from the horizontal position. If the steering wheel is not within this specified position, proceed with the front main steering angle sensor adjustment and spoke angle adjustment.

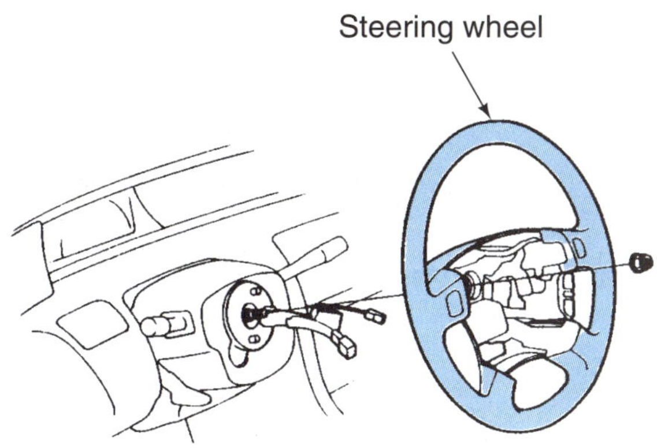


**Figure 13-31** Removing the steering wheel (Courtesy of American Honda Motor Co., Inc.)

3. Set the steering wheel so the front wheels are straight-ahead, and remove the steering wheel retaining nut. Use a steering wheel puller to remove the steering wheel (Figure 13-31).
4. Check to see if the yellow paint mark on the front main steering angle sensor is facing straight down (Figure 13-32). When this paint mark is facing down, the front main steering angle sensor is in the electronically neutral position.
5. If the yellow paint mark on the front main steering angle sensor is not facing downward, temporarily install the steering wheel with the spokes in the horizontal position. Turn the steering wheel until this yellow paint mark is facing downward.
6. Return the steering wheel to the horizontal position, and remove the steering wheel.
7. Install the steering wheel, aligning it with the serration that makes the spoke angle closest to horizontal. Be sure the steering wheel openings fit over the pins on the cable reel for the air bag system (Figure 13-33). Do not push down hard on the steering wheel until the serrations and cable reel pins are properly aligned. When the serrations and cable reel pins are properly aligned, push the steering wheel down into place, and install the retaining nut.
8. Hold the steering wheel and tighten the retaining nut to the specified torque.

The cable reel contains a conductive ribbon that connects the air bag module on top of the steering wheel to the air bag electrical system, while allowing steering wheel rotation.

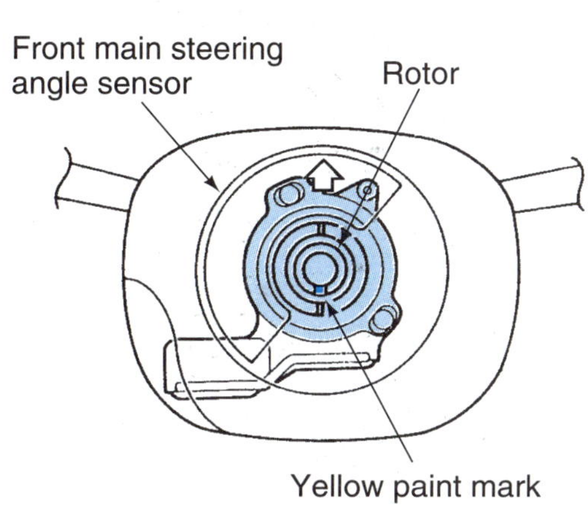


**Figure 13-32** Yellow paint mark on the front main steering angle sensor indicating the electronically neutral sensor position (Courtesy of American Honda Motor Co., Inc.)

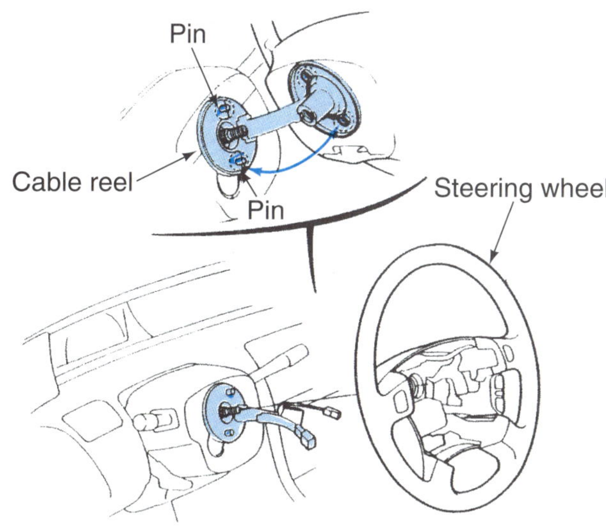


**Figure 13-33** Proper alignment of steering wheel openings and cable reel pins (Courtesy of American Honda Motor Co., Inc.)

## Front Sub Steering Angle Sensor Adjustment

Use this procedure for the front sub steering angle sensor adjustment:

1. Raise the front and rear suspension with a floor jack, and place jack stands under the proper chassis locations specified by the car manufacturer. All four wheels must be off the floor.
2. Set the steering wheel in the straight-ahead driving position.
3. Connect a jumper wire across the 4WS system service check connector terminals.
4. Pull the parking brake on fully, and turn on the ignition switch. Be sure the parking brake warning light is illuminated.
5. Turn the ignition switch off.
6. Cut the tie strap off the front sub steering angle sensor cover, and remove this cover (Figure 13-34).
7. Remove the wiring harness from the clamp, and disconnect the wiring harness connector.
8. Loosen the front sub steering angle sensor lock nut, and then tighten the lock nut fully by hand. Back this lock nut off three quarters of a turn, and connect the connector.
9. Be sure the front wheels are in the straight-ahead driving position, and turn the steering wheel until the 4WS indicator light is illuminated. Keep the steering wheel in this position.
10. Slowly turn the front sub steering angle sensor clockwise until the 4WS indicator light goes off, and mark the sensor position in relation to the housing.
11. Slowly rotate the front sub steering angle sensor counterclockwise until the 4WS indicator light begins to blink, and mark the sensor in relation to the housing (Figure 13-35). Set the front sub steering angle sensor in the center of the range from where the light went off to where the light began to blink. Hold the sensor in this position and tighten the lock nut to the specified torque.
12. Turn off the ignition switch. If the front sub steering angle sensor harness is twisted, disconnect the connector and straighten the harness. Install the harness in the clamp and install the sensor cover. Secure the cover with a new tie strap.
13. Perform the electronic neutral check described previously in this chapter.

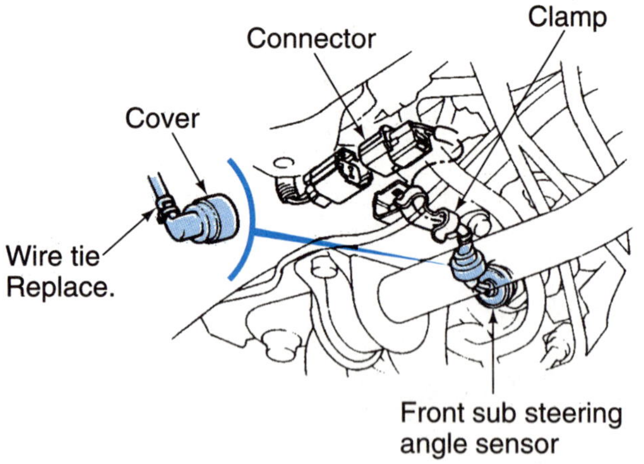


**Figure 13-34** Cutting the tie strap and removing the front sub steering angle sensor cover (Courtesy of American Honda Motor Co., Inc.)

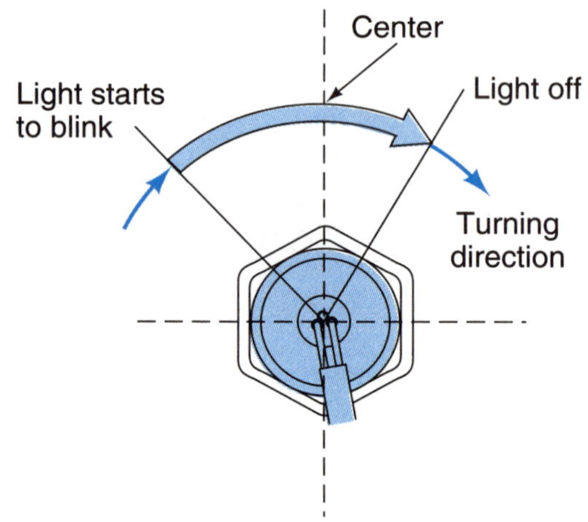


**Figure 13-35** Adjusting front sub steering angle sensor (Courtesy of American Honda Motor Co., Inc.)

## Rear Sub Steering Angle Sensor Adjustment

The rear main steering angle sensor is not adjustable.

Proceed with these steps to adjust the rear sub steering angle sensor:

1. Raise the front and rear suspension with a floor jack, and place jack stands under the proper chassis locations specified by the car manufacturer. All four wheels must be off the floor.
2. Connect a jumper wire across the terminals in the 4WS system service check connector, and be sure any trouble codes have been displayed.
3. Release the parking brake and turn the ignition switch on. Be sure the parking brake warning light goes off.
4. Turn off the ignition switch.
5. Remove the cap bolt and washer, and install the rear steering center lock pin.
6. Remove the rear sub steering angle sensor wire from the clamp, and disconnect the wiring harness connector.
7. Loosen the rear sub steering angle sensor lock nut. Tighten this lock nut fully by hand, and then back it off approximately one-half turn.
8. Connect the rear sub steering angle sensor connector and set the front wheels in the straight-ahead driving position.
9. Turn on the ignition switch.
10. Push the left rear wheel fully to the left by hand, and then push this wheel slowly to the right until the 4WS indicator light comes on. This action places the main rear steering angle sensor in the electronically neutral position.
11. Slowly turn the rear sub steering angle sensor counterclockwise until the 4WS indicator light goes off, and mark the sensor in relation to the housing.
12. Slowly rotate this sensor clockwise until the 4WS indicator light starts to blink, and mark the sensor in relation to the housing. Turn the sensor to the center position between where the indicator light went off and the light started to blink (Figure 13-36). Hold the sensor in this position and tighten the lock nut to the specified torque.
13. Turn off the ignition switch.
14. If the rear sub steering angle sensor wiring is twisted, disconnect the connector, straighten the harness, and reconnect the connector.
15. Disconnect the jumper wire from the service check connector.

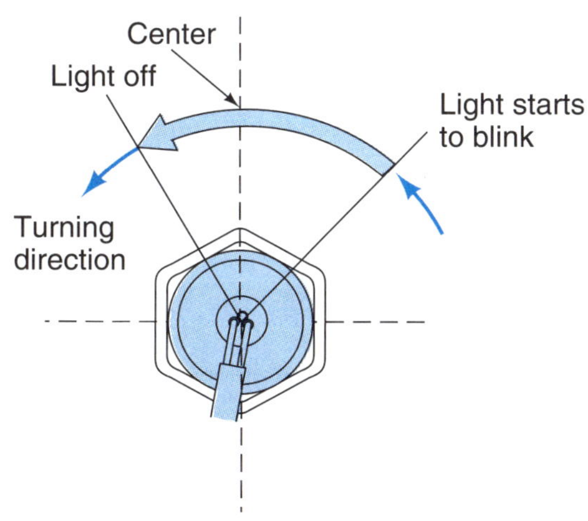


**Figure 13-36** Adjusting the rear sub steering angle sensor (Courtesy of American Honda Motor Co., Inc.)

**Classroom Manual**
Chapter 13, page 271

16. Remove the rear steering center lock pin, and install the cap bolt and washer. Tighten the cap bolt to the specified torque.
17. Install the rear steering actuator cover and perform the electronic neutral check described previously in this chapter.

## Guidelines for Four Wheel Steering Service and Adjustment

1. Prior to 4WS system diagnosis, a preliminary system inspection should be performed.
2. If the 4WS control unit detects a problem in the system, the control unit enters the fail-safe mode.
3. The 4WS system provides a damper control mode that gradually moves the rear wheels to the centered position when the fail-safe mode is entered.
4. A jumper wire may be connected across the service check connector terminals to obtain the 4WS system trouble codes.
5. If the trouble codes are obtained with the engine running, the code display informs the technician if the trouble codes are in the main or sub processor in the control unit.
6. When the control unit stores a trouble code representing the main steering angle sensor system, the clock-radio fuse must be removed to cancel the 4WS indicator light.
7. Some trouble codes represent problems caused by abnormal or harsh driving conditions. The 4WS control unit does not illuminate the 4WS indicator light when these codes are present.
8. The rear steering actuator is serviced as a unit except for sensor and tie rod service.
9. Axial impact or rotational force must not be applied to the shaft screw in the rear steering actuator.
10. During many rear steering actuator service operations a rear center lock pin must be installed in the actuator.
11. An electronic neutral check may be performed to determine if the front and rear steering sensors are adjusted properly.
12. The front main steering sensor, front sub steering sensor, and rear sub steering sensor are adjustable.
13. The rear main steering sensor is not adjustable.

Photo Sequence 11 shows a typical procedure for diagnosing an electronically controlled, four wheel steering system.

### CASE STUDY

A customer complained about the 4WS indicator light coming on intermittently on a Honda Prelude with an electronic 4WS system. The technician asked the customer about any other steering problems, and the customer reported the car steered normally. The technician road tested the car, but the 4WS light did not come on, which indicated there were no electronic problems in the system. The customer was concerned about a possible safety hazard while driving this vehicle with the 4WS indicator light illuminated. In reply to this concern, the technician explained to the customer about the fail-safe function in the 4WS system and the rear wheels being centered in this mode.

The technician asked the customer about any recent service work completed on the vehicle. In response to this question, the customer replied that the car had been in a rear end collision recently, and since the body work was completed the 4WS indicator light

# Photo Sequence 11

## Typical Procedure for Diagnosing an Electronically Controlled, Four Wheel Steering System

P11-1 Road test the vehicle to check 4WS operation and indicator light.

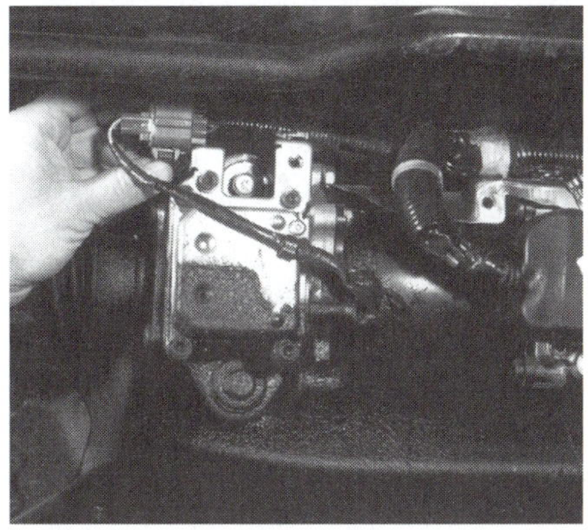

P11-2 Raise the vehicle on a lift, and check all electrical connectors on the front steering gear and rear steering actuator.

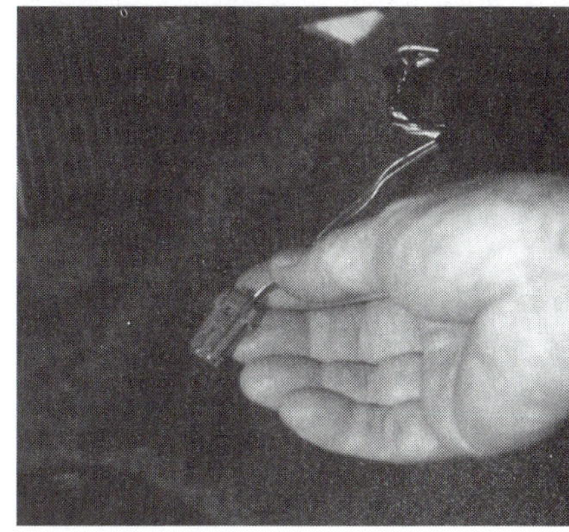

P11-3 Lower the vehicle and locate the service check connector behind the center console.

P11-4 Look up the diagnostic procedure and trouble codes in the car manufacturer's service manual.

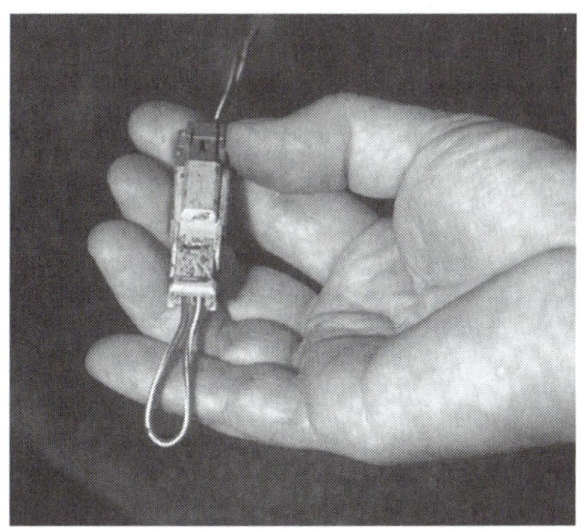

P11-5 Connect a jumper wire between the terminals in the service check connector.

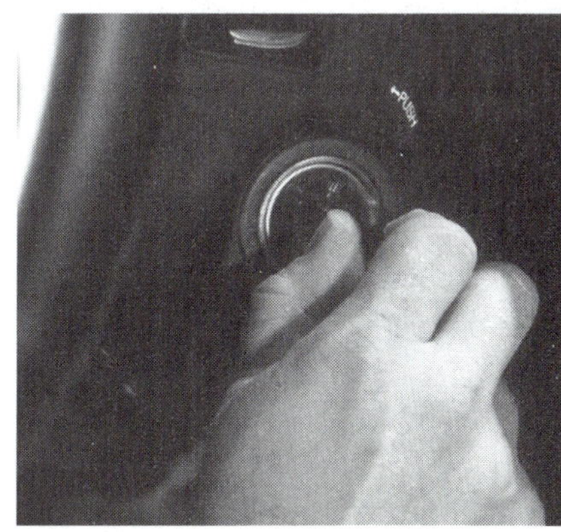

P11-6 Turn on the ignition switch.

P11-7 Observe the 4WS indicator light flashes to obtain the fault codes.

P11-8 Turn off the ignition switch.

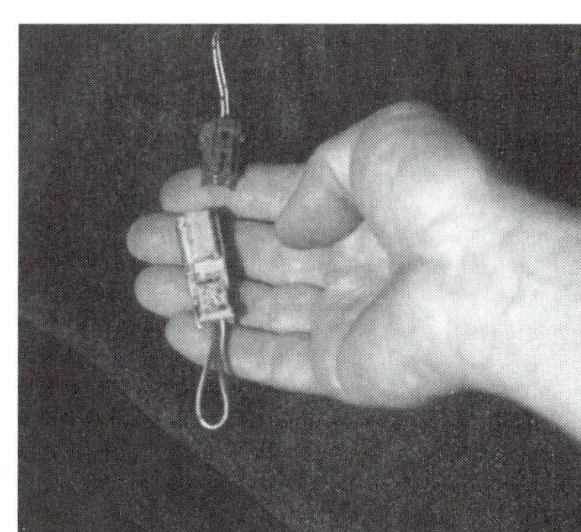

P11-9 Remove the jumper wire from the service check connector.

problem started occurring. The technician informed the customer that a 4WS system diagnosis and inspection should be performed.

Since the 4WS indicator light was not illuminated, the technician concluded that a trouble code diagnosis would probably not provide any diagnostic answers. However, the technician checked the system for codes in case there was a code caused by abnormal or harsh driving, which would not cause the indicator light to be illuminated.

When the technician raised the vehicle on a hoist, it was clearly visible that many of the rear suspension and body parts had been replaced recently. Even the rear steering actuator cover had been replaced. The technician removed the rear steering actuator cover to inspect the wiring on the actuator. All the wiring connectors were inspected, including the terminals on the rear main steering angle sensor. When the technician inspected the rear sub steering angle sensor wiring harness, he found this harness had been punctured by a sharp object near the sensor. The technician probed the sensor wires at the sensor, and connected a pair of ohmmeter leads from each wire at the sensor to the corresponding colored wire in the sensor connector. Each wire showed a normal, zero-ohm resistance. The technician repeated these ohmmeter connections and wiggled the wires at the damaged location. On one of the wires, the ohmmeter reading went to infinite while wiggling the wires, indicating an intermittent open circuit.

The technician replaced the rear sub steering angle sensor and performed the electronic neutral check and rear sub steering angle sensor adjustment. During a road test, the 4WS indicator light did not come on.

## Terms to Know

Fail-safe function
Service check connector
Front main steering angle sensor
Front sub steering angle sensor
Rear main steering angle sensor
Rear sub steering angle sensor
Rear steering center lock pin
Electronic neutral check
Cable reel

## ASE Style Review Questions

1. While discussing the fail-safe function:
*Technician A* says the 4WS indicator light is illuminated during the fail-safe function.
*Technician B* says the rear wheels steer normally when the 4WS control unit enters the fail-safe mode.
Who is correct?
**A.** A only
**B.** B only
**C.** Both A and B
**D.** Neither A nor B

2. While discussing the fail-safe function and damper control:
*Technician A* says the rear wheels move instantly to the centered position when the 4WS control unit enters the fail-safe mode.
*Technician B* says the return spring moves the rear wheels away from the centered position.
Who is correct?
**A.** A only
**B.** B only
**C.** Both A and B
**D.** Neither A nor B

3. While discussing trouble code diagnosis:
*Technician A* says the 4WS system service check connector is located under the driver's seat.
*Technician B* says the 4WS system service check connector is positioned under the center console.
Who is correct?
**A.** A only
**B.** B only
**C.** Both A and B
**D.** Neither A nor B

4. While discussing trouble code diagnosis:
*Technician A* says many 4WS system trouble codes are cancelled when the ignition switch is turned off.
*Technician B* says codes representing problems caused by abnormal or harsh driving conditions do not illuminate the 4WS indicator light.
Who is correct?
**A.** A only
**B.** B only
**C.** Both A and B
**D.** Neither A nor B

5. While discussing rear steering actuator service:
*Technician* A says the rear steering actuator is a replacement unit except for tie rods and sensors.
*Technician B* says the rear steering actuator may be overhauled.
Who is correct?
**A.** A only
**B.** B only
**C.** Both A and B
**D.** Neither A nor B

6. While discussing rear steering actuator service:
*Technician A* says axial impact on the shaft screw may damage the actuator.
*Technician B* says rotational force on the shaft screw may damage the actuator.
Who is correct?
**A.** A only
**B.** B only
**C.** Both A and B
**D.** Neither A nor B

7. While discussing rear steering actuator service:
*Technician A* says a rear steering center lock pin should be installed in the left rear tie rod for many 4WS service procedures.
*Technician B* says the engine may be started with the rear steering center lock pin in place.
Who is correct?
**A.** A only
**B.** B only
**C.** Both A and B
**D.** Neither A nor B

8. While discussing front and rear steering angle sensor service:
*Technician A* says the electronic neutral check determines if the front or rear steering sensors require adjustment.
*Technician B* says the 4WS indicator light will flash trouble codes and perform the electronic neutral check at the same time.
Who is correct?
**A.** A only
**B.** B only
**C.** Both A and B
**D.** Neither A nor B

9. While discussing front and rear steering angle sensor service:
*Technician A* says the front main steering angle sensor is not adjustable.
*Technician B* says the rear main steering angle sensor is adjustable.
Who is correct?
**A.** A only
**B.** B only
**C.** Both A and B
**D.** Neither A nor B

10. While discussing front and rear steering angle sensor service:
*Technician A* says the parking brake and the ignition switch are on prior to the front sub steering angle sensor adjustment.
*Technician B* says the parking brake and the ignition switch are on prior to the rear sub steering angle sensor adjustment.
Who is correct?
**A.** A only
**B.** B only
**C.** Both A and B
**D.** Neither A nor B

# Frame Diagnosis and Service

CHAPTER 14

Upon completion and review of this chapter, you should be able to:

- ❑ Take the necessary precautions to avoid frame damage.
- ❑ Diagnose the causes of frame damage.
- ❑ Follow safety precautions when measuring and welding frames.
- ❑ Visually inspect frames.
- ❑ Perform basic frame welding.
- ❑ Measure frames with a plumb bob, and diagnose frame damage.
- ❑ Measure frames with a tram gauge, and diagnose frame damage.
- ❑ Perform unitized body measurements with a tram gauge.
- ❑ Perform unitized body measurements with a dedicated bench system.

**Basic Tools**

Basic technician's tool set

Service manual

Tape measure

Chalk

Floor jack

Jack stands

## Frame Diagnosis

### Preventing Frame Damage

Since frame problems affect wheel alignment, technicians must be able to diagnose frame defects so they are not confused with wheel alignment problems. Follow these precautions to minimize frame damage:

1. Do not overload the vehicle.
2. Place the load evenly in a vehicle.
3. Do not operate the vehicle on extremely rough terrain.
4. Do not mount equipment such as a snow plow on a vehicle unless the frame is strong enough to carry the additional load and force.

Section modulus is a measure of the frame's strength based on height, width, thickness, and the shape of the side rails. Section modulus does not account for the type of material in the frame.

### Diagnosis of Frame Problems

**Side Sway.** The causes of frame side sway are the following:

1. collision damage
2. fire damage
3. use of equipment on the vehicle for which the frame was not designed

**Sag.** The causes of frame sag are:

1. vehicle loads that exceed the load carrying capacity of the frame
2. uneven load distribution
3. sudden changes in section modulus
4. holes drilled in the frame flange
5. too many holes drilled in the frame web
6. holes drilled too close together in the web
7. welds on the frame flange
8. cutting holes in the frame with a cutting torch
9. cutting notches in the frame rails
10. a fire involving the vehicle
11. collision damage
12. the use of equipment for which the frame was not designed

Yield strength is a measure of the strength of the material from which the frame is manufactured. Yield strength is the maximum load that may be placed on a material and still have the material retain its original shape. Yield strength is measured in pounds per square inch (psi) or kilopascals (kPa).

**Buckle.** The causes of frame buckle are:

1. collision damage
2. using equipment such as a snow plow when the frame was not designed for this type of service
3. a fire involving the vehicle

The frame web refers to the vertical side of the frame.

Frame flange is the horizontal part of the frame on the top and bottom of the web.

**Diamond-Shaped.** Diamond-shaped frame damage may be caused by:

1. collision damage
2. towing another vehicle with a chain attached to one corner of the frame
3. being towed by another vehicle with a chain attached to one corner of the vehicle frame

**Twist.** Frame twist may be caused by:

1. an accident, or collision, especially one involving a rollover
2. operating the vehicle on extremely rough terrain

# Checking Frame Alignment

## Safety Concerns

While servicing vehicle frames, always wear the proper safety clothing and safety items for the job being performed. This includes proper work clothing, safety goggles, ear protection, respirator, proper gloves, welding shield, and safety shoes (Figure 14-1).

If arc welding is necessary on a vehicle frame, follow these precautions:

1. Remove the negative battery cable before welding (Figure 14-2).
2. Remove the fuel tank before welding (Figure 14-3).
3. Protect the interior and exterior of the vehicle as necessary (Figure 14-4).

## Visual Inspection

Prior to frame measurement, the frame and suspension should be visually inspected. Check for wrinkles on the upper flange of the frame, which indicate a sag problem. Visually inspect the lower flange for wrinkles, which are definite evidence of buckle. Since suspension, or axle, problems may appear as frame problems, the suspension components should be inspected for wear and damage. For example, an offset rear axle may appear as a diamond-shaped frame. Check all suspension mounting bushings, and inspect leaf-spring shackles and center bolts.

The frame should be inspected for cracks, bends, and severe corrosion. Minor frame bends are not visible, but severe bends may be visible. Straight cracks may occur at the edge of the frame flange, and sunburst cracks may radiate from a hole in the frame web or crossmember (Figure 14-5).

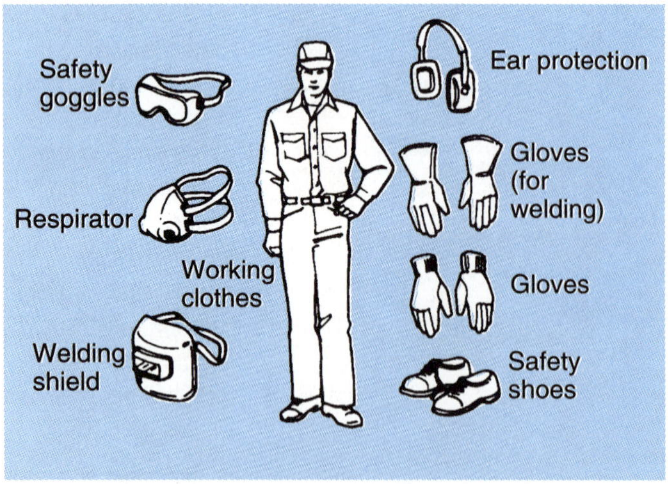


Figure 14-1 Safety items for frame service (Courtesy of Oldsmobile Division, General Motors Corporation)

Figure 14-2 Remove the negative battery cable before arc welding on a vehicle. (Courtesy of Oldsmobile Division, General Motors Corporation)

**Figure 14-3** Remove the fuel tank before arc welding on a vehicle. (Courtesy of Oldsmobile Division, General Motors Corporation)

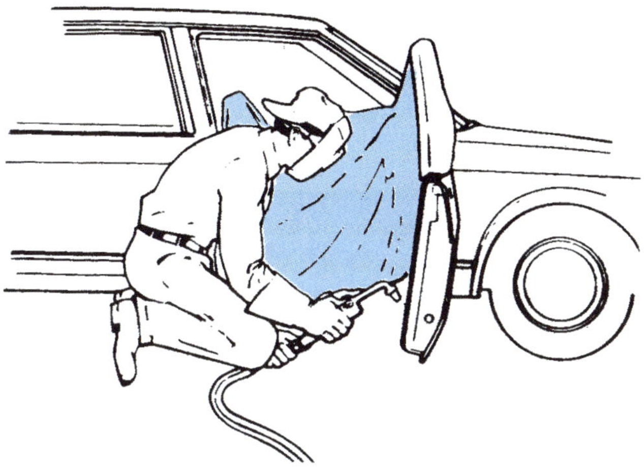

**Figure 14-4** Protect the interior and exterior of the vehicle while welding. (Courtesy of Oldsmobile Division, General Motors Corporation)

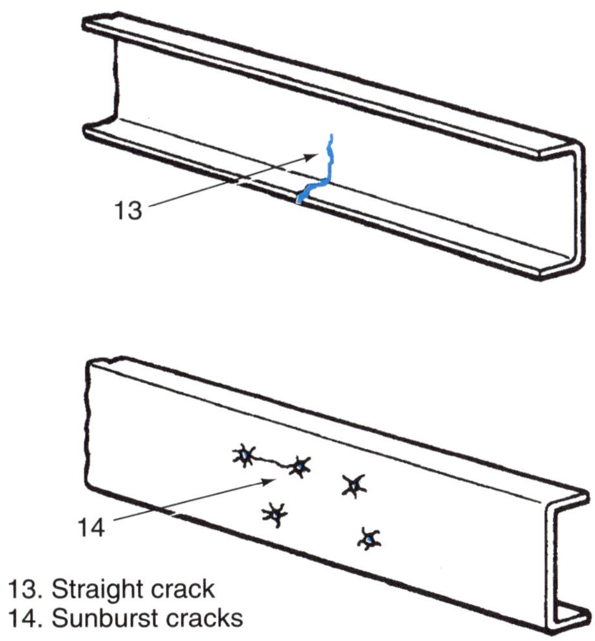


**Figure 14-5** Straight cracks may occur in the frame flange, and sunburst cracks may radiate from a hole in the frame web or crossmember. (Courtesy of Chevrolet Motor Division, General Motors Corporation)

## Frame Welding

**WARNING:** Always follow the vehicle manufacturer's recommendations in the service manual regarding frame welding and reinforcing, or mounting additional equipment on the frame.

**CAUTION:** If the frame is cracked, frame alignment is often necessary.

Follow these steps for a typical frame welding procedure:

1. Remove any components that would interfere with the weld or be damaged by heat.
2. Find the extreme end of the crack and drill a 0.250-in (6-mm) hole at this point.
3. V-grind the entire length of the crack from the starting point to the drilled hole.
4. The bottom of the crack should be opened 0.062 in (2 mm) to allow proper weld penetration. A hacksaw blade may be used to open the crack.
5. Arc weld with the proper electrode and welding procedure.

A plumb bob is a weight with a sharp, tapered point that is suspended and centered on a string. Plumbers use this tool for locating pipe openings directly above each other in the tops and bottoms of partitions.

**Special Tools**

Plumb bob

## Frame Measurement, Plumb Bob Method

Photo Sequence 12 shows a typical procedure for performing frame measurement using the plumb bob method.

**Locating the Frame Centerline.** Some vehicle manufacturers recommend measuring the frame with the plumb bob method. Follow these steps to complete a plumb bob measurement for frame damage:

1. Place the vehicle on a level area of the shop floor, and use a floor jack to raise the front and rear suspension off the floor. Support the chassis on jack stands at the manufacturer's recommended locations.
2. Suspend a plumb bob at locations 1, 2, 11, and 12 on the inside of the frame web, and allow the plumb bob to almost touch the floor surface (Figure 14-6). Points 1 and 2 are

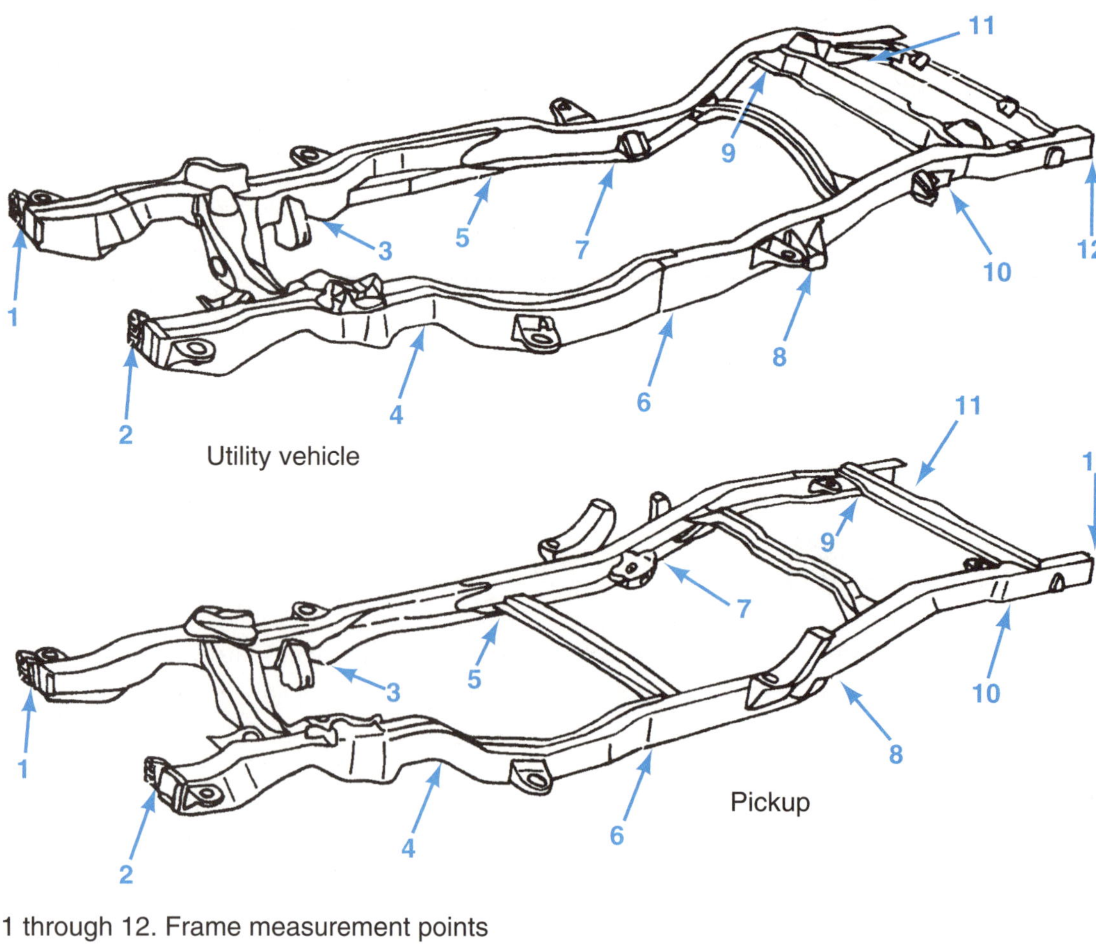


**Figure 14-6** Suspend a plumb bob at the locations shown on the frame, and mark these locations on the floor with chalk. (Courtesy of Chevrolet Motor Division, General Motors Corporation)

# Photo Sequence 12 Typical Procedure for Performing Frame Measurement, Plumb Bob Method

**P12-1** Park the vehicle on a level area of the shop floor.

**P12-2** Raise the front suspension with a floor jack, and lower the chassis onto jack stands positioned at the manufacturer's recommended lift points.

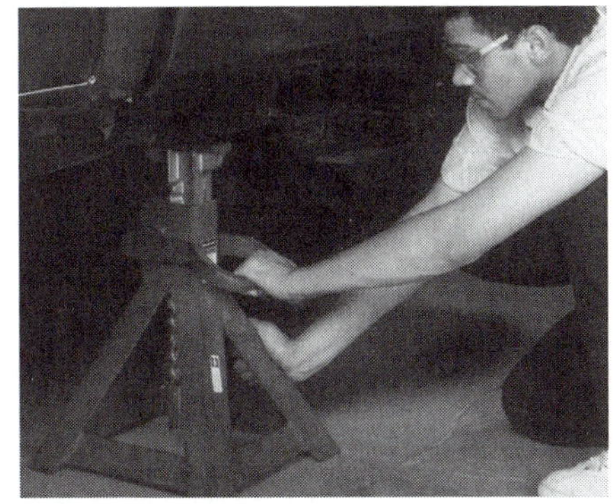

**P12-3** Raise the rear suspension with a floor jack, and lower the chassis onto jack stands positioned at the manufacturer's recommended lift points.

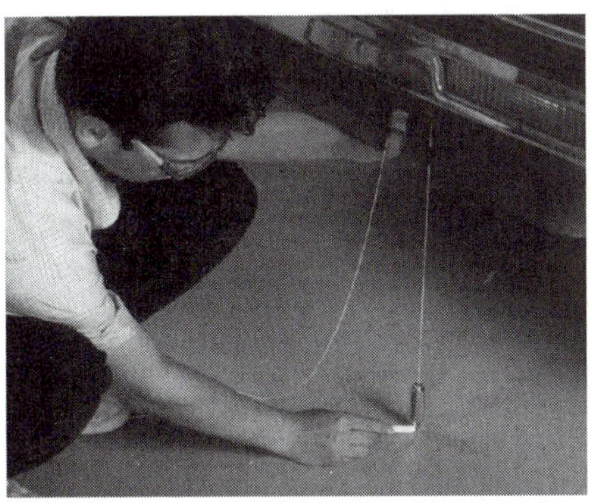

**P12-4** Suspend a plumb bob at the manufacturer's recommended frame measurement locations and place a chalk mark on the floor directly under the plumb bob.

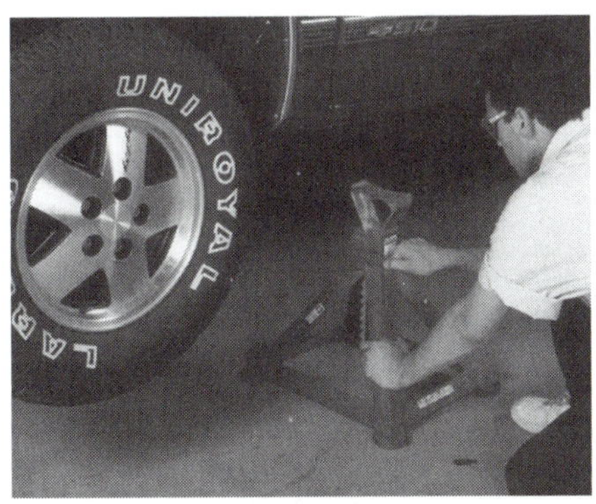


**P12-5** Use a floor jack to lift the vehicle, remove the jack stands, and lower the vehicle.

**P12-6** Drive the vehicle away from the chalk-marked area.

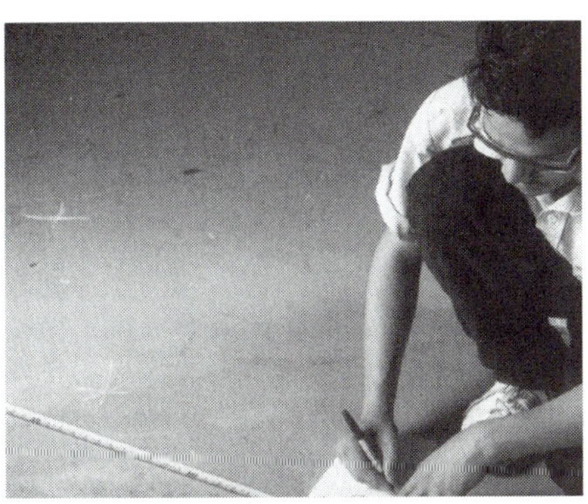

**P12-7** Use a tape measure to perform the vehicle manufacturer's recommended frame measurements between the chalk marks on the floor.

**P12-8** Compare the frame measurements obtained to the vehicle manufacturer's specifications in the service manual.

at the centerline of the slotted hole in each front bumper bracket, and points 11 and 12 are on the inside of the rear frame web. Mark these plumb bob locations on the floor with chalk.

**CAUTION:** The points on the left and right frame webs must be at the same location on each web for accurate measurements.

3. Use a plumb bob to transfer points 3 through 10 from the frame to chalk marks on the floor.
4. Raise the vehicle with the floor jack, remove the jack stands, and lower the vehicle onto the floor. Move the vehicle away from the chalk-marked area.

**SERVICE TIP:** When using a tape measure, avoid twists and bends in the tape to provide accurate readings.

5. Measure the distance between points 1 and 2 with a tape measure, and chalk mark the exact half-way point in this distance. This mark is the frame centerline at the front.
6. Measure the distance between points 11 and 12, and chalk mark the exact center of this measurement. This chalk mark is the frame centerline at the rear.
7. Draw a straight chalk line from the centerline at the front of the frame to the centerline at the rear of the frame. This chalk line is the complete centerline of the frame.

**Horizontal Frame Measurements.** The procedure used for horizontal frame measurements follows:

1. Measure the distance from the frame centerline to points 3 through 10. Each pair of these points should be equal within 0.125 in (3 mm). For example, the distance from point 5 to the centerline should be equal to the distance from point 6 to the centerline.
2. Measure diagonally from point 1 to 6 and point 2 to 5. These distances should be equal or within the manufacturer's specified tolerance. Place a straight chalk mark from points 1 to 6 and points 2 to 5. These diagonal lines should cross each other at the frame centerline.
3. Repeat step 2 at all the other diagonal frame measurements, such as points 3 to 10, 4 to 9, 5 to 12, and 6 to 11. All these diagonal measurements should be equal or within the vehicle manufacturer's specified tolerance. Each pair of diagonal lines must cross each other at the vehicle centerline (Figure 14-7).

When frame problems such as side sway, buckle, or diamond condition are present, some of the frame horizontal measurements are not within specifications, and some of the diagonal chalk lines do not cross at the frame centerline, depending on the location of the damaged area.

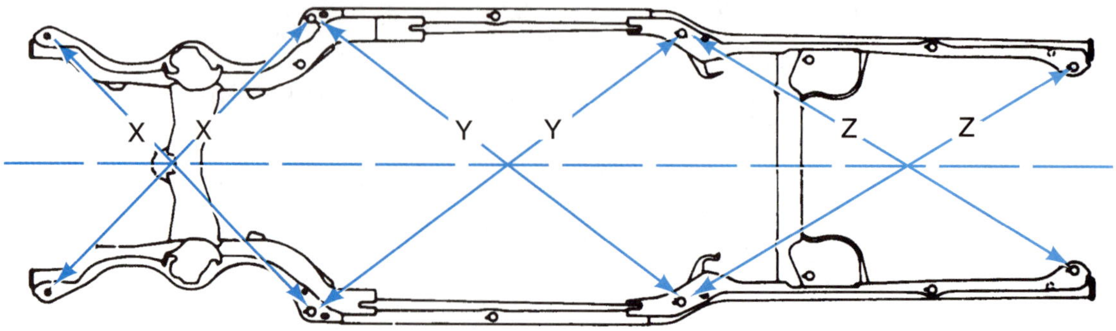


**Figure 14-7** Each pair of diagonal measurement lines on a frame must cross each other at the frame centerline. (Courtesy of Chevrolet Motor Division, General Motors Corporation)

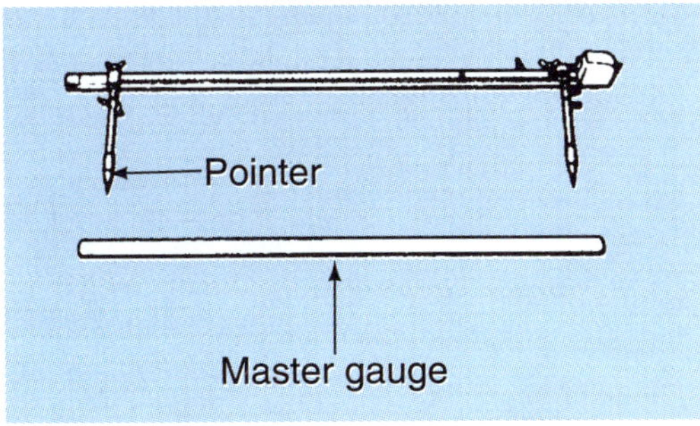


Figure 14-8 Tram gauge for frame measurements (Courtesy of Toyota Motor Corporation)

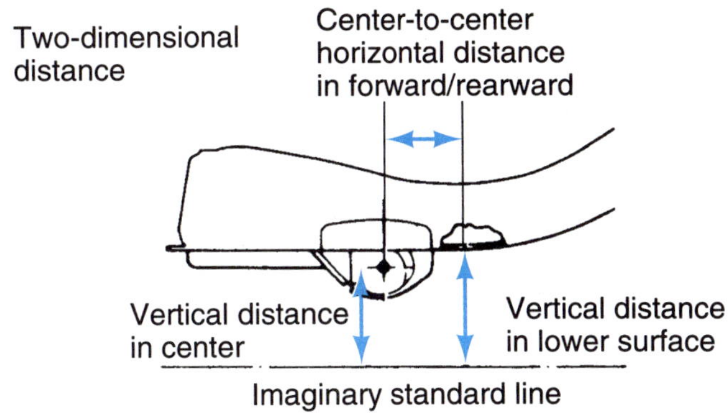


Figure 14-9 Horizontal and vertical frame measurements (Courtesy of Toyota Motor Corporation)

## Frame Measurement, Tram Gauge Method

**TRADE JARGON:** A tram gauge may be called a tracking, or track, gauge.

A tram gauge is a long, straight bar with two adjustable pointers. The distance between the pointers is adjustable, and the height of the pointers is also adjustable (Figure 14-8).

The horizontal frame measurements may be completed with a tram gauge rather than using the plumb bob method. When measuring a frame, horizontal measurements are completed straight across the frame and diagonally forward and rearward on the frame as we have discussed previously.

Vertical measurements from specific locations on the frame to an imaginary standard line are also necessary to detect such conditions as frame sag and twist (Figure 14-9). This imaginary horizontal line used for vertical frame measurements is called the datum line.

An imaginary horizontal line parallel to the frame is used for vertical frame measurements. This line is called the datum line.

**SERVICE TIP:** When the tram gauge is installed on the frame, be sure the pointers are seated properly in the frame openings (Figure 14-10).

When vertical frame measurements are completed with the tram gauge, the gauge pointers must be set at the manufacturer's specified distance for each frame location (Figure 14-11). With the tram gauge pointers set at the specified height and the gauge properly installed across the frame in the recommended frame openings, the tram gauge should be level, if the frame vertical measurements are within specifications. If the frame is twisted, the tram gauge is not level when adjusted to specifications and installed in the twisted area.

**Special Tools**

Tram gauge

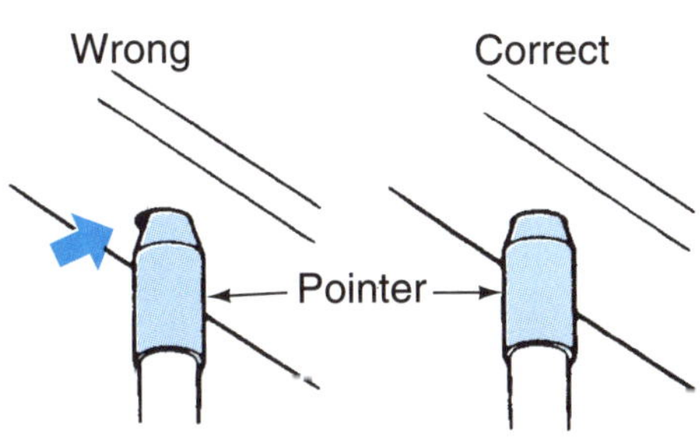


Figure 14-10 Proper and improper tram gauge pointer seating in the frame openings (Courtesy of Toyota Motor Corporation)

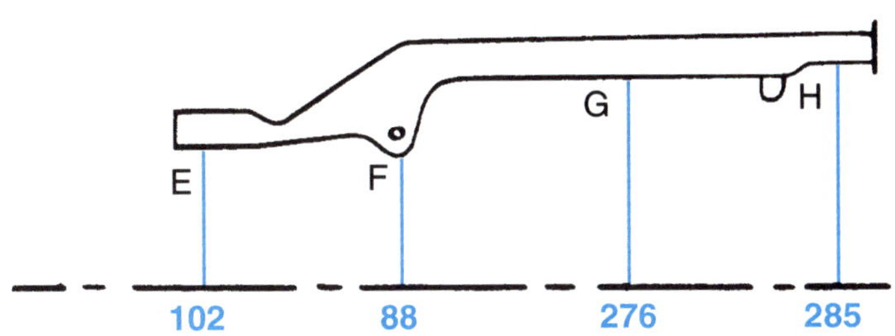


3. All tolerances ±3 mm
4. All dimensions are calculated on a horizontal plane

Figure 14-11 Tram gauge setting and locations for vertical frame measurements, rear subframe (Courtesy of Cadillac Motor Car Division, General Motors Corporation)

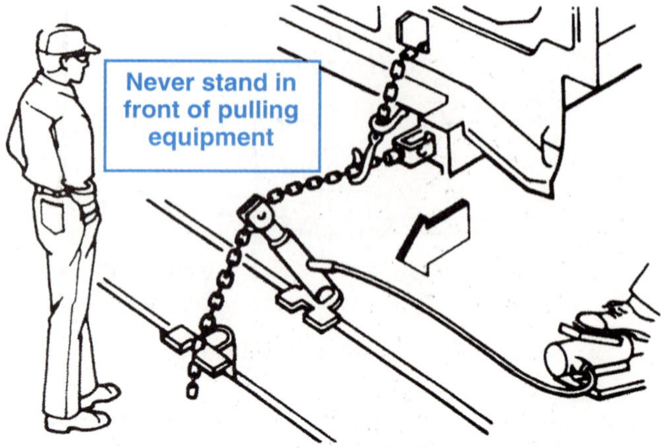


**Figure 14-12** Do not stand in front of pulling equipment while it is in operation. (Courtesy of Oldsmobile Division, General Motors Corporation)

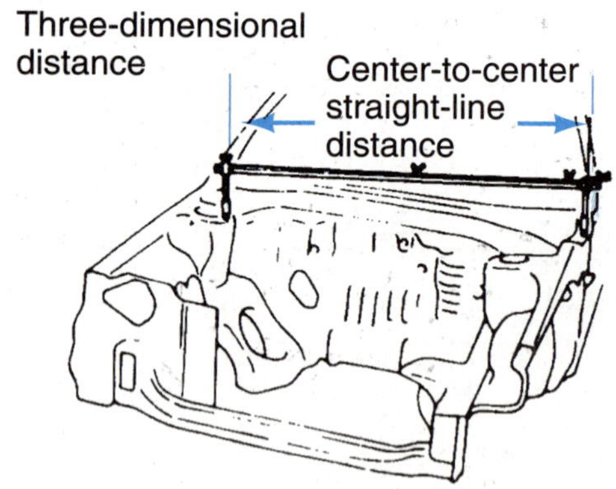


**Figure 14-13** Performing an upper strut tower measurement with a tram gauge on a unitized body vehicle (Courtesy of Toyota Motor Corporation)

To measure frame sag, three tram gauges may be installed at various locations for vertical frame measurement. When viewed from the front or the rear, the tram gauges must be level with each other. For example, if the tram gauge near the center of the frame is lower than the tram gauges at the front and rear of the frame, the frame is sagged.

### Frame Straightening

**Classroom Manual**
Chapter 14, page 284

Frame straightening is usually done with special hydraulically operated bending equipment. This equipment must be operated according to the equipment manufacturer's recommended procedures. Since frame straightening is usually done by experienced body technicians, we will not discuss this service in detail. All safety precautions must be observed while operating frame-straightening equipment. For example, never stand in front of pulling equipment when it is in operation (Figure 14-12).

## Measuring Unitized Body Alignment

### Tram Gauge

A tram gauge may be used to perform various measurements on unitized bodies such as the upper strut towers (Figure 14-13) and front crossmember (Figure 14-14).

Since the lower control arms are attached to the front crossmember, or cradle, front wheel alignment angles are affected by a bent crossmember. The engine and transaxle mounts are also connected to the front crossmember. Therefore, a bent crossmember may cause improper engine and transaxle position, which results in front drive axle vibration. Loose or worn front crossmember mounts may cause improper front wheel alignment and engine position. The tram gauge may be used to perform vertical measurements on the front and rear subframes on a unitized body vehicle.

### Bench

**Special Tools**

Dedicated bench system

A dedicated bench system is necessary to check many unitized body measurements after medium or heavy collision damage. The dedicated bench system has three main parts: the bench, transverse beams, and the dedicated fixtures. When used together, this equipment will perform many undercar unitized body measurements, including length, width, and height at the same time. The bench contains strong steel beams mounted on heavy casters (Figure 14-15). The top of the bench acts as a datum line.

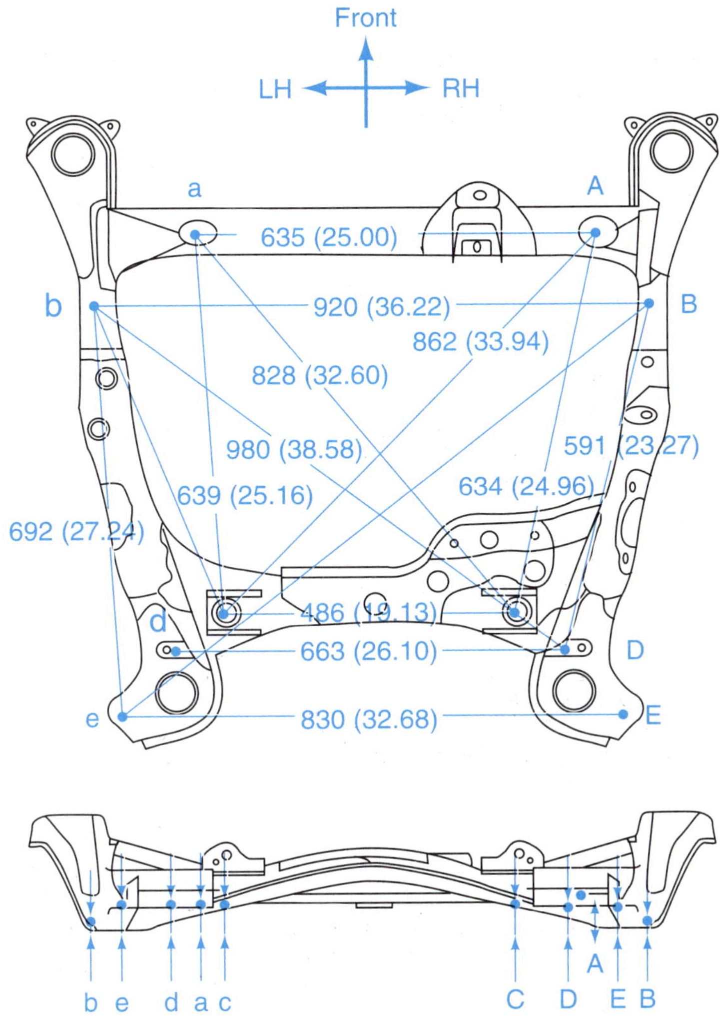


**Figure 14-14** Tram gauge measurements on the front crossmember of a unitized body vehicle (Courtesy of Toyota Motor Corporation)

## Transverse Beams

The transverse beams are mounted perpendicular to the bench. There are a variety of holes in the bench for proper transverse beam attachment (Figure 14-16). Various holes in the transverse beams provide the correct dedicated fixture positions.

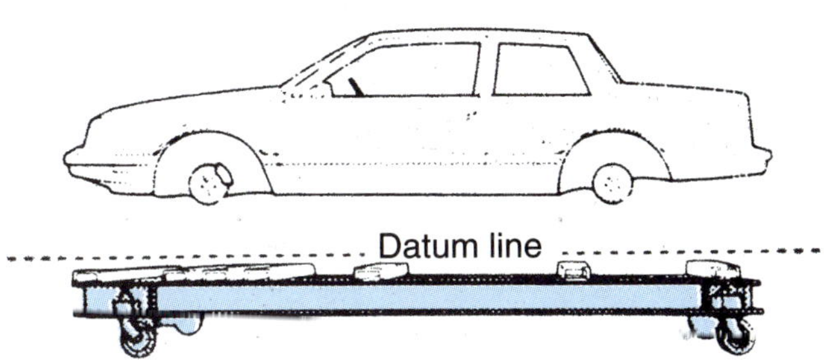


**Figure 14-15** Bench from a dedicated bench system (Courtesy of Oldsmobile Division, General Motors Corporation)

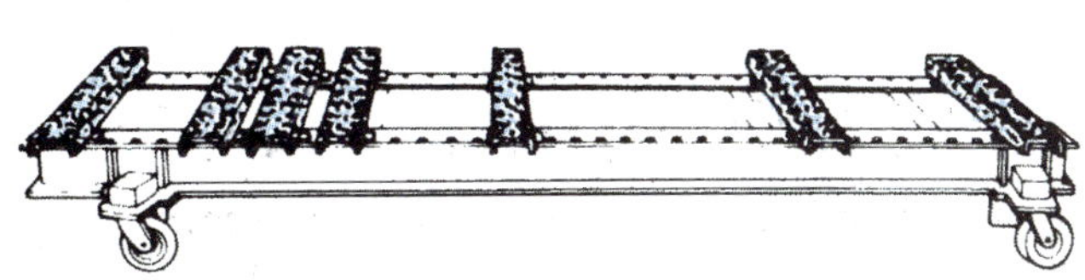

**Figure 14-16** Transverse beams placed on the bench (Courtesy of Oldsmobile Division, General Motors Corporation)

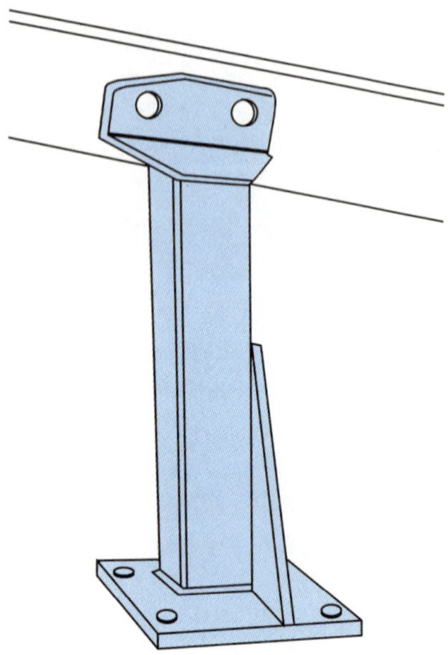

Figure 14-17 Dedicated fixtures are specific to the body style being measured. (Courtesy of Oldsmobile Division, General Motors Corporation)

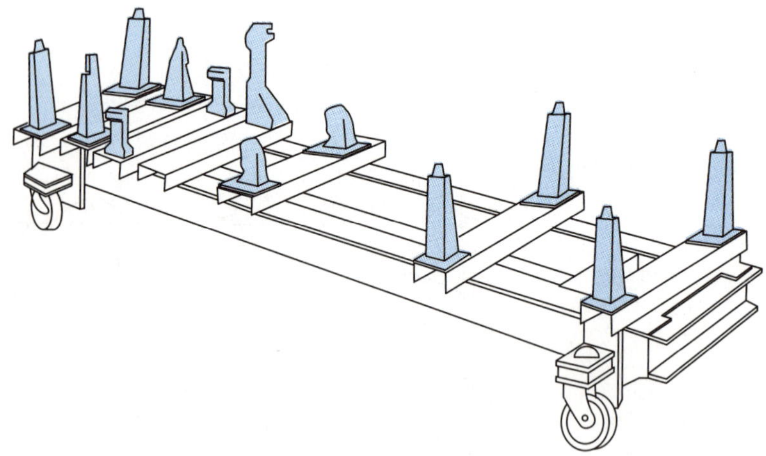

Figure 14-18 Dedicated fixtures are bolted to the transverse beams in specific locations depending on the body style. (Courtesy of Oldsmobile Division, General Motors Corporation)

## Dedicated Fixtures

The dedicated fixtures are specific to the body style being measured (Figure 14-17). These fixtures are bolted to the transverse beams, and the holes in the upper end of the fixtures must be aligned with specific body openings (Figure 14-18).

The unitized body must be straightened until the holes in the dedicated fixtures and the body openings are aligned. If the holes in the unitized body are behind and above the dedicated fixture openings, the unitized body must be pulled forward and downward in that area (Figure 14-19).

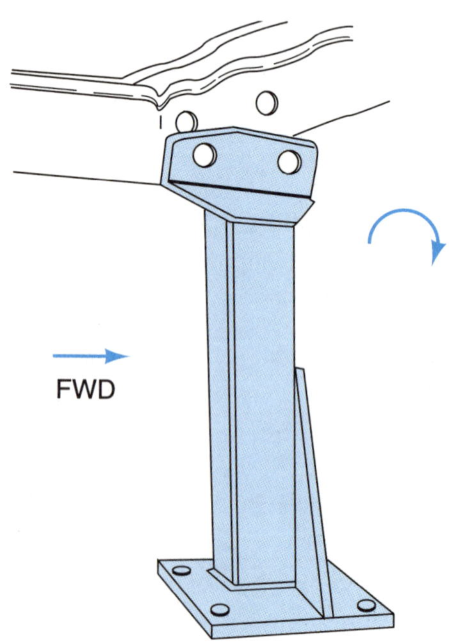


Figure 14-19 When the holes in the body are behind and above the openings in the dedicated fixture, the unitized body must be pulled forward and downward in that area. (Courtesy of Oldsmobile Division, General Motors Corporation)

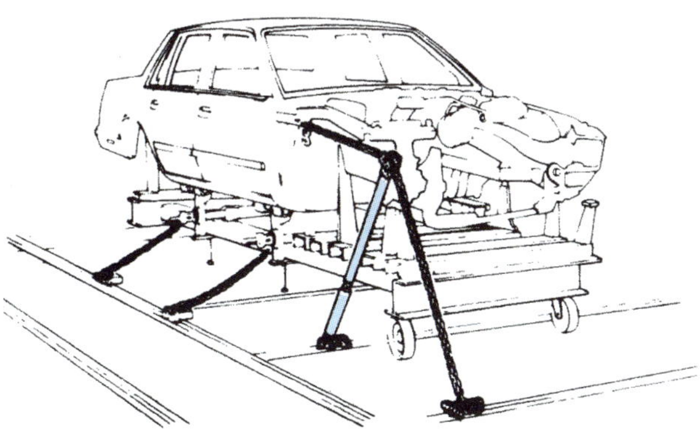

Figure 14-20 Single-pull unitized body straightening system (Courtesy of Oldsmobile Division, General Motors Corporation)

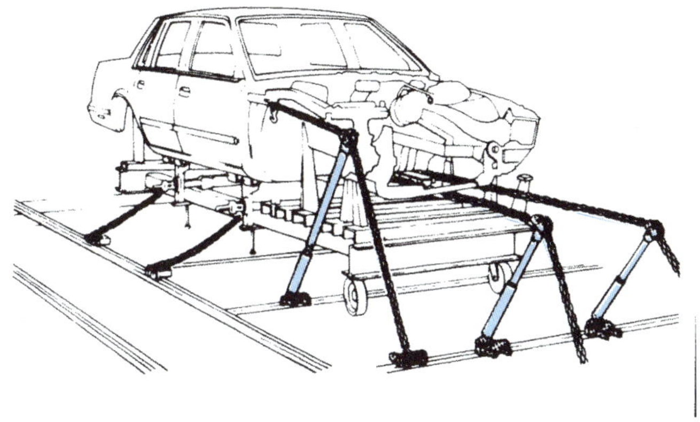

Figure 14-21 Multipull unitized body straightening system (Courtesy of Oldsmobile Division, General Motors Corporation)

### Unitized Body Straightening

**WARNING:** Always follow all the recommended precautions and procedures in the vehicle manufacturer's service manual and in the equipment manufacturer's operator's manual when straightening a unitized body. Failure to follow these precautions and procedures may result in severe personal injury and property damage.

Since unitized body straightening is usually performed by experienced body technicians, our discussion is brief on this subject. During the straightening process on a unitized body, the body is securely bolted to the dedicated bench at all the possible locations except those where misalignment is present. The dedicated bench is securely chained to special holding fixtures in the shop floor. Hydraulically operated pulling equipment is then attached securely to the unitized body to pull the body in the intended direction. Pulling equipment may be single pull in one location (Figure 14-20), or multipull in several locations at the same time (Figure 14-21). The unitized body is pulled until the holes in the body are aligned with the holes in the dedicated fixtures.

**Classroom Manual**
Chapter 14, page 280

## Guidelines for Diagnosing and Measuring Frames and Unitized Bodies

1. Frame damage may be caused by vehicle overloading or uneven loading.
2. Frame damage may result from operating the vehicle on extremely rough terrain or mounting additional equipment, such as a snow plow on the vehicle, when the frame was not designed for this type of equipment.
3. Frame side sway may be caused by collision damage, fire damage, or the use of equipment for which the frame was not designed.
4. There are many causes of frame sag, including excessive or uneven loads, drilling too many holes in the frame, cutting holes or notches in the frame with a torch, collision or fire damage, and sudden changes in section modulus.
5. Frame buckle may be caused by collision or fire damage, and additional equipment attached to the frame for which the frame was not designed.
6. A diamond-shaped frame may be caused by collision damage, or towing and/or being towed with the chain attached to one corner of the frame.
7. Frame twist may result from collision damage or operating the vehicle on extremely rough terrain.

8. When arc welding on a vehicle, follow the basic safety precautions, including wearing proper protective clothing, removing the negative battery cable, removing the fuel tank, and protecting the interior and exterior of the vehicle.
9. Frames should be visually inspected for wrinkles on the upper and lower flanges, and for cracks or bends.
10. A plumb bob may be suspended from specific points on the vehicle frame, and these points marked with chalk. The distance between these points may be measured with a tape measure to check the frame.
11. Horizontal and vertical frame measurements may be performed with a tram gauge.
12. When the tram gauge is installed to measure vertical frame measurements, the pointers should be set at the specified distance and the tram gauge should be level.
13. A tram gauge may be used for various measurements in unitized body vehicles.
14. A dedicated bench system is necessary to check many unitized body measurements after medium or heavy collision damage.
15. The holes in the dedicated bench fixtures must be aligned with the openings in the unitized body if the body measurements are correct.

## CASE STUDY

A customer complained about a vibration problem on an Oldsmobile Cutlass Supreme with front wheel drive. Further questioning of the customer revealed that the vibration occurred when accelerating and decelerating at 40 to 60 mph (64 to 96 kph). During a road test, the technician found the customer's description of the problem was accurate. The car did have a vibration problem at the speed indicated by the customer. While road testing the car, the technician thought about the causes of this problem and decided the most likely cause of the vibration would be inner front drive axle joints. The technician informed the customer that a drive axle joint inspection was necessary.

The technician raised the vehicle on a hoist and checked the front drive axle joints. He was surprised to discover the inner and outer joints were all in good condition with no cracked boots or looseness. An inspection of the engine cradle indicated the crosspiece at the rear of the cradle was severely damaged near the left side of the cradle. It appeared this crosspiece may have struck an object, such as a rock, during off-road driving. A check of the cradle mounts indicated the cradle openings were jammed into the rubber mounts on the left side of the cradle. The rubber mounts were bulged at the rear of the cradle openings, indicating the left side of the cradle was driven rearward.

The technician informed the customer that a cradle removal, measurement, and possible replacement was necessary. After the customer approved this service operation, the technician removed the cradle and measured it with a tram gauge. The diagonal measurements on the cradle were not within the manufacturer's specifications. Since the cradle was severely distorted, a replacement cradle was installed, and all damaged cradle mounting bushings were replaced. A road test indicated the vibration problem was corrected.

## Terms to Know

Section modulus
Yield strength
Frame web
Frame flange
Sunburst frame cracks
Plumb bob
Tram gauge
Datum line
Dedicated bench system
Single-pull unitized body straightening system
Multipull unitized body straightening system

## ASE Style Review Questions

1. While discussing frame sag:
*Technician A* says frame sag may be caused by holes drilled in the frame flange.
*Technician B* says frame sag may be caused by holes drilled too close together in the frame web.
Who is correct?
**A.** A only
**B.** B only
**C.** Both A and B
**D.** Neither A nor B

2. While discussing frame side sway:
*Technician A* says frame side sway may be caused by uneven load distribution.
*Technician B* says frame side sway may result from too many holes drilled in the frame web.
Who is correct?
**A.** A only
**B.** B only
**C.** Both A and B
**D.** Neither A nor B

3. While discussing a diamond-shaped frame condition:
*Technician A* says this condition may be caused when the vehicle is involved in a fire.
*Technician B* says this condition may be caused by towing another vehicle with the chain attached to one corner of the frame.
Who is correct?
**A.** A only
**B.** B only
**C.** Both A and B
**D.** Neither A nor B

4. When discussing the plumb bob method of frame measurement:
*Technician A* says the distance from the frame centerline to the same point on each side of the frame should be equal.
*Technician B* says diagonal frame measurement lines on each side of the vehicle should cross each other at the frame centerline.
Who is correct?
**A.** A only
**B.** B only
**C.** Both A and B
**D.** Neither A nor B

5. While discussing vertical frame measurements:
*Technician A* says the datum line is an imaginary line that is sloped toward the rear of the vehicle in relation to the frame.
*Technician B* says the datum line is an imaginary line that is parallel to the vehicle frame and is used for vertical frame measurements.
Who is correct?
**A.** A only
**B.** B only
**C.** Both A and B
**D.** Neither A nor B

6. While discussing vertical frame measurements:
*Technician A* says the tram gauge should be level when it is placed across the frame at the specified locations.
*Technician B* says the pointers on the tram gauge are not adjustable.
Who is correct?
**A.** A only
**B.** B only
**C.** Both A and B
**D.** Neither A nor B

7. When discussing unitized body measurements:
*Technician A* says a tram gauge may be used for many unitized body measurements.
*Technician B* says a dedicated bench system is used to check unitized body measurements after medium or heavy collision damage.
Who is correct?
**A.** A only
**B.** B only
**C.** Both A and B
**D.** Neither A nor B

8. While discussing unitized body measurements:
*Technician A* says a dedicated bench system performs measurements between the upper strut towers.
*Technician B* says the transverse beams on a dedicated bench system are specific to the body style being measured.
Who is correct?
**A.** A only
**B.** B only
**C.** Both A and B
**D.** Neither A nor B

9. While discussing unitized body measurements with a dedicated bench system:
*Technician A* says the holes in the dedicated fixtures must be aligned with matching holes in the unitized body.
*Technician B* says the dedicated fixtures are bolted to the bench.
Who is correct?
**A.** A only
**B.** B only
**C.** Both A and B
**D.** Neither A nor B

10. While discussing frame measurement with a tram gauge:
*Technician A* says three tram gauges may be installed across the frame at various locations to check for sag.
*Technician B* says tram gauge pointers should be at the specified measurement for each frame measurement location.
Who is correct?
**A.** A only
**B.** B only
**C.** Both A and B
**D.** Neither A nor B

## Table 14-1 ASE TASK

Check front wheel setback; determine needed repairs.

| Problem Area | Symptoms | Possible Causes | Classroom Manual | Shop Manual |
|---|---|---|---|---|
| **STEERING CONTROL** | Steering pull while driving straight ahead | Severe front wheel setback condition | 283 | 342 |

## Table 14-2 ASE TASK

Check front cradle (subframe) alignment; determine needed repairs or adjustments.

| Problem Area | Symptoms | Possible Causes | Classroom Manual | Shop Manual |
|---|---|---|---|---|
| **STEERING CONTROL** | Steering pull while driving straight ahead | 1. Improper camber caused by improperly positioned cradle | 287 | 346 |
| | | 2. Improper setback caused by improperly positioned cradle | 283 | 346 |

## Table 14-3 ASE TASK

Check SAI (steering axis inclination)/KPI (king pin inclination); determine needed repairs.

| Problem Area | Symptoms | Possible Causes | Classroom Manual | Shop Manual |
|---|---|---|---|---|
| **STEERING CONTROL** | Steering pull during hard acceleration | Improper SAI on one side of front suspension | 283 | 346 |
| | Steering pull while braking | Improper SAI on one side of front suspension | 283 | 346 |
| | Erratic steering on road irregularities | Improper SAI on one side of front suspension | 282 | 346 |

## Table 14-4 ASE TASK

Check included angle; determine needed repairs.

| Problem Area | Symptoms | Possible Causes | Classroom Manual | Shop Manual |
|---|---|---|---|---|
| **STEERING CONTROL** | Steering pull while braking or accelerating | Improper SAI on one side of front suspension | 283 | 346 |
| | Erratic steering on road irregularities | Improper SAI on one side of front suspension | 282 | 346 |

## Table 14-5 ASE TASK

Check and adjust front wheel toe.

| Problem Area | Symptoms | Possible Causes | Classroom Manual | Shop Manual |
|---|---|---|---|---|
| TIRE LIFE | Feathered wear on front tire treads | Improper front wheel toe | 287 | 346 |

## Table 14-6 ASE TASK

Check toe-out-on-turns (turning radius); determine needed repairs.

| Problem Area | Symptoms | Possible Causes | Classroom Manual | Shop Manual |
|---|---|---|---|---|
| TIRE LIFE | Excessive front tire tread wear, scuffing | Improper turning radius | 287 | 346 |
| TIRE NOISE | Tire squeal while cornering | Improper turning radius | 287 | 346 |
| STEERING WHEEL POSITION | Steering wheel not centered while driving straight ahead | Improper tie rod sleeve adjustment | 287 | 346 |

CHAPTER 15

# Camber and Caster Measurement and Adjustment

Upon completion and review of this chapter, you should be able to:

- ❑ Perform a prealignment inspection.
- ❑ Diagnose wheel alignment problems related to camber and caster settings.
- ❑ Diagnose tire wear problems.
- ❑ Measure front wheel camber.
- ❑ Adjust front wheel camber on various front suspension systems.
- ❑ Diagnose bent front struts.
- ❑ Measure front wheel caster.
- ❑ Adjust front wheel caster on various front suspension systems.

## Wheel Alignment Preliminary Diagnosis and Inspection

**Basic Tools**

Basic technician's tool set

Service manual

### Customer Complaints Related to Suspension or Brakes

Drivers may experience a variety of symptoms related to incorrect wheel alignment or defective brakes (Figure 15-1). All the symptoms in Figure 15-1 are not related to wheel alignment, but the customer may request an alignment to correct these problems. The technician must diagnose and correct the cause of the specific complaint. Wheel alignment angles that affect tire wear are toe and camber. The most likely symptoms of incorrect toe or camber settings are pull to one side while driving, wander, feathered tire wear, or tire wear on one side.

### Road Test

In many cases, a road test may be necessary to verify customer complaints. During a road test, the vehicle should be driven under the conditions when the customer complaint occurred. While road

Which result from wheel alignment?

- Vibration, shimmy
- Pull while driving
- Pull while braking
- Wander
- Slow returnability
- Wheel off-center
- Tire wear
- High effort

Are any symptoms really misunderstandings of normal characteristics?

Figure 15-1 Customer complaints related to wheel alignment or brakes (Courtesy of Chrysler Corporation)

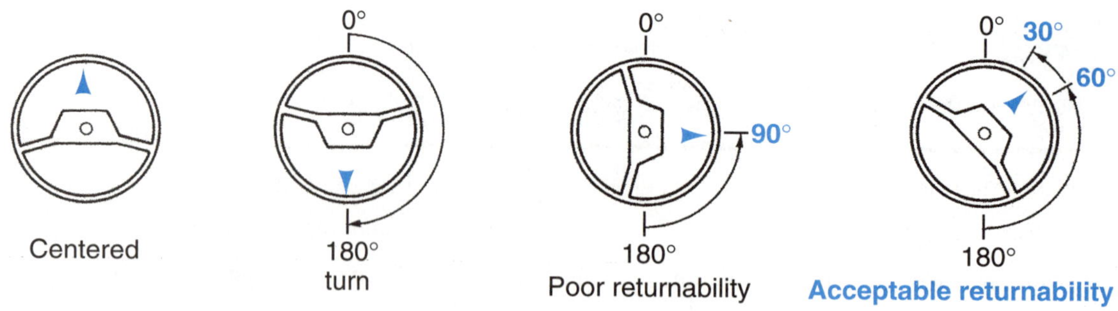


**Figure 15-2** Steering wheel returnability (Courtesy of Chrysler Corporation)

testing the vehicle, the technician should listen for any unusual noises related to suspension and steering. The technician should check for these steering problems during a road test:

> Torque steer may be defined as the tendency of the steering to pull to one side during hard acceleration.

1. excessive vertical chassis oscillations
2. chassis waddle
3. steering wander, pull, or drift
4. high steering effort and binding
5. tire squeal while cornering
6. bump steer
7. torque steer
8. memory steer
9. steering wheel return

> Bump steer is the tendency of the steering to veer suddenly in one direction when one or both of the front wheels strikes a bump.

Chassis waddle may be caused by excessive wheel or tire runout. Excessive vertical chassis oscillations are the result of defective shock absorbers or struts. (Refer to Chapter 4 for wheel and tire runout diagnosis, and Chapter 5 for shock absorber and strut diagnosis.)

Most steering systems are designed so the steering wheel returns to within 30° to 60° of center after a 180° turn (Figure 15-2). If the steering wheel returnability is not satisfactory and memory steer is evident, the steering shaft dash seal or the steering shaft universal joints may be binding. When memory steer occurs on a front wheel drive car, the upper strut mounts may be binding. High steering effort, or binding, can also be caused by a loose power steering belt, low power steering fluid level, worn front tires, or a defective steering gear.

> When memory steer occurs, the vehicle doesn't want to steer straight ahead after a turn because the steering does not return to the straight-ahead position.

A front wheel drive vehicle with unequal drive axles produces some torque steer on hard acceleration. Some front wheel drive vehicles, especially those with higher horsepower, have equal front drive axles to reduce torque steer. On a front wheel drive vehicle, torque steer is aggravated by different tire tread designs on the front tires or uneven wear on the front tires.

Bump steer occurs if the tie rods are not the same height, and therefore one of the tie rods is not parallel to the lower control arm. This condition may be caused by a bent or worn idler arm or pitman arm on a parallelogram steering linkage. On a rack and pinion steering system, worn steering gear mounting bushings may cause this unparallel condition between one of the tie rods and the lower control arm.

Steering pull may be caused by improper caster or camber adjustments.

**SERVICE TIP:** If steering pull, or drift, is evident during the road test, drive the vehicle on the same road in the opposite direction to determine if a crosswind is causing the steering pull.

## Prealignment Inspection

The technician should identify the exact suspension or steering complaint before a wheel alignment or other suspension work is performed. The technician must diagnose the cause, or causes,

| Condition | Probable Cause |
|---|---|
| Tire wear:<br>Outer shoulder<br>Inner shoulder<br>Sawtooth pattern:<br>Sharp edge toward center<br>Sharp edge away from center<br><br>Both shoulders<br>Tread roll under<br>Cupping, dishing | Too much positive camber<br>Too much negative camber<br>Too little toe-in<br>Too much toe-in<br>Under inflation, high-speed cornering,<br>overloading<br>High speed in curves<br>Wheel imbalance, radial runout |
| Shimmy:<br>With or without vehicle instability<br>Without instability | Too much positive caster, unequal caster between wheels<br>Tire imbalance, tire runout, driveline vibration |
| Vibration<br>(Caster/camber not a probable cause. The condition is listed because it is sometimes misdiagnosed as shimmy.) | Driveline misalignment, driveline imbalance, vehicle shake (accompanied by a characteristic moan), tire runout, unequal weight distribution on each side of vehicle centerline between wheels |
| Steering wander/pull | Incorrect camber<br>Unequal caster<br>Overload, which elevates front end |
| Brake pull | Too much negative caster<br>Unequal tire pressures, brake line damage that impedes hydraulic action on one side |
| Hard steering | Incorrect caster<br>Damaged steering linkage, worn steering linkage, damaged spindles, rear end overload, bent steering arm causing incorrect turning angle. |

**Figure 15-3** Diagnosis of wheel alignment problems

of the complaint, and correct this problem during the wheel alignment (Figure 15-3). Since collision damage may affect wheel alignment angles, an inspection for collision damage on the vehicle should be completed prior to a wheel alignment (Figure 15-4).

Worn suspension and steering components must be replaced before a wheel alignment is performed. A wheel alignment should be performed after suspension components, such as struts, have been replaced. The following components and measurements should be checked in a prealignment inspection:

1. curb weight
2. tires
3. suspension height
4. steering wheel play
5. shock absorbers or struts
6. wheel bearing adjustment
7. ball joint condition
8. control arms and bushings

Look at the complete vehicle

**Note:** If you suspect collision damage to the inner body panels, measure the reference points of body and frame alignment as described in the service manual.

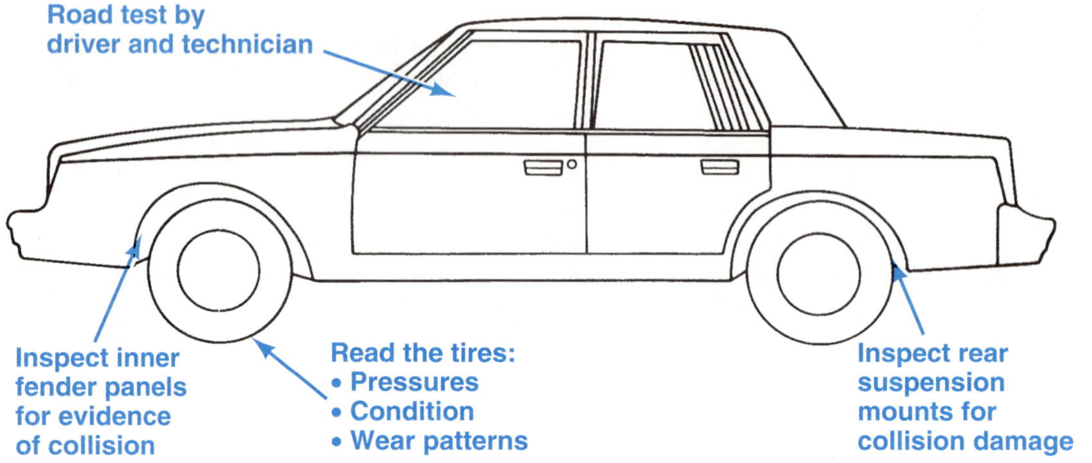


**Figure 15-4** Collision damage inspection (Courtesy of Chrysler Corporation)

9. steering linkages and tie rod ends
10. stabilizers and bushings
11. full fuel tank

Components to be checked in a prealignment inspection, and wheel alignment measurements are summarized as shown (Figure 15-5).

The following checks should be completed with the car on the floor:

1. Check for excessive mud adhered to the chassis. Remove heavy items from the trunk and passenger compartment that are not considered in the vehicle curb weight. If heavy items such as tool display cases or merchandise are normally carried in a vehicle, these items should be left in the car during a wheel alignment.
2. Inflate tires to the recommended pressure and note any abnormal tread wear or damage on each tire. Be sure all tires are the same size.
3. Check the front tires and wheels for radial runout. (Refer to Chapter 4 for this measurement.)
4. Check the suspension height. If this measurement is not within specifications, check for broken or sagged springs. On a torsion bar suspension system, check the torsion bar adjustment.
5. When the wheels are in the straight-ahead position, rotate the steering wheel back and forth to check for play in the steering column, steering gear, or linkages.
6. Check the shock absorbers or struts for loose mounting bushings and bolts. Examine each shock absorber or strut for leakage.
7. Check the condition of each shock absorber or strut with a bounce test at each corner of the vehicle. (This test is explained in Chapter 5.)

The following checks should be performed with the vehicle raised and the suspension supported:

1. Check the front wheel bearings for lateral movement. (This check is described in Chapter 3.) On front wheel drive cars, perform this check on all four wheel bearings. Wheel bearings must be properly adjusted prior to a wheel alignment. Clean, repack, or adjust the wheel bearings as necessary.
2. Measure the ball joint radial and axial movement. If excessive movement exists in either direction, ball joint replacement is required. (Ball joint diagnosis and replacement is

## PREALIGNMENT INSPECTION CHECKLIST

Owner ______________ (LAST) ______________ (FIRST) Phone __________ Date __/__/__

Address ______________________________ Ser. No. __________

Make ________ Model ________ Yr. ________ Lic. No. ________ Mileage ________

| **1. Road Test Disclosed** | | | | | |
|---|---|---|---|---|---|
| Vehicle Pulls | | Yes | No | Right | Left |
| | Above 30 MPH | | | | |
| | Below 30 MPH | | | | |
| | Bump Steer | | | | |
| | When Braking | | | | |
| Steering Wheel Movement<br>When Stopping From 2-3 MPH—Front Brakes | | | | | |
| Vehicle Steers Hard | | | | | |
| Steering Wheel Returns Normally | | | | | |
| Steering Wheel Position | | | | | |
| Vibration | | Yes | No | Front | Rear |

| **2. Tire Pressure** | SPECS Front ____ Rear ____ |
|---|---|
| RECORD PRESSURE FOUND<br>RF ____ LF ____ RR ____ LR ____ | |

| **3. Chassis Height** | SPECS Front ____ Rear ____ | |
|---|---|---|
| RECORD HEIGHT FOUND<br>RF ____ LF ____ RR ____ LR ____ | | |
| Springs Sagged | YES | NO |
| Torsion Bars Adjusted | | |

| **4. Rubber Bushings** | **OK** |
|---|---|
| Upper Control Arm | |
| Lower Control Arm | |
| Sway Bar/Stabilizer Link | |
| Strut Rod | |
| Rear Bushing | |

| **5. Shock Absorbers/Struts** | Front | Rear |
|---|---|---|
| | | |

| **6. Steering Linkage** | Rear OK | Front OK |
|---|---|---|
| Tie-Rod Ends | | |
| Idler Arm | | |
| Center Link | | |
| Sector Shaft | | |
| Pitman Arm | | |
| Gearbox/Rack Adjustment | | |
| Gearbox/Rack Mounting | | |

| **7. Ball Joints** | | **OK** |
|---|---|---|
| Load Bearings | | |
| SPECS | READINGS | |
| Right ____ Left ____ | Right ____ Left ____ | |
| Follower | | |
| Upper Strut Bearing Mount | | |
| Rear | | |

| **8. Power Steering** | **OK** |
|---|---|
| Belt Tension | |
| Fluid Level | |
| Leaks/Hose Fittings | |
| Spool Valve Centered | |

| **9. Tires/Wheels** | **OK** |
|---|---|
| Wheel Runout | |
| Condition | |
| Equal Tread Depth | |
| Wheel Bearing | |

| **10. Brakes Operating Properly** | |
|---|---|

| **11. Alignment** | Specs | | Initial Readings | | Adj. Readings | |
|---|---|---|---|---|---|---|
| | Right | Left | Right | Left | Right | Left |
| Camber | | | | | | |
| Caster | | | | | | |
| Toe | | | | | | |

| Bump Steer | TOE CHANGE Right Wheel | | TOE CHANGE Left Wheel | |
|---|---|---|---|---|
| | Amount | Direction | Amount | Direction |
| Chassis DOWN 3" | | | | |
| Chassis UP 3" | | | | |

| | Specs | | Initial Readings | | Adj. Readings | |
|---|---|---|---|---|---|---|
| | Right | Left | Right | Left | Right | Left |
| Toe-Out On Turns | | | | | | |
| SAI | | | | | | |
| Rear Camber | | | | | | |
| Rear Total Toe | | | | | | |
| Rear Indiv. Toe | | | | | | |
| Wheel Balance | Front ____ | | | Rear ____ | | |
| Radial Tire Pull | Yes ____ | | | No ____ | | |

**FLAG READINGS**

LR RR LF RF

COMMENTS:

Figure 15-5 Prealignment inspection and wheel alignment checklist

explained in Chapter 6.) Be sure the suspension is supported correctly during ball joint diagnosis.

3. Inspect the control arms for damage, and check the control arm bushings for wear.
4. Check all the steering linkages and tie rod ends for looseness.
5. Check for worn stabilizer mounting links and bushings.
6. Check for loose steering gear mounting bolts and worn mounting brackets and bushings.

| Condition | Rapid wear at shoulders | Rapid wear at center | Cracked treads | Wear on one side | Feathered edge | Bald spots | Scalloped wear |
|---|---|---|---|---|---|---|---|
| Effect | 1. 2. | | | | | | |
| Cause | Under-inflation or lack of rotation | Over-inflation or lack of rotation | Under inflation or excessive speed | Excessive camber | Incorrect toe | Unbalanced wheel<br>Or tire defect* | Lack of rotation of tires or worn or out-of-alignment suspension |
| Correction | Adjust pressure to specifications when tires are cool rotate tires | | | Adjust camber to specifications | Adjust toe-in to specifications | Dynamic or static balance wheels | Rotate tires and inspect suspension see group 2 |

**Figure 15-6** Tire tread wear patterns and causes (Courtesy of Chrysler Corporation)

### Tire Wear Diagnosis

Various types of tire tread wear patterns indicate specific alignment or balance defects (Figure 15-6).

## Camber Measurement and Adjustment

### Camber Measurement Procedure

A front suspension alignment may be performed with the vehicle on an alignment ramp and the front wheels positioned on turn tables (Figure 15-7). The alignment rack must be maintained in a level position on the shop floor.

There are many different types of equipment available to measure wheel alignment angles. All types of wheel alignment equipment must be used according to the equipment manufacturer's directions. Since magnetic gauges are one of the most basic types of alignment equipment, we will

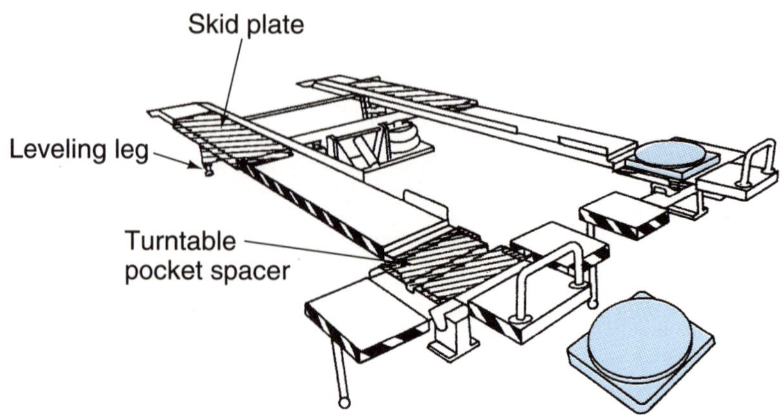


**Figure 15-7** Alignment ramp with turn tables (Courtesy of Snap-On Tools Corporation)

**Figure 15-8** Magnetic gauge mounted on wheel hub (Courtesy of Snap-On Tools Corporation)

**Figure 15-9** Magnetic gauge mounted on a rim clamp (Courtesy of Snap-On Tools Corporation)

discuss the use of typical magnetic gauges. The magnetic gauge may be mounted directly to the wheel hub (Figure 15-8) or attached to a rim clamp (Figure 15-9).

**Special Tools**

Magnetic wheel alignment gauge set

Rim clamps

**WARNING:** When rim clamps are used to mount magnetic gauges on aluminum or magnesium rims, be sure the clamps are designed for this type of rim. Rim clamps designed for steel rims only may damage aluminum or magnesium rims.

If the rim clamps are used to mount the magnetic gauges, the rims must be checked for runout. This is done by checking the camber reading, and then rotating the wheel 180° and taking another camber reading. A variation in the two readings indicates a bent rim.

Use the following procedure to measure front wheel camber:

1. Perform a prealignment inspection.
2. Position the vehicle with the front wheels on the turn tables.
3. Install the magnetic gauges on the front wheels and remove the turn table locking pins.

**SERVICE TIP:** Clean the outer lips on the front wheel hubs and check these surfaces for metal burrs before installing magnetic gauges. Any dirt or irregularities on these surfaces causes inaccurate magnetic gauge readings.

**WARNING:** If the magnetic gauges can be rocked vertically or horizontally after they are installed on the wheel hubs or rim clamps, recheck the gauge mounting surfaces for dirt or metal irregularities. Any gauge rocking action causes inaccurate camber or caster readings.

4. Place the front wheels in the straight position with the turn table pointers in the zero position.
5. Push up and down on the front bumper to jounce the vehicle several times.
6. Read the camber setting on each front wheel. The bubble on the gauge indicates the camber on each wheel.
7. Compare the camber setting on each front wheel to the manufacturer's specifications.

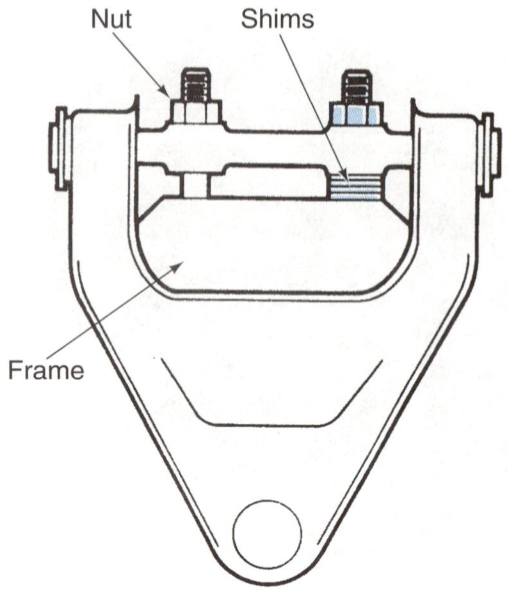


**Figure 15-10** Shim-type camber adjustment between the upper control arm mounting and the inside of the frame (Courtesy of Hunter Engineering Company)

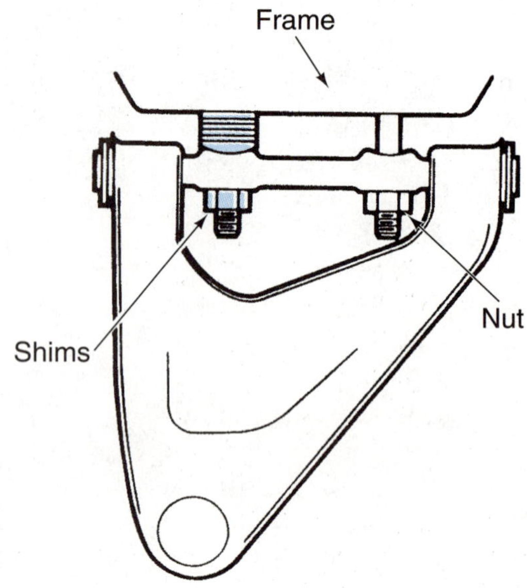


**Figure 15-11** Shim-type camber adjustment between the upper control arm mounting and the outside of the frame (Courtesy of Hunter Engineering Company)

## Camber Adjustment Procedure

**Shims.** Various methods are provided for camber adjustment by the car manufacturers. Some car manufacturers provide a shim-type camber adjustment between the upper control arm mounting and the inside of the frame (Figure 15-10). In this type of camber adjustment, increasing the shim thickness moves the camber setting to a more negative position, whereas decreasing shim thickness changes the camber toward a more positive position. Shims of equal thickness should be added or removed on both upper control arm mounting bolts to change the camber setting without affecting the caster setting.

Other vehicle manufacturers provide a shim-type camber adjustment between the upper control arm mounting and the outside of the frame (Figure 15-11). On this shim-type camber adjustment, increasing the shim thickness increases positive camber.

Eccentric cams are out-of-round pieces of metal mounted on a retaining bolt with the shoulder of the cam positioned against a component. When the cam is rotated, the component contacting the cam shoulder is moved.

**Eccentric Cams.** Some car manufacturers provide a cam-type camber adjustment on the inner ends of the upper control arm (Figure 15-12). Other suspension systems have a cam adjustment on the inner ends of the lower control arms (Figure 15-13). The suspension adjustments shown in Figures 15-10 through 15-13 are used on rear wheel drive cars with long and short arm front suspension systems.

Some MacPherson strut-type front suspension systems on front wheel drive cars have a cam on one of the steering knuckle-to-strut bolts to adjust camber (Figure 15-14).

**Slotted Strut Mounts and Frames.** On some MacPherson strut front suspension systems, the upper strut mounts may be loosened and moved inward or outward to adjust camber (Figure 15-15). Other MacPherson strut-type front suspension systems do not provide a camber adjustment. Automotive parts manufacturers sell a number of replacement suspension parts that provide adjustment capabilities when the original components did not have such capabilities. If the front wheel camber adjustment is not correct, and the suspension does not have a camber adjustment, check for worn components such as ball joints or upper strut mounts. Also check the engine cradle mounts. Since the inner ends of the lower control arms are attached to the engine cradle on many front wheel drive cars, a cradle that is out of position may cause improper camber readings.

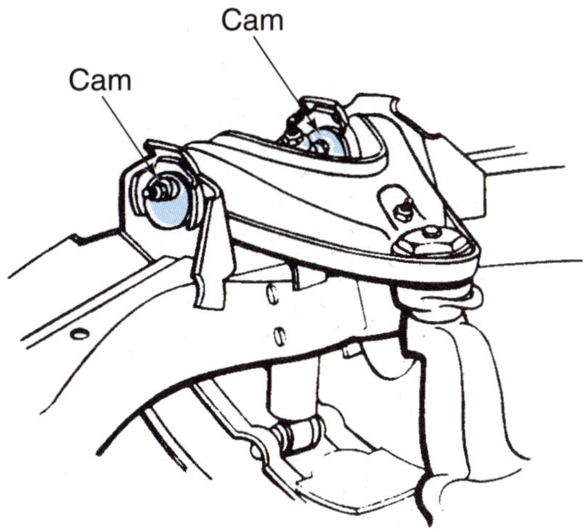


Figure 15-12 Cam adjustment on the inner ends of the upper control arms (Courtesy of Hunter Engineering Company)

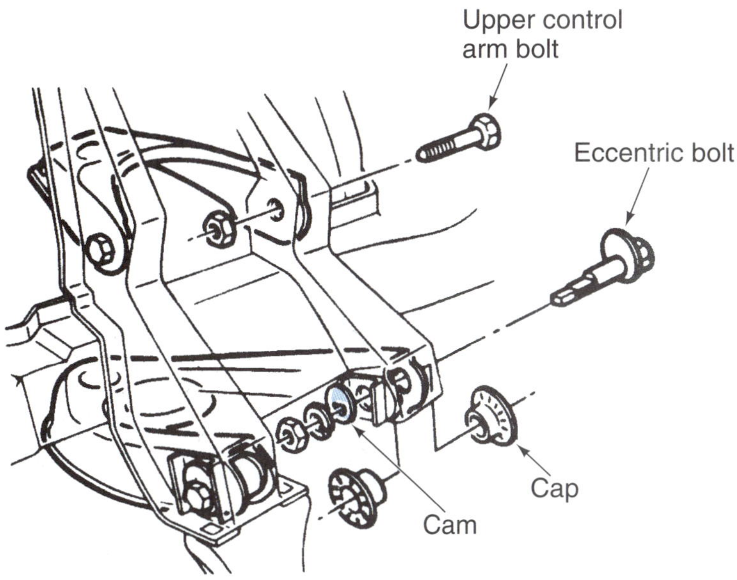


Figure 15-13 Cam adjustment on the inner ends of the lower control arms (Courtesy of Hunter Engineering Company)

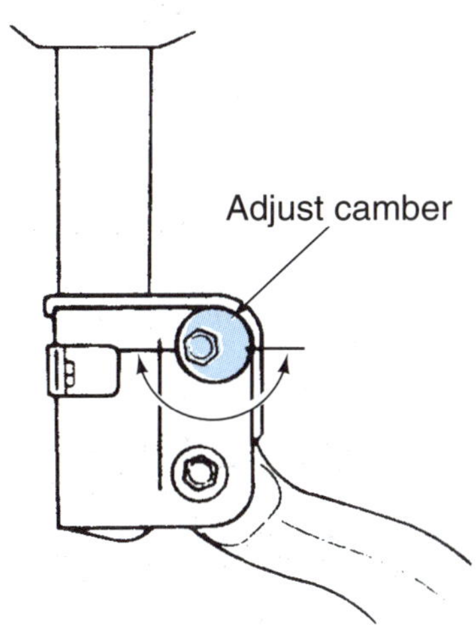


Figure 15-14 Eccentric steering knuckle-to-strut camber adjusting bolt, MacPherson strut front suspension (Courtesy of Hunter Engineering Company)

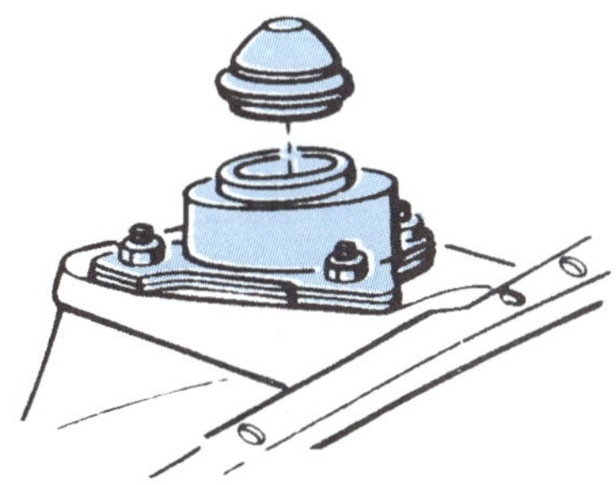

Figure 15-15 The upper strut mounting bolts may be loosened and the mount moved inward or outward to adjust camber on some MacPherson strut front suspension systems. (Courtesy of Hunter Engineering Company)

A camber plate is locked in place with a pop rivet on some upper strut mounts. The pop rivet may be removed to release the camber plate (Figure 15-16). After the upper strut mount bolts are loosened, the mount may be moved inward or outward to adjust the camber. When the camber is adjusted to the vehicle manufacturer's specification, the strut mount bolts should be tightened to the correct torque and a new pop rivet should be installed.

On some long and short arm front suspension systems, the bolt holes in the frame are elongated for the inner shaft on the upper control arm. When these bolts are loosened, the upper control arm may be moved inward or outward to adjust camber (Figure 15-17).

**Eccentric Bushing or Ball Joint Eccentric in Upper Control Arm.** In some long and short arm front suspension systems, an eccentric bushing in the upper end of the steering knuckle may

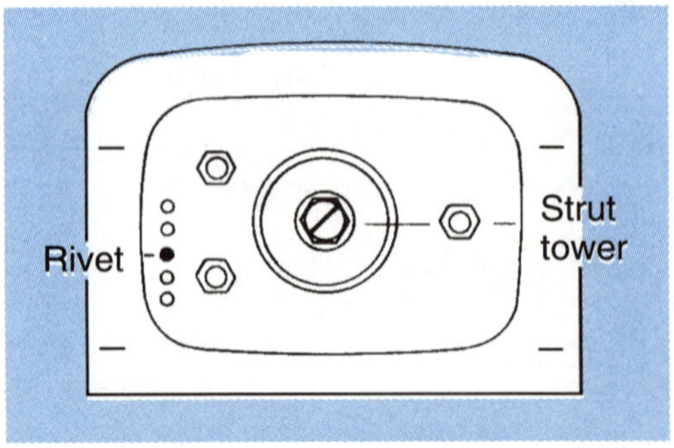


**Figure 15-16** Upper strut mount with camber plate and rivet, modified MacPherson strut front suspension (Courtesy of Hunter Engineering Company)

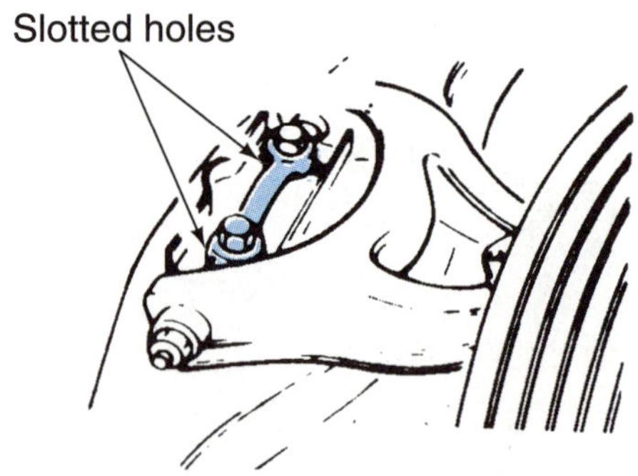


**Figure 15-17** Elongated bolt holes for upper control arm shaft bolts provide camber adjustment on some long and short arm front suspension systems. (Courtesy of Hunter Engineering Company)

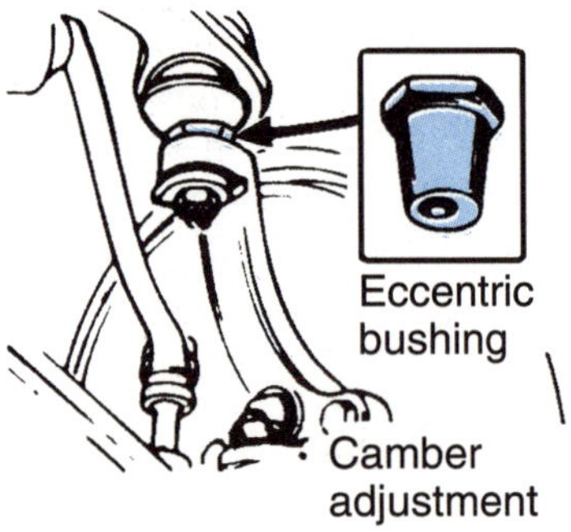


**Figure 15-18** An eccentric bushing in upper end of the steering knuckle may be used to adjust camber on some long and short arm front suspension systems. (Courtesy of Hunter Engineering Company)

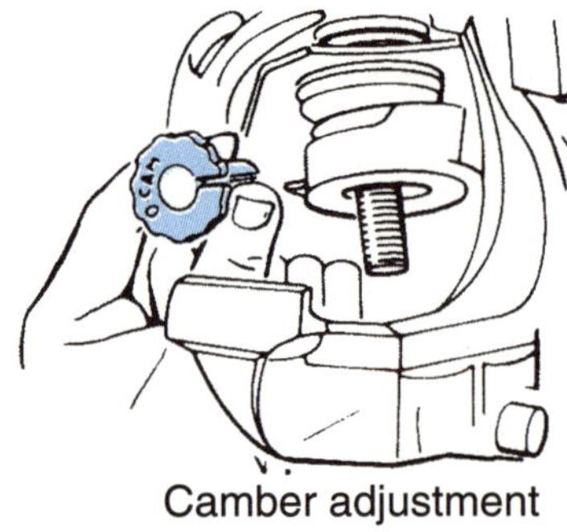


**Figure 15-19** Upper ball joint eccentric may be used to adjust camber. (Courtesy of Hunter Engineering Company)

be used to adjust camber (Figure 15-18). On other long and short arm front suspension systems, a ball joint eccentric may be rotated to adjust camber (Figure 15-19).

### Camber Adjustment on Twin I-Beam Front Suspension System

**Special Tools**

Camber-caster adjusting equipment, twin I-beam front suspension

On twin I-beam front suspension systems, a special clamp-type bending tool with a hydraulic jack is used to bend the I-beams and change the camber setting. On a twin I-beam front suspension, the clamp and hydraulic jack must be positioned properly for adjusting the camber to a more positive setting (Figure 15-20) and a more negative setting (Figure 15-21).

## Bent Front Strut Diagnosis

### Diagnostic Procedure

**Classroom Manual**
Chapter 15, page 293

Front struts that are bent forward or rearward may be diagnosed as follows:

1. With the vehicle on the aligning rack and the magnetic gauges in position for the camber measurement, swing one front wheel out 10° and read the camber.
2. Turn the same wheel inboard 10° and read the camber. The difference between the two readings is usually less than 4 1/2°. If the strut is bent forward or rearward, the difference in the camber readings will be excessive. A 10° difference is not uncommon.
3. Repeat steps 1 and 2 at the other front wheel.

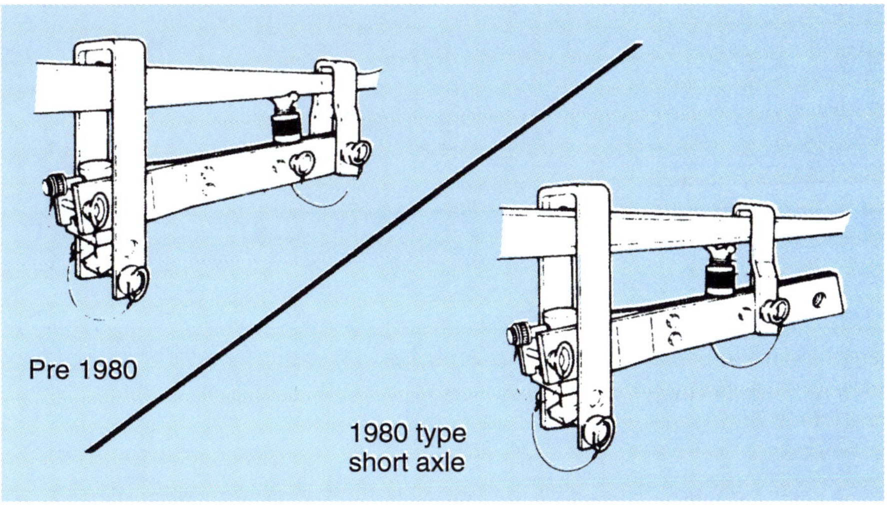


**Figure 15-20** Adjusting camber to a more positive setting, twin I-beam front suspension (Courtesy of Hunter Engineering Company)

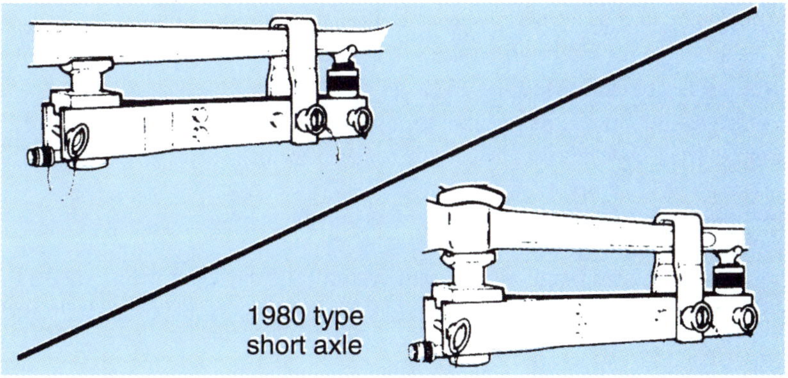


**Figure 15-21** Adjusting camber to a more negative setting, twin I-beam front suspension (Courtesy of Hunter Engineering Company)

Front struts could also be bent inboard or outboard. To diagnose this condition, use the following procedure:

1. With the wheel aligner in operation, sit on the front fender to load the suspension downward and read the camber.
2. Unload the suspension and lift up on the car while the camber is recorded. The two camber readings should be within 1/2°. If the strut is bent inboard or outboard, the difference in the two camber readings will be excessive. A 4° to 6° camber change is not uncommon.
3. Repeat steps 1 and 2 on the other front wheel.

# Caster Measurement and Adjustment

## Caster Measurement Procedure

Use the following procedure to measure the front wheel caster:

1. Install a brake pedal jack between the front seat and the brake pedal. Operate this jack to apply the brakes (Figure 15-22). If the vehicle has power brakes, the engine must be running when the brakes are applied.

**Special Tools**

Brake pedal jack

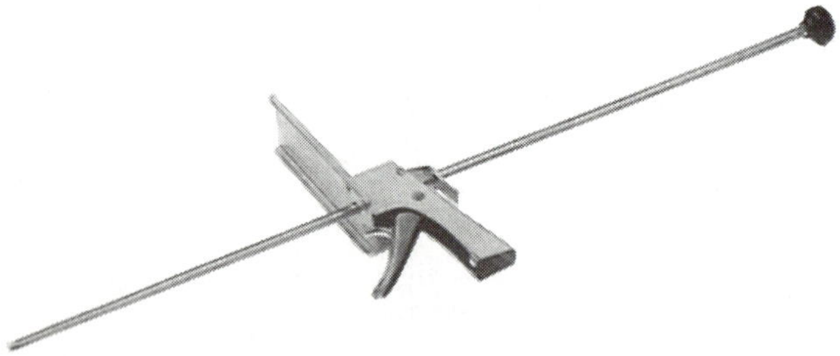

**Figure 15-22** Brake pedal jack is used to apply the brakes during caster measurement. (Courtesy of Snap-On Tools Corporation)

2. Install the magnetic gauges on the front wheel hubs and remove the turn table locking pins. If the wheel hubs are inaccessible for the magnetic gauges, use the rim clamps to mount the magnetic gauges.
3. Be sure the turn table pointers are in the zero position, and then turn the wheel being checked 20° outward as indicated by the turn table pointer.
4. Adjust the caster vial to zero on the wheel being checked (Figure 15-23).
5. Turn the wheel being checked to the 20° inward position on the turn table dial and record the caster reading (Figure 15-24).
6. Measure the caster on the opposite front wheel with the same procedure.

Photo Sequence 13 shows a typical procedure for measuring front wheel camber and caster.

## Caster Adjustment Procedure

**Shims.** On a long and short arm front suspension system, the same shims are used for camber and caster adjustment. If the shims are located between the upper control arm shaft and the inside of

**Figure 15-23** Adjusting caster vial to zero with wheel turned 20° outward (Courtesy of Snap-On Tools Corporation)

**Figure 15-24** Measuring caster with front wheel turned 20° inward (Courtesy of Snap-On Tools Corporation)

**Photo Sequence 13**

# Typical Procedure for Measuring Front Wheel Camber and Caster

P13-1 Position the vehicle on the alignment rack with the front tires located properly on the turn tables, and remove the turn table locking pins.

P13-2 Attach the magnetic gauges to the front wheel hubs.

P13-3 Jounce the front bumper several times to stabilize the suspension.

P13-4 Read the camber on each front wheel.

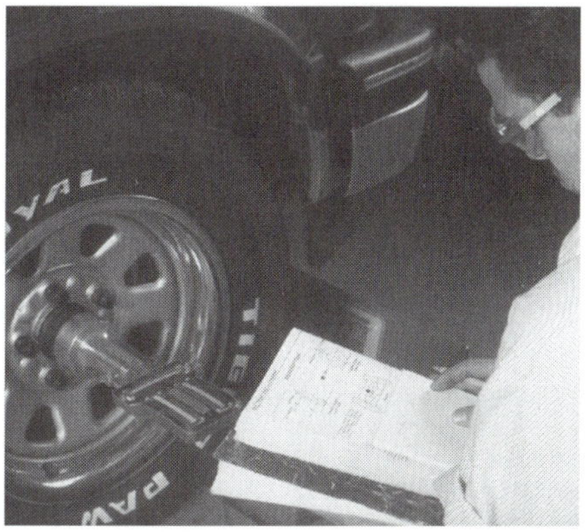

P13-5 Compare the camber reading to specifications.

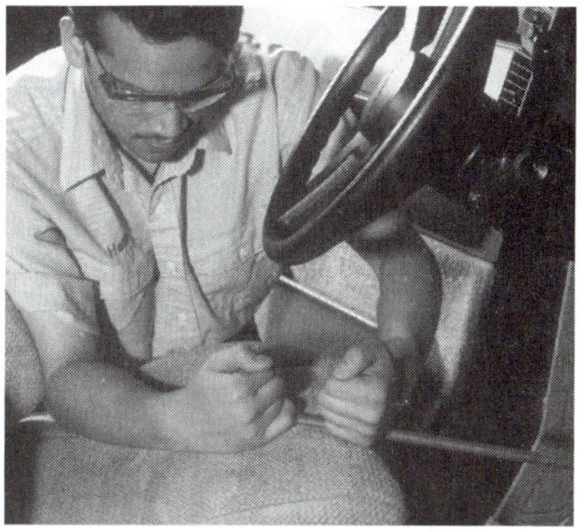

P13-6 Install a brake pedal jack between the brake pedal and the front seat.

P13-7 Turn the wheel being checked 20° outward and adjust the caster vial to zero.

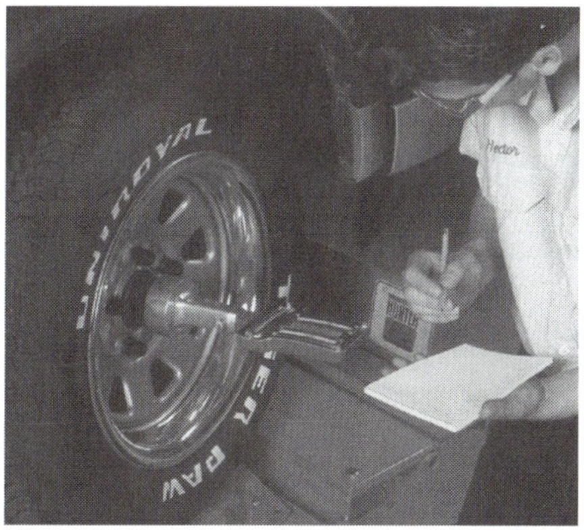

P13-8 Turn the wheel being checked 20° inward and record the caster reading.

P13-9 Measure the caster on the opposite front wheel with the same procedure.

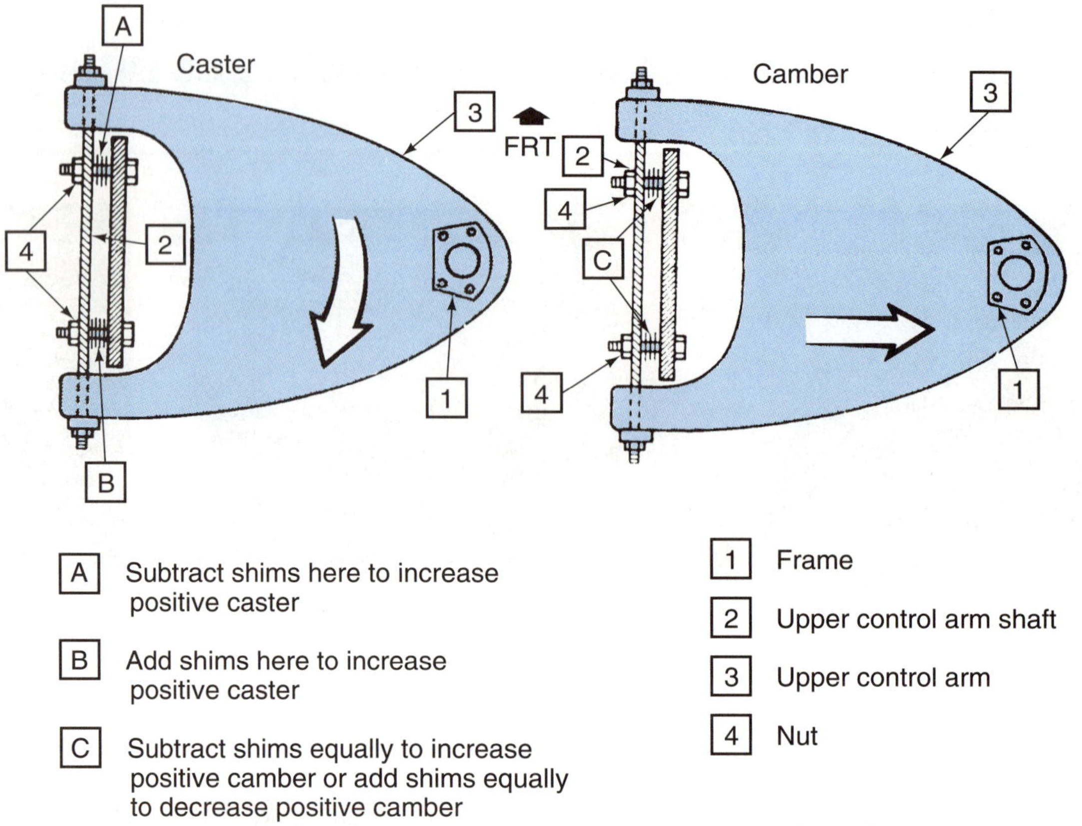


**Figure 15-25** Subtract shims from the front upper control arm mounting bolt and add shims to the rear mounting bolt to increase positive caster. (Courtesy of Chevrolet Motor Division, General Motors Corporation)

the frame, subtract shims from the front mounting bolt and add shims to the rear mounting bolt to increase positive caster (Figure 15-25). After the caster is adjusted, the camber should be re-checked.

**Strut Rod Length Adjustment.** On some suspension systems, the nuts on the forward end of the strut rod may be adjusted to lengthen or shorten the strut rod, which changes the caster setting (Figure 15-26). Shorten the strut rod to increase the positive caster.

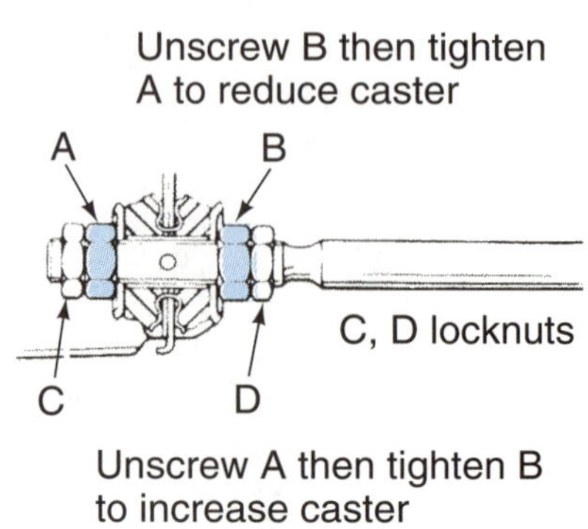


**Figure 15-26** Strut rod nuts may be adjusted to lengthen or shorten the strut rod and adjust caster. (Courtesy of Hunter Engineering Company)

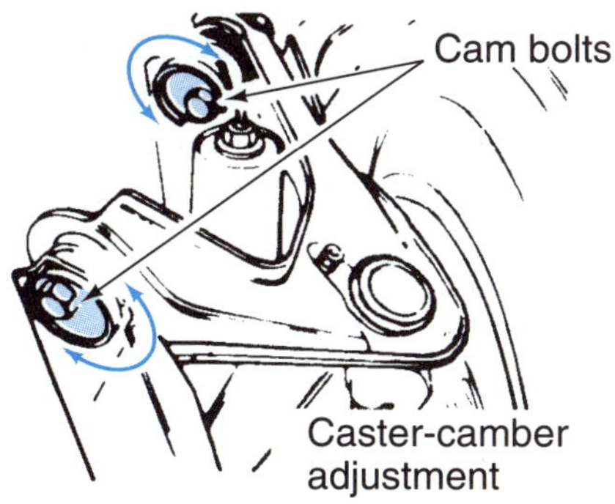


**Figure 15-27** Eccentric cams on the inner ends of the upper or lower control arms may be used to adjust camber and caster. (Courtesy of Hunter Engineering Company)

**Eccentric Cams.** The same eccentric cams on the inner ends of the upper or lower control arms may be used to adjust camber or caster (Figure 15-27). If an eccentric bushing in the outer end of the upper control arm is rotated to adjust camber, this same eccentric also adjusts caster.

**Slotted Strut Mounts and Frames.** If a caster adjustment is required and the suspension does not have an adjustment for this purpose, check for worn components such as upper strut mounts and ball joints. Also check the engine cradle mounts. Since the inner ends of the lower control arms are attached to the engine cradle on many front wheel drive cars, a cradle that is out of position may cause improper caster readings.

The slots in the frame at the upper control arm shaft mounting bolts provide camber and caster adjustment. If the upper strut mount is adjustable on a MacPherson strut suspension, the mount retaining bolts may be loosened and the mount moved forward or rearward to adjust caster. Rearward movement of the mount increases positive caster. In some suspension systems, one of the upper strut mount bolt holes has to be elongated with the proper size drill and a round file prior to caster adjustment (Figure 15-28).

**CAUTION:** After any alignment angle adjustment, always be sure the adjustment bolts are tightened to the vehicle manufacturer's specified torque. Loose suspension adjustment bolts will result in improper alignment angles, steering pull, and tire wear. Loose suspension adjustment bolts may cause reduced directional stability, which could result in a vehicle accident and possible personal injury.

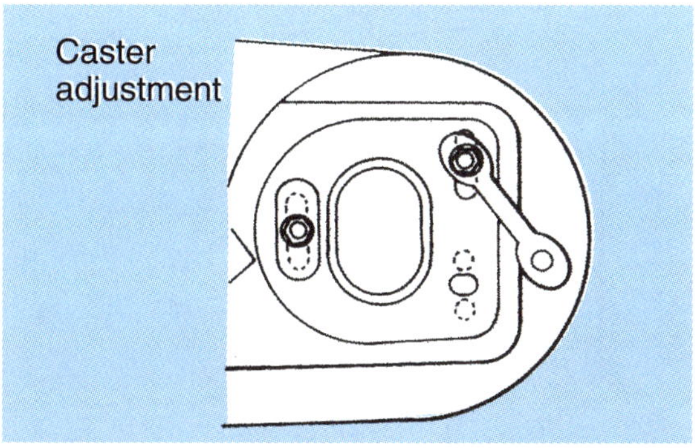


**Figure 15-28** Some adjustable upper strut mounts may be moved forward or rearward to adjust caster. (Courtesy of Hunter Engineering Company)

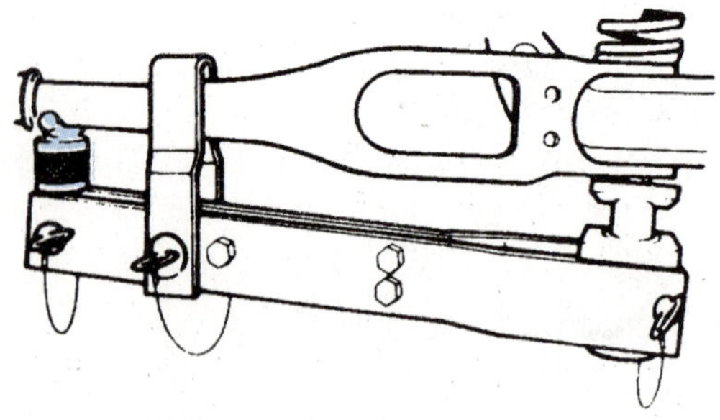

**Figure 15-29** Hydraulic jack and bending tool positioned on the radius rod to increase positive caster on a twin I-beam front suspension (Courtesy of Hunter Engineering Company)

## Caster Adjustment on Twin I-Beam Front Suspension Systems

**CAUTION:** Improper use of bending equipment may result in personal injury. Always follow the equipment and vehicle manufacturer's recommended procedures.

The same bending tool and hydraulic jack may be used to adjust camber and caster on twin I-beam front suspension systems. This equipment is installed on the radius rods connected from the I-beams to the chassis to adjust caster. If the hydraulic jack is placed near the rear end of the radius rod and the clamp arm is positioned over the center area of this rod, the rod is bent downward in the center when the jack is operated (Figure 15-29). This bending action tilts the I-beam and caster line rearward to increase positive caster.

**Classroom Manual**
Chapter 15, page 300

If the hydraulic jack is placed in the center area of the radius arm and the clamps are positioned at the ends of the arm, the arm is bent upward when the jack is operated (Figure 15-30). When this bending action occurs, the I-beam and caster line are tilted forward to decrease positive caster.

**CUSTOMER CARE:** Always concentrate on quality workmanship and customer satisfaction. Most customers do not mind paying for vehicle repairs if the work is done properly, and their vehicle problem is corrected. A follow-up phone call to determine customer satisfaction a few days after repairing a car indicates that you are interested in the customer's car and that you consider quality work and satisfied customers a priority.

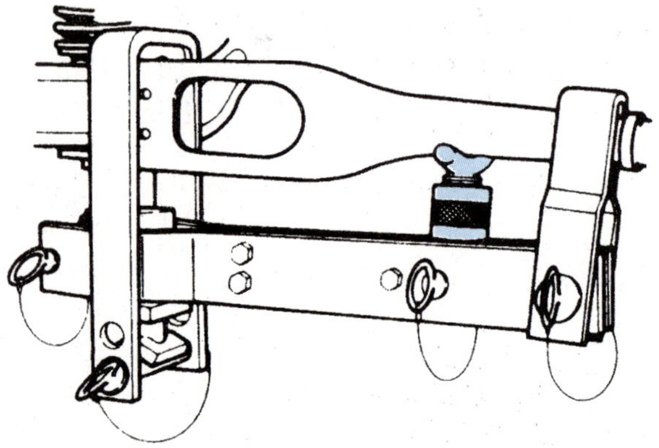

**Figure 15-30** Hydraulic jack and clamp positioned on radius rod to decrease positive caster on a twin I-beam front suspension system (Courtesy of Hunter Engineering Company)

# Guidelines for Measuring and Adjusting Camber and Caster

1. A road test may be necessary to verify customer complaints prior to a wheel alignment.
2. Worn suspension and steering components must be replaced before a wheel alignment is performed.
3. After suspension components such as struts have been replaced, a wheel alignment should be performed.
4. A prealignment inspection of all steering and suspension components must be completed before a wheel alignment procedure.
5. Improper camber settings cause wear on one side of the front tire treads.
6. Magnetic wheel alignment gauges may be mounted directly to the wheel hubs or on rim clamps.
7. The magnetic gauge mounting surfaces on the wheel hub lips or rim clamps must be clean and smooth to prevent vertical or horizontal magnetic gauge movement.
8. Front wheel camber is measured with the front wheels in the straight-ahead position.
9. Different methods for adjusting camber include shims, eccentric cams, slotted strut mounts and frames, eccentric bushings or ball joints, and upper strut mount camber plates.
10. A special tool is available to bend the I-beams and adjust camber on a twin I-beam front suspension.
11. To measure front wheel caster, the front wheel being measured is turned 20° outward and then 20° inward.
12. Caster adjustments on various suspension systems include shims, strut rod length, slotted upper strut mounts and frames, and eccentric cams.
13. On a twin I-beam front suspension system, the radius rods may be bent with a special tool to adjust caster.

## CASE STUDY

A customer complained about memory steer on a front wheel drive Olds 98. During a road test, the technician noticed the car did not have the directional controls he had experienced on other similar cars. The steering had a slight tendency to wander and the vehicle did not steer straight ahead after a turn, as the customer had indicated. The technician had to return the steering wheel to the straight-ahead position after a turn. The customer was advised that a wheel alignment and complete check of the front suspension was required to correct the problem.

A prealignment inspection did not reveal any suspension or steering problems, and the front tires did not indicate any abnormal wear. During the wheel alignment, the technician found the caster setting on each front wheel was +1°, but the caster specification was +3° plus or minus 0.5°. The technician assumed the lack of positive caster was causing the improper steering wheel return and memory steer as well as the wander problem.

On this vehicle, front wheel caster is adjustable by removing the crossbrace between the front struts and elongating two of the upper strut mounting bolt holes after the strut and mount are dropped downward until these mounting bolts pull out of the holes (Figure 15-31).

When the technician removed the crossbrace between the struts, she found the holes had been elongated previously, indicating someone had adjusted the caster at an earlier date. The technician adjusted the caster to the car manufacturer's specifications on both front wheels and checked all the other alignment angles on the front and rear suspension. All the

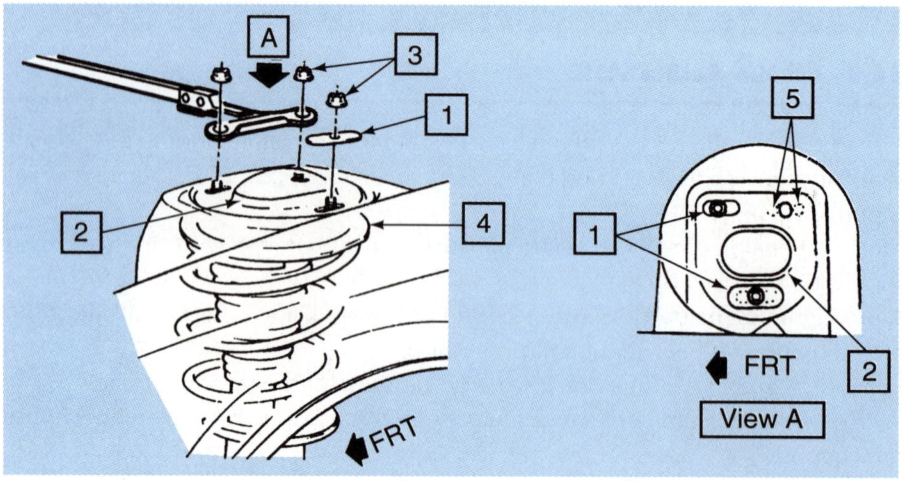


**Figure 15-31** Elongating upper strut mount bolt holes to adjust caster (Courtesy of Oldsmobile Division, General Motors Corporation)

other front and rear wheel alignment angles were within specifications. The technician thought the caster adjustment should correct the wander and memory steer problems, but she decided to road test the car to be sure the customer's complaint was corrected. A road test revealed that the car now had excellent directional control once the steering wheel was placed in the straight-ahead position, but the lack of steering wheel return and the memory steer problem were still present.

The technician thought about the possible causes and realized that a binding condition in an upper strut mount could result in the memory steer condition. With the vehicle parked on the shop floor and the engine running, the technician asked an assistant to turn the steering wheel fully in both directions. During this front wheel movement, the technician felt both front coil springs and found the right spring had a chattering and jerking action, which indicated a binding problem in the upper strut mount.

The customer was contacted to obtain permission to replace the upper strut mount. Once this approval was obtained, the technician removed the right front strut and replaced the upper mount. After the strut was reinstalled, a road test indicated the lack of steering wheel return and the memory steer problem were corrected.

## Terms to Know

Prealignment inspection
Alignment ramp
Bump steer
Torque steer
Memory steer
Turn tables
Camber adjustment
Caster adjustment
Magnetic gauges
Rim clamps
Eccentric cams or bushings
Brake pedal jack

# ASE Style Review Questions

1. While discussing camber adjustments on a MacPherson strut front suspension system:
   *Technician A* says an eccentric strut-to-steering knuckle bolt is used to adjust camber.
   *Technician B* says shims may be inserted between the lower control arm mounting bolts and the chassis to adjust camber.
   Who is correct?
   **A.** A only
   **B.** B only
   **C.** Both A and B
   **D.** Neither A nor B

2. While discussing camber adjustment on a long and short arm front suspension system with adjusting shims positioned between the upper control arm mounting shaft and the outside of the frame:
   *Technician A* says adding more shims on both mounting bolts increases negative camber.
   *Technician B* says adding more shims on both mounting bolts increases positive camber.
   Who is correct?
   **A.** A only
   **B.** B only
   **C.** Both A and B
   **D.** Neither A nor B

3. While discussing a front suspension height that is 1 in (2.54 cm) less than specified:
   *Technician A* says the suspension height must be correct before a wheel alignment is performed.
   *Technician B* says a satisfactory wheel alignment may be performed without correcting the suspension height.
   Who is correct?
   **A.** A only
   **B.** B only
   **C.** Both A and B
   **D.** Neither A nor B

4. While discussing tire pressure and wheel bearing adjustment as they relate to wheel alignment:
   *Technician A* says the front wheel bearing adjustment should be adjusted after a wheel alignment.
   *Technician B* says a tire pressure check is not required before a wheel alignment.
   Who is correct?
   **A.** A only
   **B.** B only
   **C.** Both A and B
   **D.** Neither A nor B

5. While discussing front wheel camber adjustment:
   *Technician A* says front wheel camber is measured with the front wheels straight ahead.
   *Technician B* says front wheel camber is measured with the front wheels turned 20° outward.
   Who is correct?
   **A.** A only
   **B.** B only
   **C.** Both A and B
   **D.** Neither A nor B

6. While discussing steering pull on a front suspension system in which the right front wheel has 2° positive camber and the left front wheel has 1/2° positive camber:
   *Technician A* says when the vehicle is driven straight ahead, the steering will pull to the left.
   *Technician B* says these camber settings will not affect steering quality.
   Who is correct?
   **A.** A only
   **B.** B only
   **C.** Both A and B
   **D.** Neither A nor B

7. When discussing the use of camber adjustment to compensate for road crown:
   *Technician A* says the left front wheel should have more positive camber than the right front wheel.
   *Technician B* says the right front wheel should have more positive camber than the left front wheel.
   Who is correct?
   **A.** A only
   **B.** B only
   **C.** Both A and B
   **D.** Neither A nor B

8. When discussing unsatisfactory steering wheel returnability:
   *Technician A* says one of the tie rod ends may be worn.
   *Technician B* says this problem may be caused by interference between the dash seal and the steering shaft.
   Who is correct?
   **A.** A only
   **B.** B only
   **C.** Both A and B
   **D.** Neither A nor B

9. When discussing a front suspension system with an adjustable radius rod to correct the caster setting:
*Technician A* says the radius rod should be shorted to decrease positive caster.
*Technician B* says the radius rod should be lengthened to decrease positive caster.
Who is correct?
**A.** A only
**B.** B only
**C.** Both A and B
**D.** Neither A nor B

10. While discussing caster adjustment on a twin I-beam front suspension:
*Technician A* says the I-beams are bent to adjust camber.
*Technician B* says the radius rods are bent to adjust caster.
Who is correct?
**A.** A only
**B.** B only
**C.** Both A and B
**D.** Neither A nor B

## Table 15-1 ASE TASK

Diagnose vehicle wander, drift, pull, hard steering, bump steer, memory steer, torque steer, and steering return problems; determine needed repairs.

| Problem Area | Symptoms | Possible Causes | Classroom Manual | Shop Manual |
|---|---|---|---|---|
| **STEERING QUALITY** | Wander, pull, drift | 1. Incorrect camber | 298 | 360 |
| | | 2. Unequal caster | 303 | 365 |
| | | 3. Rear suspension overload | 303 | 357 |
| | Excessive steering effort | Excessive positive caster | 302 | 360 |
| | Bump steer | 1. Worn or bent idler arm or pitman arm | 304 | 356 |
| | | 2. Worn steering gear mounting bushings | 304 | 356 |
| | Torque steer | 1. Unequal wear on front tires | 304 | 356 |
| | | 2. Different types of front tires or tire treads | 304 | 356 |
| | Steering wheel return problems | 1. Improper caster setting | 302 | 356 |
| | | 2. Binding in steering column or universal joints | 302 | 356 |

## Table 15-2 ASE TASK

Check and adjust front wheel camber on suspension systems with a camber adjustment.

| Problem Area | Symptoms | Possible Causes | Classroom Manual | Shop Manual |
|---|---|---|---|---|
| TIRE LIFE | Excessive wear on the edges of tire tread | Excessive positive or negative camber | 297 | 357 |
| STEERING CONTROL | Steering pull | Excessive positive camber, one front wheel | 298 | 357 |

## Table 15-3 ASE TASK

Check and adjust front wheel camber on nonadjustable suspension systems; determine needed repairs.

| Problem Area | Symptoms | Possible Causes | Classroom Manual | Shop Manual |
|---|---|---|---|---|
| TIRE LIFE | Excessive wear on the edges of the tire tread | Excessive positive or negative camber | 297 | 357 |
| STEERING CONTROL | Steering pull | Excessive positive camber, one front wheel | 298 | 357 |

## Table 15-4 ASE TASK

Check and adjust caster on suspension systems with a caster adjustment.

| Problem Area | Symptoms | Possible Causes | Classroom Manual | Shop Manual |
|---|---|---|---|---|
| STEERING CONTROL | Steering pull | Unequal caster on front wheels | 302 | 357 |
| STEERING EFFORT | Excessive steering effort | Excessive positive caster on both front wheels | 302 | 357 |
| STEERING WHEEL RETURN | Excessive steering wheel returning force | Excessive positive caster on both front wheels | 302 | 357 |

# Table 15-5 ASE TASK

Check and adjust caster on nonadjustable suspension systems; determine needed repairs.

| Problem Area | Symptoms | Possible Causes | Classroom Manual | Shop Manual |
|---|---|---|---|---|
| **STEERING CONTROL** | Steering pull | Unequal caster on front wheels | 302 | 365 |
| **STEERING EFFORT** | Excessive steering effort | Excessive positive caster on both front wheels | 302 | 365 |
| **STEERING WHEEL RETURN** | Excessive steering wheel returning force | Excessive positive caster on both front wheels | 302 | 365 |

CHAPTER 16

# Steering Axis Inclination, Setback, and Toe Measurement and Adjustment

Upon completion and review of this chapter, you should be able to:

- ❑ Correct setback conditions.
- ❑ Check and correct front engine cradle position.
- ❑ Measure steering axis inclination (SAI).
- ❑ Correct SAI angles that are not within specifications.
- ❑ Measure front wheel toe.
- ❑ Adjust front wheel toe.
- ❑ Measure turning radius.
- ❑ Correct a turning radius that is not within specifications.
- ❑ Measure toe change during front wheel jounce and rebound to check steering linkage height.
- ❑ Center steering wheel.

**Basic Tools**

Basic technician's tool set

Service manual

## Setback Measurement and Correction Procedure

Computer wheel aligners have setback measuring capabilities on the front wheels. These aligners automatically display setback angles with the other front suspension alignment angles as explained in the following chapter. If a front wheel drive car has experienced collision damage on the left front, the left side of the cradle, or subframe, may be driven rearward (Figure 16-1). This type of collision damage also moves the bottom of the left front strut rearward to a setback condition and shifts the caster to a negative position.

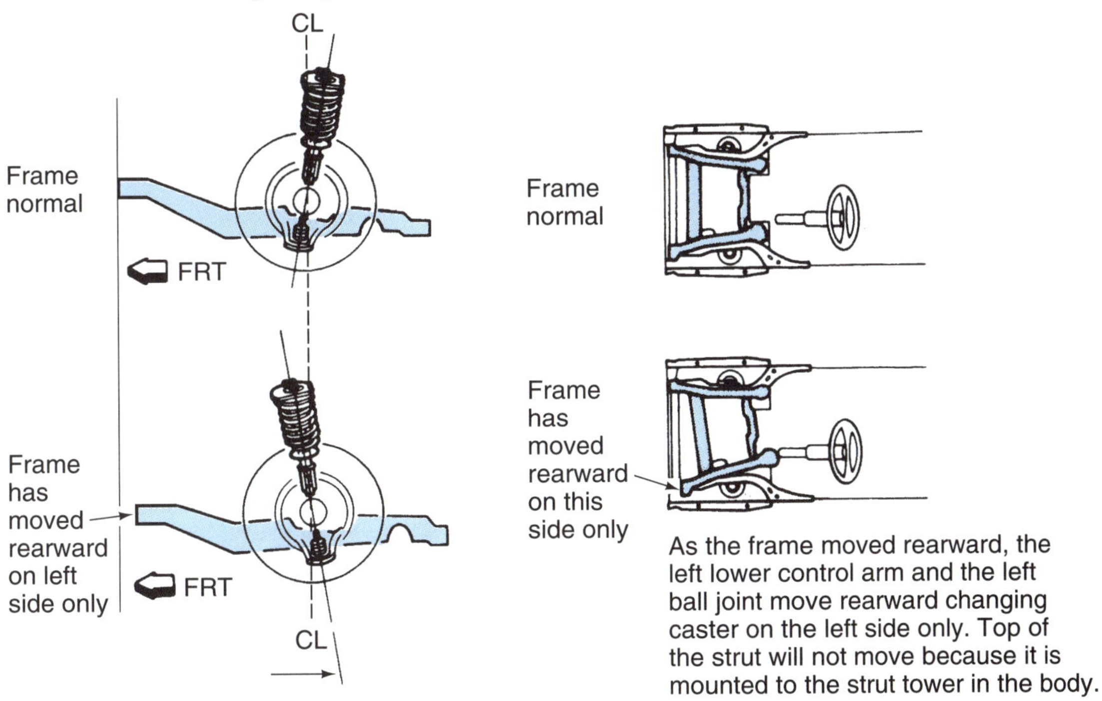


**Figure 16-1** Collision damage on the left front moves the cradle and strut rearward to a setback condition and shifts the caster to a negative position. (Courtesy of Oldsmobile Division, General Motors Corporation)

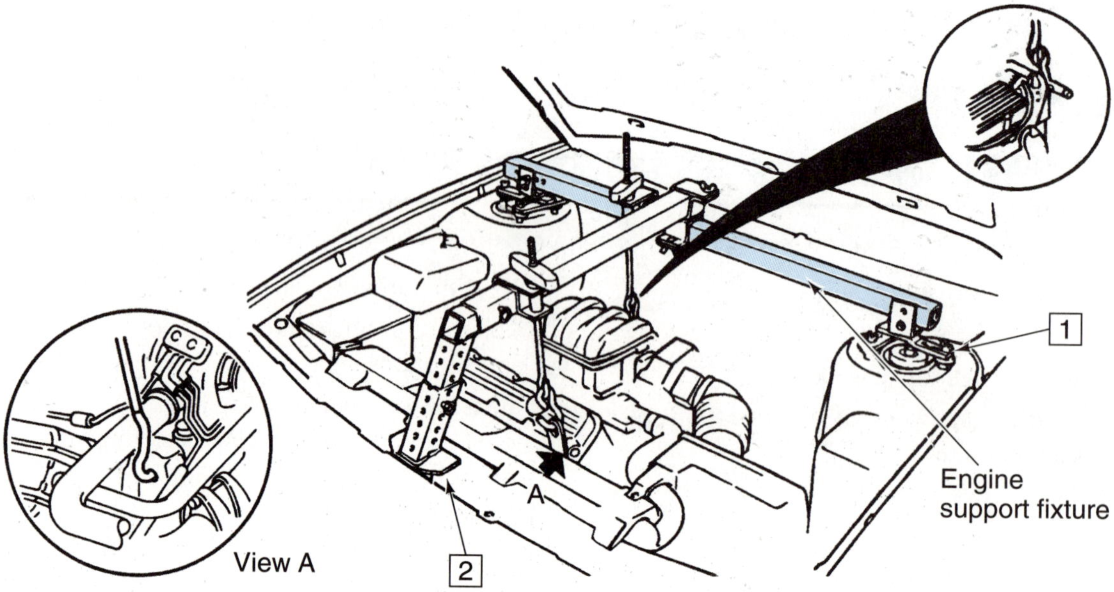


1 Thread onto strut attaching bolts (3) per side

2 Rubber pad rests on radiator support

**Figure 16-2** Engine support fixture installed to support engine and transaxle prior to cradle removal (Courtesy of Oldsmobile Division, General Motors Corporation)

**WARNING:** If the engine cradle has to be removed or loosened, the engine support fixture must hold the engine in the normal position to prevent damage to the fan shroud, power steering hoses, and other components.

**CAUTION:** Prior to any service operation where the body is going to be moved in relation to the cradle, or subframe, the intermediate shaft on the rack and pinion gear must be disconnected. If this shaft is not disconnected, it may be damaged, resulting in loss of steering control and serious vehicle damage and personal injury.

A tram gauge is a long, straight, adjustable bar with an adjustable, graduated pointer at each end of the bar. A tram gauge is used to perform measurements on automotive bodies and frames.

**Special Tools**

Tram gauge

With this type of collision damage, it may be possible to adjust the caster back to the specified setting, but that does not correct the setback problem. If the caster is adjusted and the setback problem is ignored, the steering may pull to one side while driving straight ahead, especially if the setback condition is severe. The cradle must be straightened and the unitized body must be pulled back to the original measurements to correct this problem. A severely bent, distorted cradle should be replaced. Prior to cradle removal, an engine support fixture must be installed to support the engine and transaxle (Figure 16-2). The engine support fixture must hold the engine in the original position to prevent damage to the power steering hoses or other components.

Before the body is moved in relation to the cradle, or subframe, the intermediate shaft on the rack and pinion steering gear must be disconnected to prevent damage to this shaft. Remove all the engine and transaxle mount bolts in the cradle, plus the cradle retaining bolts and insulators to remove the cradle. Perform the front underbody measurements with a tram gauge (Figure 16-3), and pull the unitized body back to these specifications. This pulling operation is usually done in an automotive body shop rather than an automotive repair shop.

The cradle must be restored to the specified measurements (Figure 16-4). Check all cradle insulators and replace any that are damaged, oil soaked, or worn. When the cradle is replaced, all the insulators must be installed properly, and all the cradle bolts should be aligned with the bolt

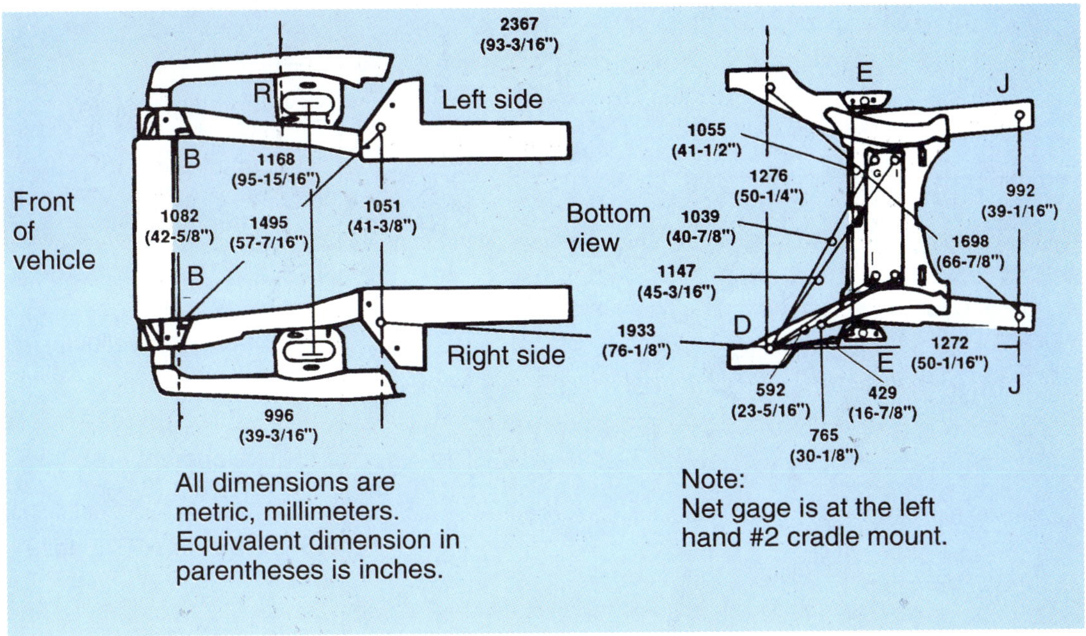


Figure 16-3 Front and rear underbody measurements (Courtesy of Oldsmobile Division, General Motors Corporation)

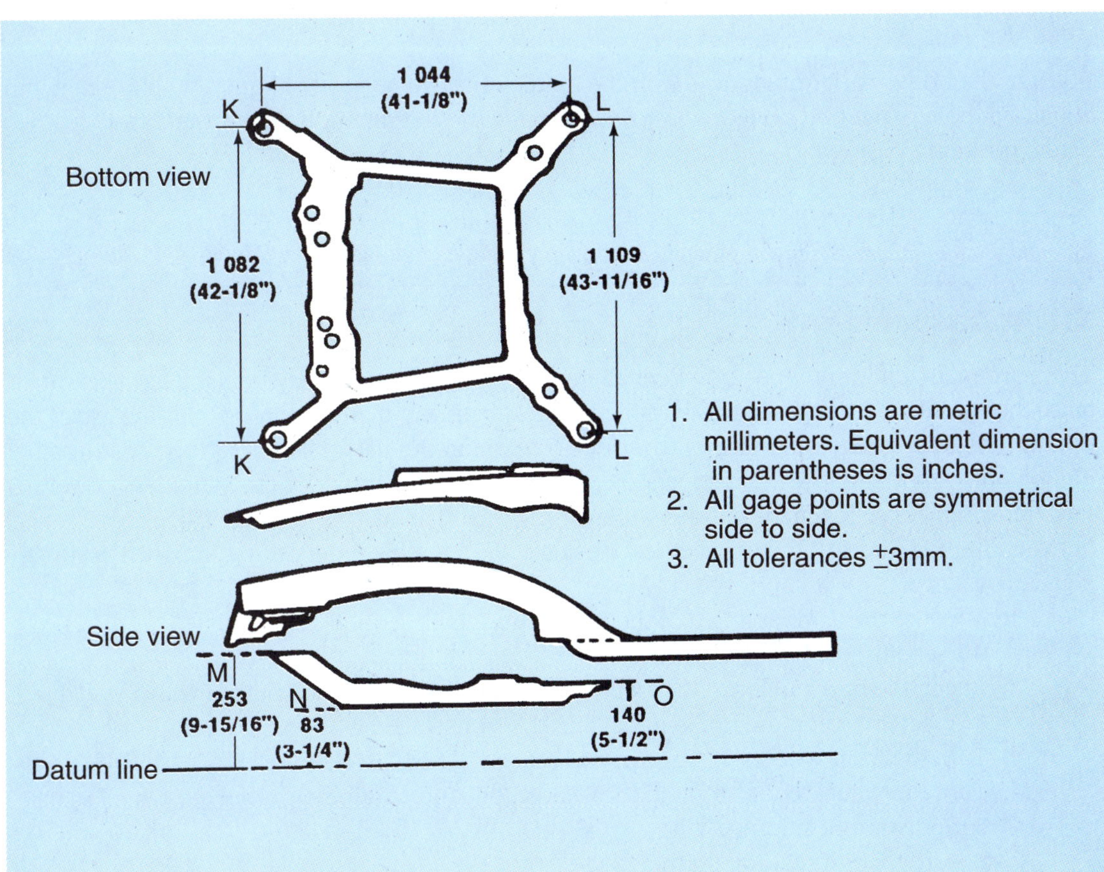


Figure 16-4 Front cradle measurements (Courtesy of Oldsmobile Division, General Motors Corporation)

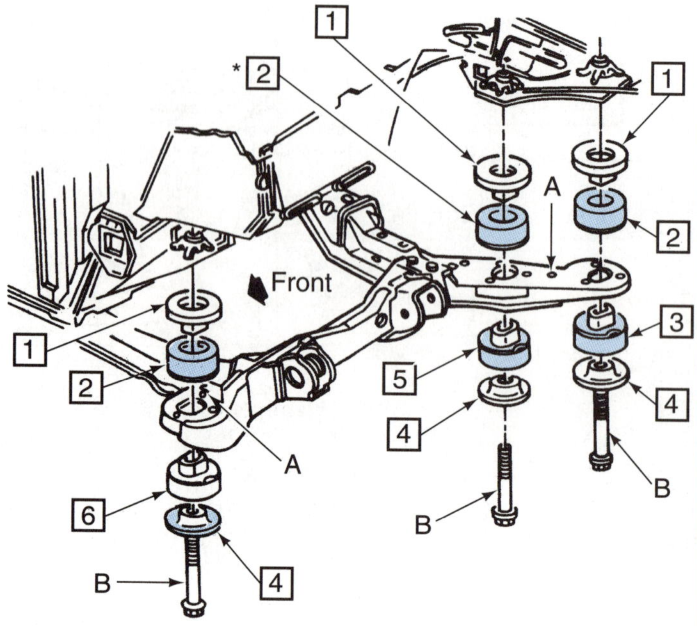


**Figure 16-5** Cradle insulators and alignment holes in cradle and body (Courtesy of Oldsmobile Division, General Motors Corporation)

**Classroom Manual**
Chapter 16, page 309

holes in the body. Alignment hole A in the cradle must be perfectly aligned with the matching alignment hole in the body (Figure 16-5). Observing this alignment hole is a quick way to check cradle position.

The steering axis inclination (SAI) line is an imaginary line through the centers of the upper and lower ball joints, or through the centers of the lower ball joint and the upper strut mount.

The SAI angle is the angle between the SAI line and the true vertical centerline of the tire and wheel.

**Special Tools**

Magnetic gauge with SAI capabilities

Brake pedal jack

# SAI Measurement and Correction Procedure

## SAI Measurement

There are many different types of wheel alignment equipment that may be used to measure SAI. Since the operator's manual supplied by each wheel alignment manufacturer details the use of their equipment, it is not the objective of this book to explain the use of different types of alignment equipment. Always follow the instructions in the operator's manual for the alignment equipment you are using. Some magnetic gauges will measure SAI. The magnetic gauge calculates the angle between the SAI line and the true vertical position. Follow this procedure to measure SAI with a magnetic gauge:

1. Measure the camber angle of the front wheel as described in Chapter 15, and record the reading.
2. Attach the hub gauges to the front wheel hubs, and apply the brakes with a pedal jack installed between the brake pedal and the front seat.
3. Turn the front wheel 20° inward, and set the SAI or KPI scale to 0°.
4. Turn the wheel to the straight-ahead position, and continue turning the wheel to the 20° outward position.
5. Record the SAI reading on the gauge.
6. If the camber reading is positive, add this reading to the SAI reading to obtain the included angle. A negative camber reading must be subtracted from the SAI angle.
7. Compare the SAI, or included, angle readings to the manufacturer's specifications.

## SAI Correction Procedure

SAI is not considered to be adjustable. If the SAI is not within specifications, the upper strut tower may be out of position, the lower control arm may be bent, or the center crossmember, or cradle, could be shifted. In most cases, these defects are caused by collision damage. When the SAI angle is correct, but the camber and included angle are less than specified, the strut or spindle may be bent. The difference in the included angles on the front wheels must not exceed 1 1/2°.

When the SAI is not within specifications, steering pull may occur while braking or accelerating, and steering may be erratic on road irregularities.

The included angle is the sum of the SAI angle and the camber angle if the camber is positive. If the camber is negative, the camber setting must be subtracted from the SAI angle to obtain the included angle.

# Front Wheel Toe Measurement and Adjustment

**Classroom Manual**
Chapter 16, page 315

## Front Wheel Toe Measurement Procedure

On older wheel aligners, toe is measured with a toe gauge placed between the front wheels. The toe gauge is an adjustable rigid bar that may be referred to as a trammel. Some alignment equipment manufacturers recommended toe measurement between chalk lines placed on each tire tread near the center. Other suggested points for toe measurement are at the inside of the tire sidewalls or between the inside edges of the rims. Toe is measured at the front and rear inside edges of the tire or rim at spindle height. Many alignment equipment manufacturers specify that a steering wheel locking device be installed during the toe measurement. The toe-in or toe-out setting is the difference in the distance between the inside front edges of the tires or rims, and the distance between the inside rear edges of the tires and rims. If the toe setting is not within specifications, feathered tread wear occurs on the front tires.

An optical toe gauge is available from some alignment equipment manufacturers. This type of toe gauge attaches magnetically to the front wheel hubs and projects a light beam to a screen on the opposite gauge. The optical toe gauge changes angular deflection of the wheels to inches of toe (Figure 16-6).

Always follow the toe gauge manufacturer's recommended procedure to measure the front wheel toe. Following is a typical front wheel toe measurement procedure:

1. Perform a prealignment inspection as outlined previously in Chapter 15.
2. Check the camber and caster, and adjust to specifications if necessary.

Toe-in occurs when the distance between the front edges of the front wheels is less than the distance between the rear edges of the front wheels.

Toe-out is present when the distance between the front edges of the front wheels is more than the distance between the rear edges of the front wheels.

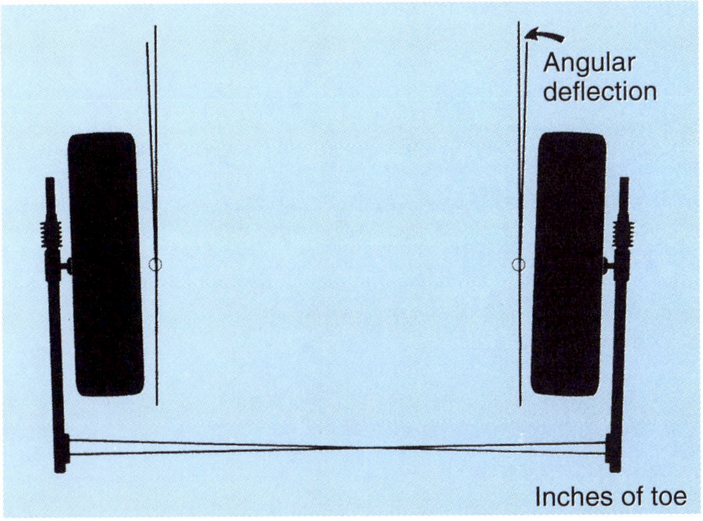


Figure 16-6 Optical toe gauge (Courtesy of Ammco Tools, Inc.)

**Special Tools**
Steering wheel clamp

3. Position the vehicle with the front wheels centered on the turn tables and remove the turn table locking pins. Be sure the front wheels are in the straight-ahead position.
4. Use a steering wheel clamp to lock the steering wheel in the straight-ahead position. If the vehicle has power steering, operate the engine when the steering wheel is locked.
5. Install the brake pedal jack between the front seat and the brake pedal. If the vehicle has power brakes, operate the engine while the brakes are locked.
6. Remove the wheel covers and grease caps. Clean the surfaces on the hub faces and the optical gauge magnetic faces. These mating surfaces must be free from grease, dirt, and metal burrs.
7. Jounce the front suspension several times and allow it to stabilize. Be sure the front wheels are still in the straight-ahead position.
8. If the steering linkage is mounted at the rear edge of the front wheels, use hand pressure to push the rear edge of the front wheels outward simultaneously. When the steering linkage is positioned at the front edge of the front wheels, use hand pressure to push the front edge of the front wheels outward simultaneously.

**Special Tools**
Toe gauge

9. Mount the optical gauges on the hub faces with the optical reflector facing toward the front of the vehicle. If the hub face is not accessible, use a rim clamp to attach the optical gauge to the hub face.
10. Use the level indicators on each optical toe gauge to locate the gauges in the level position.
11. Read the toe on the screen at the outer end of the toe gauges, and add the toe on each front wheel to obtain the total toe (Figure 16-7). Compare the total toe reading to the total toe specifications, and adjust the toe as required.

Total toe is the sum of the toe settings on the front wheels.

Changes in other alignment angles will alter the toe setting. For example, if the positive caster is increased, the top of the spindle is moved rearward. If the steering linkage is at the rear edge of the front wheels, this action rotates the rear of the steering arm downward, which moves the wheel toward a toe-out position. Therefore, other alignment angles, such as camber, caster, and SAI, must be adjusted within the manufacturer's specifications before toe is checked. If the toe setting is not within the manufacturer's specifications, a toe adjustment is required. Photo Sequence 14 shows a typical procedure for measuring front wheel toe.

## Toe Adjustment

**WARNING:** Do not heat tie rod sleeves to loosen them. This action may weaken the sleeves, resulting in sudden sleeve failure, loss of steering control, personal injury, and vehicle damage.

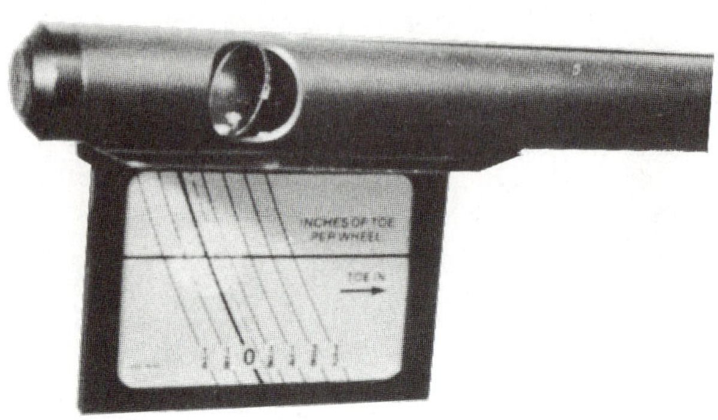


**Figure 16-7** Optical toe gauge screen (Courtesy of Ammco Tools Inc.)

**Photo Sequence 14**

# Typical Procedure for Measuring Front Wheel Toe

P14-1 Position the front tires on the turn tables, remove the locking pins, and be sure the front wheels are straight ahead and the turn tables are in the zero position.

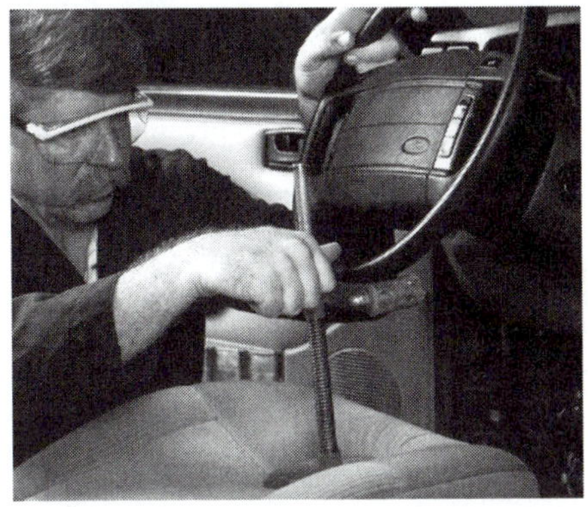

P14-2 Install a steering wheel clamp. If the car has power steering, run the engine when the clamp is installed.

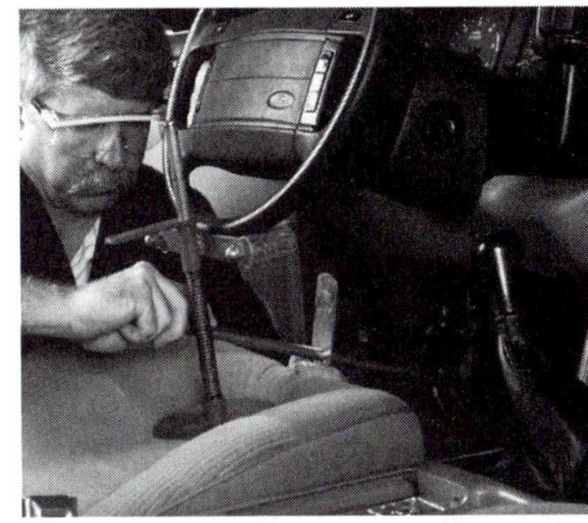

P14-3 Install a brake pedal jack between the brake pedal and the front seat. If the car has power brakes, run the engine while installing the jack.

P14-4 Remove the wheel covers and grease caps. Be sure the outer hub surfaces are free from grease, dirt, and metal burrs.

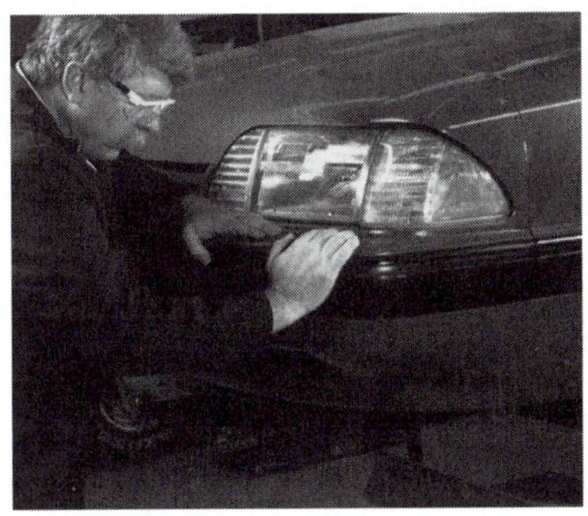

P14-5 Jounce the front suspension several times and allow it to stabilize.

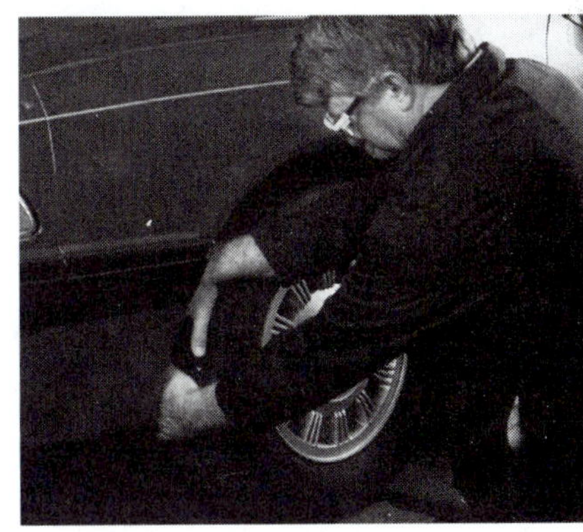

P14-6 If the steering linkage is at the rear of the front wheels, push outward simultaneously on the rear of the front tires. When the steering linkage is at the front of the front wheels, perform this action at the front edge of the front tires.

P14-7 Mount the optical toe gauges on the front wheel hubs.

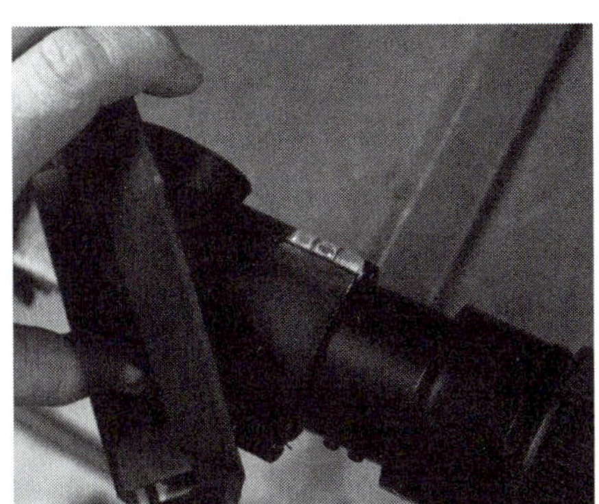

P14-8 Use the level indicators on the gauges to be sure the gauges are level.

P14-9 Read the front wheel toe on the toe gauges, and compare the reading to specifications.

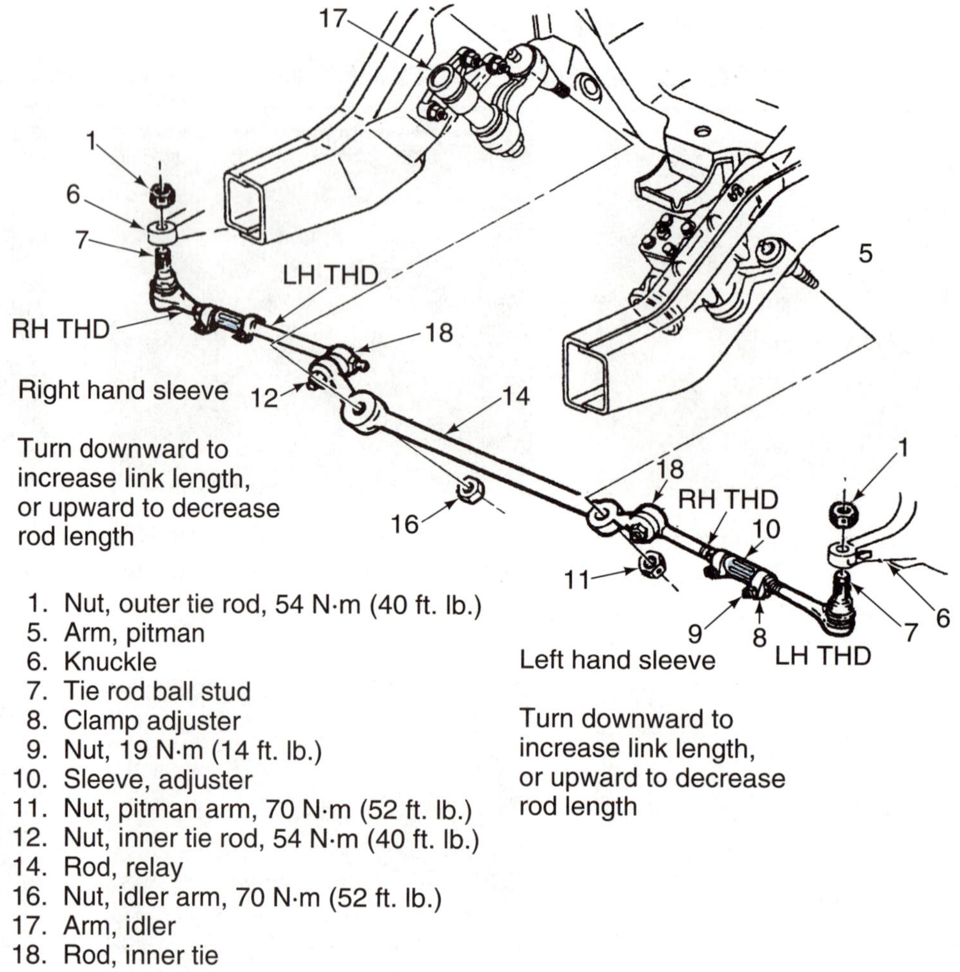


**Figure 16-8** Rotating tie rod sleeves to adjust front wheel toe (Courtesy of Chevrolet Motor Division, General Motors Corporation)

**WARNING:** Do not use a pipe wrench to loosen tie rod sleeves. This tool may crush and weaken the sleeve, causing sleeve failure, loss of steering control, personal injury, and vehicle damage. A tie rod rotating tool must be used for this job.

**Special Tools**

Tie rod sleeve rotator

When a front wheel toe adjustment is required, apply penetrating oil to the tie rod adjusting sleeves and sleeve clamp bolts. Loosen the tie rod adjusting sleeve clamp bolts enough to allow the clamps to rotate partially. One end of each tie rod sleeve contains a right-hand thread, and the opposite end contains a left-hand thread. These threads match the threads on the tie rod and outer tie rod end. When the tie rod sleeve is rotated, the complete tie rod, sleeve, and tie rod end assembly is lengthened or shortened. Use a tie rod sleeve rotating tool to turn the sleeves until the toe-in on each front wheel is equal to one-half the total toe specification (Figure 16-8).

After the toe-in adjustment is completed, the adjusting sleeve clamps must be turned so the openings in the clamps are positioned away from the slots in the adjusting sleeves (Figure 16-9). Replace clamp bolts that are rusted, corroded, or damaged, and tighten these bolts to the specified torque.

On some rack and pinion steering gears, the tie rod lock nut is loosened and the tie rod is rotated to adjust toe on each front wheel (Figure 16-10). Prior to rotating the tie rod, the small outer bellows boot clamp should be removed to prevent boot twisting during tie rod rotation. After the toe-in adjustment, the tie rod nut must be torqued to specifications, and the small bellows boot clamp must be reinstalled.

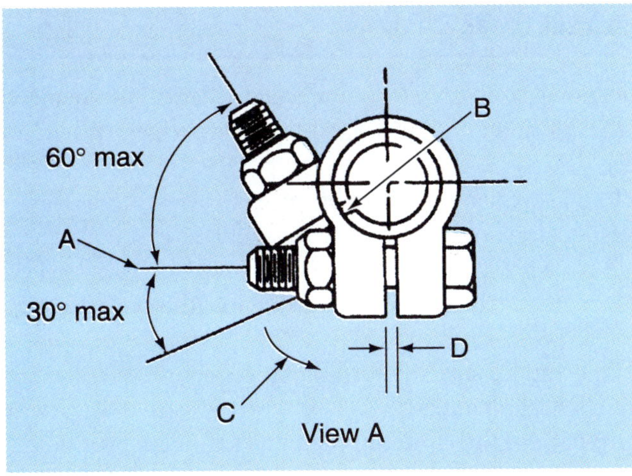


Figure 16-9 Proper tie rod adjusting sleeve clamp position (Courtesy of Chevrolet Motor Division, General Motors Corporation)

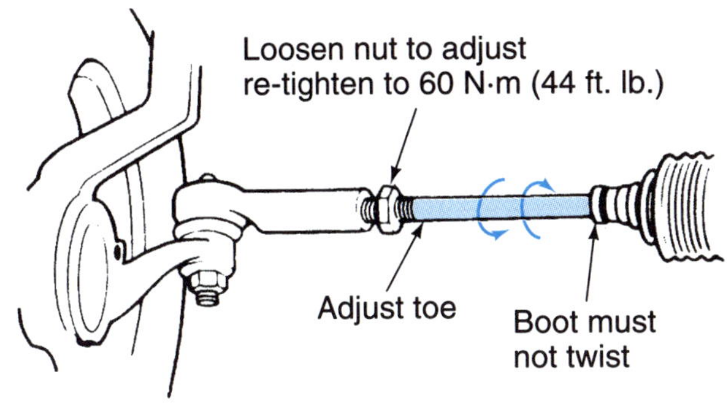


Figure 16-10 Rotating tie rod to adjust the toe on a rack and pinion steering gear (Courtesy of Chevrolet Motor Division, General Motors Corporation)

On other rack and pinion steering gears, the tie rod and outer tie rod end have internal threads, and a threaded adjuster is installed in these threads. One end of the adjuster has a right-hand thread, and the opposite end has a left-hand thread. Matching threads are located in the outer tie rod end and tie rod. Clamps on the outer tie rod end and tie rod secure these components to the adjuster. A hex-shaped nut is designed into the center of the adjuster (Figure 16-11). After the clamps are loosened, an open end wrench is placed on this hex nut to rotate the adjuster and change the toe setting.

When the toe-in adjustment is completed, the toe must be set within the manufacturer's specifications with the steering wheel in the centered position. If the steering wheel is not centered, a centering adjustment is required.

**Classroom Manual**
Chapter 16, page 316

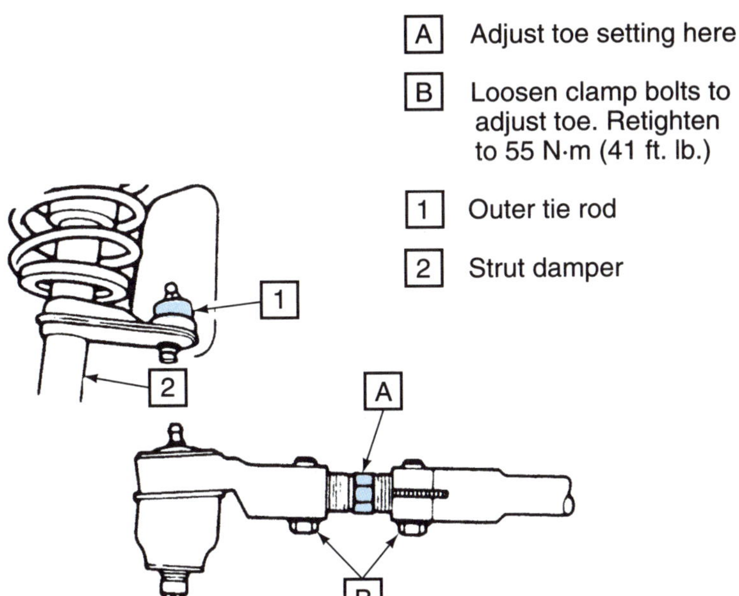


Figure 16-11 Rack and pinion steering gear with externally threaded toe adjuster and internal threads on the tie rod and outer tie rod end (Courtesy of Oldsmobile Division, General Motors Corporation)

## Turning Radius Measurement

Turning radius may be referred to as cornering angle.

Turning radius, or cornering angle, is the amount of toe-out on turns.

If the turning radius is not within specifications, excessive front tire tread wear occurs while cornering. Tire squeal may also occur while cornering if the turning radius is improper. When turning radius is measured, follow these steps:

1. Perform a prealignment inspection as explained in Chapter 15.
2. Measure and adjust the other suspension angles in this order: camber, caster, steering axis inclination (SAI), and toe.
3. Position the front wheels on the center of the turn tables, and apply the brakes with a brake pedal jack.
4. Remove the turn table locking pins and be sure the turning radius gauges are in the zero position with the front wheels straight ahead.
5. Turn the right front wheel inward toward the center of the vehicle until the turning radius gauge indicates 20° (Figure 16-12). Read the turning radius on the left front wheel and record the reading. On many steering systems, the turning radius on the wheel on the inside of the turn is 2° more than the turning radius on the wheel on the outside of the turn when the inside wheel is in the 20° outward position.
6. Turn the left front wheel inward toward the center of the vehicle until the turning radius gauge reads 20°. Read the turning radius on the right front turn table radius gauge and record the readings.
7. Compare the turning radius readings on each front wheel to the manufacturer's specifications. Turning radius or toe-out on turns is not adjustable. If the toe-out on turns is not within the manufacturer's specifications, the steering arms are bent or components such as tie rod ends are worn.

## Checking Toe Change and Steering Linkage Height

Bump steer is the tendency of the steering to veer suddenly in one direction when one or both front wheels strike a bump.

If the steering linkage on each side is not the same height, a condition called bump steer may occur. When this steering linkage condition is present, abnormal toe changes occur during wheel

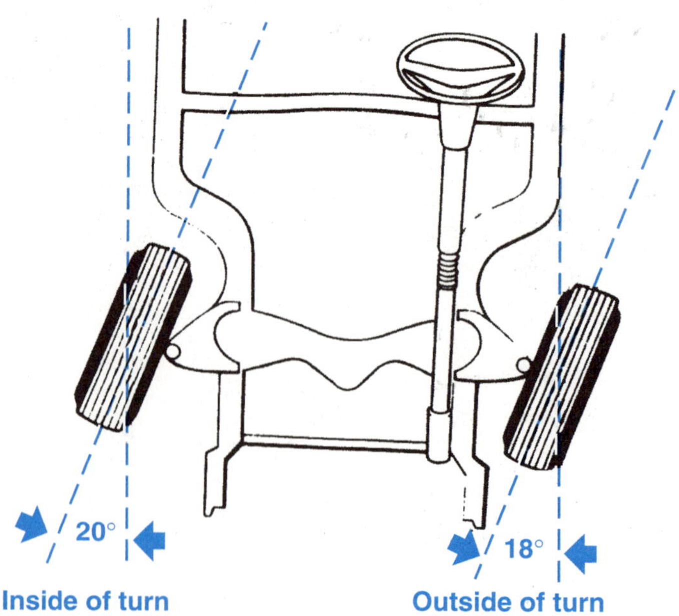


**Figure 16-12** Toe-out on turns (Courtesy of Hunter Engineering Company)

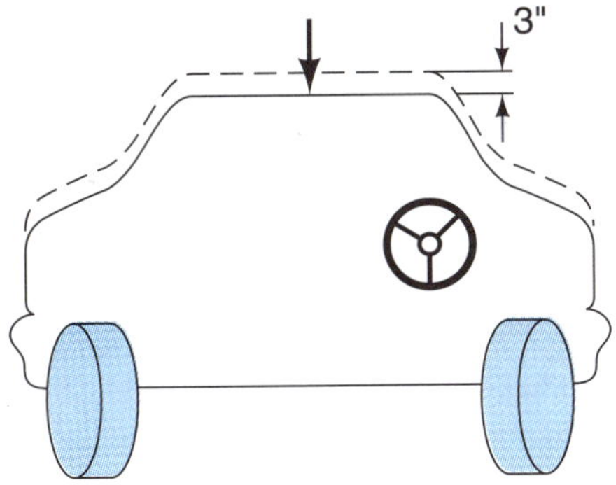


**Figure 16-13** Normal toe change during wheel jounce (Courtesy of Ammco Tools Inc.)

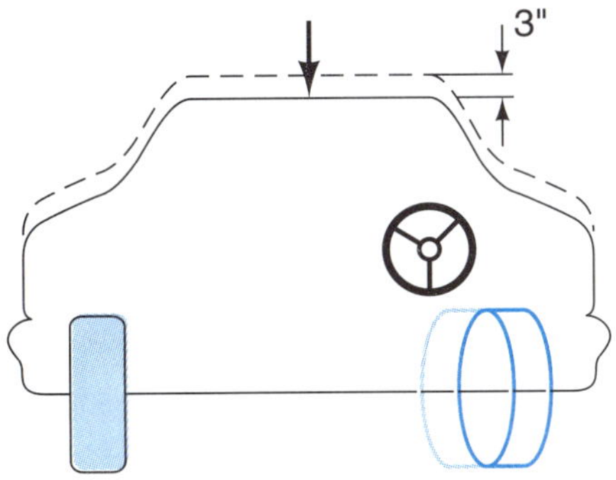


**Figure 16-14** If the toe on one front wheel remains at the original setting while the toe on the opposite wheel toes in or out during wheel jounce, unequal steering linkage height is indicated. (Courtesy of Ammco Tools Inc.)

jounce and rebound. Follow this procedure to check for abnormal toe changes during wheel jounce and rebound:

1. Be sure the toe is adjusted to specifications with the front wheels centered and the steering wheel locked.
2. Pull the chassis downward approximately 3 inches (in), 7.6 centimeters (cm), and observe the toe change. It is acceptable if the toe setting on each front wheel remains at the original reading, or if each front wheel toes in or toes out an equal small amount (Figure 16-13).
3. During wheel jounce, if one front wheel toes inward or outward while the opposite wheel remains at the original setting, the steering linkage height is not equal and must be corrected (Figure 16-14).
4. During wheel jounce, if one front wheel toes in and the opposite front wheel toes out, the steering linkage height is unequal and must be corrected (Figure 16-15).
5. Push the chassis upward 3 in (7.6 cm) and check the toe change (Figure 16-16). If the steering linkage height is equal, the toe setting on each front wheel remains the same or moves the same amount to a toe-in or toe-out position.

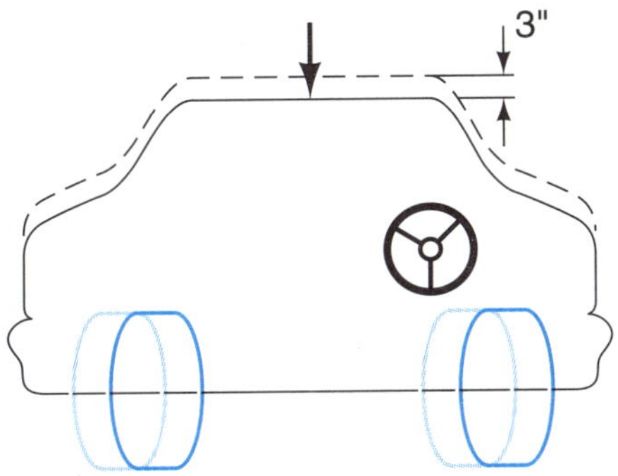


**Figure 16-15** If the toe on one front wheel toes in and the toe on the opposite front wheel toes out during wheel jounce, unequal steering linkage height is indicated. (Courtesy of Ammco Tools Inc.)

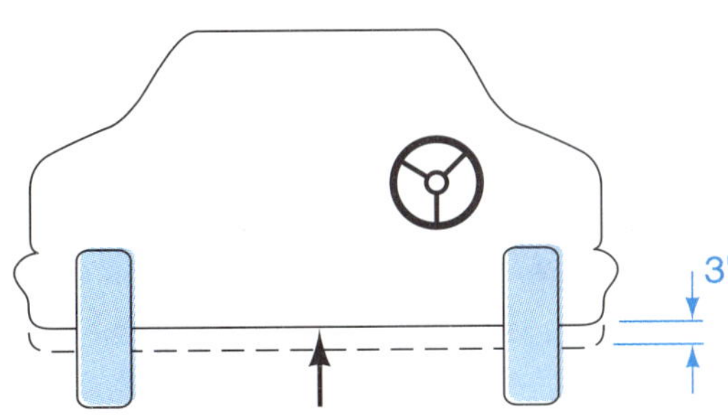


**Figure 16-16** Pull the chassis upward 3 in (7.6 cm) to check for improper toe change indicating unequal steering linkage height. (Courtesy of Ammco Tools Inc.)

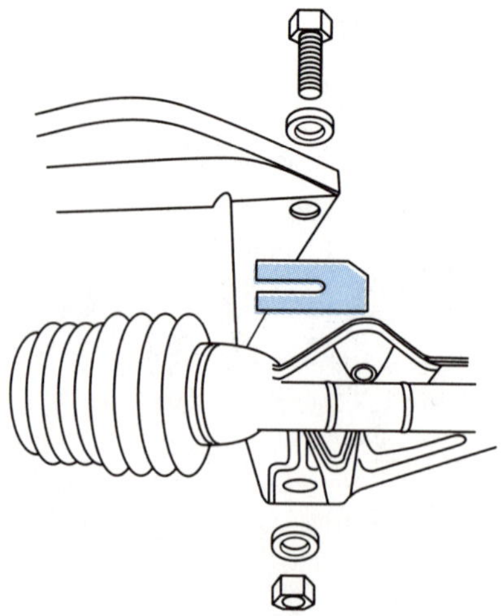

Figure 16-17 On some imported vehicles, shims between the rack and pinion steering gear and the chassis adjust the steering linkage height. (Courtesy of Ammco Tools Inc.)

6. If one front wheel remains at the original toe setting while the opposite front wheel changes to a toe-in or toe-out setting, the steering linkage height is unequal. When one front wheel moves to a toe-in position and the opposite front wheel moves to a toe-out setting, unequal steering linkages are indicated.

On some imported cars, the steering linkage height can be adjusted with shims between the rack and pinion steering gear and the chassis (Figure 16-17). The idler arm may be moved upward or downward on some domestic vehicles to adjust the steering linkage height.

On other vehicles, if the steering linkage height is unequal, steering components are worn or bent. These components include tie rods, tie rod ends, idler arms, pitman arms, and rack and pinion steering gear mounting bushings.

## Steering Wheel Centering Procedure

**SERVICE TIP:** The most accurate check of a properly centered steering wheel is while driving the vehicle straight ahead during a road test.

Road test the vehicle and determine if the steering wheel spoke is centered when the vehicle is driven straight ahead. If steering wheel centering is necessary, follow this procedure:

1. Lift the front end of the vehicle with a hydraulic jack and position safety stands under the lower control arms. Lower the vehicle onto the safety stands, and place the front wheels in the straight-ahead position.
2. Use a piece of chalk to mark each tie rod sleeve in relation to the tie rod, and loosen the sleeve clamps (Figure 16-18).

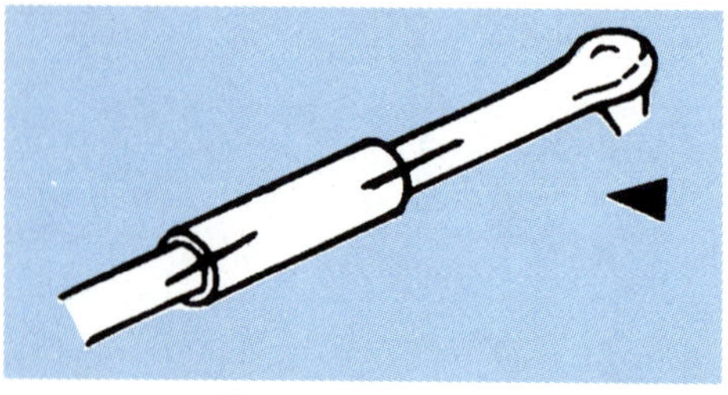

Figure 16-18 Marking tie rod sleeves in relation to the tie rods and outer tie rod ends (Courtesy of Ammco Tools, Inc.)

3. Position the steering wheel spoke in the position it was in while driving straight ahead during the road test. Turn the steering wheel to the centered position and note the direction of the front wheels.
4. If the steering wheel spoke is low on the left side while driving the vehicle straight ahead, use a tie rod sleeve rotating tool to shorten the left tie rod and lengthen the right tie rod (Figure 16-19). One-quarter turn on a tie rod sleeve moves the steering wheel position approximately one inch. Turn the tie rod sleeves the proper amount to bring the steering wheel to the centered position. For example, if the steering wheel spoke was two inches off center, turn each tie rod sleeve one-half turn.
5. If the steering wheel spoke is low on the right side while driving the vehicle straight ahead, lengthen the left tie rod and shorten the right tie rod.
6. Mark each tie rod sleeve in its new position in relation to the tie rod. Be sure the sleeve clamp openings are positioned properly as indicated previously in this chapter. Tighten the clamp bolts to the specified torque.
7. Lift the front chassis with a floor jack, and remove the jack stands. Lower the vehicle onto the shop floor, and check the steering wheel position during a road test.

When all the front suspension alignment angles are adjusted within the manufacturer's specifications, road test the vehicle. When the vehicle is driven on a relatively smooth straight road, there should be directional stability with no drift to the right or left, and the steering wheel must be centered.

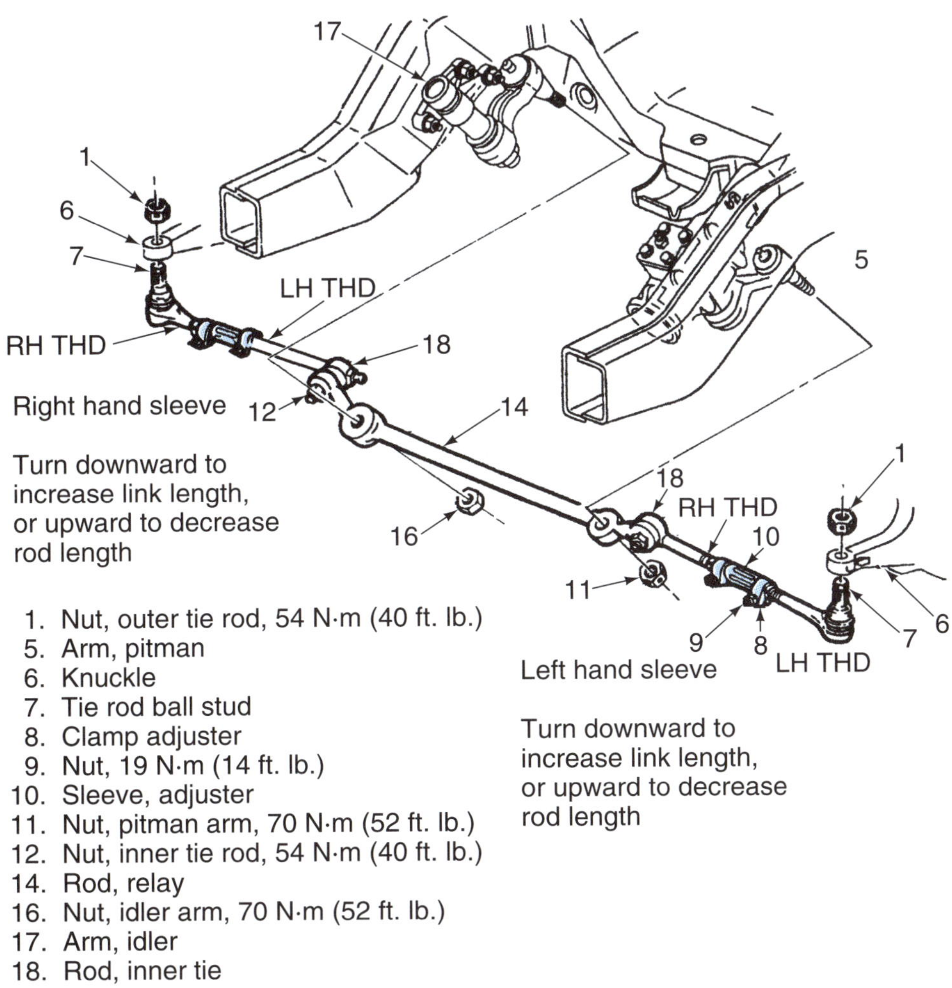


**Figure 16-19** Rotating tie rod sleeves to center steering wheel (Courtesy of Chevrolet Motor Division, General Motors Corporation)

**CUSTOMER CARE:** Like everyone else, individuals involved in the automotive service industry do make mistakes. If you make a mistake that results in a customer complaint, always be willing to admit your mistake and correct it. Do not try to cover up the mistake or blame someone else. Customers are usually willing to live with an occasional mistake that is corrected quickly and efficiently.

## Guidelines for SAI and Toe Measurement and Adjustment

1. SAI is not adjustable.
2. If the SAI is not within specifications, the upper strut tower may be out of position, the lower control arm may be bent, or the front cradle may be out of position.
3. The difference between the included angles on the two front wheels must not exceed 1.5°.
4. If the SAI angle is not correct on one front wheel, steering pull may occur while braking or during hard acceleration, and steering may be erratic.
5. The camber and caster must be within specifications before checking the front wheel toe.
6. While checking the front wheel toe, the wheels must be straight ahead with the steering wheel locked and the brakes applied.
7. The tie rod sleeves must be rotated with a sleeve rotating tool to adjust front wheel toe.
8. If heat is used to loosen tie rod sleeves, the sleeves may break, causing loss of steering control.
9. When a pipe wrench is used to rotate tie rod sleeves, the sleeves may be crushed and weakened, resulting in sleeve failure and loss of steering control.
10. Before tightening the tie rod sleeve clamps, the opening in the clamps must be positioned away from the slots in the sleeves.
11. The small outer bellows boot clamp must be removed to prevent boot twisting before the tie rod is rotated to adjust toe on a rack and pinion steering system.
12. Steering wheel centering may be necessary after a toe adjustment.
13. During a turn, the turning radius on the inside wheel is more than the turning radius on the outside wheel.
14. During wheel jounce and rebound, if the toe on both front wheels remains the same or both front wheels move an equal amount toward a toe-in or toe-out position, the tie rod height is equal.
15. During wheel jounce and rebound, if one front wheel remains at the original toe setting and the opposite front wheel moves toward a toe-in or toe-out position, the steering arm height is unequal.
16. During wheel jounce and rebound, if one front wheel moves toward a toe-in position and the opposite front wheel moves toward a toe-out setting, the tie rod height is unequal.
17. Unequal tie rod height may be caused by loose rack and pinion steering gear mounting bushings, worn or bent idler arms and pitman arms or tie rods and tie rod ends.
18. If the tie rod height is unequal, one tie rod is no longer parallel to the lower control arm. This condition causes improper toe changes during front wheel jounce and rebound, which causes the steering to suddenly veer to one side. The term bump steer is applied to this steering problem.

## CASE STUDY

A customer complained about erratic steering on a front wheel drive Dodge Dynasty. A road test revealed the car steered reasonably well on a smooth road surface, but the steering would suddenly swerve to the right or left while driving on irregular road surfaces.

The technician performed a preliminary wheel alignment inspection and found the right tie rod end was loose, but all the other suspension and steering components were in satisfactory condition. The technician replaced the loose tie rod end, but a second road test indicated that the bump steer problem was still present. After advising the customer that a complete wheel alignment was necessary, the technician drove the vehicle on the wheel aligner and carefully checked all front and rear alignment angles. Each front and rear wheel alignment angle was within specifications. The technician realized that somehow this problem had not been diagnosed correctly.

While thinking about this problem, the technician remembered a general diagnostic procedure he learned while studying Automotive Technology. This procedure stated: listen to the customer complaints, be sure the complaint is identified, think of the possible causes, test to locate the exact problem, and be sure the complaint is eliminated. The technician realized he had not thought much about the causes of the problem, and so he began to recall the wheel alignment theory he learned in college. He remembered that the tie rods must be parallel to the lower control arms, and if the tie rod height is unequal, this parallel condition no longer exists. The technician also recalled that unequal tie rod height causes improper toe changes during wheel jounce and rebound, which result in bump steer.

A check of the toe during front wheel jounce and rebound indicated the toe on the right front wheel remained the same during wheel jounce and rebound, but the toe on the left front wheel moved to a toe-out position. Since the tie rods had been inspected during the preliminary alignment inspection, the technician turned his attention to the rack and pinion steering gear mounting. He found the bushing on the right end of the steering gear was worn and loose. This bushing was replaced and all the steering gear mounting bolts were tightened to the specified torque. A check of the toe change during wheel jounce and rebound revealed a normal toe change.

From this experience, the technician learned two things:

1. Understanding wheel alignment theory is very important in diagnosing steering problems.
2. Always be thorough! During a prealignment inspection, check all suspension and steering components, including rack and pinion steering gear mountings.

## Terms to Know

Tram gauge
Optical toe gauge
Toe-in
Toe-out
Total toe
Turning radius
Cornering angle
Toe-out on turns
Bump steer

## ASE Style Review Questions

1. While discussing an SAI angle that is more than specified on the left side of a MacPherson strut front suspension:
   *Technician A* says that the camber adjustment bolt should be adjusted to correct the problem.
   *Technician B* says that the left front steering arm is bent.
   Who is correct?
   **A.** A only
   **B.** B only
   **C.** Both A and B
   **D.** Neither A nor B

2. While discussing unequal SAI angles on the left and right sides of the front suspension:
   *Technician A* says this problem causes tread wear on the front tires.
   *Technician B* says this problem causes brake pull during sudden stops.
   Who is correct?
   **A.** A only
   **B.** B only
   **C.** Both A and B
   **D.** Neither A nor B

3. While discussing an SAI angle that is 3° greater on the right front suspension than the SAI angle on the left front suspension:
   *Technician A* says this SAI angle variation must be corrected.
   *Technician B* says this SAI angle variation is satisfactory.
   Who is correct?
   **A.** A only
   **B.** B only
   **C.** Both A and B
   **D.** Neither A nor B

4. While discussing toe adjustment:
   *Technician A* says the tie rod sleeves may be heated with an acetylene torch to loosen them.
   *Technician B* says the tie rod sleeves should be loosened with a pipe wrench.
   Who is correct?
   **A.** A only
   **B.** B only
   **C.** Both A and B
   **D.** Neither A nor B

5. While discussing toe adjustment in relation to other front suspension angles:
   *Technician A* says if the positive caster is increased on the right front wheel, this wheel moves toward a toe-out position.
   *Technician B* says that caster change does not affect the toe-in setting.
   Who is correct?
   **A.** A only
   **B.** B only
   **C.** Both A and B
   **D.** Neither A nor B

6. While discussing toe measurement:
   *Technician A* says the front wheel toe should be checked with the right front wheel turned 20° outward.
   *Technician B* says the front wheel toe should be checked with the front wheels straight ahead.
   Who is correct?
   **A.** A only
   **B.** B only
   **C.** Both A and B
   **D.** Neither A nor B

7. While discussing front wheel toe adjustment:
   *Technician A* says the right tie rod sleeve has a right-hand thread inside each end of the sleeve.
   *Technician B* says the opening in the sleeve clamp must be positioned above one of the slots in the tie rod sleeve.
   Who is correct?
   **A.** A only
   **B.** B only
   **C.** Both A and B
   **D.** Neither A nor B

8. While discussing turning radius measurement:
   *Technician A* says a bent steering arm will cause the turning radius to be out of specification.
   *Technician B* says if the turning radius is not within specifications, tire tread wear is excessive while cornering.
   Who is correct?
   **A.** A only
   **B.** B only
   **C.** Both A and B
   **D.** Neither A nor B

9. While discussing toe change during wheel jounce and rebound:
*Technician A* says if both front wheels toe-in or toe-out a small equal amount during front wheel jounce and rebound, the tie rods are parallel to the lower control arms.
*Technician B* says if one front wheel toes-in and the opposite front wheel toes-out during front wheel jounce and rebound, the tie rod height is unequal.
Who is correct?
**A.** A only
**B.** B only
**C.** Both A and B
**D.** Neither A nor B

10. While discussing toe change during wheel jounce and rebound:
*Technician A* says the improper toe change during front wheel jounce and rebound may cause bump steer.
*Technician B* says improper toe change during front wheel jounce and rebound may be caused by a worn upper strut mount.
Who is correct?
**A.** A only
**B.** B only
**C.** Both A and B
**D.** Neither A nor B

## Table 16-1 ASE TASK

Check front wheel setback; determine needed repairs.

| Problem Area | Symptoms | Possible Causes | Classroom Manual | Shop Manual |
|---|---|---|---|---|
| **STEERING CONTROL** | Steering pull while driving straight ahead | Severe front wheel setback condition | 328 | 377 |

## Table 16-2 ASE TASK

Check front cradle (subframe) alignment; determine needed repairs or adjustments.

| Problem Area | Symptoms | Possible Causes | Classroom Manual | Shop Manual |
|---|---|---|---|---|
| **STEERING CONTROL** | Steering pull while driving straight ahead | **1.** Improper camber caused by improperly positioned cradle | 312 | 378 |
| | | **2.** Improper setback caused by improperly positioned cradle | 312 | 378 |

## Table 16-3 ASE TASK

Check SAI (steering axis inclination)/KPI (king pin inclination); determine needed repairs.

| Problem Area | Symptoms | Possible Causes | Classroom Manual | Shop Manual |
|---|---|---|---|---|
| STEERING CONTROL | Steering pull during hard acceleration | Improper SAI on one side of front suspension | 311 | 380 |
| | Steering pull while braking | Improper SAI on one side of front suspension | 311 | 380 |
| | Erratic steering on road irregularities | Improper SAI on one side of front suspension | 311 | 380 |

## Table 16-4 ASE TASK

Check included angle; determine needed repairs.

| Problem Area | Symptoms | Possible Causes | Classroom Manual | Shop Manual |
|---|---|---|---|---|
| STEERING CONTROL | Steering pull while braking or accelerating | Improper SAI on one side of front suspension | 312 | 380 |
| | Erratic steering on road irregularities | Improper SAI on one side of front suspension | 312 | 381 |

## Table 16-5 ASE TASK

Check and adjust front wheel toe.

| Problem Area | Symptoms | Possible Causes | Classroom Manual | Shop Manual |
|---|---|---|---|---|
| TIRE LIFE | Feathered wear on front tire treads | Improper front wheel toe | 315 | 381 |

## Table 16-6 ASE TASK

Check toe-out on turns (turning radius); determine needed repairs.

| Problem Area | Symptoms | Possible Causes | Classroom Manual | Shop Manual |
|---|---|---|---|---|
| **TIRE LIFE** | Excessive front tire tread wear, scuffing | Improper turning radius | 317 | 386 |
| **TIRE NOISE** | Tire squeal while cornering | Improper turning radius | 317 | 386 |
| **STEERING WHEEL POSITION** | Steering wheel not centered while driving straight ahead | Improper tie rod sleeve adjustment | 318 | 386 |

CHAPTER 17

# Computer Wheel Alignment and Rear Wheel Alignment Procedures

Upon completion and review of this chapter, you should be able to:

- ❑ Recognize the symptoms of improper rear wheel alignment.
- ❑ Use a track gauge to measure rear wheel tracking.
- ❑ Diagnose rear wheel tracking problems from the track gauge measurements.
- ❑ Position a vehicle on the alignment rack and connect the wheel units on a computer wheel aligner to the wheel rims.
- ❑ Select the specifications in the computer wheel aligner for the vehicle being aligned.
- ❑ Perform a preliminary wheel alignment inspection with the preliminary inspection screen.
- ❑ Perform a ride height inspection and measurement with the ride height screen.
- ❑ Perform an automatic or manual wheel runout compensation procedure.
- ❑ Use the turning angle screen, and measure turning radius.
- ❑ Obtain all the front and rear wheel alignment angles on the alignment angle screen.
- ❑ Use the adjustment screens to perform the necessary suspension adjustments to set all the front and rear alignment angles within specifications.
- ❑ Use the diagnostic drawing and text screens and the shim select screens to assist in the necessary front and rear suspension adjustments.
- ❑ Diagnose the causes of improper rear wheel alignment.
- ❑ Perform rear wheel camber and toe adjustments.

**Basic Tools**

Basic technician's tool set

Service manual

## Four Wheel Alignment Diagnosis

Conditions that indicate a four wheel alignment is necessary are :

1. The front wheel toe-in is set correctly with the steering wheel centered on the alignment rack, but the steering wheel is not centered when the vehicle is driven straight ahead.
2. There are no worn suspension parts or defective tires, and all front suspension angles are within specifications, but the vehicle wanders, or drifts, to one side.
3. The vehicle is not overloaded, all front suspension angles are within specifications, and there are no worn suspension parts, but tire wear re-occurs.

The geometric centerline is an imaginary line through the exact center between the front and rear wheels.

## Rear Wheel Tracking Measurement with a Track Gauge

Prior to the introduction of computer alignment systems, wheel alignment equipment basically performed a two wheel alignment on the front wheels, and a track gauge was used to measure rear wheel tracking. A track gauge is connected between the front and rear wheels to determine the rear wheel tracking, or thrust line. When all four wheels are parallel to the geometric centerline of the vehicle, the thrust line is positioned at the geometric centerline. Under this condition, the rear wheels track directly behind the front wheels (Figure 17-1).

The thrust line is an imaginary line at a 90° angle to the centerline between the rear wheels and projected forward.

**Special Tools**

Track gauge

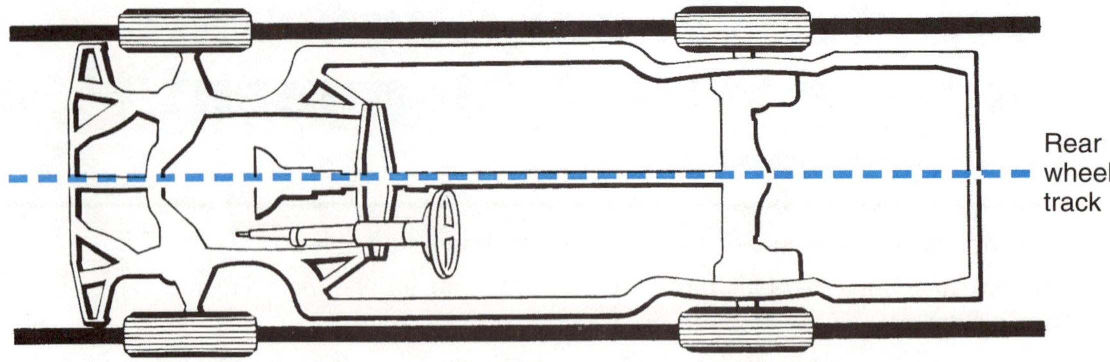


**Figure 17-1** When all four wheels are parallel to the geometric centerline, the thrust line is positioned at the geometric centerline and the rear wheels track directly behind the front wheels. (Courtesy of Hunter Engineering Company)

Setback occurs when one wheel is driven rearward in relation to the opposite wheel.

If the left front of the vehicle is involved in a severe collision, the left front wheel may be driven rearward to a setback condition, and the rear wheels may be moved straight sideways. Under this condition, the rear wheels no longer track directly behind the front wheels (Figure 17-2). The term sideset is applied to the condition where the rear axle is moved straight sideways.

A track gauge is a straight metal bar that is graduated in inches or millimeters. The scale on the outside edge of the track bar is similar to a tape measure. Three straight metal pointers are mounted in brackets on the track gauge. These pointers are also graduated much like a tape measure. When the wing nuts in the brackets are loosened, the brackets may be moved lengthwise on the track bar. The metal pointers may be adjusted inward or outward in the brackets.

To check rear wheel tracking, one of the track gauge pointers is positioned against the rear outside edge of the front wheel rim at spindle height. The other two pointers are positioned on the outside edges of the rear wheel rim at spindle height. The track gauge must be kept parallel to the side of the vehicle (Figure 17-3).

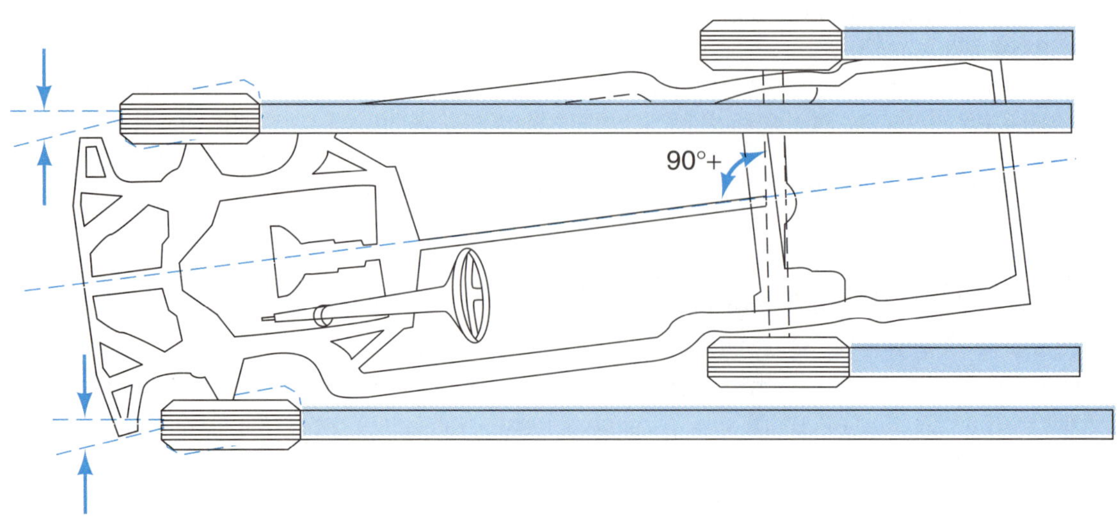


**Figure 17-2** Collision damage causes front wheel setback and improper rear wheel tracking with rear axle sideset. (Courtesy of Hunter Engineering Company)

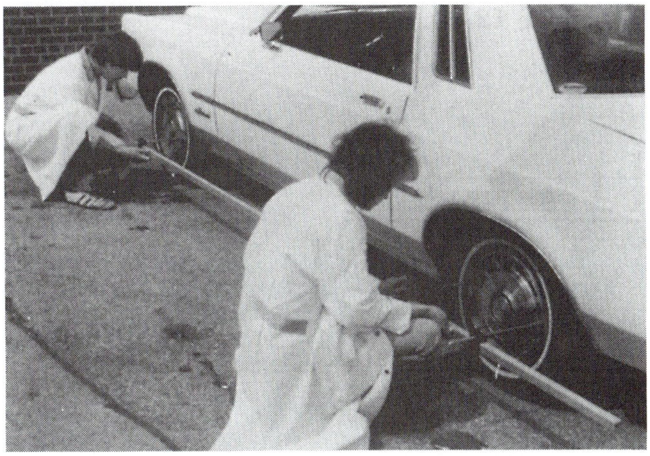

Figure 17-3 Track bar positioned to measure rear wheel tracking

## Rear Wheel Track Measurement Procedure

Follow these steps to measure rear wheel tracking:

1. Lift one side of the front suspension with a floor jack and lower the suspension onto a jack stand. Be sure the tire is lifted off the shop floor.
2. Place a lateral runout gauge against the outside edge of the rim lip and rotate the wheel to check wheel runout. (Refer to Chapter 4 for wheel runout measurement.) Place a chalk mark at the location of the maximum wheel runout. If the wheel runout is excessive, replace the wheel.
3. Rotate the chalk mark to the exact top or bottom of the wheel, and remove the jack stand and floor jack to lower the wheel onto the shop floor.
4. Repeat steps 1, 2, and 3 at the other three wheels.
5. Hold the single pointer on the track gauge so it contacts the rear edge of the front rim.
6. Adjust the other two pointers so they contact the outer edges of the rear wheel rim.
7. Adjust the pointers on the rear wheel so the track gauge is parallel to the side of the vehicle, and be sure all three pointers are clamped securely.
8. Move the track gauge to the opposite side of the vehicle and position the pointers in the same location on these wheels. If the vehicle has proper tracking, all three pointers on the track gauge will contact the front and rear rims at exactly the same location on both sides of the vehicle, and the track gauge will be parallel to the sides of the vehicle.

## Examples of Improper Rear Wheel Tracking Measured with a Track Gauge

**Offset Rear Axle.** When the track gauge is positioned on the left side of the vehicle, the rear pointer on the rear wheel must be moved inward further than the pointer at the front edge of the rear rim. On the right side of the vehicle, the front pointer is positioned forward, ahead of the rim edge, when the two pointers are positioned on the edges of the rear rim. This indicates the wheelbase is less between the wheels on the right side of the vehicle than on the left side. The front pointer on the rear rim is some distance from the rim face (Figure 17-4). These pointer positions indicate the right rear wheel has moved forward and the left rear wheel has moved rearward. This rear axle offset condition moves the thrust line to the left of the geometric centerline.

Rear axle offset occurs when the rear axle is turned so one rear wheel is moved forward and the opposite rear wheel is moved rearward.

**Setback.** If a vehicle has been subjected to a severe right front sideways impact, the left front wheel may be driven outward and rearward. The track gauge and pointers are positioned on the

A track gauge may be called a track bar.

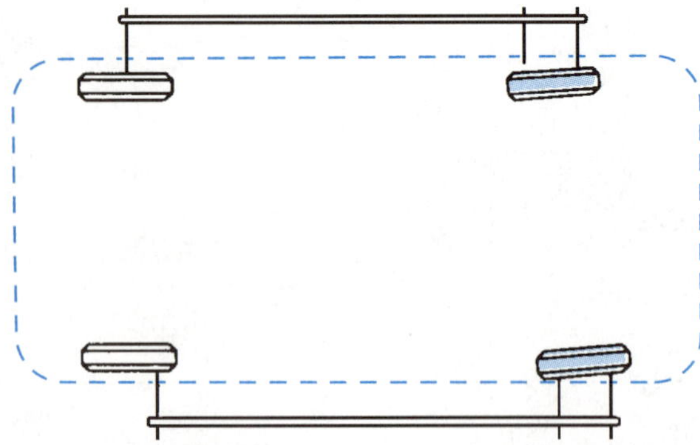

**Figure 17-4** Track bar measurement of an offset rear axle

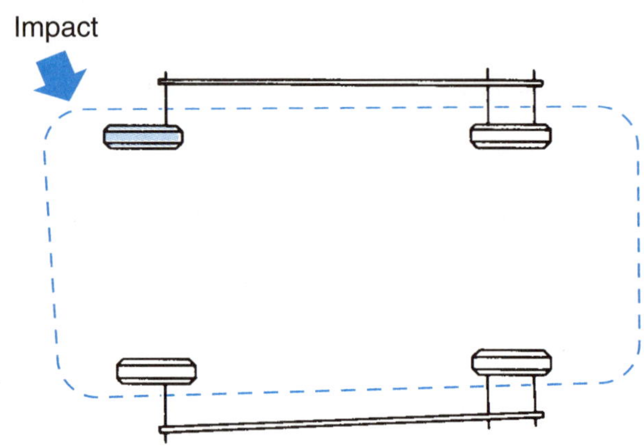


**Figure 17-5** Track bar measurement of a left front wheel that is moved outward and rearward to a setback position

right side of the vehicle and then moved to the left side. On the left side, the front pointer is positioned ahead of the rim edge, and the track gauge is not parallel to the side of the vehicle (Figure 17-5).

The track gauge must be moved inward on the front pointer to position the bar parallel to the side of the vehicle. When this adjustment is made, the distance on the front pointer is less on the left side of the vehicle than this distance on the right side of the vehicle. These track gauge measurements indicate the left front wheel is moved rearward to a setback position, and this wheel is also moved outward.

**Chassis Diamond-Shaped Condition.** When a vehicle experiences severe collision damage on the left front, the entire left side of the vehicle may be driven rearward in relation to the right side of the chassis. Under this condition, the left front wheel is moved to a setback condition and the rear axle is offset, which moves the thrust line to the left of the geometric centerline. This particular type of damage is most likely to occur on a rear wheel drive vehicle with a frame and a one-piece rear axle.

Since the entire left side of the vehicle is driven rearward, the wheel base is the same on both sides of the vehicle. Therefore, the track gauge pointers indicate the wheelbase is the same between the front and rear wheels on both sides of the vehicle. However, the two rear track gauge

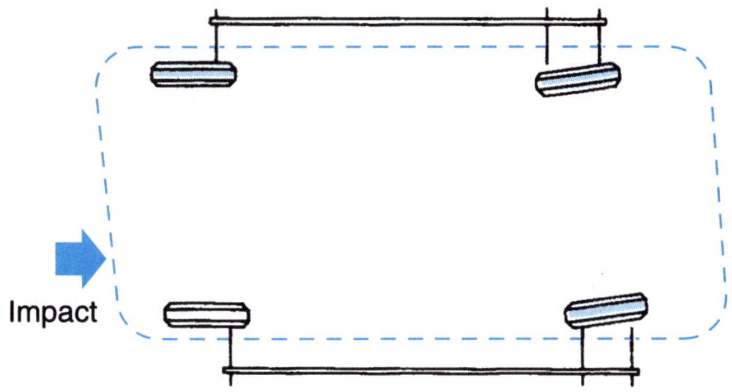


Figure 17-6 A severe left front collision impact moves the entire left side of the vehicle rearward, resulting in front wheel setback and rear axle offset, which provides a diamond-shaped chassis.

pointers must be adjusted to different lengths to contact the rear rim edges. On the left side of the vehicle, the rear track gauge pointer is some distance from the rear rim edge if these two pointers are set at the same measurement (Figure 17-6).

When the track gauge is positioned on the right side of the vehicle, there is some distance between the front pointer and the edge of the rear rim if both rear pointers are set at the same distance.

Measuring vehicle tracking with a track gauge requires two people and a considerable amount of time to perform the wheel runout and track bar measurements. The track gauge also lacks precision accuracy. Computer alignment systems perform the same checks by measuring the thrust line angle, setback, and all the other front and rear suspension alignment angles with speed and accuracy.

# Front and Rear Wheel Alignment with Computer Alignment Systems

## Preliminary Procedure

Photo Sequence 15 shows a typical procedure for performing four wheel alignment with a computer wheel aligner.

**CAUTION:** When using a computer wheel aligner, always follow the equipment manufacturer's recommended procedures to provide accurate readings and avoid equipment and vehicle damage, or personal injury.

**Special Tools**

Computer wheel aligner with rim clamps

The vehicle should be on an alignment rack with a turning plate under each front tire and conventional slip plates under the rear tires. Slip plates under the rear tires allow unrestricted movement of the rear wheels during rear wheel camber and toe adjustments. The center of the front wheel spindles should be in line with the zero mark on the turning plates. The locking pins must be left in these plates, and the transmission selector must be in the park position.

Mount the rim clamps on each wheel. It may be necessary to remove the chrome discs, or hub caps, before these clamps can be installed. The adjustment knob on the rim clamp should be pointing toward the top of the wheel, and the bubble on the wheel unit should be centered. A set knob on the wheel unit allows the service technician to lock the wheel unit in this position. Each wheel unit contains a microprocessor with a preheat circuit, which stabilizes the readings in relation to temperature. Touch keypads on some wheel units allow the entry of commands from each

# Photo Sequence 15
## Typical Procedure for Performing Four Wheel Alignment with a Computer Wheel Aligner

**P 15-1** Position the vehicle on the alignment ramp.

**P15-2** Be sure the front tires are positioned properly on the turn tables.

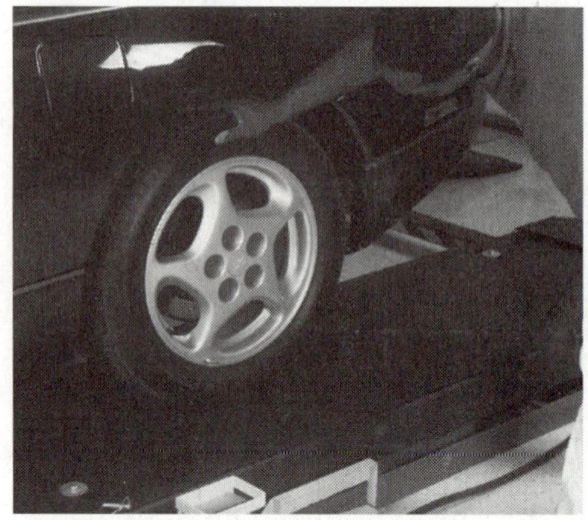

**P15-3** Position the rear wheels on slip plates.

**P15-4** Attach the wheel units.

**P15-5** Select the vehicle make and model year.

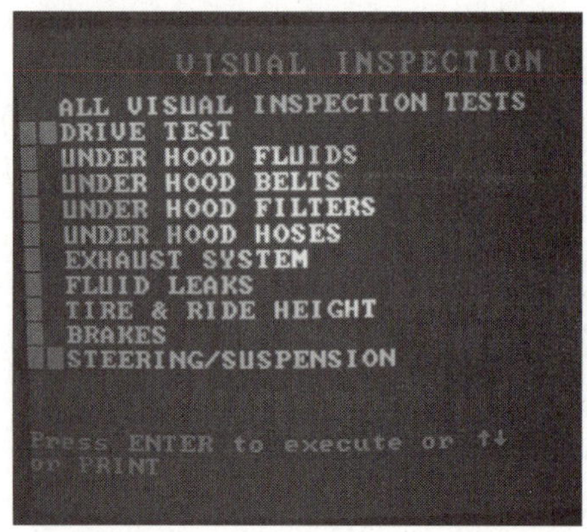


**P15-6** Check items on the screen during preliminary inspection.

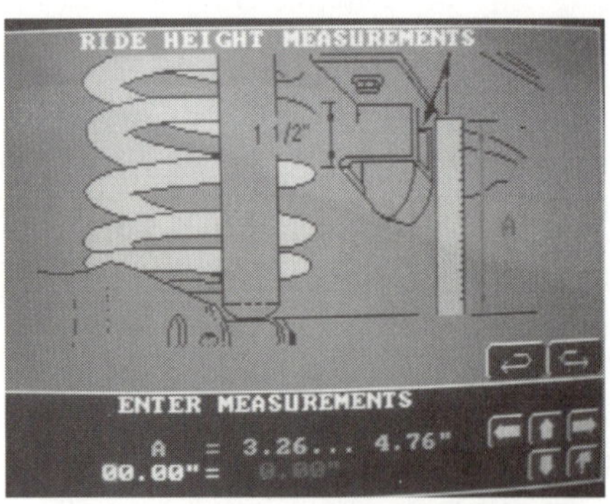


**P15-7** Display the ride height screen.

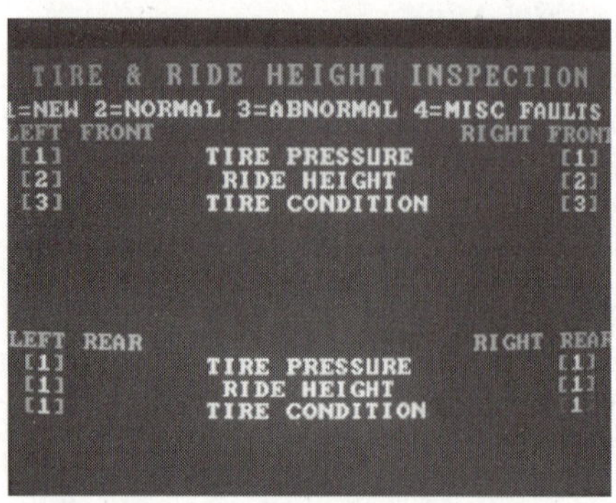


**P15-8** Check tire condition for each tire on the tire condition screen.

**P15-9** Display wheel runout compensation screen.

## Photo Sequence 15
## Typical Procedure for Performing Four Wheel Alignment with a Computer Wheel Aligner (continued)

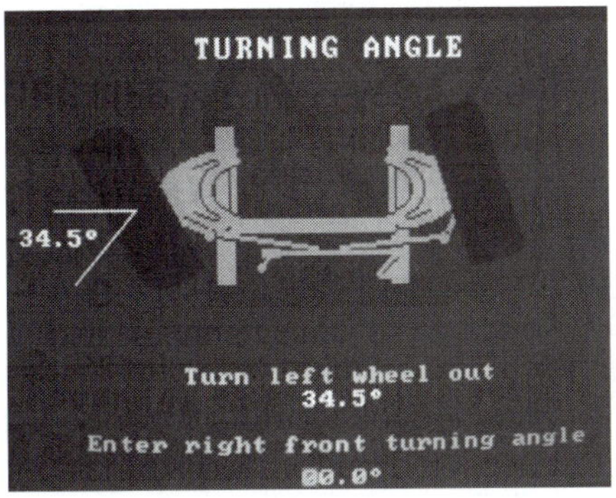


**P15-10** Display turning angle screen and perform turning angle check.

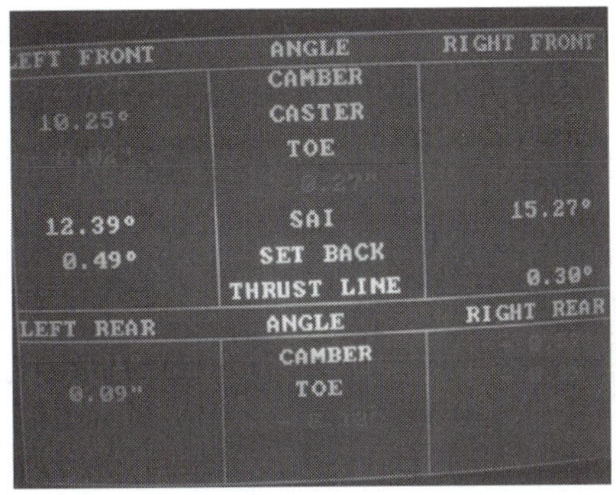


**P15-11** Display front and rear wheel alignment angle screen.

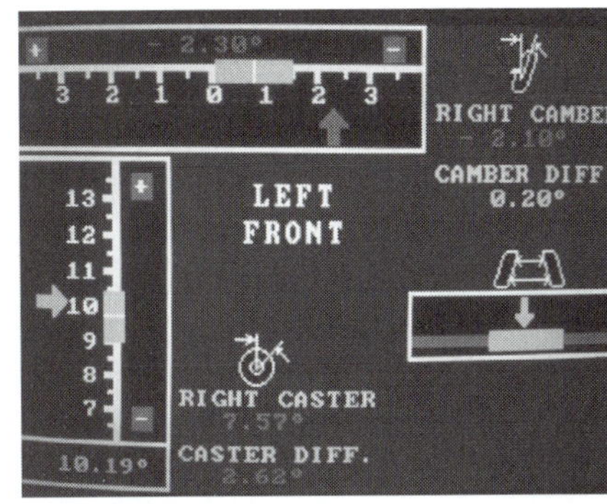


**P15-12** Display adjustment screens.

unit. Light emitting diodes (LEDs), or electronic signals, between the front wheel units and the front and rear wheel units are used to perform alignment measurements. The front wheel units have arms that project toward the front of the vehicle to transmit signals between these units. When a blocked beam prompt appears on any screen, the beam between two wheel units is blocked. This could be caused by a person standing in the beam, an open car door, or a suspension jack. The blocked beam prompt must be eliminated before any tests are completed.

Since the manufacturers of various types of computer alignment systems publish detailed operator's manuals for their specific equipment, our objective is to discuss some of the general screens that a technician uses while measuring wheel alignment angles with a typical computer aligner.

### Main Menu

One of the first items displayed on the computer aligner screen is the main menu screen (Figure 17-7). From this screen, the technician makes a selection by pressing the number on the key pad that matches the number beside the desired procedure. On some main menus, a cursor is used to select the type of alignment.

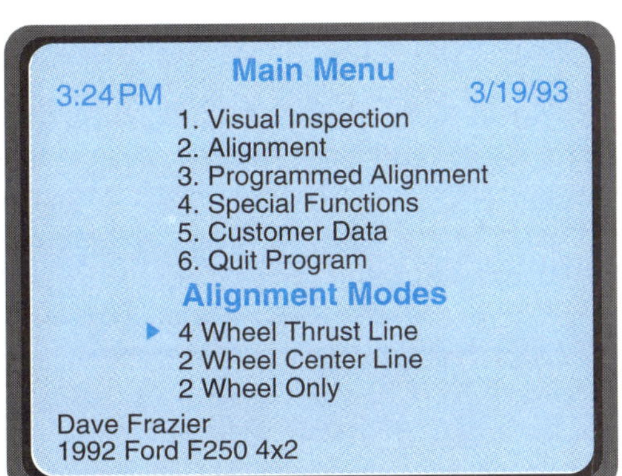


**Figure 17-7** Typical main menu (Courtesy of Bear Automotive Service Equipment Company)

Figure 17-8 Bar graph and wand used to enter vehicle specifications

## Specifications Menu

On some of the early computer alignment systems, the vehicle specifications were contained on bar graphs. The technician moved a wand across the appropriate bar graph to enter the specifications for the vehicle being aligned (Figure 17-8). On other computer wheel aligners, the specifications are contained on a floppy disc, and the technician selects the appropriate make, model, and year of vehicle with the cursor on the screen (Figures 17-9 and 17-10).

## Preliminary Inspection Screen

The preliminary inspection screen allows the technician to enter the condition of all front and rear suspension components, which must be checked during a preliminary alignment inspection. On

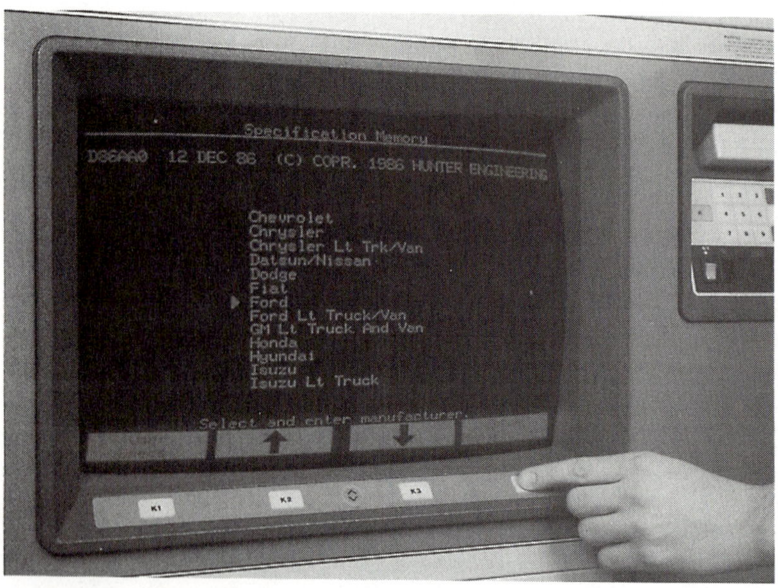


Figure 17-9 Selecting vehicle make with cursor (Courtesy of Hunter Engineering Company)

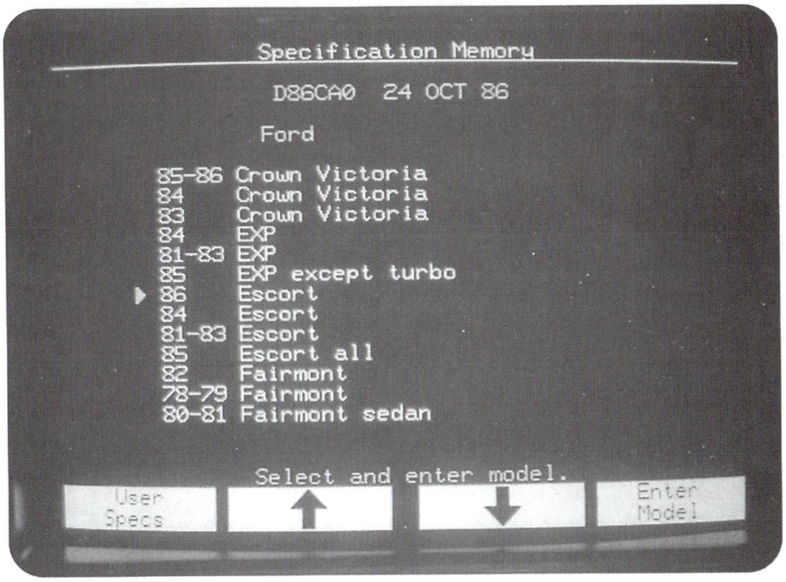


**Figure 17-10** Selecting vehicle model and year with cursor (Courtesy of Hunter Engineering Company)

some computer alignment systems, the technician may enter checked, marginal, and repair or replace for each component (Figure 17-11). These entered conditions may be printed out at any time or with the complete wheel alignment results.

## Ride Height Screen

Some computer wheel aligners provide a ride height screen with a graphic display indicating the exact location where the ride height should be measured (Figure 17-12). The computer aligner compares the ride height measurements entered by the technician to the specifications.

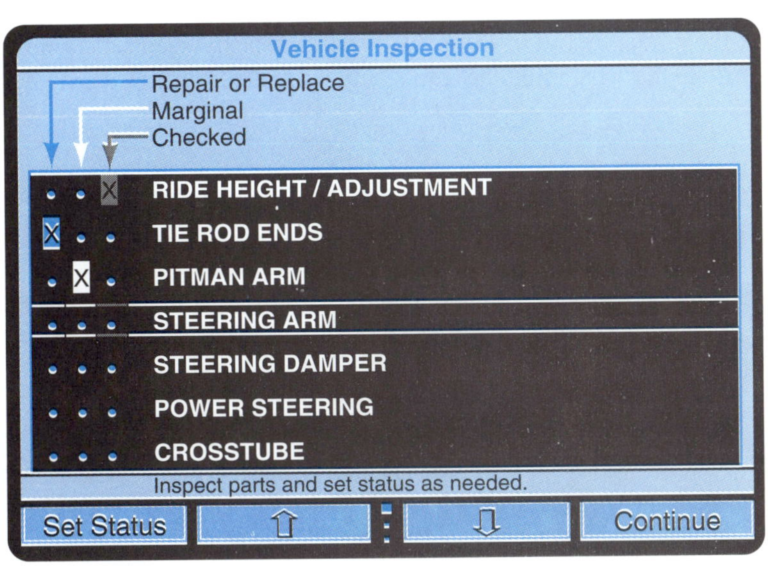


**Figure 17-11** Preliminary inspection screen (Courtesy of Hunter Engineering Company)

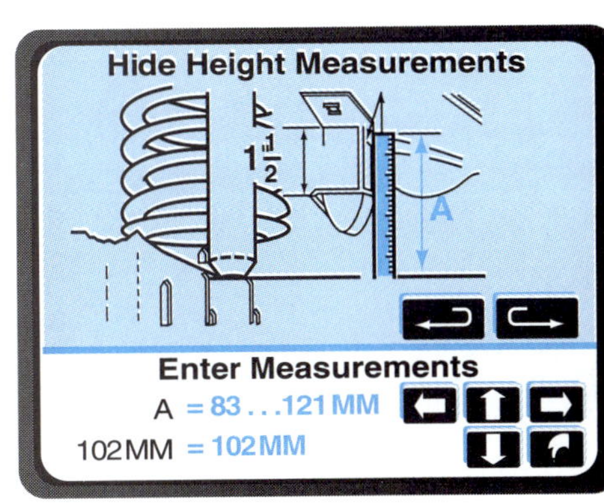


**Figure 17-12** Ride height screen (Courtesy of Bear Automotive Service Equipment Company)

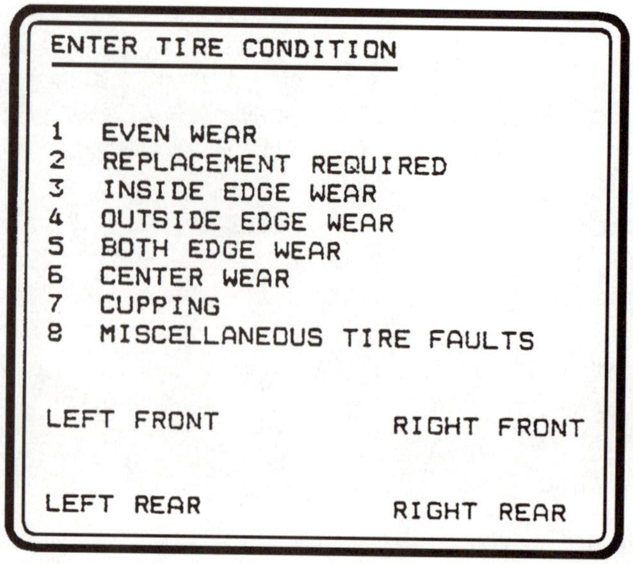


**Figure 17-13** Tire condition screen (Courtesy of Bear Automotive Service Equipment Company)

## Tire Condition Screen

When the tire inspection screen is displayed, the technician may enter various tire wear conditions for each tire (Figure 17-13). The cursor on the screen is moved to the tire being inspected. The tire conditions are printed out with the preliminary inspection results and the alignment report.

## Wheel Runout Compensation Screen

When the wheel runout screen is displayed, the technician is directed to level and lock each wheel unit, and then press the compensation button to provide automatic wheel runout compensation (Figure 17-14). If the computer aligner does not have automatic wheel runout compensation, a manual wheel runout procedure must be followed. This type of computer aligner displays a wheel runout measurement screen. During this procedure, the wheel being checked for runout is lifted with the hydraulic jack on the alignment rack, and the wheel is rotated until the rim clamp knob faces downward. Level and lock the wheel unit in this position, and then push yes on the wheel unit as instructed on the screen. Rotate the wheel until the rim clamp knob faces upward, and then

**Figure 17-14** Wheel runout compensation screen (Courtesy of Bear Automotive Service Equipment Company)

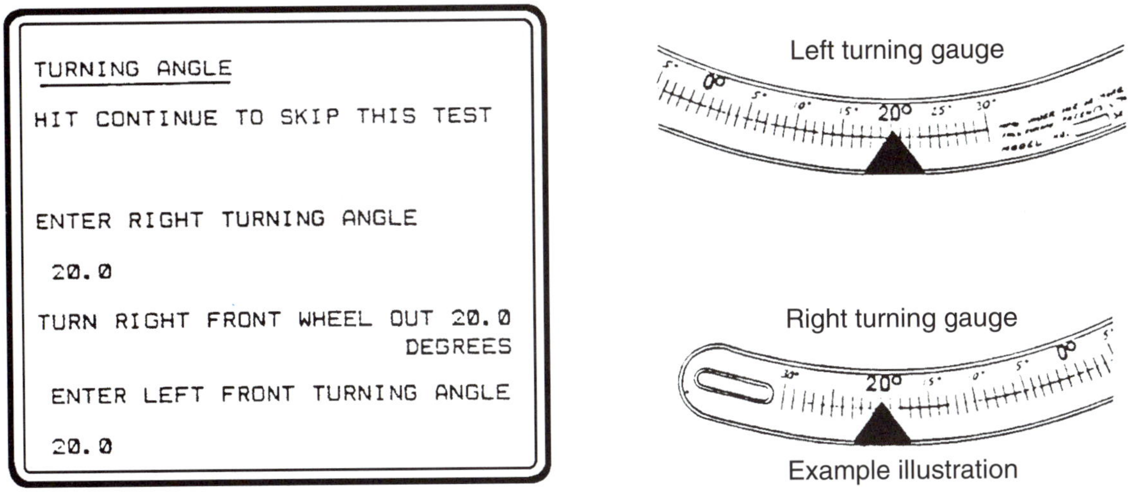


**Figure 17-15** Turning angle screen (Courtesy of Bear Automotive Service Equipment Company)

level and lock the wheel unit. After this procedure, press yes on the wheel unit. The wheel being checked is selected on the screen, and this same basic procedure is followed at each wheel.

## Wheel Alignment Screens

### Turning Angle Screen

When the turning angle screen is displayed, the technician removes the locking pins from the turning plates. Each front wheel must be turned outward a specified amount, and the turning angle on the opposite front wheel must be entered with the keypad as directed on the screen (Figure 17-15).

### Front and Rear Wheel Alignment Angle Screen

**SERVICE TIP:** Many computer wheel aligners have the capability to print out any single screen. It may be helpful to print out the wheel alignment angle screen before and after the alignment angles are adjusted to the vehicle manufacturer's specifications. Many customers appreciate receiving these printouts.

Prior to a display of the wheel alignment angle screen, the screen directs the technician to position the front wheels straight ahead, apply the brake pedal depressor, and level and lock the wheel units. After this procedure, the screen directs the technician to swing the front wheels outward a specific amount (Figure 17-16). This front wheel swing may be referred to as a caster/SAI swing. The amount of front wheel swing varies depending on the make of computer aligner.

After the front wheel swing procedure, the front and rear wheel alignment angles are displayed (Figure 17-17). Some computer wheel aligners highlight any wheel alignment angles that are not within specifications. If cross camber and cross caster are displayed on the screen, these readings indicate the maximum difference allowed between the right and left side readings.

## Adjustment Screens

The technician may select adjustment screens that provide bar graph readings of the alignment angles. A highlighted bar on the graph indicates the specified range for the alignment angle, and an arrow shows the exact position of the alignment angle. As the alignment angle is adjusted, the

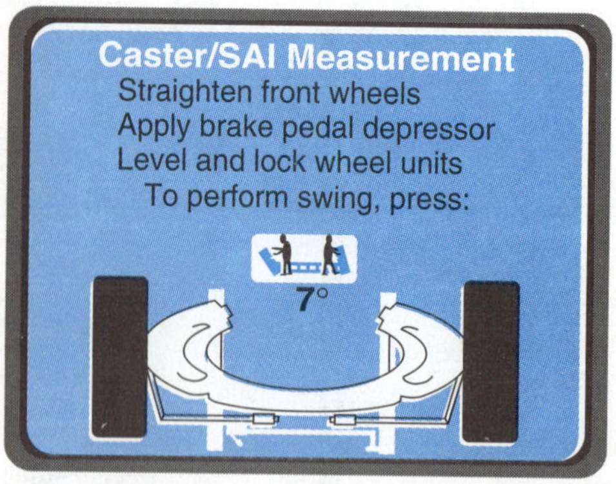


**Figure 17-16** Screen instructions prior to wheel alignment angle display screen (Courtesy of Bear Automotive Service Equipment Company)

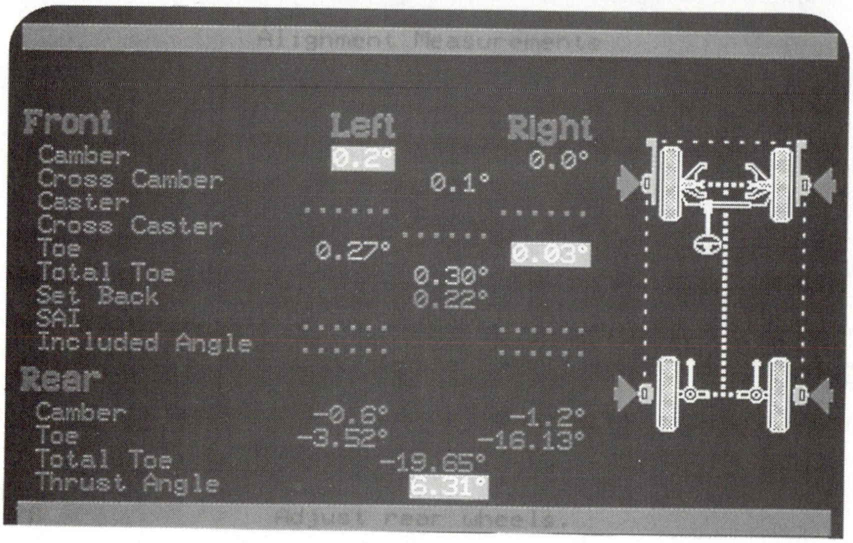


**Figure 17-17** Front and rear wheel alignment angle screen (Courtesy of Hunter Engineering Company)

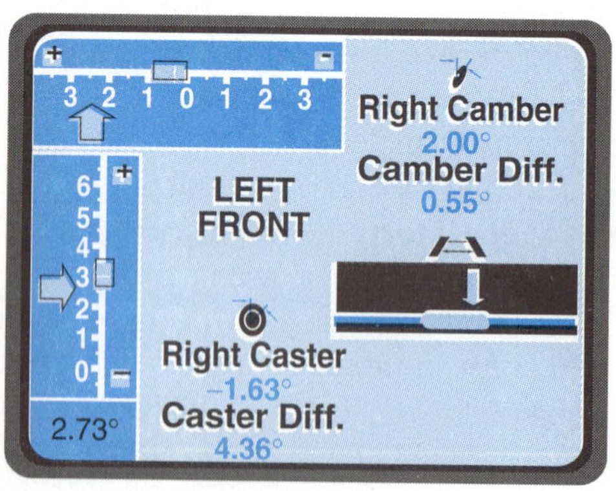


**Figure 17-18** Wheel alignment adjustment screen with bar graphs and digital readings (Courtesy of Bear Automotive Service Equipment Company)

arrow moves to indicate the present setting. If the alignment angle is within specifications, the arrow is pointing to the highlighted bar on the graph. Some adjustment screens provide bar graphs and digital readings of alignment angles (Figure 17-18).

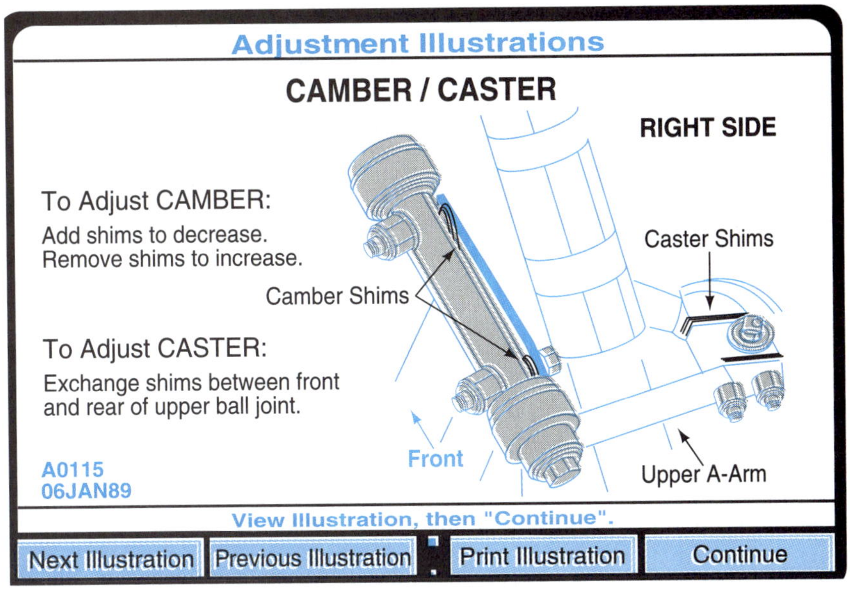


**Figure 17-19** Diagnostic drawings illustrate adjustment locations and instructions (Courtesy of Hunter Engineering Company)

Some computer aligners have a jack and hold feature that allows the suspension to be lifted on the alignment rack to perform an adjustment, while maintaining accurate displays on the adjustment screen.

### Shim Select Screens

On many front wheel drive cars, the rear wheel camber and toe are adjusted with shims. Some computer aligners have shim display screens, which indicate the thickness of shim required and the proper position of the shim.

## Diagnostic Drawings and Text Screens

The technician may select diagnostic drawings of the suspension on the vehicle being aligned. These drawings illustrate adjustment locations and instructions (Figure 17-19).

On some computer wheel aligners, the technician may select print at any time and print out the displayed screen, including diagnostic drawings. Another optional procedure is to print out the wheel alignment report before and after the adjustment of wheel alignment angles.

**Classroom Manual**
Chapter 17, page 332

## Causes of Improper Rear Wheel Alignment

These suspension or chassis defects will cause incorrect rear wheel alignment:

1. collision damage that results in a bent frame or distorted unitized body
2. a leaf-spring eye that is unwrapped or spread open
3. leaf-spring shackles that are broken, bent, or worn
4. broken leaf springs, or leaf-spring center bolts
5. worn rear upper or lower control arm bushings
6. worn trailing arm bushings, or dislocated trailing arm brackets
7. bent components such as radius rods, control arms, struts, and rear axles

# Rear Suspension Adjustments

## Rear Wheel Camber Adjustment

**WARNING:** Always follow the vehicle manufacturer's recommended rear suspension adjustment procedure in the service manual to avoid improper alignment angles and reduced steering control.

Improper rear wheel camber causes excessive wear on the edges of the rear tire treads. Cornering stability may be affected by improper rear wheel camber, especially while cornering at high speeds. Improper rear wheel camber on a front wheel drive car may change the understeer or oversteer characteristics of the vehicle. Since a tilted wheel rolls in the direction of the tilt, rear wheel camber may cause steering pull. For example, excessive positive camber on the right rear wheel causes the rear suspension to drift to the right and causes steering pull to the left.

Understeer is the tendency of a vehicle not to turn as much as desired by the driver, and to move toward the outside of the turn, especially at high speed. Most front wheel drive cars have a slight amount of understeer because of the large amount of weight on the front suspension.

Oversteer is the tendency of a vehicle to turn sharper than the turn selected by the driver. This action is caused by the rear of the car wanting to move to the outside of the turn, which results in the front end of the car turning more sharply. Rear wheel drive cars have a tendency to oversteer because of the increased weight on the rear suspension.

**Special Tools**

Special rear wheel camber adjustment tools for the vehicle being aligned

Rear suspension adjustments vary depending on the type of suspension system. Some manufacturers of wheel alignment equipment and tools provide detailed diagrams for front and rear wheel alignment adjustments on all makes of domestic and imported vehicles. Our objective is to show a few of the common methods of adjusting rear wheel alignment angles on various types of rear suspension systems.

On some semi-independent rear suspension systems, camber and toe are adjusted by inserting different sizes of shims between the rear spindle and the spindle mounting surface. These shims are retained by the spindle mounting bolts. The shim thickness is changed between the top or bottom of the spindle to adjust camber (Figure 17-20).

Many rear camber shims are now circular. As mentioned previously, some computer wheel aligners indicate the thickness of shim required and the proper shim position. These same shims also adjust rear wheel toe.

On some transverse leaf rear suspension systems, such as those on a later model Oldsmobile Cutlass Supreme, the spindle must be removed from the strut, and the lower strut bolt hole elongated to allow a camber adjustment (Figure 17-21). Once this bolt hole is elongated, assemble the knuckle and strut, and leave the retaining bolts loose. The top of the tire and the spindle may be moved outward or inward to adjust camber. After the camber is adjusted to specifications, the strut-to-knuckle nuts must be tightened to the specified torque.

On some independent rear suspension systems with a lower control arm and ball joint, the strut-to-knuckle bolts are loosened and a special camber tool must be installed on the lower strut-to-knuckle bolt and the rear side of the strut (Figure 17-22). Rotate the adjusting bolt in the back of the tool to adjust the camber, and then tighten the strut-to-knuckle nuts to the specified torque.

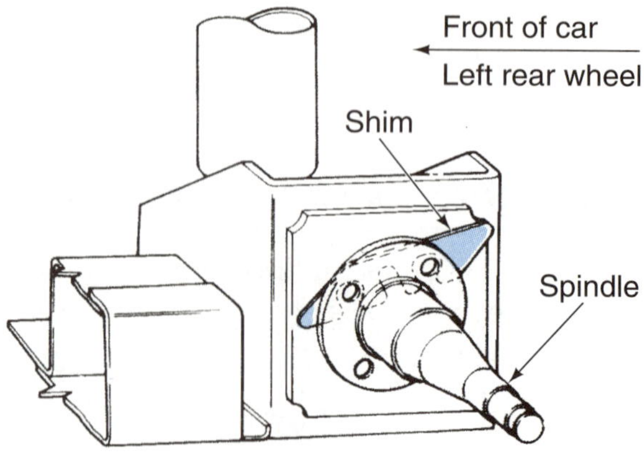


**Figure 17-20** Shim adjustment for rear wheel camber, semi-independent rear suspension system (Courtesy of Chrysler Corporation)

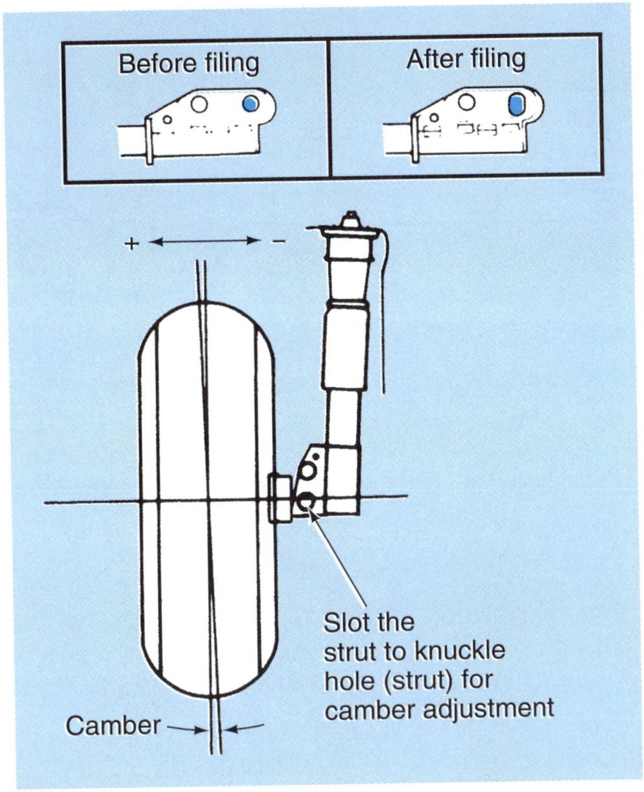


Figure 17-21 Elongating the lower strut bolt hole provides camber adjustment. (Courtesy of Oldsmobile Division, General Motors Corporation)

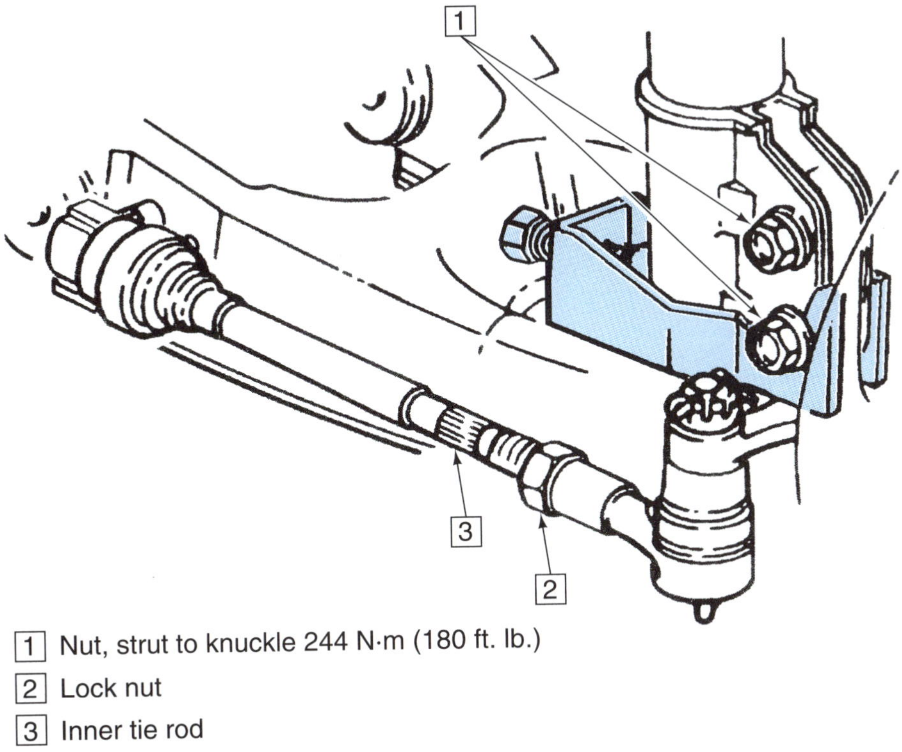


Figure 17-22 Camber adjustment with special tool inserted on rear lower strut-to-knuckle bolt (Courtesy of Oldsmobile Division, General Motors Corporation)

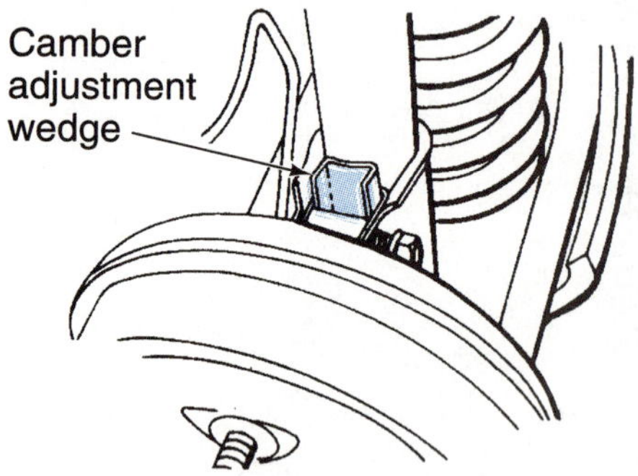


Figure 17-23 Camber adjusting wedge installed between the top of the rear knuckle and strut (Courtesy of Hunter Engineering Company)

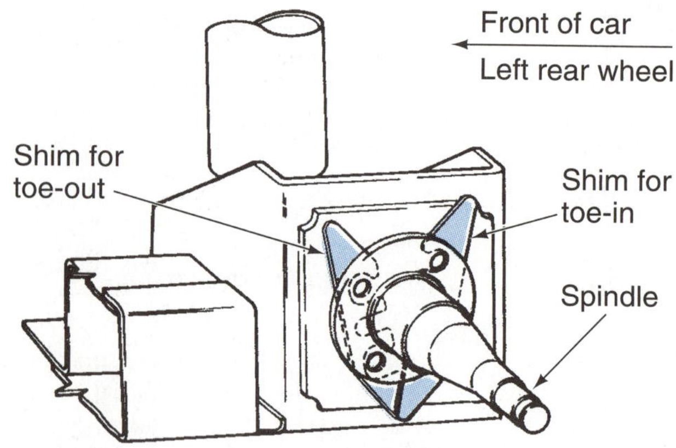


Figure 17-24 Adjusting rear wheel toe by changing shim thickness between the front and rear edges of the rear spindle mounting flange (Courtesy of Chrysler Corporation)

On other rear suspension systems, a camber adjustment wedge may be installed between the top of the knuckle and the strut to adjust camber (Figure 17-23). The nuts on the strut-to-knuckle bolts must be loosened prior to wedge installation, and these nuts must be tightened to the specified torque after the camber is adjusted to specifications.

## Rear Wheel Toe Adjustment

Improper rear wheel toe moves the thrust line away from the geometric centerline, which causes steering pull and feathered tire tread wear. If the left rear wheel has excessive toe-out, the thrust line is moved to the left of the geometric centerline. Excessive toe-in on the left rear wheel moves the thrust line to the right of the the geometric centerline. When toe-in or toe-out is excessive on the right rear wheel, it has the opposite effect on the thrust line compared to the left rear wheel. When the thrust line is moved to the left of the geometric centerline, the steering pulls to the right, whereas a thrust line positioned to the right of the geometric centerline results in steering pull to the left.

The toe may be adjusted on many semi-independent rear suspension systems by changing the shim thickness between the front and rear edges of the rear spindle mounting flange (Figure 17-24).

On some transverse leaf rear suspension systems, a special toe adjusting tool is inserted in openings in the jack pad and the rear lower control rod (Figure 17-25).

**Special Tools**

Special rear wheel toe adjusting tools for the vehicle being aligned

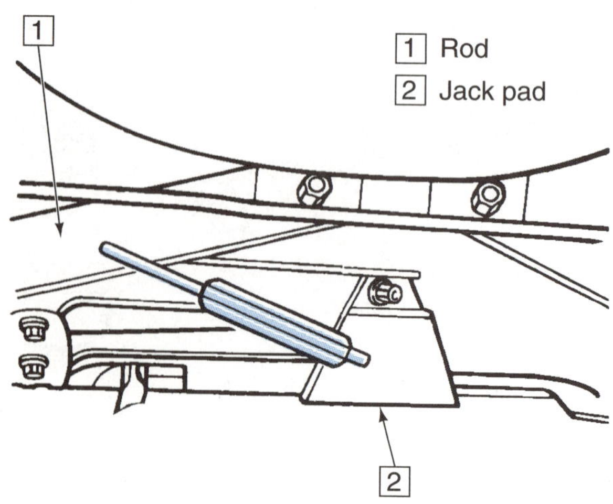


Figure 17-25 Special toe adjusting tool for transverse leaf-spring rear suspension (Courtesy of Oldsmobile Division, General Motors Corporation)

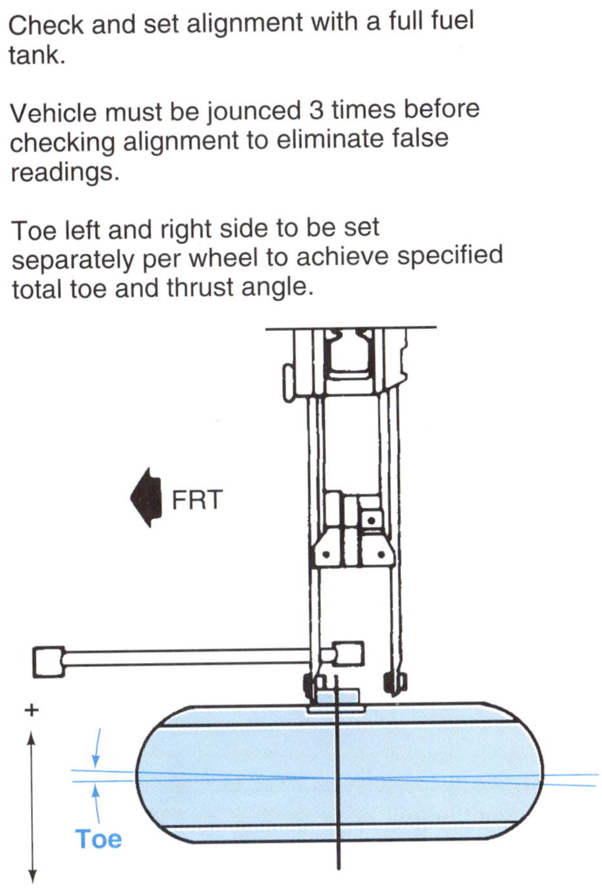


**Figure 17-26** Moving lower rear control rod position to adjust rear wheel toe on a transverse leaf rear suspension (Courtesy of Oldsmobile Division, General Motors Corporation)

This tool has a turnbuckle in the center of the tool. When the nut on the inner end of the rear lower control rod is loosened, the turnbuckle is rotated to lengthen or shorten the tool and to move the rear lower control rod to adjust the rear wheel toe (Figure 17-26). After the toe is adjusted to specifications, the nut on the rear lower control rod bolt must be tightened to the specified torque.

On some independent rear suspension systems, an eccentric star wheel on the rear lower control rod may be rotated to adjust the rear wheel toe (Figure 17-27). On some rear suspension systems with a lower control arm and ball joint, the nut on the tie rod is loosened, and the tie rod is

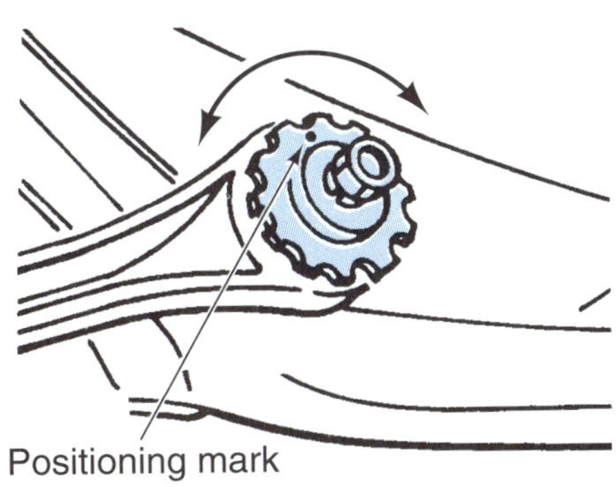


**Figure 17-27** Eccentric star wheel on the inner end of the rear lower control rod for toe adjustment (Courtesy of Hunter Engineering Company)

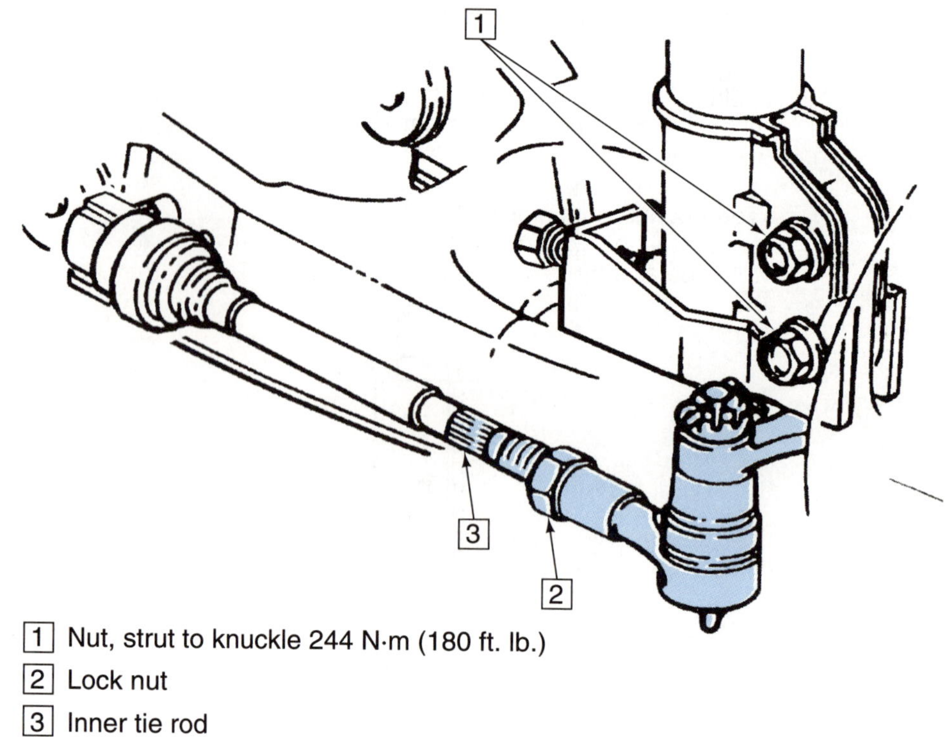


**Figure 17-28** Rotating rear tie rod to change tie rod length and adjust rear wheel toe (Courtesy of Oldsmobile Division, General Motors Corporation)

**Classroom Manual**
Chapter 17, page 327

rotated to adjust the rear wheel toe (Figure 17-28). After the toe is adjusted to specifications, the tie rod nut must be tightened to the specified torque.

**CUSTOMER CARE:** Some computer wheel aligners can operate a remote monitor. A remote monitor may be placed in the customer waiting room so customers may observe the suspension service and adjustments on their vehicle. Present the customer with a printout of the wheel alignment angles before and after the alignment angles were adjusted to the vehicle manufacturer's specifications. Take time to explain briefly to the customer the alignment angle adjustments performed on the car and the cause and correction of any customer complaints.

## Guidelines for the Use of Track Bars, Computer Alignment Systems, and Rear Wheel Alignment

1. When all four wheels are parallel to the geometric centerline and the thrust line is positioned on the geometric centerline, the rear wheels track directly behind the front wheels.
2. Prior to the introduction of computer wheel aligners, a two-wheel alignment was performed on the front wheels, and a track gauge was used to measure rear wheel tracking.
3. If the track gauge measurements are the same on both sides of the vehicle, the rear wheel tracking is satisfactory.

4. The main menu on a computer wheel aligner allows the technician to select the desired procedure and the type of wheel alignment.
5. The specifications menu allows the technician to select specifications for the vehicle being aligned with a cursor on the screen or a wand and bar graph.
6. The preliminary inspection menu allows the technician to select all suspension components as satisfactory or unsatisfactory from a list on the screen.
7. On a ride height screen, the ride height measurement location is displayed, and the technician enters the ride height measurement. The computer compares this measurement to specifications.
8. An automatic or manual wheel runout compensation procedure must be performed at each wheel, because wheels with excessive runout affect the position of the rim clamps and the alignment readings.
9. A turning angle screen allows the technician to check the turning radius and enter this reading.
10. The front and rear wheel alignment angle screen displays all the front and rear wheel alignment angles.
11. Adjustment screens display bar graphs of the alignment angles with the specified angle highlighted. An arrow points to the actual position of the alignment angle. When the alignment angle is adjusted, the arrow moves to indicate the new setting.
12. Shim select screens show the proper location and thickness of shims required to bring the alignment angles within specifications.
13. Diagnostic drawings and text screens illustrate adjustment locations and provide instructions regarding adjustments.
14. On many semi-independent rear suspension systems, shims are positioned between the spindle and the spindle mounting surface to adjust rear wheel camber and toe.
15. On some transverse leaf rear suspension systems, the lower bolt hole in the rear strut must be elongated to provide a camber adjustment.
16. On some rear suspension systems with a lower control arm and ball joint, a special tool is installed on the strut and lower strut bolt to adjust camber.
17. A camber wedge may be installed between the strut and knuckle to adjust camber on some rear suspension systems.
18. On some transverse leaf rear suspension systems, the toe is adjusted with a special turnbuckle tool installed in openings in the jack pad and the rear lower control rod. When the inner nut on the lower control rod is loosened, the tool is lengthened or shortened to adjust the rear wheel toe.
19. A star wheel is located on the inner end of the lower control rod to adjust the toe on some rear suspension systems.
20. On other rear suspension systems, the tie rod nut may be loosened and the tie rod rotated to adjust rear wheel toe.

## CASE STUDY

A customer complained that the left rear tire on her Cutlass Calais had suffered severe tread wear in the last 2,000 miles. The technician examined this tire and found the customer was not exaggerating. All the tread wear indicators were showing, and the other tires showed very little wear. The customer indicated that the left rear tire tread looked like the tread on the other tires 2,000 miles previously. While checking the tread wear on the left rear tire, the technician noticed the car had been painted recently. Further discussion with the customer revealed the car had been purchased recently, and there was no warranty on the vehicle. During a road test, the technician found that the steering pulled to the left while driving straight ahead. The technician suggested a thorough prealignment inspection and complete four wheel alignment with a computer alignment system.

During the prealignment inspection, all the steering and suspension components appeared to be in normal condition, except the left rear tire.. However, the technician noticed the left rear quarter panel and other body components in this area had been replaced. The car had apparently suffered severe collision damage in this area.

The four wheel alignment results were satisfactory except the left rear wheel had 5° toe-in and the thrust angle was 8°. This severe rear toe-in condition had moved the thrust line to the right of the geometric centerline. When the technician referred to the manufacturer's service manual, this information stated the rear wheel toe is not adjustable, and if the rear wheel camber and toe are not within specifications, the rear underbody and suspension should be inspected for damage. The technician located the underbody dimension specifications in the service manual (Figures 17-29 and 17-30).

**Special Tools**

Tram gauge

A tram gauge is similar to a track gauge, but the tram gauge has two adjustable pointers at the ends of the gauge. The tram gauge is used to measure distances between two locations on the frame, or chassis.

The technician began checking the rear suspension underbody measurements with a tram gauge. When dimension M was measured between the center of the bolt heads in the trailing arm brackets, this measurement was 1 inch (25.4 millimeters) less than specified. This measurement indicated the left rear subframe had been forced inward toward the center of the vehicle, which resulted in the improper toe and thrust angle. Further inspection of the underbody revealed the floor pan was buckled.

When the rear subframe was forced inward, the left rear trailing arm was also bent inward. Further inspection of the rear axle channel with a long straight edge indicated this channel was bent rearward in the center. The car was sent to the body shop to pull the rear

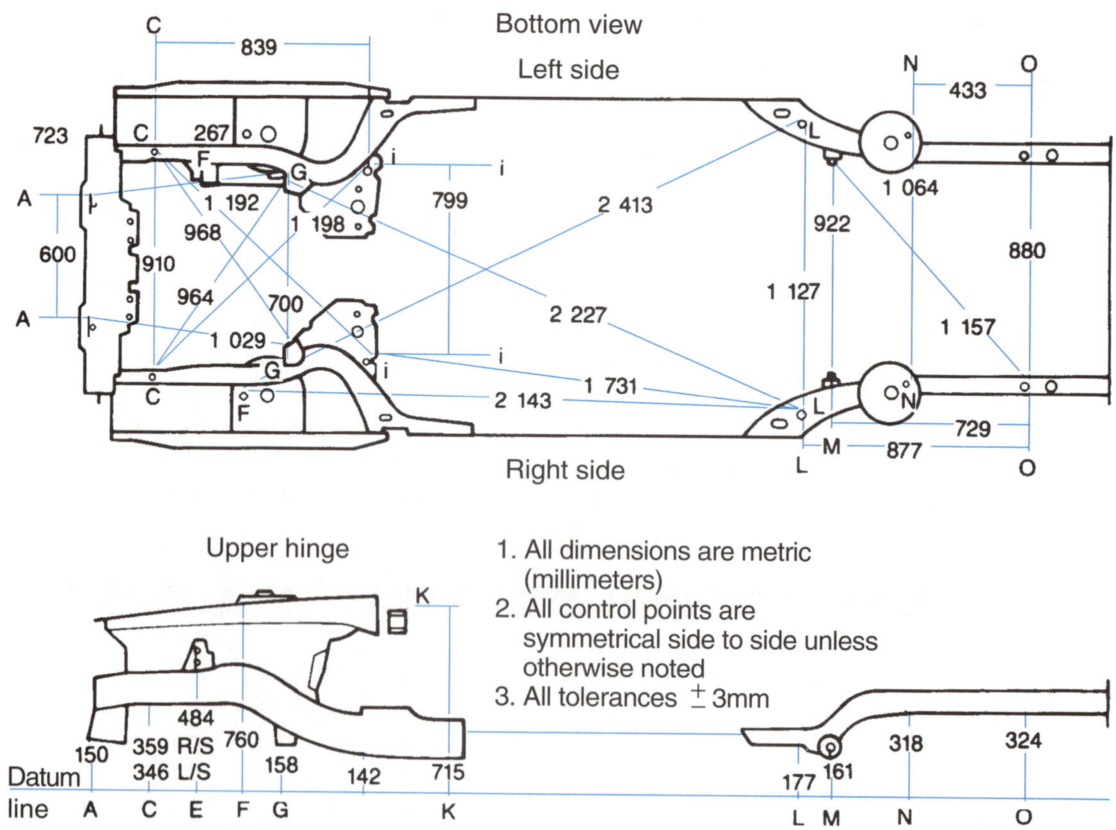


Right side gage hole "C", relative to left side gage hole "C", should be 13 mm above, 23 mm rearward, and 10 mm more outboard from center of vehicle.

**Figure 17-29** Underbody measurement specifications and locations (Courtesy of Oldsmobile Division, General Motors Corporation)

| REF. | HORIZONTAL | VERTICAL | LOCATION |
|---|---|---|---|
| A | Leading edge of 24 mm gage hole* | Lower surface at gage hole to datum line | Front end lower tie bar |
| B | Center of 16 mm gage hole | None | Front end upper tie bar |
| C | Leading edge of 20 mm gage hole | Lower surface at gage hole to datum line | Motor compartment side rail, forward of transaxle anchor support plate on left side rail, and at rear of engine mounting support on right side rail. For access on right side with air conditioning, remove air compressor |
| D | Center of 16 mm gage hole | None | Front upper surface of motor compartment side rail |
| E | Center of lower attaching hole in transaxle support anchor plate | Datum line to horizontal centerline of transaxle support anchor plate lower attaching hole | Transaxle support anchor plate on left side rail |
| F | Center of shock tower strut front attaching hole | Upper surface at shock tower strut front attaching hole | Shock tower |
| G | Leading edge of oblong master gage hole | Datum line to horizontal leading edge of oblong master gage hole | Front suspension lower control arm mount support |
| H | Center of 16 mm gage hole | None | Gage hole in upper surface of motor compartment side rail |
| I | Center of 19 mm gage hole | Datum line to lower surface at gage hole | Outboard 19 mm gage hole in front suspension control arm mounting reinforcement support |
| J | Center of hood hinge pivot pin head | None | Hood hinge |
| K | Center of front upper hinge pin hole | Upper surface at hinge pin hole | Front upper door hinge, body side |
| L | Leading edge of 22 mm gage hole | Lower surface at leading edge of gage hole to datum line | Compartment pan longitudinal rail, forward of rear suspension spring seat support |
| M | Center of attaching bolt head | Datum line to horizontal centerline of attaching bolt head of rear suspension control arm support. Dimension is calculated by placing side surface of pointer at center of bolt head | Rear suspension control arm support forward of rear suspension spring seat support |
| N | Center of 16 mm gage hole | Datum line to lower surface at gage hole | Outboard gage hole of rear suspension spring seat support. (Not accessible with rear suspension assembly in place.) |
| O | Leading edge of 21 mm gage hole | Lower surface at gage hole to datum line | Compartment pan longitudinal rail. For access on right side, remove muffler |
| P (Right) | Center of lower stud attaching hole of right side steering unit rack and pinion support mount plate | Centerline of lower stud attaching hole of right side steering unit rack and pinion support mount plate to datum line | Steering unit rack and pinion support mount plate attached to lower right side of dash panel |
| P (Left) | Center of lower stud attaching hole of left side steering unit rack and pinion support mount plate | Centerline of lower stud attaching hole of left side steering unit rack and pinion support mount plate to datum line | Steering unit rack and pinion support mount plate attached to lower left side of dash panel |
| Q (Right) | Center of upper stud attaching hole to right side steering unit rack and pinion support mount plate | Centerline of upper stud attaching hole of right side steering unit rack and pinion support mount plate to datum line | Steering unit rack and pinion support mount plate attached to lower right side of dash panel |
| Q (Left) | Center of upper stud attaching hole to left side steering unit rack and pinion support mount plate | Centerline of upper stud attaching hole of left side steering unit rack and pinion support mount plate to datum line | Steering unit rack and pinion support mount plate attached to lower left side of dash panel |
| R | Center of outboard attaching hole for front suspension — lower control arm support | Lower surface at gage hole | Front end lower tie bar assembly |

*Leading edge refers to the most forward edge of the hole

**Figure 17-30** Explanation of underbody horizontal and vertical measurement locations (Courtesy of Oldsmobile Division, General Motors Corporation)

subframe and unitized body back to the original position and provide the specified distance between the center of bolt heads on the trailing arms. Since the rear axle channel is a critical component in providing rear suspension ride quality and alignment, the customer was advised this component should be replaced. A new rear tire was also installed.

When the rear suspension was reassembled, all the alignment angles were rechecked. The rear wheel toe and thrust line were now within specifications. During a road test, there was no evidence of steering pull or any other steering problems.

From this experience, the technician learned the importance of four wheel alignment. The technician also discovered that accurate underbody measurement specifications are absolutely essential.

## Terms to Know

Geometric centerline
Thrust line
Track gauge
Rear wheel tracking
Setback
Rear axle sideset
Rear axle offset
Diamond-shaped chassis
Main menu
Specifications menu
Preliminary inspection screen
Ride height screen
Tire condition screen
Wheel runout compensation screen
Turning angle screen
Front and rear wheel alignment angle screen
Adjustment screen
Shim select screen
Diagnostic drawing and text screen
Oversteer
Understeer
Tram gauge

## ASE Style Review Questions

1. While discussing a vehicle on which all the suspension components are in satisfactory condition, all the front suspension alignment angles are within specifications, and the steering wheel is centered on the alignment rack, but the steering wheel is not centered while driving straight ahead:
*Technician A* says that the rear wheel toe-in and thrust line may not be within the manufacturer's specifications.
*Technician B* says that a four wheel alignment should be performed on the vehicle.
Who is correct?
**A.** A only
**B.** B only
**C.** Both A and B
**D.** Neither A nor B

2. While discussing these track gauge measurements: On the right side of the vehicle, the front pointer is aligned with the rear edge of the front rim, and the rear pointer on the rear wheel must be moved inward more than the forward pointer on the rear wheel. On the left side of the vehicle, the front pointer is ahead of the rear edge on the front rim, and the forward rear pointer must be moved inward more than the rear pointer.
*Technician A* says this condition indicates rear axle sideset condition.
*Technician B* says this condition indicates front wheel setback.
Who is correct?
**A.** A only
**B.** B only
**C.** Both A and B
**D.** Neither A nor B

3. While discussing thrust line and geometric centerline:
*Technician A* says the vehicle thrust line should be located on the geometric centerline of the vehicle.
*Technician B* says the thrust line position is acceptable if it is within 8° on either side of the geometric centerline.
Who is correct?
**A.** A only
**B.** B only
**C.** Both A and B
**D.** Neither A nor B

4. While discussing computer wheel aligners:
*Technician A* says the technician selects defective suspension components from a list on the screen.
*Technician B* says the defective suspension components selected by the technician are printed out in the alignment report.
Who is correct?
**A.** A only
**B.** B only
**C.** Both A and B
**D.** Neither A nor B

5. While discussing computer wheel aligners:
*Technician A* says it is not necessary to check or compensate for wheel runout.
*Technician B* says an automatic or manual wheel runout checking procedure must be followed.
Who is correct?
**A.** A only
**B.** B only
**C.** Both A and B
**D.** Neither A nor B

6. While discussing computer wheel aligners:
*Technician A* says the adjustment screens provide bar graph readings of the alignment angles, and an arrow points to the exact position of the alignment angle.
*Technician B* says on the front and rear wheel alignment angle screen, the cross camber reading indicates the difference between the camber setting on each front wheel and the specified camber.
Who is correct?
**A.** A only
**B.** B only
**C.** Both A and B
**D.** Neither A nor B

7. While discussing computer wheel aligners:
*Technician A* says on many computer wheel aligners the technician may print out any screen at any time including diagnostic drawings.
*Technician B* says some computer wheel aligners have a jack and hold feature that allows the suspension to be lifted with a jack while still maintaining accurate alignment angle readings on the screen.
Who is correct?
**A.** A only
**B.** B only
**C.** Both A and B
**D.** Neither A nor B

8. While discussing a rear suspension system with a shim-type camber and toe adjustment that requires an increase in positive camber:
*Technician A* says that a thicker shim must be installed on the upper spindle mounting bolts and a thinner shim positioned on the lower spindle mounting bolts.
*Technician B* says that a thicker shim must be positioned on the front spindle mounting bolts and a thinner shim on the rear spindle mounting bolts.
Who is correct?
**A.** A only
**B.** B only
**C.** Both A and B
**D.** Neither A nor B

9. While discussing a vehicle thrust line that is 6° to the left of the geometric centerline:
*Technician A* says this problem may be caused by excessive toe-out on the right rear wheel.
*Technician B* says this defect may be caused by excessive toe-out on the left rear wheel.
Who is correct?
**A.** A only
**B.** B only
**C.** Both A and B
**D.** Neither A nor B

10. While discussing a vehicle on which the front suspension alignment angles are within the manufacturer's specifications, but the thrust line is 7° to the right of the vehicle centerline:
*Technician A* says this vehicle will have satisfactory directional control.
*Technician B* says this vehicle will tend to steer to the left when driven straight ahead.
Who is correct?
**A.** A only
**B.** B only
**C.** Both A and B
**D.** Neither A nor B

## Table 17-1 ASE TASK

Check and adjust front and rear wheel camber on suspension systems with a camber adjustment.

| Problem Area | Symptoms | Possible Causes | Classroom Manual | Shop Manual |
|---|---|---|---|---|
| TIRE LIFE | Wear on the outside edges of rear tire treads | Excessive negative or positive camber | 330 | 410 |
| STEERING CONTROL | Steering pull | Improper rear wheel camber | 330 | 410 |
| | Oversteer or understeer while cornering at high speed | Improper rear wheel camber | 330 | 410 |

## Table 17-2 ASE TASK

Check front and rear wheel camber on nonadjustable suspension systems; determine needed repairs.

| Problem Area | Symptoms | Possible Causes | Classroom Manual | Shop Manual |
|---|---|---|---|---|
| TIRE LIFE | Wear on outside edges of rear tire treads | Excessive negative or positive camber | 330 | 411 |
| STEERING CONTROL | Steering pull | Improper rear wheel camber | 330 | 411 |

## Table 17-3 ASE TASK

Check rear wheel toe; determine needed repairs or adjustments.

| Problem Area | Symptoms | Possible Causes | Classroom Manual | Shop Manual |
|---|---|---|---|---|
| TIRE LIFE | Feathered wear on rear tires | Improper rear wheel toe setting | 330 | 412 |
| STEERING CONTROL | Steering pull | Improper toe setting on one rear wheel | 330 | 412 |

## Table 17-4 ASE TASK

Check rear wheel thrust angle; determine needed repairs or adjustments.

| Problem Area | Symptoms | Possible Causes | Classroom Manual | Shop Manual |
|---|---|---|---|---|
| STEERING CONTROL | Steering pull | Thrust line positioned to the left or right of the geometric centerline | 326 | 408 |

# APPENDIX A

## Automotive Suspension and Steering Systems Special Tool Suppliers

Automotive Group
Kent-Moore Division, SPX Corporation
Roseville, MI

Bear Automotive Service Equipment Company
Milwaukee, WI

Easco/KD Tools
Lancaster, PA

Hennessy Industries, Inc.
LaVerge, TN

Hunter Engineering Company
Bridgeton, MO

Mac Tools, Inc.
Washington Courthouse, OH

OTC Division, SPX Corporation
Owatonna, MN

Sears Industrial Sales
Cincinnati, OH

Snap-On Tools Corporation
Kenosha, WI

Specialty Products Company
Longmont, CO

Sun Electric Corporation
Crystal Lake, IL

# APPENDIX B

## Metric Conversions

| | to convert these | to these, | multiply by: |
|---|---|---|---|
| *TEMPERATURE* | Centigrade Degrees | Fahrenheit Degrees | 1.8 then + 32 |
| | Fahrenheit Degrees | Centigrade Degrees | 0.556 after − 32 |
| *LENGTH* | Millimeters | Inches | 0.03937 |
| | Inches | Millimeters | 25.4 |
| | Meters | Feet | 3.28084 |
| | Feet | Meters | 0.3048 |
| | Kilometers | Miles | 0.62137 |
| | Miles | Kilometers | 1.60935 |
| *AREA* | Square Centimeters | Square Inches | 0.155 |
| | Square Inches | Square Centimeters | 6.45159 |
| *VOLUME* | Cubic Centimeters | Cubic Inches | 0.06103 |
| | Cubic Inches | Cubic Centimeters | 16.38703 |
| | Cubic Centimeters | Liters | 0.001 |
| | Liters | Cubic Centimeters | 1000 |
| | Liters | Cubic Inches | 61.025 |
| | Cubic Inches | Liters | 0.01639 |
| | Liters | Quarts | 1.05672 |
| | Quarts | Liters | 0.94633 |
| | Liters | Pints | 2.11344 |
| | Pints | Liters | 0.47317 |
| | Liters | Ounces | 33.81497 |
| | Ounces | Liters | 0.02957 |
| *WEIGHT* | Grams | Ounces | 0.03527 |
| | Ounces | Grams | 28.34953 |
| | Kilograms | Pounds | 2.20462 |
| | Pounds | Kilograms | 0.45359 |
| *WORK* | Centimeter-Kilograms | Inch-Pounds | 0.8676 |
| | Inch-Pounds | Centimeter-Kilograms | 1.15262 |
| | Meter-Kilograms | Foot-Pounds | 7.23301 |
| | Foot-Pounds | Newton-Meters | 1.3558 |
| *PRESSURE* | Kilograms/Square Centimeter | Pounds/Square Inch | 14.22334 |
| | Pounds/Square Inch | Kilograms/Square Centimeter | 0.07031 |
| | Bar | Pounds/Square Inch | 14.504 |
| | Pounds/Square Inch | Bar | 0.06895 |

# GLOSSARY

*Note: Terms are highlighted in color, followed by Spanish translation in bold.*

Accidental air bag deployment An unintended air bag deployment caused by improper service procedures.
**Despliegue accidental del Airbag** Despliegue imprevisto del Airbag ocasionado por procedimientos de reparación inadecuados.

Adjustment screen A display on a computer wheel aligner that allows the technician to see the results of certain suspension adjustments.
**Pantalla para la visualización de ajustes** Representación visual en una computadora para alineación de ruedas que le permite al técnico ver los resultados de ciertos ajustes en la suspensión.

Air bag deployment module The air bag and deployment canister assembly that is mounted in the steering wheel for the driver's side air bag, or in the dash panel for the passenger's side air bag.
**Unidad de despliegue del Airbag** El conjunto del Airbag y elemento de despliegue montado en el volante de dirección para proteger al conductor, o en el tablero de instrumentos para proteger al pasajero.

Air bleeding The process of bleeding air from a hydraulic system such as the power steering system.
**Muestra de aire** Proceso a través del cual se extrae el aire de un sistema hidraúlico, como por ejemplo un sistema de dirección hidraúlica.

Air conditioning programmer (ACP) A control unit that may contain the computer, solenoids, motors, and vacuum diaphragms for air conditioning control.
**Programador del acondicionamiento de aire** Unidad de control que puede incluir la computadora, los solenoides, los motores y los diafragmas de vacío para el control del acondicionamiento de aire.

Alignment ramp A metal ramp positioned on the shop floor on which vehicles are placed during wheel alignment procedures.
**Rampa de alineación** Rampa de metal ubicada en el suelo del taller de reparación de automóviles sobre la que se colocan los vehículos durante los procedimientos de alineación de ruedas.

Anhydrous calcium grease A special lubricant used in manual rack and pinion steering gears.
**Grasa de calcio anhidro** Lubricante especial utilizado en mecanismos de dirección de cremallera y piñón manuales.

Anti-theft lug nuts Locking lug nuts that help to prevent wheel theft.
**Tuercas de orejetas anti-robo** Tuercas de orejetas autobloqueantes que ayudan a evitar el robo de las ruedas.

Anti-theft wheel covers Locking wheel covers that help to prevent wheel theft.
**Cubrerruedas anti-robo** Cubrerruedas autobloqueantes que ayudan a evitar el robo de las ruedas.

ASE blue seal of excellence An ASE logo displayed by automotive service shops that employ ASE certified technicians.
**Sello azul de excelencia de la ASE** Logotipo exhibido en talleres de reparación de automóviles donde se emplean mecánicos certificados por la ASE.

ASE technician certification ASE provides certification of automotive technicians in eight different areas of expertise.
**Certificación de mecánico de la ASE** Certificación de mecánico de automóviles otorgada por la ASE en ocho áreas diferentes de especialización.

Auto/manual diagnostic check A test procedure in diagnosing a Ford computer-controlled suspension system.
**Revisión diagnóstica automática/manual** Procedimiento de prueba llevado a cabo para diagnosticar un sistema de suspensión controlado por computadora de la Ford.

Auto test A test procedure in diagnosing a Ford computer-controlled suspension system.
**Prueba automática** Procedimiento de prueba llevado a cabo para diagnosticar un sistema de suspensión controlado por computadora de la Ford.

Axle pullers A special puller required for axle shaft removal.
**Extractores del eje** Extractor especial requerido para la remoción del árbol motor.

Backup power supply A voltage source, usually located in the air bag computer, that is used to deploy an air bag if the battery cables are disconnected in a collision.
**Alimentación de reserva** Fuente de tensión, por lo general localizada en la computadora del Airbag, que se utiliza para desplegar el Airbag si se desconectan los cables de la batería a consecuencia de una colisión.

Ball joint axial movement Vertical movement in a ball joint because of internal joint wear.
**Movimiento axial de la junta esférica** Movimiento vertical en una junta esférica ocasionado por el desgaste interno de la misma.

Ball joint radial movement Horizontal movement in a ball joint because of internal joint wear.
**Movimiento radial de la junta esférica** Movimiento horizontal en una junta esférica ocasionado por el desgaste interno de la misma.

Ball joint removal and pressing tools Special tools required for ball joint removal and replacement.
**Herramientas para la remoción y el ajuste de la junta esférica** Herramientas especiales requeridas para la remoción y el reemplazo de la junta esférica.

Ball joint unloading Removing the ball joint tension supplied by the vehicle weight prior to ball joint diagnosis.
**Descarga de la junta esférica** Remoción de la tensión de la junta esférica provista por la carga del vehículo antes de llevarse a cabo la diagnosis de la junta esférica.

Ball joint wear indicator A visual method of checking ball joint wear.
**Indicador de desgaste de la junta esférica** Método visual de revisar el desgaste interior de una junta esférica.

Bearing abrasive roller wear Fine scratches on the bearing surface.
**Desgaste abrasivo del rodillo del cojinete** Rayados finos en la superficie del cojinete.

Bearing abrasive step wear A fine circular wear pattern on the ends of the rollers.
**Desgaste abrasivo escalonado del cojinete** Desgaste circular fino en los extremos de los rodillos.

**Bearing brinelling** Straight line indentations on the races and rollers.

**Acción de Brinnell en un cojinete** Hendiduras en línea recta en los anillos y los rodillos.

**Bearing etching** A loss of material on the bearing rollers and races.

**Corrosión del cojinete** Pérdida de material en los rodillos y los anillos del cojinete.

**Bearing fatigue spalling** Flaking of the surface metal on the rollers and races.

**Escamación y fatiga del cojinete** Condición que ocurre cuando el metal de la superficie de los rodillos y los anillos comienza a escamarse.

**Bearing frettage** A fine, corrosive wear pattern around the races and rollers, with a circular pattern on the races.

**Cinceladura del cojinete** Desgaste corrosivo y fino alrededor de los anillos y los rodillos que se hace evidente a través de figuras circulares en los anillos.

**Bearing galling** Metal smears on the ends of the rollers.

**Desgaste por rozamiento en un cojinete** Ralladuras metálicas en los extremos de los rodillos.

**Bearing heat discoloration** A dark brown or bluish discoloration of the rollers and races caused by excessive heat.

**Descoloramiento del cojinete ocasionado por el calor** Descoloramiento marrón oscuro o azulado de los rodillos y los anillos ocasionado por el calor excesivo.

**Bearing preload** A tension placed on the bearing rollers and races by an adjustment or assembly procedure.

**Carga previa del cojinete** Tensión aplicada a los rodillos y los anillos del cojinete a través de un procedimiento de ajuste o de montaje.

**Bearing pullers** Special tools designed for bearing removal.

**Extractores de cojinetes** Herramientas especiales diseñadas para la remoción del cojinete.

**Bearing smears** Metal loss from the races and rollers in a circular, blotched pattern.

**Ralladuras en los cojinetes** Pérdida del metal de los anillos y los rodillos que se hace evidente a través de una figura circular oxidada.

**Bearing stain discoloration** A light brown or black discoloration of the rollers and races caused by incorrect lubricant or moisture.

**Descoloramiento del cojinete** Descoloramiento marrón claro o negro de los rodillos y los anillos ocasionado por la humedad o la utilización de un lubricante incorrecto.

**Bellows boots** Accordion-style boots that provide a seal between the tie rods and the housing on a rack and pinion steering gear.

**Botas de fuelles** Botas en forma de acordeón que proveen una junta de estanqueidad entre las barras de acoplamiento y el alojamiento en un mecanismo de dirección de cremallera y piñón.

**Belt tension gauge** A special gauge used to measure drive belt tension.

**Calibrador de tensión de la correa de transmisión** Calibrador especial que se utiliza para medir la tensión de una correa de transmisión.

**Body sway** Excessive body movement from side to side.

**Oscilación de la carrocería** Movimiento lateral excesivo de la carrocería.

**Brake pedal jack** A special tool installed between the front seat and the brake pedal to apply the brakes during certain wheel alignment measurements.

**Gato del pedal de freno** Herramienta especial instalada entre el asiento delantero y el pedal de freno que se utiliza para aplicar los frenos durante ciertas medidas de alineación de ruedas.

**Bump steer** The tendency of the steering to veer suddenly in one direction when one or both front wheels strike a bump.

**Cambio de dirección ocacionado por promontorios en el terreno** Tendencia de la dirección a cambiar repentinamente de sentido cuando una o ambas ruedas delanteras golpea un promontorio.

**Cable reel** A conductive ribbon mounted on top of the steering column to maintain electrical contact between the air bag inflator module and the air bag electrical system. This component may be called a clock spring electrical connector.

**Bobina de cable** Cinta conductiva montada sobre la columna de dirección que se utiliza para mantener el contacto eléctrico entre la unidad infladora y el sistema eléctrico del Airbag. Dicho componente se conoce también como conector eléctrico de cuerda de reloj.

**Camber adjustment** A method of adjusting the inward or outward tilt of a front or rear wheel in relation to the true vertical centerline of the tire.

**Ajuste de la combadura** Método de ajustar la inclinación hacia adentro o hacia afuera de una rueda delantera o trasera con relación a la línea central vertical real del neumático.

**Camber angle** The inward or outward tilt of a line through the center of a front or rear tire in relation to the true vertical centerline of the tire and wheel.

**Ángulo de combadura** Inclinación hacia adentro o hacia afuera de una línea a través del centro de una rueda delantera o trasera con relación a la línea central vertical real del neumático y la rueda.

**Carbon monoxide** A poisonous gas present in vehicle exhaust in small quantities.

**Monóxido de carbono** Gas mortífero presente en pequeñas cantidades en el escape de los motores.

**Caster adjustment** A method of adjusting the forward or rearward tilt of a line through the center of the upper and lower ball joints, or lower ball joint and upper strut mount, in relation to the true vertical centerline of the tire and wheel viewed from the side.

**Ajuste de comba de eje** Método de ajustar la inclinación hacia adelante o hacia atrás de una línea a través del centro de las juntas esféricas superior e inferior, o la junta esférica inferior y el montaje del montante superior, con relación a la línea central vertical real del neumático y la rueda vista desde la parte lateral.

**Caster angle** A line through the center of the upper and lower ball joints, or lower ball joint and upper strut mount, in relation to the true vertical centerline of the tire and wheel viewed from the side.

**Ángulo de comba de eje** Línea a través del centro de las juntas esféricas superior e inferior, o la junta esférica inferior y el montaje del montante superior, con relación a la línea central vertical real del neumático y la rueda vista desde la parte lateral.

**Circlip** A round circular clip used as a locking device on components such as front drive axles.

**Grapa circular** Grapa circular esférica que se utiliza como dispositivo de bloqueo en algunos componentes, como por ejemplo ejes de mando de tracción delantera.

**Claw washer** A special locking washer used to retain the tie rods to the rack in some rack and pinion steering gears.

**Arandela de garra** Arandela de bloqueo especial que se utiliza para sujetar las barras de acoplamiento a la cremallera en algunos mecanismos de dirección de cremallera y piñón.

**"C" locks** A thick metal locking device used to lock components in place, such as the rear drive axles.

**Retenedores en forma de C** Dispositivo de bloqueo metálico grueso que se utiliza para ajustar componentes en su posición, como por ejemplo ejes de mando de tracción trasera.

**Clock spring electrical connector** A conductive ribbon in a plastic container mounted on top of the steering column that maintains electrical contact between the air bag inflator module and the air bag electrical system.

**Conector eléctrico de cuerda de reloj** Cinta conductiva envuelta en una cubierta plástica montada sobre la parte superior de la columna de dirección, que mantiene el contacto eléctrico entre la unidad infladora y el sistema eléctrico del Airbag.

**Coil spring compressing tool** A special tool required to compress a coil spring prior to removal of the spring from the strut.

**Herramienta para la compresión del muelle helicoidal** Herramienta especial requerida para comprimir un muelle helicoidal antes de remover el muelle del montante.

**Collapsible steering column** A steering column that is designed to collapse when impacted by the driver in a collision to help reduce driver injury.

**Columna de dirección plegable** Columna de dirección diseñada para plegarse al ser impactada por el conductor durante una colisión con el propósito de reducir las lesiones que el conductor pueda recibir.

**Computer command ride (CCR) system** A computer-controlled suspension system that controls strut or shock absorber firmness in relation to driving and road conditions.

**Sistema de viaje ordenado por computadora** Sistema de suspensión controlado por computadora que controla la firmeza de los montantes o de los amortiguadores de acuerdo a las condiciones de viaje o del camino.

**Computer wheel aligner** A type of wheel aligner that uses a computer to measure wheel alignment angles at the front and rear wheels.

**Computadora para alineación de ruedas** Tipo de alineador de ruedas que utiliza una computadora para medir los ángulos de alineación de las ruedas delanteras y traseras.

**Continual strut cycling** A test procedure for a computer command ride suspension system that allows continual cycling of the strut armatures.

**Funcionamiento cíclico continuo del montante** Procedimiento de prueba para un sistema de suspensión de viaje ordenado por computadora que permite el funcionamiento cíclico continuo de las armaduras de los montantes.

**Control arm bushing tools** Special tools required for control arm bushing removal and replacement.

**Herramientas para el buje del brazo de mando** Herramientas especiales requeridas para la remoción y el reemplazo del buje del brazo de mando.

**Cornering angle** The turning angle of one front wheel in relation to the opposite front wheel during a turn. This angle may be called turning radius.

**Ángulo de viraje** Ángulo de giro de una rueda delantera con relación a la rueda delantera opuesta durante un viraje. Dicho ángulo se conoce también como radio de giro.

**Crocus cloth** A very fine paper for polishing or removing small abrasions from metal.

**Tela fina de esmeril** Papel sumamente fino que se utiliza para pulir o remover pequeñas abrasiones del metal.

**Curb riding height** The distance between the vehicle chassis and the road surface measured at specific locations.

**Altura del cotén del viaje** Distancia entre el chasis del vehículo y la superficie del camino medida en puntos específicos.

**Current code** A fault code in a computer that is present at all times.

**Código actual** Código de fallo siempre presente en una computadora.

**Cylinder end stopper** A circular bushing in the end of a rack and pinion steering gear housing.

**Tapador en el extremo del cilindro** Buje circular en el extremo del alojamiento de un mecanismo de dirección de cremallera y piñón.

**Data link connector (DLC)** An electrical connector for computer system diagnosis mounted under the instrument panel or in the engine compartment.

**Conector de enlace de datos** Conector eléctrico montado debajo del tablero de instrumentos o en el compartimiento del motor, que se utiliza para la diagnosis del sistema informático.

**Datum line** A straight reference line such as the top of a dedicated bench system.

**Línea de datos** Línea recta de referencia, como por ejemplo la parte superior de un sistema de banco dedicado.

**Dedicated bench system** A heavy steel bed with special fixtures for aligning unitized bodies.

**Sistema de banco dedicado** Asiento pesado de acero con aparatos especiales que se utiliza para la alineación de carrocerías unitarias.

**Diagnostic drawing and text screen** A display on a computer wheel aligner that provides illustrations and written instructions for the technicians.

**Pantalla para la visualización de textos y diagramas diagnósticos** Representación visual en una computadora para alineación de ruedas que le provee a los mecánicos ilustraciones e instrucciones escritas.

**Dial indicator** A precision measuring device with a stem and a rotary pointer.

**Indicador de cuadrante** Dispositivo para medidas precisas con vástago y aguja giratoria.

**Diamond-shaped chassis** A vehicle chassis that is shaped like a diamond from collision damage.

**Chasis en forma de diamante** Chasis que ha adquirido la forma de un diamante a causa del impacto recibido durante una colisión.

**Diesel particulates** Carbon particles emitted in diesel engine exhaust.

**Partículas de diesel** Partículas de carbón presentes en el escape de un motor diesel.

**Directional stability** The tendency of a vehicle steering to remain in the straight-ahead position when driven straight ahead on a smooth, level road surface.

**Estabilidad direccional** Tendencia de la dirección del vehículo a permanecer en línea recta al ser así conducido en un camino cuya superficie es lisa y nivelada.

**Drive cycle diagnosis** A test procedure when diagnosing computer-controlled suspension.

**Diagnosis del ciclo propulsor** Procedimiento de prueba llevado a cabo durante la diagnosis de una suspensión controlada por computadora.

**Dynamic imbalance** The imbalance of a wheel in motion.

**Desequilibrio dinámico** Desequilibrio de una rueda en movimiento.

**Eccentric camber bolt** A bolt with an out-of-round metal cam on the bolt head that may be used to adjust camber.

**Perno de combadura excéntrica** Perno con una leva metálica con defecto de circularidad en su cabeza, que puede utilizarse para ajustar la combadura.

**Eccentric cams or bushings** Out-of-round metal cams mounted on a retaining bolt with the shoulder of the cam positioned against a component. When the cam is rotated, the component position is changed.

**Bujes o levas excéntricas** Levas metálicas con defecto de circularidad montadas sobre un perno de retenida; el apoyo de la leva está colocado contra un componente. Al girar la leva, cambia la posición del componente.

**Electronically erasable programmable read only memory (EEPROM)** A chip in a computer that may be erased easily with special equipment.
**Memoria de solo lectura borrable y programable electrónicamente (EEPROM)** Pastilla en una computadora que puede borrarse fácilmente con equipo especial.

**Electronic neutral check** A check while servicing an electronically controlled four wheel steering system.
**Revisión electrónica neutra** Revisión llevada a cabo durante la reparación de un sistema de dirección en las cuatro ruedas controlado electrónicamente.

**Fail-safe function** A mode entered by a computer if the computer detects a fault in the system.
**Función de autoprotección** Modo adaptado por una computadora si la misma detecta un fallo en el sistema.

**Fault codes** Numeric codes stored in a computer memory representing specific computer system faults.
**Códigos de fallo** Códigos numéricos almancenados en la memoria de una computadora que representan fallos específicos en el sistema informático.

**Floor jack** A hydraulically operated lifting device for vehicle lifting.
**Gato de pie** Dispositivo activado hidráulicamente que se utiliza para levantar un vehículo.

**Flow control valve** A special valve that controls fluid movement in relation to system demands.
**Válvula de control de flujo** Válvula especial que controla el movimiento del fluido de acuerdo a las exigencias del sistema.

**Frame flange** The upper or lower horizontal edge on a vehicle frame.
**Brida del armazón** Borde horizontal superior o inferior en el armazón del vehículo.

**Frame web** The vertical side of a vehicle frame.
**Malla del armazón** Lado vertical del armazón del vehículo.

**Front and rear wheel alignment angle screen** A display on a computer wheel aligner that provides readings of the front and rear wheel alignment angles.
**Pantalla para la visualización del ángulo de alineación de las ruedas delanteras y traseras** Representación visual en una computadora para alineación de ruedas que provee lecturas de los ángulos de alineación de las ruedas delanteras y traseras.

**Front main steering angle sensor** An input sensor mounted in the front steering gear in an electronically controlled four wheel steering system.
**Sensor principal del ángulo de la dirección delantera** Sensor de entrada montado en el mecanismo de dirección delantera en un sistema de dirección en las cuatro ruedas controlado electrónicamente.

**Front sub steering angle sensor** An input sensor mounted in the front steering gear in an electronically controlled four wheel steering system.
**Sensor auxiliar del ángulo de la dirección delantera** Sensor de entrada montado en el mecanismo de dirección delantera en un sistema de dirección en las cuatro ruedas controlado electrónicamente.

**Functional test 211** A diagnostic procedure in a computer-controlled suspension system.
**Prueba funcional 211** Procedimiento diagnóstico en un sistema de suspensión controlado por computadora.

**Gear backlash, or lash** Movement between gear teeth that are meshed with each other.
**Contragolpe o juego en los engranajes** Movimiento entre los dientes del engranaje que están endentados entre sí.

**Geometric centerline** An imaginary line through the exact center of the front and rear wheels.
**Línea central geométrica** Línea imaginaria a través del centro exacto de las ruedas delanteras y traseras.

**Hard fault code** A fault code in a computer that represents a fault that is present at all times.
**Código de fallo duro** Código de fallo en una computadora que representa un fallo que está siempre presente.

**Heavy spots** A spot in a tire casing that is heavier than the rest of the tire.
**Puntos pesados** Punto en la cubierta del neumático más pesado que el resto del neumático.

**History code** A fault code in a computer that represents an intermittent defect.
**Código histórico** Código de fallo en una computadora que representa un defecto intermitente.

**Hydraulic press** A hydraulically operated device for disassembling and assembling components that have a tight press-fit.
**Prensa hidráulica** Dispositivo activado hidráulicamente que se utiliza para desmontar y montar componentes con un fuerte ajuste en prensa.

**Instrument panel cluster (IPC)** A display in the instrument panel that may include gauges, indicator lights, or digital displays to indicate system functions.
**Instrumentos agrupados en el tablero de instrumentos** Representación visual en el tablero de instrumentos que puede incluir calibradores, luces indicadoras, o visualizaciones digitales para indicar las funciones del sistema.

**Integral reservoir** A reservoir that is joined with another component such as a power steering pump.
**Tanque integral** Tanque unido a otro componente, como por ejemplo una bomba de la dirección hidráulica.

**Intermittent fault code** A fault code in a computer memory that is caused by a defect that is not always present.
**Código de fallo intermitente** Código de fallo en la memoria de una computadora ocasionado por un defecto que no está siempre presente.

**International system (SI)** A system of weights and measures.
**Sistema internacional** Sistema de pesos y medidas.

**Jack stand** A metal stand that may be used to support a vehicle.
**Soporte de gato** Soporte de metal que puede utilizarse para apoyar un vehículo.

**Lateral movement** Movement from side to side.
**Movimiento lateral** Movimiento de un lado a otro.

**Lateral runout** The variation in side-to-side movement.
**Desviación lateral** Variación del movimiento de un lado a otro.

**Light emitting diode (LED)** A special diode that emits light when current flows through the diode.
**Diodo emisor de luz (LED)** Diodo especial que emite luz cuando una corriente fluye a través del mismo.

**Lithium-based grease** A special lubricant that is used on the rack bearing in manual rack and pinion steering gears.
**Grasa con base de litio** Lubricante especial utilizado en el cojinete de la cremallera en mecanismos de dirección de cremallera y piñón manuales.

**Machinist's rule** A steel ruler used for measuring short distances. These rulers are available in USC or metric measurements.
**Regla para mecánicos** Regla de acero que se utiliza para medir distancias cortas. Dichas reglas están disponibles en medidas del USC o en medidas métricas.

**Magnetic wheel alignment gauge** A wheel alignment gauge that is held on each front hub with a magnet.

**Calibrador magnético para la alineación de ruedas** Calibrador para la alineación de ruedas que se fija a cada uno de los cubos delanteros con un imán.

**Main menu** A display on a computer wheel aligner from which the technician selects various test procedures.

**Menú principal** Representación visual en una computadora para alineación de ruedas de la que el mecánico puede eligir varios procedimientos de prueba.

**Memory steer** The tendency of the vehicle steering not to return to the straight-ahead position after a turn, but to keep steering in the direction of the turn.

**Dirección de memoria** Tendencia de la dirección del vehículo a no regresar a la posición de línea recta después de un viraje, sino a continuar girando en el sentido del viraje.

**Molybdenum disulphide lithium-based grease** A special lubricant containing molybdenum disulphide and lithium that may be used on some steering components.

**Grasa de bisulfuro de molibdeno con base de litio** Lubricante especial compuesto de bisulfuro de molibdeno y litio que puede utilizarse en algunos componentes de la dirección.

**Multipull unitized body straightening system** A hydraulically operated system that pulls in more than one location when straightening unitized bodies.

**Sistema de tiro múltiple para enderezar la carrocería unitaria** Sistema activado hidráulicamente que tira hacia más de una dirección al enderezar carrocerías unitarias.

**National Institute for Automotive Service Excellence (ASE)** An organization that provides voluntary automotive technician certification in eight areas of expertise.

**Instituto Nacional para la excelencia en la reparación de automóviles** Organización que provee una certificación voluntaria de mecánico de automóviles en ocho áreas diferentes de especialización.

**Optical toe gauge** A piece of equipment that uses a light beam to measure front wheel toe.

**Calibrador óptico del tope** Equipo que utiliza un rayo de luz para medir el tope de la rueda delantera.

**Oversteer** The tendency of a vehicle to turn sharper than the turn selected by the driver.

**Sobreviraje** Tendencia de un vehículo a girar más de lo que el conductor desea.

**Pilot bearing** A bearing that supports the end of a shaft, such as a pinion shaft in a rack and pinion steering gear.

**Cojinete piloto** Cojinete que apoya el extremo de un árbol, como por ejemplo el árbol de piñón en un mecanismo de dirección de cremallera y piñón.

**Pinion and bearing assembly** The pinion gear and supporting bearings in a rack and pinion steering gear.

**Conjunto del piñón y cojinete** Engranaje del piñón y cojinetes de soporte en un mecanismo de dirección de cremallera y piñón.

**Pinpoint tests** Specific test procedures to locate the exact cause of a fault code in a computer system.

**Pruebas llevadas a cabo con precisión** Procedimientos de prueba específicos que se llevan a cabo para localizar la causa exacta de un código de fallo en un sistema informático.

**Pitman arm puller** A special puller required to remove a pitman arm.

**Extractor del brazo pitman** Extractor especial requerido para la remoción de un brazo pitman.

**Plumb bob** A weight with a sharp, tapered point that is suspended and centered on a string.

**Plomada** Balanza con un extremo cónico puntiagudo, suspendida y centrada en una hilera.

**Ply separation** A parting of the plies in a tire casing.

**Separación de estrias** División de las estrias en la cubierta de un neumático.

**Pneumatic tools** Tools operated by air pressure.

**Herramientas neumáticas** Herramientas activadas por aire a presión.

**Power steering pump pressure gauge** A gauge and valve with connecting hoses for checking power steering pump pressure.

**Calibrador de presión de la bomba de la dirección hidráulica** Calibrador y válvula con mangueras de conexión utilizadas para verificar la presión de la bomba de la dirección hidráulica.

**Powertrain control module (PCM)** A computer that controls engine and possibly transmission functions.

**Unidad de control del tren transmisor de potencia** Computadora que controla las funciones del motor y posiblemente las de la transmisión.

**Prealignment inspection** A check of steering and suspension components prior to a wheel alignment.

**Inspección antes de una alineación** Verificación de los componentes de la dirección y de la suspensión antes de llevarse a cabo una alineación de ruedas.

**Preliminary inspection screen** A display on a computer wheel aligner that allows the technician to check and record the condition of many suspension and steering components.

**Pantalla para la visualización durante una inspección preliminar** Representación visual en una computadora para alineación de ruedas que le permite al mecánico revisar y anotar la condición de un gran número de componentes de la suspensión y de la dirección.

**Pressure relief valve** A valve designed to limit pump pressure, such as a power steering pump.

**Válvula de alivio de presión** Válvula diseñada para limitar la presión de una bomba, como por ejemplo una bomba de la dirección hidráulica.

**Programmed ride control (PRC) system** A computer-controlled suspension system in which the computer operates an actuator in each strut to control strut firmness.

**Sistema de control programado del viaje** Sistema de suspensión controlado por computadora en el que la computadora opera un accionador en cada uno de los montantes para controlar la firmeza de los mismos.

**Rack bushing** A bushing that supports the rack in the rack and pinion steering gear housing.

**Buje de la cremallera** Buje que apoya la cremallera en el alojamiento del mecanismo de dirección de cremallera y piñón.

**Radial runout** The variations in diameter of a round object such as a tire.

**Desviación radial** Variaciones en el diámetro de un objeto circular, como por ejemplo un neumático.

**Real time damping (RTD)** The real time damping module controls the road-sensing suspension system.

**Amortiguamiento en tiempo real** La unidad del amortiguamiento en tiempo real controla el sistema de suspensión con equipo sensor.

**Rear axle offset** A condition in which the complete rear axle assembly has turned so one rear wheel has moved forward and the opposite rear wheel has moved rearward.

**Desviación del eje trasero** Condición que ocurre cuando todo el conjunto del eje trasero ha girado de manera que una de las ruedas traseras se ha movido hacia adelante y la opuesta se ha movido hacia atrás.

**Rear axle sideset** A condition in which the rear axle assembly has moved sideways from its original position.

**Resbalamiento lateral del eje trasero** Condición que ocurre cuando el conjunto del eje trasero se ha movido lateralmente desde su posición original.

**Rear main steering angle sensor** An input sensor in the rear steering actuator of an electronically controlled four wheel steering system.

**Sensor principal del ángulo de la dirección trasera** Sensor de entrada en el accionador de la dirección trasera de un sistema de dirección en las cuatro ruedas controlado electrónicamente.

**Rear steering center lock pin** A special pin used to lock the rear steering during specific service procedures in an electronically controlled four wheel steering system.

**Pasador de cierre central de la dirección trasera** Pasador especial que se utiliza para bloquear la dirección trasera durante procedimientos de reparación específicos en un sistema de dirección en las cuatro ruedas controlado electrónicamente.

**Rear sub steering angle sensor** An input sensor in the rear steering actuator of a four wheel steering system.

**Sensor auxiliar del ángulo de la dirección trasera** Sensor de entrada en el accionador de la dirección trasera de un sistema de dirección en las cuatro ruedas.

**Rear wheel tracking** Refers to the position of the rear wheels in relation to the front wheels.

**Encarrilamiento de las ruedas traseras** Se refiere a la posición de las ruedas traseras con relación a las ruedas delanteras.

**Remote reservoir** A container containing fluid that is mounted separately from the fluid pump.

**Tanque remoto** Recipiente lleno de líquido que se monta separado de la bomba del fluido.

**Ribbed V-belt** A belt containing a series of small "V" grooves on the underside of the belt.

**Correa nervada en V** Correa con una serie de ranuras pequeñas en forma de "V" en la superficie inferior de la misma.

**Ride height screen** A display on a computer wheel aligner that illustrates the locations for ride height measurement.

**Pantalla para la visualización de la altura del viaje** Representación visual en una computadora para alineación de ruedas que muestra los puntos donde se mide la altura del viaje.

**Rim clamps** Special clamps designed to clamp on the wheel rims and allow the attachment of alignment equipment such as the wheel units on a computer wheel aligner.

**Abrazaderas de la llanta** Abrazaderas especiales diseñadas para sujetar las llantas de las ruedas y así permitir la fijación del equipo de alineación, como por ejemplo las unidades de la rueda en una computadora para alineación de ruedas.

**Road feel** A feeling experienced by a driver during a turn when the driver has a positive feeling that the front wheels are turning in the intended direction.

**Sensación del camino** Sensación experimentada por un conductor durante un viraje cuando está completamente seguro de que las ruedas delanteras están girando en la dirección correcta.

**Road sensing suspension (RSS)** A computer-controlled suspension system that senses road conditions and adjusts suspension firmness to match these conditions in a few milliseconds.

**Suspensión con equipo sensor** Sistema de suspensión controlado por computadora que advierte las condiciones actuales del camino y ajusta la firmeza de la suspensión para equiparar dichas condiciones en un período de milisegundos.

**Scan tester** A digital computer system tester used to read trouble codes and perform other diagnostic functions.

**Verificador de exploración** Verificador digital del sistema informático que se utiliza para leer códigos indicativos de problemas y llevar a cabo otras funciones diagnósticas.

**Seal drivers** Designed to maintain even contact with a seal case and prevent seal damage during installation.

**Empujadores para sellar** Diseñados para mantener un contacto uniforme con el revestimiento de la junta de estanqueidad y evitar el daño de la misma durante el montaje.

**Section modulus** The measurement of a frame's strength based on height, width, thickness, and the shape of the side rails.

**Coeficiente de sección** Medida de la resistencia de un armazón, basada en la altura, el ancho, el espesor y la forma de las vigas laterales.

**Selective shim** Shims of different thicknesses that are used to provide an adjustment such as bearing preload.

**Laminillas selectivas** Laminillas de diferentes espesores que se utilizan para proveer un ajuste, como por ejemplo la carga previa del cojinete.

**Self-locking nut** A special nut designed so it will not loosen because of vibration.

**Tuerca de cierre automático** Tuerca con un diseño especial que evita su aflojamiento a causa de la vibración.

**Serpentine belt** A ribbed V-belt drive system in which all the belt-driven components are on the same vertical plane.

**Correa serpentina** Sistema de transmisión con correa nervada en V en el que todos los componentes accionados por una correa se encuentran sobre el mismo plano vertical.

**Service bay diagnostics** A diagnostic system supplied by Ford Motor Company to their dealers that diagnoses vehicles and provides communication between the dealer and manufacturer.

**Diagnóstico para el puesto de reparación** Sistema diagnóstico que les suministra la Ford a sus distribuidores de automóviles; dicho sistema diagnostica vehículos y permite una buena comunicación entre el distribuidor y el fabricante.

**Service check connector** A diagnostic connector used to diagnose computer systems. A scan tester may be connected to this connector.

**Conector para la revisión de reparaciones** Conector diagnóstico que se utiliza para diagnosticar sistemas informáticos. A dicho conector se le puede conectar un verificador de exploración.

**Setback** Occurs when one front wheel is driven rearward in relation to the opposite front wheel.

**Retroceso** Condición que ocurre cuando una de las ruedas delanteras se mueve hacia atrás con relación a la rueda delantera opuesta.

**Sheared injected plastic** Is inserted into steering column shafts and gear shift tubes to allow these components to collapse if the driver is thrown against the steering column in a collision.

**Plástico cortado inyectado** Insertado en los árboles de la columna de dirección y en los tubos del cambio de engranajes de velocidad para permitir que estos componentes se pleguen si la columna de dirección es impactada por el conductor durante una colisión.

**Shim select screen** A display on a computer wheel aligner that shows the technician the thickness and location of the proper shims for rear wheel alignment.

**Pantalla para la visualización durante la selección de laminillas** Representación visual en una computadora para alineación de ruedas que le muestra al mecánico el espesor y la ubicación de las laminillas correctas para la alineación de las ruedas traseras.

**Shock absorber manual test** A test in which the lower end of the shock absorber is disconnected and the shock absorber is operated manually to determine its condition.

**Prueba manual del amortiguador** Prueba en la que se desconecta el extremo inferior del amortiguador con el fin de poder operarlo manualmente y lograr determinar su condición.

**Shock absorber or strut bounce test** A test in which the vehicle is bounced by leaning on the bumper to determine shock absorber condition.

**Prueba de rebote del montante o del amortiguador** Prueba en la que se ejerce presión sobre el parachoques con el fin de hacer rebotar el vehículo y lograr determinar la condición del amortiguador.

**Shop layout** The location of all shop facilities including service bays, equipment, safety equipment, and offices.

**Arreglo del taller de reparación** Ubicación de todas las instalaciones del taller, incluyendo los puestos de reparación, el equipo, el equipo de seguridad, y las oficinas.

**Single-pull unitized body straightening system** Hydraulically operated equipment that pulls in one location while straightening unitized bodies.

**Sistema enderezador de tiro único para la carrocería unitaria** Sistema activado hidráulicamente que tira hacia una dirección al enderezar carrocerías unitarias.

**Soft-jaw vise** A vise equipped with soft metal, such as copper, in the jaws.

**Tornillo con tenacilla maleable** Tornillo equipado con un metal blando, como por ejemplo cobre, en las tenacillas.

**Specifications menu** A display on a computer wheel aligner that allows the technician to select, enter, alter, and display vehicle specifications.

**Menú de especificaciones** Representación visual en una computadora para alineación de ruedas que le permite al mecánico seleccionar, introducir datos, cambiar e indicar especificaciones referentes al vehículo.

**Spiral cable** A conductive ribbon mounted in a plastic container on top of the steering column that maintains electrical contact between the air bag inflator module and the air bag electrical system. A spiral cable may be called a clock spring electrical connector.

**Cable espiral** Cinta conductiva montada en un recipiente plástico sobre la columna de dirección que mantiene el contacto eléctrico entre la unidad infladora y el sistema eléctrico del Airbag. El cable espiral se conoce también como conector eléctrico de cuerda de reloj.

**Spring fill diagnostics** A diagnostic procedure in a computer-controlled air suspension system.

**Diagnóstico con relleno para muelles** Procedimiento diagnóstico en un sistema de suspensión de aire controlado por computadora.

**Spring insulators, or silencers** Rings usually made from plastic and positioned on each end of a coil spring to reduce noise and vibration transfer to the chassis.

**Aisladores de muelles, o silenciadores** Anillos, por lo general fabricados de plástico y colocados a ambos extremos de un muelle helicoidal, que se utilizan para disminuir la transferencia de ruido y de vibración al chasis.

**Spring sag** Occurs when a spring becomes weak, and the curb riding height is reduced compared to the original height.

**Agotamiento de los muelles** Condición que ocurre cuando el muelle se debilita; la altura del cotén del viaje disminuye si se la compara con la altura orginal.

**Stabilizer bar, or sway bar** A round steel bar connected between the front or rear lower control arms that reduces body sway.

**Barra estabilizadora o barra de oscilación lateral** Barra circular de acero contectada entre los brazos de mando delantero o trasero que disminuye la oscilación lateral de la carrocería.

**Static imbalance** Refers to the imbalance of a wheel and tire at rest.

**Desequilibrio estático** Se refiere al desequilibrio de una rueda y de un neumático cuando se ha detenido la marcha del vehículo.

**Steering effort** The amount of effort required by the driver to turn the steering wheel.

**Esfuerzo de dirección** Amplitud de esfuerzo requerido por parte del conductor para girar el volante de dirección.

**Steering pull** The tendency of the steering to pull to the right or left when the vehicle is driven straight ahead on a smooth, straight road surface.

**Tiro de la dirección** Tendencia de la dirección a desviarse hacia la derecha o hacia la izquierda mientras se conduce el vehículo en línea recta en un camino cuya superficie es lisa y nivelada.

**Steering wheel freeplay** The amount of steering wheel movement before the front wheels begin to turn.

**Juego libre del volante de dirección** Amplitud de movimiento del volante de dirección antes de que las ruedas delanteras comiencen a girar.

**Steering wheel locking tool** A special tool used to lock the steering wheel during certain wheel alignment procedures.

**Herramienta para el cierre del volante de dirección** Herramienta especial que se utiliza para bloquear el volante de dirección durante ciertos procedimientos de alineación de ruedas.

**Stethoscope** A special tool that amplifies sound to help diagnose noise location.

**Estetoscopio** Herramienta especial que amplifica el sonido para ayudar a diagnosticar la procedencia de los ruidos.

**Strikeout, or rebound, bumper** A rubber block that prevents the control arm from striking the chassis when the wheel hits a large road irregularity.

**Parachoques de rebote** Bloque de caucho que evita que el brazo de mando choque contra el chasis cuando la rueda golpea una irregularidad en el camino.

**Strut cartridge** The inner components in a strut that may be replaced rather than replacing the complete strut.

**Cartucho del montante** Componentes internos de un montante que pueden ser reemplazados en vez de tener que reemplazarse todo el montante.

**Strut chatter** A chattering noise as the steering wheel is turned often caused by a binding upper strut mount.

**Vibración del montante** Rechinamiento producido mientras se gira el volante de dirección. A menudo ocasionado por el trabamiento del montaje del montante superior.

**Strut rod, or radius rod** A rod connected from the lower control arm to the chassis to prevent forward and rearward control arm movement.

**Varilla del montante o varilla radial** Varilla conectada del brazo de mando inferior al chasis que se utiliza para evitar el movimiento hacia adelante o hacia atrás del brazo de mando.

**Strut tower** A circular, raised, reinforced area inboard of the front fenders that supports the upper strut mount and strut assembly.

**Torre del montante** Área circular, elevada y reinforzada en la parte interior de los guardafangos delanteros que apoya el conjunto del montante y montaje del montante superior.

**Sulfuric acid** A very corrosive acid used in automotive batteries.

**Ácido sulfúrico** Ácido sumamente corrosivo utilizado en las baterías de automóviles.

**Sunburst frame cracks** Radiate outward from an opening in the vehicle frame.

**Grietas del armazón en forma de rayos de sol** Grietas que se proyectan hacia afuera desde una abertura en el armazón del vehículo.

**Supplemental inflatable restraint (SIR)** An air bag system.
**Sistema de seguridad inflable suplementario** Sistema del Airbag.

**Tapered-head bolts** A bolt with a taper on the underside of the head.
**Pernos de cabeza cónica** Perno con una cabeza cuya superficie inferior es cónica.

**Teflon ring compressing tool** A special tool used to compress the Teflon rings on the pinion prior to installation in a rack and pinion steering gear.
**Herramienta para la compresión de anillos de teflón** Herramienta especial que se utiliza para comprimir los anillos de teflón en el piñón antes de ser instalados en un mecanismo de dirección de cremallera y piñón.

**Teflon ring expander** A special tool required to expand the Teflon rings prior to installation on the pinion in a rack and pinion steering gear.
**Expansor para anillos de teflón** Herramienta especial requerida para expandir los anillos de teflón antes de ser instalados en un mecanismo de dirección de cremallera y piñón.

**Thrust line** A line positioned at a 90° angle to the rear axle and projected toward the front of the vehicle.
**Línea de empuje** Línea colocada a un ángulo de 90° con relación al eje trasero y proyectada hacia la parte frontal del vehículo.

**Tie rod end and ball joint puller** A special puller designed for removing tie rod ends and ball joints.
**Extractor para la junta esférica y el extremo de la barra de acoplamiento** Extractor especial diseñado para la remoción de las juntas esféricas y los extremos de barras de acoplamiento.

**Tie rod sleeve adjusting tool** A special tool required to rotate tie rod sleeves without damaging the sleeve.
**Herramienta para el ajuste del manguito de la barra de acoplamiento** Herramienta especial requerida para girar los manguitos de las barras de acoplamiento sin estropearlos.

**Tire changer** Equipment used to demount and mount tires on wheel rims.
**Cambiador de neumáticos** Equipo que se utiliza para desmontar y montar los neumáticos en las llantas de las ruedas.

**Tire condition screen** A display on a computer wheel aligner that allows the technician to check and enter tire condition.
**Pantalla para la visualización de la condición del neumático** Representación visual en una computadora para alineación de ruedas que le permite al mecánico revisar e introducir datos referentes a la condición del neumático.

**Tire conicity** Occurs when the tire belt is wound off-center in the manufacturing process creating a cone-shaped belt, which results in steering pull.
**Conicidad del neumático** Condición que ocurre cuando la correa del neumático se devana fuera del centro durante el proceso de fabricación creando así una correa en forma cónica y trayendo como resultado el tiro de la dirección.

**Tire rotation** Involves moving each wheel and tire to a different location on the vehicle to increase tire life.
**Cambio de posición de los neumáticos** Se refiere a la rotación de cada rueda y neumático a una posición diferente en el vehículo para aumentar la vida útil del neumático.

**Tire thump** A pounding noise as the tire and wheel rotate usually caused by improper wheel balance.
**Ruido sordo del neumático** Ruido similar al de un golpe pesado que se produce mientras el neumático y la rueda están girando. Por lo general este ruido lo ocasiona el desequilibrio de la rueda.

**Tire tread depth gauge** A special tool required to measure tire tread depth.
**Calibrador de la profundidad de la huella del neumático** Herramienta especial requerida para medir la profundidad de la huella del neumático.

**Tire vibration** Vertical or sideways tire oscillations.
**Vibración del neumático** Oscilaciones verticales o laterales del neumático.

**Toe gauge** A special tool used to measure front or rear wheel toe.
**Calibrador del tope** Herramienta especial que se utiliza para medir el tope de las ruedas delanteras o traseras.

**Toe-in** A condition in which the distance between the front edges of the tires is less than the distance between the rear edges of the tires.
**Convergencia** Condición que ocurre cuando la distancia entre los bordes frontales de los neumáticos es menor que la distancia entre los bordes traseros de los mismos.

**Toe-out** A condition in which the distance between the front edges of the tires is more than the distance between the rear edges of the tires.
**Divergencia** Condición que ocurre cuando la distancia entre los bordes delanteros de los neumáticos es mayor que la distancia entre los bordes traseros de los mismos.

**Toe-out on turns** The turning angle of the wheel on the inside of a turn compared to the turning angle of the wheel on the outside of the turn.
**Divergencia durante un viraje** El ángulo de giro de la rueda en el interior de un viraje comparado con el ángulo de giro de la rueda en el exterior del viraje.

**Torque steer** The tendency of the steering to pull to one side during hard acceleration on front wheel drive vehicles with unequal length front drive axles.
**Dirección de torsión** Tendencia de la dirección a desviarse hacia un lado durante una aceleración rápida en un vehículo de tracción delantera con ejes de mando desiguales.

**Total toe** The sum of the toe angles on both wheels.
**Tope total** Suma de los ángulos del tope en ambas ruedas.

**Track gauge** A long straight bar with adjustable pointers used to measure rear wheel tracking in relation to the front wheels.
**Calibrador del encarrilamiento** Barra larga y recta con agujas ajustables que se utiliza para medir el encarrilamiento de las ruedas traseras con relación a las ruedas delanteras.

**Traction control system (TCS)** A computer-controlled system that prevents one drive wheel from spinning more than the opposite drive wheel on acceleration.
**Sistema para el control de la tracción** Sistema de suspensión controlado por computadora que evita que una rueda motriz gire más que la opuesta durante la aceleración del vehículo.

**Tram gauge** A long, straight bar with adjustable pointers used to measure unitized bodies.
**Calibrador del tram** Barra larga y recta con agujas ajustables que se utiliza para medir carrocerías unitarias.

**Tread wear indicators** Raised portions near the bottom of the tire tread that are exposed at a specific tread wear.
**Indicadores del desgaste del neumático** Secciones elevadas cerca de la parte inferior de la huella del neumático que quedan expuestas cuando la huella alcanza una cantidad de desgaste específica.

**Trim height** The normal chassis riding height on a computer-controlled air suspension system.
**Altura equilibrada** Altura normal de viaje del chasis en un sistema de suspensión controlado por computadora.

**Turning angle screen** A display on a computer wheel aligner that displays the turning angle.

**Pantalla para la visualización del ángulo de giro** Representación visual en una computadora para alineación de ruedas que muestra el ángulo de giro.

**Turning imbalance** A condition in which more effort is required to turn the steering wheel in one direction compared to the opposite direction.

**Desequilibrio de giro** Condición que ocurre cuando se requiere más esfuerzo para girar el volante en una dirección que en la dirección opuesta.

**Turning radius** The turning angle on the front wheel on the inside of a turn compared to the front wheel turning angle on the outside of the turn.

**Radio de giro, o círculo de giro** Ángulo de giro de la rueda delantera en el interior de un viraje con relación al ángulo de giro de la rueda delantera en el exterior del viraje.

**Turning radius gauge** A gauge with a degree scale mounted on turn tables under the front wheels.

**Calibrador del radio de giro** Calibrador con una escala de grados montado sobre plataformas giratorias debajo de las ruedas delanteras.

**Turn tables** Heavy steel plates placed under the front wheels during a wheel alignment that allow front wheel turning.

**Plataformas giratorias** Placas pesadas de acero que se colocan debajo de las ruedas delanteras durante una alineación de ruedas para permitir el giro de las mismas.

**Understeer** The tendency of a vehicle not to turn as much as desired by the driver.

**Dirección pobre** Tendencia de un vehículo a no girar tanto como lo desea el conductor.

**United States Customary (USC)** A system of weights and measures patterned after the British system.

**Sistema usual estadounidense (USC)** Sistema de pesos y medidas desarrollado según el modelo del Sistema Imperial Británico.

**Vacuum hand pump** A mechanical pump with a vacuum gauge and hose used for testing vacuum-operated components.

**Bomba de vacío manual** Bomba mecánica con un calibrador de vacío y una manguera que se utiliza para probar componentes a depresión.

**V-belt** A drive belt with a V-shape.

**Correa en V** Correa de transmisión en forma de V.

**Vehicle lift** A hydraulically or air-operated mechanism for lifting vehicles.

**Levantamiento del vehículo** Mecanismo activado hidráulicamente o por aire que se utiliza para levantar vehículos.

**Vehicle wander** The tendency of the steering to pull to the right or left when the vehicle is driven straight ahead on a straight road.

**Desviación de la marcha del vehículo** Tendencia de la dirección a desviarse hacia la derecha o hacia la izquierda cuando se conduce el vehículo en línea recta en un camino cuya superficie es lisa.

**Wear indicator** A device that visually indicates ball joint wear.

**Indicador de desgaste** Dispositivo que indica de manera visual el desgaste de una junta esférica.

**Wheel balancer** Equipment for wheel balancing that is usually computer controlled.

**Equilibrador de ruedas** Equipo para la equilibración de ruedas que por lo general es controlado por computadora.

**Wheel runout compensation screen** A display on a computer wheel aligner that is shown during the wheel runout compensation procedure.

**Pantalla para la visualización de la compensación de desviación de la rueda** Representación visual en una computadora para alineación de ruedas que aparece durante el procedimiento de compensación de desviación de la rueda.

**Woodruff key** A half-moon shaped metal key used to retain a component, such as a pulley, on a shaft.

**Chaveta woodruff** Chaveta de metal en forma de media luna que se utiliza para sujetar un componente, como por ejemplo una roldana, en un árbol.

**Yield strength** A measurement of the material strength from which a frame is manufactured.

**Límite para fluencia** Medida de la resistencia del material del que está fabricado un armazón.

# INDEX

Note: Page numbers in **bold** print reference non-text material.